Inverse Methods in Electromagnetic Imaging

Part 2

This Advanced Research Workshop was co-sponsored by:
NATO, Scientific Affairs Division
DFVLR, Institute for Radio Frequency Technology
U.S.-ARO, European Branch Office
SIEMENS A.G., Medical Division
UIC/EECS, Communications Laboratory

NATO ASI Series

Advanced Science Institutes Series

A series presenting the results of activities sponsored by the NATO Science Committee, which aims at the dissemination of advanced scientific and technological knowledge, with a view to strengthening links between scientific communities.

The series is published by an international board of publishers in conjunction with the NATO Scientific Affairs Division

A	Life Sciences	Plenum Publishing Corporation
B	Physics	London and New York
C	Mathematical and Physical Sciences	D. Reidel Publishing Company Dordrecht, Boston and Lancaster
D	Behavioural and Social Sciences	Martinus Nijhoff Publishers
E	Engineering and Materials Sciences	The Hague, Boston and Lancaster
F	Computer and Systems Sciences	Springer-Verlag
G	Ecological Sciences	Berlin, Heidelberg, New York and Tokyo

Series C: Mathematical and Physical Sciences Vol. 143 - Part 2

Inverse Methods in Electromagnetic Imaging

Part 2

edited by

Wolfgang-M. Boerner
Communications Laboratory, EECS Department,
University of Illinois at Chicago, U.S.A.

Hans Brand
High Frequency Engineering Laboratories,
Universität Erlangen-Nürnberg, F.R.G.

Leonard A. Cram
Thorn EMI, Electronics Ltd.,
U.K.

Dag T. Gjessing
ESTP, Royal Norwegian Council for Industrial
and Scientific Research, Kjeller, Norway

Arthur K. Jordan
Space Sciences Division,
Naval Research Laboratory, U.S.A.

Wolfgang Keydel
Institüt für HF-Technik,
DFVLR, F.R.G.

Günther Schwierz
Siemens Medical Division,
Siemens AG, Erlangen, F.R.G.

Martin Vogel
Institüt für HF-Technik,
DFVLR, F.R.G.

Springer-Science+Business Media, B.V.

Proceedings of the NATO Advanced Research Workshop on
Inverse Methods in Electromagnetic Imaging
Bad Windsheim, Franconia, F.R.G.
18-24 September 1983

Library of Congress Cataloging in Publication Data

NATO Advanced Research Workshop on Inverse Methods in Electromagnetic Imaging
(1983: Bad Windsheim, Germany)
Inverse methods in electromagnetic imaging.

(NATO ASI series. Series C, Mathematical and physical sciences; vol. 143)
"Proceedings of the NATO Advanced Research Workshop on Inverse Methods in Electromagnetic Imaging, Bad Windsheim, Franconia, F.R.G., September 18–24, 1983"–T.p. verso.
"Published in cooperation with NATO Scientific Affairs Division."
Bibliography: p.
Includes index.
1. Imaging systems–Congresses. 2. Electromagnetic waves–Congresses.
I. Boerner, Wolfgang M. II. North Atlantic Treaty Organization, Scientific Affairs Divison. III. Title. IV. Series: NATO ASI series. Series C, Mathematical and physical sciences; no. 143.
TK8315.N36 1983 621.36'7 84-22843

ISBN 978-94-010-9446-7 ISBN 978-94-010-9444-3 (eBook)
DOI 10.1007/978-94-010-9444-3

Originally published by D. Reidel Publishing Company, Dordrecht, Holland 1985
Softcover reprint of the hardcover 1st edition 1985

TABLE OF CONTENTS (Part 2)

TOPIC V - HOLOGRAPHIC AND TOMOGRAPHIC IMAGING AND RELATED PHASE PROBLEMS

(Papers are cross-referenced with the **Final Technical Program Outline.** Given are session and sequence of presentation, i.e., OS.3)

CONTENTS (Part 1)

TOPIC II - NUMERICAL INVERSION METHODS

(Papers are cross-referenced with the **Final Technical Program Outline.** Given are session and sequence of presentation, i.e., OS.3)

TOPIC III - POLARIZATION UTILIZATION IN THE ELECTROMAGNETIC VECTOR INVERSE PROBLEM

(Papers are cross-referenced with the **Final Technical Program Outline.** Given are session and sequence of presentation, i.e., OS.3)

III.10 (SR.4)

STUDY OF TWO SCATTERER INTERFERENCE WITH A POLARIMETRIC FM/CW RADAR

P.S.P. Wei, M. M. Rackson, T. C. Bradley,
T. H. Meyer, and J. D. Kelly

Boeing Aerospace Company, Mail Stop 8H-51
P. O. Box 3999, Seattle, WA 98124, U.S.A.

ABSTRACT The coherence zone for a single photon, which is related to the range resolution for a radar, is defined in terms of physical optics and the uncertainty principle. Two scatterers located approximately within this zone give rise to interference in the scattered wavefield. In this paper we study the two-scatterer interference profiles measured with a 45 GHz FM/CW radar. Downrange translations of one scatterer with respect to another for a distance of $\lambda/4$ produces a phase shift of π which alters the lineshapes from constructive to destructive interference. In a computer simulation, the back-scattered wave from each scatterer is modeled by a sin X/X type function with associated amplitude and phase. Excellent agreement with measured data is obtained. The resolution conditions for two coherent scatterers have been analyzed in detail.

I. Introduction

In inverse scattering, one wishes to determine the vector potential from measured cross sections. For radar phenomenology, we are interested in extracting information about the size, shape, distribution, and motion of the targets. The problem is not easily solvable except for simple geometric shapes. With incomplete measurements, we can only get a partial solution. Prior knowledge about the target is usually helpful.

For a given radar system, it is of interest to study its coherent property by the interference effect of two or more scatterers under controlled conditions. We believe that understanding the radar response is important for developing the algorithms for detection, discrimination, and identification of targets. The

W.-M. Boerner et al. (eds.), Inverse Methods in Electromagnetic Imaging, Part 2, 673–682.

eventual goal will be to recognize complex targets in a random environment of clutter.

In this paper we present some preliminary results on the two-scatterer interference profiles measured with a polarimetric 45 GHz FM/CW radar. Depending on the conditions, the observed range profile may have one or two major peaks. The lineshape, the peak amplitude, and apparent peak separation vary with the parameters such as the radar band-width, the coarse and the fine separations between scatterers. Using a simple model of the sinX/X type sampling spectrum function to represent each scatterer, we find that a linear superposition gives a remarkably good simulation of the measured profile. The resolution conditions for two coherent scatterers are evaluated. A quantitative calibration of both the amplitude and range scales is obtained.

II. Theory of Coherence

Photons or electromagnetic waves exhibit both particle and wave natures (1,2). The former is manifested in its energy and momentum, whereas the latter shows up in interference and diffraction. According to Dirac, each photon interferes only with itself (3). The wave phenomena of light can best be demonstrated by the Young's double slit experiment (4).

The fundamentals of coherent optics have been reviewed by Thompson (5). From optics textbooks (1,2), it is convenient to describe the coherence effects in terms of temporal and spatial parts. The coherence time t is the duration over which we can reasonably predict the phase of the wave at a given point in space. From the uncertainty principle, t is just the inverse of the frequency bandwidth w. The longitudinal coherence length is defined as $L = ct = c/w$, where c is the speed of light. We shall see that L is related to the down-range resolution for a radar. For $w = 75$ MHz, we have $L = 4$ meters.

The spatial (lateral or transverse) coherence is related to the finite extent of the source. At a range R from a circular aperture of diameter D at wavelength λ, the transverse coherence length may be defined as $H = R\lambda/D$. In a plane containing the target and normal to the line-of-sight from the source, the transverse coherence varies with an Airy disk type function, where H is the separation between the first two nulls on two sides of the maximum. According to a visibility criterion of van Cittert and Zernike, two slits located at a separation of H/π or closer would produce readily observable interference fringes of the Young's type (1,2).

From the longitudinal and transverse coherence lengths defined above, we may regard a cylindrical volume with base $\pi H^2/4$ and length L as the coherence zone for a single photon, as defined by the source. Two scatterers located roughly within a coherence zone will be sensed by the incident wave simultaneously. A summation of the scattered waves, each with its amplitude and phase, yields a resultant wave. The coherence interference effect becomes small when two scatterers are separated by distances much larger than L or H, respectively, in the down-range or cross-range directions.

III. Simulation

In a simple monostatic case, we transmit linearly vertical polarized waves and measure the backscattering amplitudes parallel to the transmitted. Downrange displacement of one scatterer with respect to another by $\lambda/4$ gives rise to a phase change of π, while a cross-range displacement does not produce a measurable change. For scatterers each located at x_i, with a scattered amplitude of A_i and phase angle of Φ_i, the scattered intensity (square of the resultant amplitude) as a function of down-range position x may be written as

$$I = A^2 = (\sum A_i \cos\Phi_i \sin X_i/X_i)^2 + (\sum A_i \sin\Phi_i \sin X_i/X_i)^2,$$

where $X_i = 2\pi(x - x_i)/L$, and $L = c/w$ is the down-range resolution length for a given bandwidth w. Note that $L >> \lambda$. In general, when the full polarization and the 3-D coordinates are taken into consideration, then A_i, Φ_i, and x_i would be vectors rather than scalars as used here.

For two scatterers, we find that the range profile is symmetrical with respect to the middle position at $(x_1 + x_2)/2$. When there is only one scatterer, the intensity reduces to the familiar $(\sin X_1/X_1)^2$ form which is just the Fraunhofer far-field diffraction pattern of a narrow slit or the Fourier transform of a square pulse.

Using the simple formula given above, codes have been developed on a VAX 11/780 computer system to calculate both the intensity I and amplitude A with w, Φ_i, A_i, x_i as adjustable parameters. The results are plotted on an Evans & Sutherland PS-300 graphic system for easy comparison with measured data.

IV. Experimental

The 50-mW frequency-modulated continuous wave (FM/CW) radar is similar to those described in a standard textbook (6). In brief, the transmitted frequency is linearly swept from 45 (λ = 6.67 mm) to 45.35 GHz at a repetition rate of 40 kHz. A time-gate of 7 μsec is applied during which the received signal is heterodyned against a small part of the transmitted signal to produce an intermediate frequency signal to be analyzed by a spectrum analyzer. Scatterers located down range are distinguished by the round-trip time delays and hence the frequency shifts. The two-scatterer interference profiles are measured from an outdoor quiet range where fences have been erected to eliminate the multipath effect.

The radar antenna has a diameter of 0.457 m. It produces a beam of 1.1° angular divergence at 3 db which is close to the diffraction limited value of λ/D = 0.84°. Two trihedrals (corner

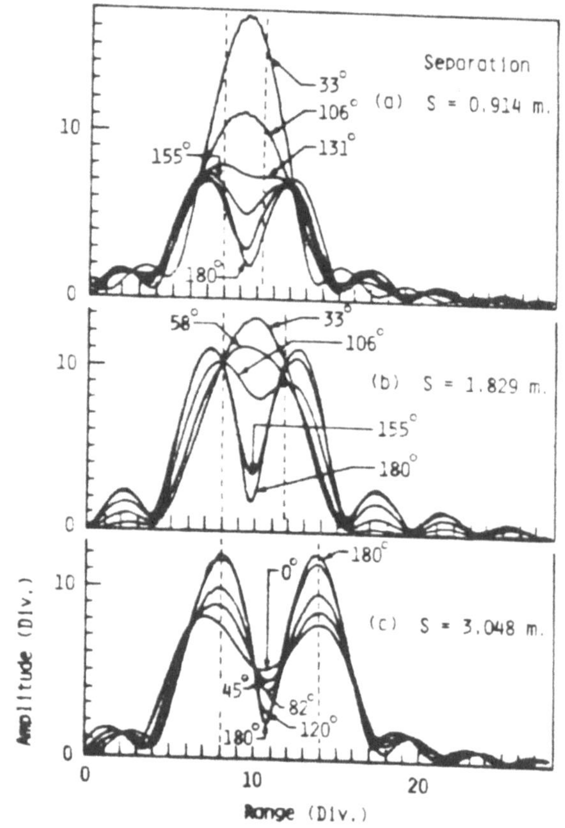

Figure 1 . Measured interference profiles from two equal scatterers at three down-range separations. The relative phase angles are assigned with reference to the simulation described in text. The approximate locations of the two scatterers are designated by dashed lines. Note that whether or not the profile resolves into two peaks, their separation and amplitude are all a function of the relative phase.

reflectors) of 50 - m^2 each are placed at a nominal range of 400 m, at about 1.8 m apart in cross-range, and at several down-range separations. The antenna position has been carefully adjusted so that the returns from each scatterer alone are about the same. With one trihedral fixed, the second one is placed on a screw-adjusted mount, where each turn of a (¼) -28 screw corresponds to 0.907 mm of down-range displacement which is equivalent to 98° of phase angle. Under a set of constant conditions the received amplitude versus range is plotted on an x-y recorder. In this paper, we are concerned with transmitting vertically polarized

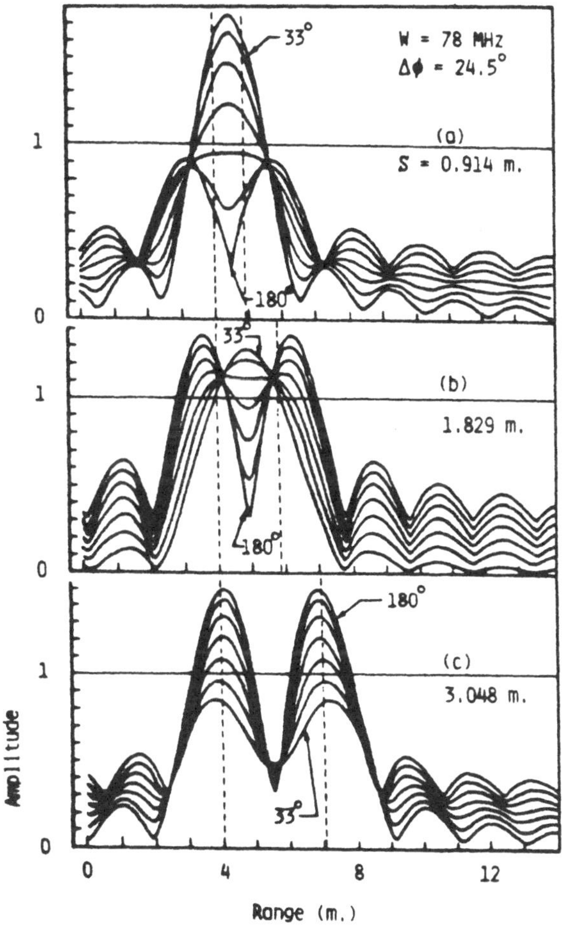

Figure 2 . Simulated interference profiles from two equal scatterers located by dashed lines. At each separation, the curves are displaced upward for each increment of relative phase to facilitate a comparison with the measured curve in Figure 1.

and receiving vertical only. Generalization to targets other than trihedrals and receiving polarization other than parallel to the transmitting will be described in a later report.

V. Results

Figure 1 shows some typical measured interference profiles from two equal trihedrals at several down-range separation S. The experiments are made by first finding a minimum or null between the two scatterers, whose approximate positions are indicated by dashed lines. At null, the two scatterers are 180° out of phase to each other. The adjustable screw is turned in small steps to record a new profile so that at 1.5 turns away from null the relative phase is at about 33°. The maximum error in phase angle is $\pm$ 12°, or 1/8 turn of the screw. An interesting observation is that at S $\leq$ 2.5 m, the profile may be either resolved into two peaks or not resolved depending on the phase angle. It is obvious that the peak height, the apparent peak separation, and the middle valley are all influenced by the relative phase. The slightly unbalanced peak heights in some scans are due to error.

Figure 2 shows some simulated interference profiles from two equal scatterers of unit amplitude for a comparison with the measured ones just described. A bandwidth of w = 78 ($\pm$1) MHz yields the best fit although the experimental value is 98 MHz. Presumably, there is some broadening due to the filtering network. Starting from Φ = 33°, each successive curve is calculated for $\Delta\Phi$ = 24.5° and displaced upward for clarity. Note that the increment corresponds to ¼ turn of the screw. With the prior knowledge about the screw position and using the simulation, we can assign the phase for the measured profiles in Figure 1. Judging from the possible error in relative phase, the excellent agreements between Figures 2 and 1 in lineshapes, peak height, and peak separation must be considered remarkable.

VI. Discussions

It is desirable to explore the conditions at which two scatterers under coherent illumination may be resolved into two major peaks. Further, we see from Figure 1 and 2 that the two peaks, if resolved, appear to be separated by a distance that may be insensitive to the actual separation. The answers to these ideas are summarized in Figure 3. For optical imaging systems dealing with incoherent light, two overlapping images are resolved according to the Rayleigh or Sparrow criteria when the valley between the two peaks is at $8/\pi^2$ or 1, respectively, of the peak height (2). The profiles for experiment and simulation at S = 0.914 m and Φ = 131° (Figures 1a and 2a) show a close encounter with a Sparrow-type condition when the central part of the profile becomes a flat top.

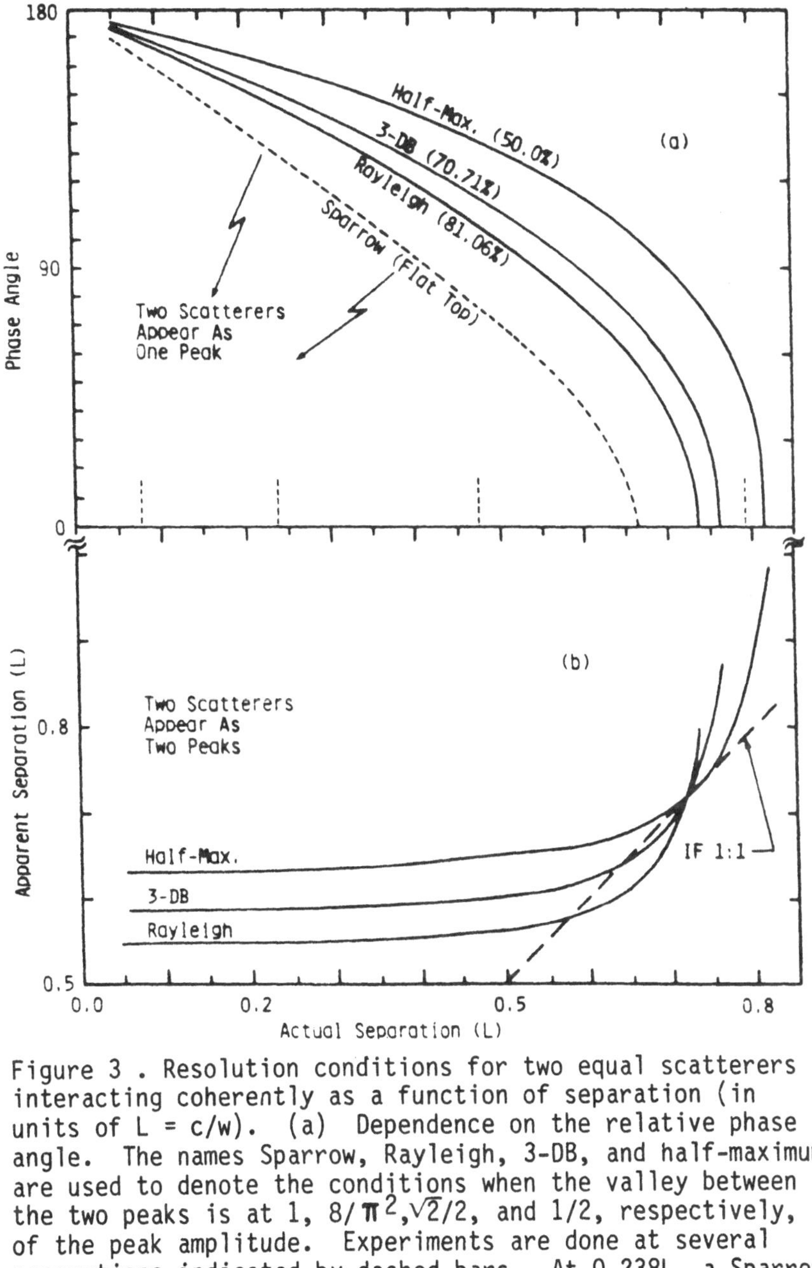

Figure 3 . Resolution conditions for two equal scatterers interacting coherently as a function of separation (in units of L = c/w). (a) Dependence on the relative phase angle. The names Sparrow, Rayleigh, 3-DB, and half-maximum are used to denote the conditions when the valley between the two peaks is at 1, $8/\pi^2$, $\sqrt{2}/2$, and 1/2, respectively, of the peak amplitude. Experiments are done at several separations indicated by dashed bars. At 0.238L, a Sparrow condition occurs near $\Phi = 131^0$ which is readily discernable in Figure 2a. (b) The apparent separation versus actual separation. Note that for actual separation smaller than 0.7L, the apparent separation tends to an asymptotic value depending on the resolution criteria.

From the interference profiles in Figures 1 and 2, we may choose to study the systematic variations of the peak amplitude and the peak displacement (apparent separation minus actual separation)

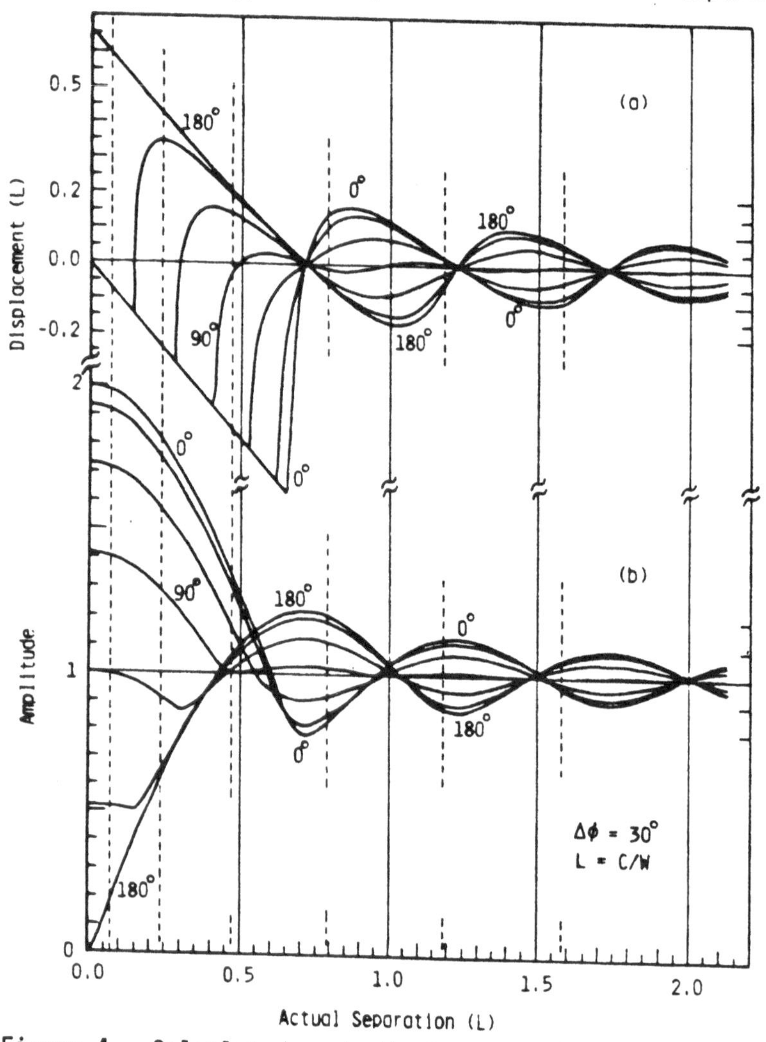

Figure 4 . Calculated peak displacement (apparent separation minus actual) and amplitude for two equal scatterers interacting coherently as a function of actual separation and relative phase angles from in-phase (0°) to out-of-phase (180°) in steps of 30°. Dashed lines denote the separations at which interference profiles are measured. Note that the curves exhibit an oscillatory behavior with a period of L. The amplitude curves in (b) are L/4 (or $\pi/2$) phase-shifted with respect to the displacement curves in (a).

as a function of the actual separation and in steps of relative phases. For the sinX/X model, the plots in Figure 4 are universal because the actual separation, expressed in L, is inversely scaled with the bandwidth w. Several separations at which the interference profiles have been measured are indicated by dashed lines. Note that the curves exhibit an oscillatory

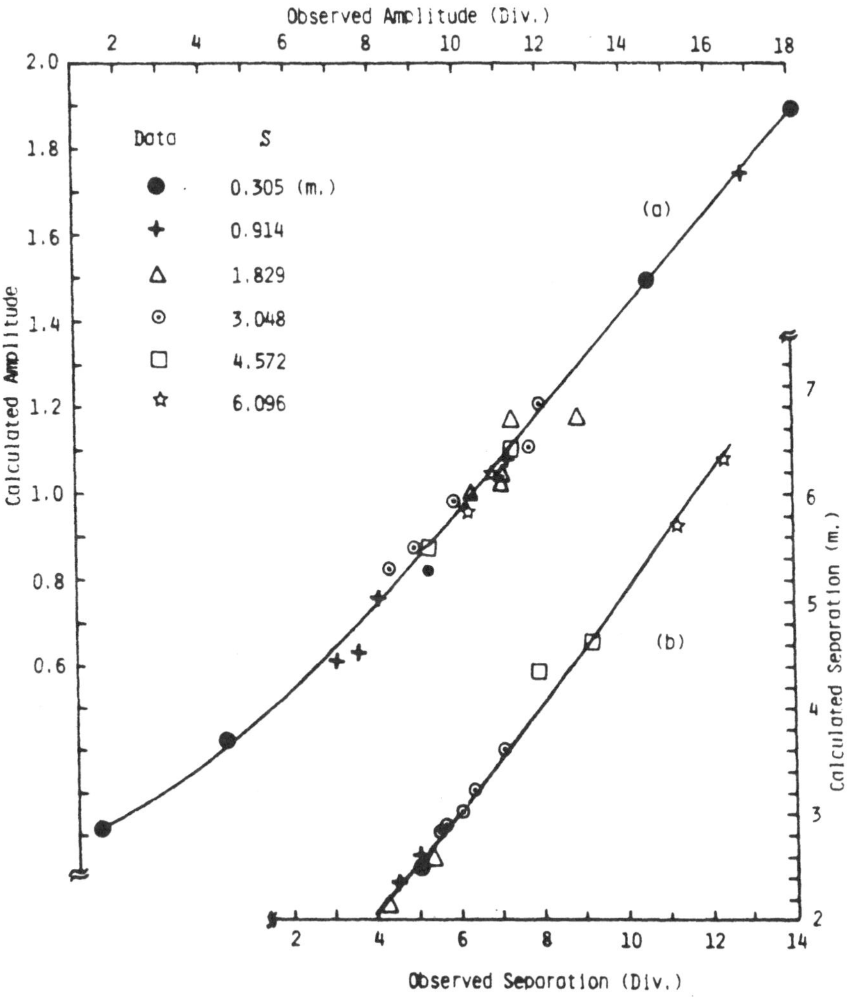

Figure 5 . Correlations of measurements and calculations. At six different down-range settings the experimentally observed peak separations (whenever resolved) and amplitudes are plotted, respectively, against the calculated ones. The smooth curves are least-squared fitted which provide a self-consistent calibration of the scales in Figure 1.

behavior of period L at separations larger than 0.5 L, and that the nodes in amplitude curves are phase-shifted by L/4 (or $\pi/2$) with respect to those in displacements. Thus, the interference effect will always be present either in amplitude or in displacement, or both.

At a larger down-range separation, the interference effect becomes small (but not zero). For example, at separations of 5 and 10, the displacements would vary between ±3.05% and ±1.52%, respectively, in units of L. The amplitudes would in turn vary between ±3.0% and ±1.6% about unity. This implies that distributed scatterers from many resolution lengths L apart would be coupled together coherently.

In Figure 5 we show the correlations of the measured peak amplitudes and separations from six sets of interference profiles with respect to the calculated values. The smooth curves are least-square fitted each with a 3rd order polynomial. The amplitude scale departs from linear at low level of signal as expected. For the separations, a straight line can be drawn near the data points with an average deviation of only ±1%, except one point. Thus, a quantitative calibration of both the amplitude and range scales has been achieved in a self-consistent fashion.

VII. Conclusions

The nice agreement between experiment and simulation provides a strong support to the single-photon coherence zone concept. At mm wavelength, the opportunity to fine tune the relative phase by a handscrew is unique. The two-scatterer interference is basic to all interference effects involving multiple scatterers. Further work on combinations of simple targets such as trihedrals and dihedrals at different polarizations should lead us to better understand the radar signatures from complex targets and targets in clutter. We wish to thank J. G. Bull, E. R. Rosenthal, M. E. Lavelle, and M. McMahon for helpful discussions and technical assistance.

VIII. References

1. M. Born and E. Wolf, "Principles of Optics" (Pergamon, Oxford, 1959).
2. E. Hecht and A. Zajac, "Optics" (Addison-Wesley, Reading, Mass., 1974), Chapter 12, and p. 360.
3. P.A.M. Dirac, "The Principle of Quantum Mechanics" (Oxford, Clarendon Press, 4th edition, 1958) p.9.
4. M. Born, "Atomic Physics" (Hafner Co., New York, 7th edition, 1962) p. 81.
5. B. J. Thompson, in "Optics in Radar Systems" SPIE Vol. 128, (1977) p. 30.
6. M. I. Skolnik, "Introduction to Radar Systems" (McGraw-Hill, New York, 2nd edition 1980), Chapter 3.

III.11 (SP.5)

POLARIZATION DEPENDENCE IN ANGLE TRACKING SYSTEMS

Daniel D. Carpenter

TRW Electronics and Defense Group
Redondo Beach, CA 90278

ABSTRACT. An analytical model is developed for a class of multimode antenna tracking feeds. This class includes certain single aperture multimode waveguide feeds and multiarm — multimode spiral feeds. The discrete Fourier transform of the mode sampling voltages is utilized to provide error and reference mode outputs for both senses of circular polarization. This is accomplished by modeling the vector heights of the mode samples as tilted beams with linear polarization for small pointing errors. The effect of implementation errors is derived and shown to produce undesirable modes that degrade the tracking performance.

The amplitude-phase implementation errors produce modes that cause cross-polarized tracking errors and a polarization dependent boresight shift. The tracking error trajectories for these effects are derived by solving the eigenvalue problem for the error measurement matrix. The decoupled equations produce the ideal weightless antenna tracking properties: first, the trajectories terminate on the boresight shift position; second, a phase shift causes a spiral-in trajectory; third, cross-polarization effects cause the target to approach asymptotically the minor axis of the polarization ellipse; fourth, adding a phase shift modifies this approach angle until the trajectory overshoots and starts to spiral in.

1. INTRODUCTION

An important feature of performance for many communications and radar systems is the end-game angle tracking property that follows acquisition. The small angle properties are important in

W.-M. Boerner et al. (eds.), Inverse Methods in Electromagnetic Imaging, Part 2, 683–694.

determining the antenna control system's ability to point continuously at the source or target. For a beacon or fixed target, the antenna feed may be required only to extract a single linear or circularly-polarized component from the incident field. For a target that has unknown or changing position and orientation, a polarization diversity feed may be required to avoid a total loss of signal. This can be accomplished by extracting two orthogonally-polarized components such as the horizontal and vertical, or the right and left circularly-polarized components.

In this paper we consider a class of circularly-polarized multimode feeds often used in practice. This includes the single aperture multimode waveguide feeds and multiarm-multimode spiral feeds. For the waveguide feeds, the single aperture permits simultaneous optimization of the error or "difference" (Δ) mode and reference or "sum" (Σ) mode channels. Implementation with corrugated waveguide permits the use of hybrid modes capable of wide bandwidth operation [2,4,8,9]. The circularly-polarized single aperture multiarm-multimode spirals, and other arrays of linear radiators, are naturally included due to the similarity of their radiation patterns with those of the TE_{11} and TE_{21} circular waveguide modes [3].

A dual-mode feed system is characterized by the unique polar form of the pointing errors. For circular polarization, the Δ mode output is proportional to the target small angle θ off boresight, and its phase is proportional to its azimuth orientation angle ϕ. The dominant Σ mode provides the proper amplitude and phase reference to extract the pointing errors [5]. This normalization can be expressed in the complex form

$$\frac{\Delta}{\Sigma} = K\,\theta\,e^{i\phi} = K\,(\alpha + i\beta) \qquad (1.1)$$

where K is a scale factor [1]. A multimode feed utilizes additional Δ and Σ mode for other received polarizations, such as linear and elliptical polarization [2,7,8,9]. Many original ideas about the use of these modes appear in the classic Chadwick and Shelton paper [3].

The usual assumption made in multimode analysis is that the ideal Δ and Σ modes are extracted. This results in expressions similar to (1.1). However, on boresight (θ=0) no constant boreshift type term is produced by this form. The Δ and Σ modes must be extracted by some finite number of mode samples to approximate the continuous fields. The mode sampling (including arm) voltages and any mode forming networks determine the overall amplitude-phase imbalances. An important effect is to produce a boresight shift term.

A multimode feed model is developed to represent the approximate effects of amplitude-phase implementation errors of the mode sampling and mode forming networks. The mode sampling outputs are assumed, by reciprocity, to produce a linearly-polarized tilted beam near boresight. The notion of a tilted beam for spiral feeds is used in [6]. The discrete Fourier transform of these "voltages" is used to produce Δ and Σ modes for right and left circular polarization. The implementation errors are shown to produce additional modes causing cross-polarization effects and a polarization dependent boresight shift. Properties of the error trajectories are obtained by extending the results of Carpenter [1] to include the additional boresight shift effects. The study in [1] extended the results of Cook and Lowell [5] by adding phase implementation errors (with ideal modes) and providing mathematical arguments to prove their observations on the minor axes seeking effect.

2. BASIC PROPERTIES OF MULTIMODE FEEDS

In general, a target or source to be tracked emits an elliptically-polarized electromagnetic field. This field arrives from the direction (θ,ϕ) at the receiver as indicated in Fig. 1. Assuming a small arrival angle θ off boresight, the target's azimuth and elevation pointing errors (α,β) are defined by

$$\alpha = \theta\cos\phi \quad , \quad \beta = \theta\sin\phi \tag{2.1}$$

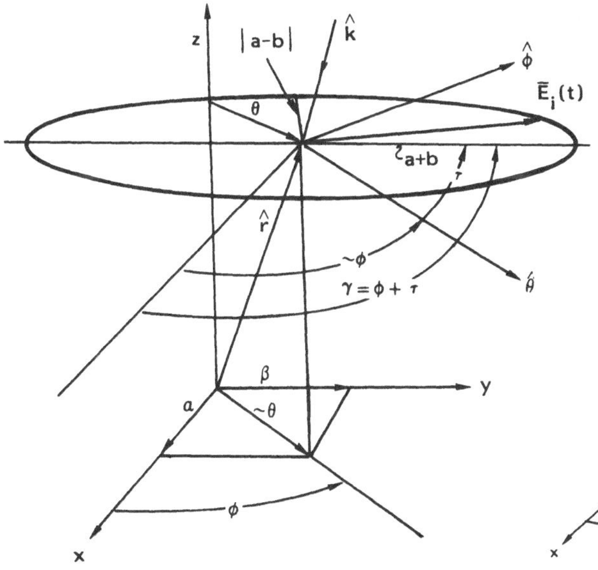

Fig. 1. Small Angle Geometry

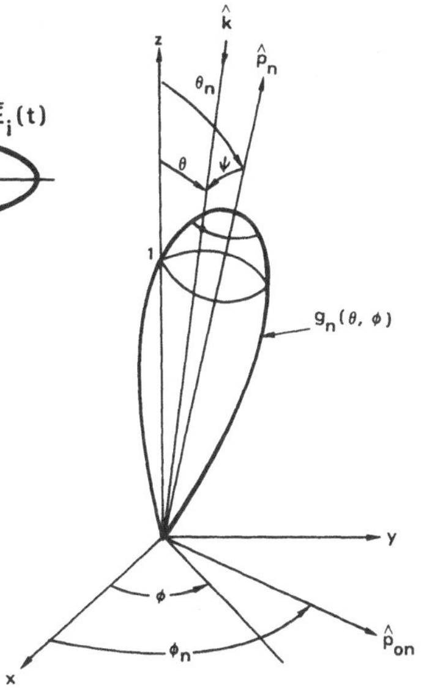

Fig. 2. Squinted Beam Geometry

The incident field can be expressed in the form

$$\overline{E}_i(t) = A(t)[ae^{-i\tau}\,\hat{e}_R^* + be^{i\tau}\,\hat{e}_L^*]\; e^{i[\omega_o t+m(t)]} \tag{2.2}$$

where τ is the tilt angle of the polarization ellipse, with semiminor and semimajor axes a and b representing the amplitudes of the right and left circularly-polarized (RCP,LCP) components. The unit base vectors are defined by the set

$$\hat{e}_R = (\hat{\theta} - i\,\hat{\phi})/\sqrt{2} = \hat{e}_L^* = \hat{e}_r\; e^{-i\phi} \tag{2.3}$$

$$\hat{e}_r = (\hat{x} - i\,\hat{y})/\sqrt{2} = \hat{e}_\ell^* \tag{2.4}$$

where we have used, for the small angle geometry,

$$\gamma = \tau + \phi \tag{2.5}$$

The tracking system must provide a means of estimating the target's pointing errors (α,β) from the incident field. The basic property of multimode feed can be obtained from the circularly-polarized TE_{11} - TM_{01} dual mode case, considered in the classic paper of Cook and Lowell [5]. Near boresight it is assumed that the far-field pattern of the feed and antenna system preserves the polarization properties of the two modes. The TE_{11} pattern is assumed constant and the TM_{01} mode pattern is linear with θ in this region. Some specific studies in support of this assumption can be found in [1,4]. The model demonstrates that for a circularly-polarized incident field the Δ mode produces a polar form for the tracking errors, and the Σ mode can be used as an amplitude and phase reference to produce the form of (1.1). The analysis in [5,10] was based on the assumption that the four mode samples produced the ideal Δ and Σ modes near boresight. For an elliptically polarized field, (1.1) is modified by a cross-polarized term. This effect will be considered for the TM_{01} mode and for the TE_{21} and TE_{01} modes used in circular waveguide tracking feeds [2,3,7,8,9].

The far-field patterns of the Δ and Σ modes are given by their effective vector heights $\overline{h}_n(\theta,\phi)$. The open-circuited voltages produced by the incident field represent the nth mode voltage.

$$\overline{M}_n(t) = \overline{h}_n(\theta,\phi)\cdot\overline{E}_i(t) \tag{2.6}$$

For small angles these expressions for the effective heights and the corresponding Δ and Σ modes and circularly-polarized TE_{11} mode outputs are, for different values of K,

$$TM_{01}:\ \overline{h} = \hat{\theta}K\theta \qquad M_0 = \frac{K}{2}\,\theta[ae^{i(\phi-\gamma)} + be^{-i(\phi-\gamma)}] \tag{2.7}$$

$$TE^o_{21}:\ \overline{h} = iK\theta[\hat{e}_R e^{-2i\phi} - \hat{e}_L e^{2i\phi}]\ ,\quad M^o_{21} = iK\theta[ae^{-i(\phi+\gamma)} - be^{i(\phi+\gamma)}] \tag{2.8}$$

$$TE^{e}_{21}: \overline{h}=K\theta\ [\hat{e}_R e^{-2i\phi} + \hat{e}_L e^{2i\phi}]\ ,\ M^{e}_{21} = K\theta[ae^{-i(\phi+\gamma)} + be^{i(\phi+\gamma)}] \quad (2.9)$$

$$TE_{01}: \overline{h} = \hat{\phi}K\theta \qquad M_{01} = i\frac{K}{2}\theta[ae^{i(\phi-\gamma)} - be^{-i(\phi+\gamma)}] \quad (2.10)$$

$$TE_{11}(RCP): \overline{h} = \hat{e}_R e^{-i\phi} \qquad M_1 = ae^{-i\gamma} \quad (2.11)$$

$$TE_{11}(LCP): \overline{h} = \hat{e}_L e^{i\phi} \qquad M_{-1} = be^{i\gamma} \quad (2.12)$$

Selecting RCP outputs, the system tracking errors (e_α, e_β) for these feeds can be put in the complex form

$$\frac{\Delta}{aK\Sigma} = \varepsilon = e_\alpha + ie_\beta \quad (2.13)$$

to yield the results

$$\varepsilon_o = \theta e^{i\phi} + (\frac{b}{a})\ \theta e^{-i(\phi-2\gamma)} \quad (2.14)$$

$$\varepsilon^{o}_{21} = i[\theta e^{i\phi} - (\frac{b}{a})\ \theta e^{-i(\phi+2\gamma)}] \quad (2.15)$$

$$\varepsilon^{e}_{21} = \theta e^{i\phi} + (\frac{b}{a})\ \theta e^{-i(\phi+2\gamma)} \quad (2.16)$$

$$\varepsilon_{01} = i[\theta e^{i\phi} - (\frac{b}{a})\ \theta e^{i(\phi-2\gamma)}] \quad (2.17)$$

These basic RCP feeds (similarly for LCP versions) respond to both senses of polarization. The cross-polarization term is proportional to b and tends to deteriorate the performance. In [5], for the TM_{01} case, it was observed that (2.14) caused the target to approach the minor axis of the polarization ellipse. When b approaches a, linear polarization occurs, and when τ became $\pm\ \pi/2$, the value ε_0 becomes zero. This null polarization produces linear trajectories with slope $\pi/2-\gamma_o$ that approaches the minor axis of ellipse.

For the TE_{11}-TM_{01} dual mode case, the null polarization occurs for linear polarization when $\tau=0$. For linear-polarized signals, the trimode feed TE_{11}-TM_{01}-TE_{01} can be used [3,4]. Combining the output modes in phase quadrature produces

$$\varepsilon_3 = \varepsilon_o - i\varepsilon_{01} = \theta e^{i\phi} \quad (2.18)$$

the null polarization is now for LCP (a=0). A similar set of arguments holds for the even and odd TE_{21} modes.

These ideal feed cases do not account for the important implementation effects of amplitude-phase imbalances. A model of the feed system that includes these effects is considered next.

3. MULTIMODE FEED MODEL

Assuming reciprocity, a mode sample V_n represents a far-field pattern described by its complex vector effective height, $\bar{h}_n(\theta,\phi)$. When receiving, this voltage is obtained from the expression.

$$V_k(t) = \bar{h}_k(\theta,\phi) \cdot \bar{E}_i(t) \tag{3.1}$$

For the multimode feed we want to extract Δ and Σ modes for right and left circular polarization. The modes must be capable of representing the lower ordered waveguide modes and related ones for multiarm-multimode spirals. This can be accomplished by considering the discrete Fourier transform of the N voltages V_n

$$M_m = \sum_{n=1}^{N} V_n e^{im\phi_n} \quad , \phi_n = -(n-1)\ 2\ \pi/N \tag{3.2}$$

For m=0, ±1, ±2, the modes M_1, M_2, $M_{-1} = M_{N-1}$, M_{-3}, and $M_0 = M_N$ represent RCΣ, RCΔ, LCΣ, LCΔ and Δ for the TM_{01} mode respectively. We note that $m \neq 0$. $N \geq 5$ elements are required for a unique set of Σ and Δ modes for both senses of circular polarization.

In order to obtain the desired feed model, (3.2) must represent the ideal multimode equations for the basic feeds considered in the previous section. It must then be capable of representing implementation errors including mode-forming networks. However, this can be obtained by using the properties of the inverse Fourier transform as follows. The overall scattering matrix transmittances $\eta_n(m)$ represent the net amplitude-phase errors imparted to each V_n for the mth mode the actual mode outputs can be expanded in terms of the ideal modes created:

$$\tilde{M}_m = \sum_{n=1}^{N} \eta_n(m)\ V_n e^{im\phi_n} = \sum_{r=1}^{N} c_r(m)\ M_r \tag{3.3}$$

with ideal mode content

$$c_r(m) = \frac{1}{N} \sum_{n=1}^{N} \eta_n(m) e^{-i(r-m)\phi_n} \tag{3.4}$$

The necessary step required to demonstrate the desired results can be accomplished by obtaining an approximation for h_n. It is assumed that the individual vector heights can be represented by a linearly-polarized tilted beam in the

$$\bar{h}_n(\theta,\phi) = g_n(\theta,\phi)\ \hat{p}_n(\theta,\phi) \tag{3.5}$$

The unit polarization vector is taken as oriented at the angle ϕ_n in the (x,y) plane and can be written in the form

$$\hat{p}_n(\theta,\phi) = \sqrt{2}\,[\hat{e}_R\, e^{-i(\phi-\phi_n)} + \hat{e}_L\, e^{i(\phi-\phi_n)}] \tag{3.6}$$

The voltage pattern is represented by a gaussian beam tilted in the direction (θ_n,ϕ_n), where ψ is the small angle between the incident field and the beam axis as shown in Fig. 2. For a linear variation with θ near boresight, neglecting higher-order terms, we have the approximation

$$g_n(\theta,\phi) \cong [1 + K\,\theta\cos(\phi-\phi_n)] \tag{3.7}$$

The resulting expression for $\overline{h}_n$ becomes

$$\overline{h}_n(\theta,\phi) = \hat{e}_R\, e^{-i(\phi-\phi_n)} + \hat{e}_L\, e^{i(\phi-\phi_n)} + \hat{\theta}K\theta + \frac{K}{2}\theta\,[\hat{e}_R\, e^{-2i(\phi-\phi_n)} + \hat{e}_L\, e^{2i(\phi-\phi_n)}] \tag{3.8}$$

From (2.11) it is observed that only the TE_{01} mode is not included. However, this does appear necessary because the orthogonal modes of (3.8) represent the TM_{01} mode and the modes necessary for polarization diversity for waveguide and multimode spirals. The specific outputs are obtained from (3.1) and (3.8) as

$$M_m = \sum_{n=1}^{N} \overline{h}_n(\theta,\phi)e^{im\phi_n} \cdot \overline{E}_i(t) = \overline{h}(m)\cdot\overline{E}_i(t) \tag{3.9}$$

with the results

$$\overline{h}(1) = \hat{e}_R e^{-i\phi} \qquad M_1 = ae^{-i\gamma} \tag{3.10}$$

$$\overline{h}(2) = \hat{e}_R K\theta e^{-2i\phi} \qquad M_2 = aK\theta e^{-i(\phi+\gamma)} \tag{3.11}$$

$$\overline{h}(-1) = \hat{e}_L e^{i\phi} \qquad M_{-1} = be^{i\gamma} \tag{3.12}$$

$$\overline{h}(-2) = \hat{e}_L K\theta e^{2i\phi} \qquad M_{-2} = bK\theta e^{i(\phi+\gamma)} \tag{3.13}$$

$$\overline{h}(0) = \hat{\theta}K\theta \qquad M_0 = \frac{K}{2}\theta[ae^{i(\phi-\gamma)} + be^{-i(\phi-\gamma)}] \tag{3.14}$$

From (2.13) we have the complex tracking error expressions (RCP)

$$\frac{M_0}{KM_1} = \theta e^{i\phi} + (b/a)\theta e^{-i(\phi-2\gamma)} \tag{3.15}$$

$$\frac{M_2^*}{KM_1^*} = \frac{M_{-2}}{KM_{-1}} = \theta e^{i\phi} \tag{3.16}$$

Eq. (3.15) represents the TM01 case (m=0), and the last two indicate ideal mode extraction for right and left circular polariztion.

The effect of implementation errors for m=0 is to add the m=±1(Σ) modes, because the m=±2 modes would be beyond cutoff. For the m=±2(Δ) mode we assume that the m=0 mode is not supported or desired. Therefore, for m=0, 2, we have from (2.13)

$$\tilde{\varepsilon}_0 = \tilde{K}(0)\ [\theta e^{-i(\phi+\Phi)} + (b/a)\ \theta e^{-i(\phi-\Phi-2\gamma)} - \tilde{B}(0)] \tag{3.17}$$

$$\tilde{\varepsilon}_2 = \tilde{K}(2)\ [\theta e^{i(\phi+\Phi)} + (b/a)\ \theta e^{-i(\phi-\Phi+2\gamma)} - \tilde{B}(2)] \tag{3.18}$$

where K, r and B represent normalizations due to c_1 (m)M_1 and Φ is a phase error not removed by calibration, defined by the equations

$$\tilde{B}(0) = \frac{1}{K}\ [\tilde{B}_1(0) + \tilde{B}_{-1}(0)\ \frac{b}{a}\ e^{2i\gamma}]\ e^{i\Phi} \tag{3.19}$$

$$\tilde{B}(2) = \frac{1}{K}\ [\tilde{B}_1(2) + \tilde{B}_{-1}(2)\ \frac{b}{a}\ e^{-2i\gamma}]\ e^{i\Phi} \tag{3.20}$$

$$K(0) = c_0(0)/c_1(1)e^{-i\Phi}, \quad K^*(2) = c_2(2)/c_1(1)\ e^{i\Phi} \tag{3.21}$$

$$B_m(0) = c_m(0)/c_1(1) \quad , \quad B_m^*(2) = c_m(2)/c_2(2) \tag{3.22}$$

$$r^*(2) = c_{-2}(2)/c_2(2) \tag{3.23}$$

The ideal m=0 mode responds to both senses of polarization with cross-polarization proportional to b. For the m=±2 ideal Δ modes, there are no cross-polarization effects. However, implementation errors create cross-polarized terms due to the m=±2 modes which are proportional to r(m)b. Eq. (3.19) can be used as a representative case by adjusting (b/a) to account for the usually small value of r(m) and (Φ-2γ) to correspond to -(Φ+2γ).

4. MULTIMODE FEED TRACKING PROPERTIES

The basic tracking equations can be obtained from (4.11) and written in component form as

$$e_\alpha = K_\alpha\ [\theta\cos(\phi+\Phi) + (b/a)\ \theta\cos(\phi-2\gamma-\Phi) - B_\alpha] \tag{4.1}$$

$$e_\beta = K_\beta\ [\theta\sin(\phi+\Phi) - (b/a)\ \theta\sin(\phi-2\gamma-\phi) - B_\beta] \tag{4.2}$$

Assuming the amplitude scale factors are calibrated out, we can express (4.1) in matrix form as

$$E = MX - B \tag{4.3}$$

where X is the target's position, E is the position as measured by the system, and B is the constant term, defined by

$$X = \begin{pmatrix} x \\ y \end{pmatrix} = \begin{pmatrix} \alpha \\ \beta \end{pmatrix}, \quad E = \begin{pmatrix} e_\alpha \\ e_\beta \end{pmatrix}, \quad B = \begin{pmatrix} B_\alpha \\ B_\beta \end{pmatrix} \tag{4.4}$$

and M is the system error measurement matrix

$$M = \begin{pmatrix} \cos\Phi & -\sin\Phi \\ \sin\Phi & \cos\Phi \end{pmatrix} + \left(\frac{b}{a}\right) \begin{pmatrix} \cos(\Phi+2\gamma) & \sin(\Phi+2\gamma) \\ \sin(\Phi+2\gamma) & -\cos(\Phi+2\gamma) \end{pmatrix} \tag{4.5}$$

When the system errors are zero (E=0), the indicated tracking errors are defined as the boresight shift

$$X_B = \begin{pmatrix} \alpha_B \\ \beta_B \end{pmatrix} = M^{-1} B \tag{4.6}$$

The ideal error correction path can be obtained by extending the basic idea used in [1,5] to include Φ and B. This process reduces the initial error X_o by the amount δE_o, and the new position as seen in Fig. 3 is

$$X_1 = X_0 - \delta E_o = [I - \delta M] X_o + \delta B = M_e X_o - \delta B \tag{4.7}$$

After n steps the target is at

$$X_n = M_e^n X_o + \delta \sum_{k=0}^{n-1} M_e^k B = M_e^n [X_0 - X_B] + X_B \tag{4.8}$$

A smooth curve is obtained by adjusting δ as seen in Fig. 4 for the cross-polarization case. Centering (4.8) yields

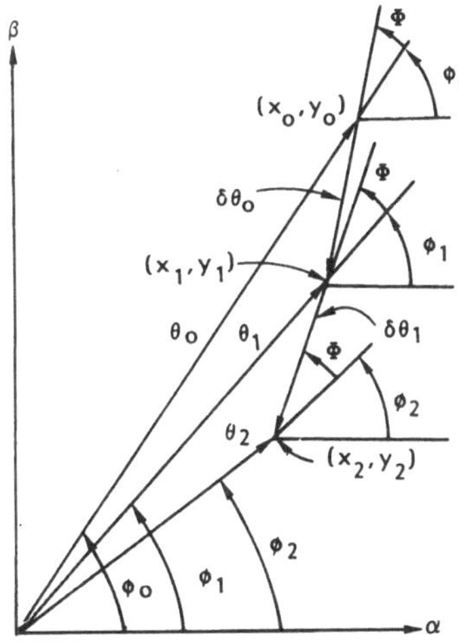

Fig. 3. Iterative Corrections

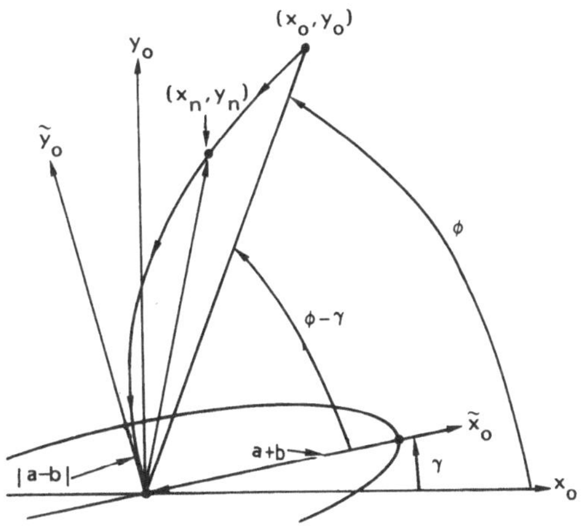

Fig. 4. Cross-Polarization Effect

$$X_{nc} = [X_n - X_B] = M_e^n[X_0 - X_B] = M_e^n X_{oc} \tag{4.9}$$

The tracking errors can be uncoupled by diagonalizing M

$$S^{-1} MS = D \tag{4.10}$$

where S is the modal matrix with columns equal to the eigenvectors of M, and D is the diagonal matrix of the eigenvalues $\lambda_1 > \lambda_2$. The equations can be written in uncoupled form

$$\tilde{X}_{nc} = S^{-1} X_{nc} = S^{-1} M_e^n X_{oc} = [S^{-1} M_e S]^n S^{-1} X_{oc}$$

$$= [I - \delta S^{-1} MS]^n S^{-1} X_{oc} = [I - \delta D]^n S^{-1} X_{oc} \tag{4.11}$$

where

$$[I - \delta D]^n = \begin{bmatrix} (1-\delta\lambda_1)^n & 0 \\ 0 & (1-\delta\lambda_2)^n \end{bmatrix} \tag{4.12}$$

For $\delta\lambda_1 < 1$ this matrix approaches zero for large n so that (4.1) approaches zero. Thus $X_n = X_B$ eventually, so that the trajectory terminates on the boresight shift position,

$$X_B = M^{-1} B = (S D^{-1} S^{-1}) B \tag{4.13}$$

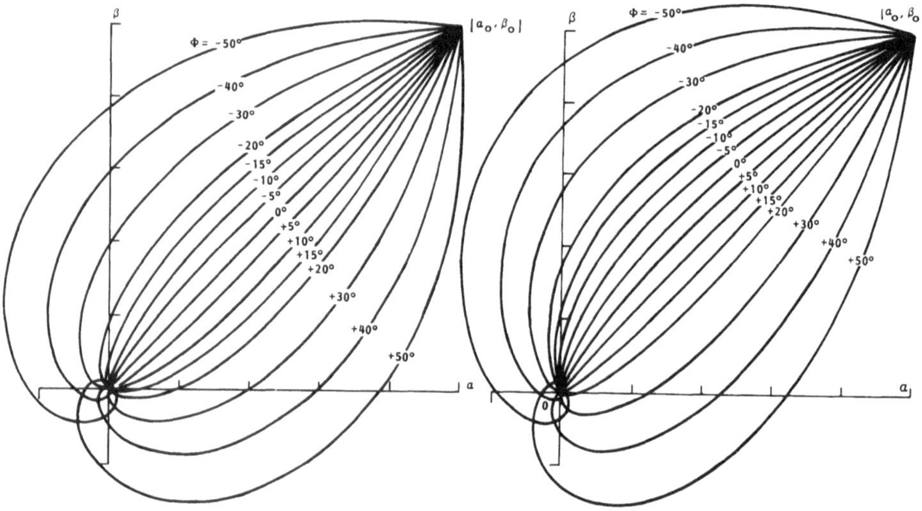

a) Phase b) Phase with Cross-Pol (b/a)=0.2,γ=0

Fig. 5. Trajectories

The special case of b=0 corresponds to a trajectory that spirals in towards the boresight point X_B. This has been graphed in Fig. 5a. For the other trajectories we have two cases. For $(b/a)^2 > \sin^2\Phi$ the eigenvalues are given by

$$\lambda_1, \lambda_2 = \cos\Phi \pm \sqrt{(b/a)^2 - \sin^2\Phi} \tag{4.14}$$

The inverse modal matrix is a rotation matrix, and the new coordinates are

$$\tilde{X}_{nc} = \begin{pmatrix} \tilde{x}_{nc} \\ \tilde{y}_{nc} \end{pmatrix} = \begin{bmatrix} \cos\gamma_0 & \sin\gamma_0 \\ -\sin\gamma_0 & \cos\gamma_0 \end{bmatrix} \begin{pmatrix} x_{nc} \\ y_{nc} \end{pmatrix} \tag{4.15}$$

where

$$\gamma_0 = \gamma + \frac{1}{2}\left[\Phi + \mathrm{Tan}^{-1}(\lambda_1 - \lambda_2)\right] \tag{4.16}$$

The uncoupled equations become

$$\tilde{X}_{nc} = [I - \delta D]^n \tilde{X}_{oc} \tag{4.17}$$

The rotation by γ changes the orientation angle ϕ_0 of X_{oc} to $\phi_0 - \gamma_0$, for X_{oc}, (Fig. 4). Therefore, we have the result

$$\frac{\tilde{y}_{nc}}{\tilde{x}_{nc}} = \left(\frac{1 - \delta\lambda_2}{1 - \delta\lambda_1}\right)^2 \frac{\tilde{y}_{oc}}{\tilde{x}_{oc}} > \mathrm{Tan}\,(\phi - \gamma_0) \tag{4.18}$$

This indicates that the trajectory at the nth step always lies above the line (x_0, y_0) due to $\lambda_1 > \lambda_2$. Therefore, the trajectory must approach the origin (X_B) asymptotically along a line inclined at angle λ_0 from the minor axis of the polarization ellipse of the incident field. When Φ is zero $(\gamma_0 = \gamma)$ the trajectory approaches X_B asymptotically along the minor axis of the polarization ellipse. For pure mode extraction $(X_B = \Phi = 0)$ Cook and Lowell pointed out this minor axis seeking effect from computer plots of (3.8).

For the phase-dominated case $(b/a)^2 < \sin^2\Phi$, the eigenvalues and eigenvectors are complex, and the uncoupled equations do not have simple interpretations. A computerized plot of a typical case is shown in Fig. 5b. The curves for $|\Phi| > 10°$ correspond to the phase dominated cases that start to overshoot the minor axis of the ellipse at $\pi/2$.

5. CONCLUSIONS

A useful class of multimode feed systems has been modeled to account for amplitude-phase implementation errors. This class includes lower order modes of circular waveguide and multiarm-multimode spirals. The implementation errors were shown to produce additional modes that produced a polarization-dependent boresight shift. They also produce curvilinear trajectories whose properties were determined as a function of phase errors and induced cross-polarization.

REFERENCES

1. Carpenter, D.D., Fleming, A.N., Vandervoet, D.B., "Monopulse RF Sensing and Station Keeping," Final technical report, No. 28668-6001-TU-00, TRW Systems, Redondo Beach, CA, Feb. 1976.

2. Clarricoats, P.J.B., Elliot, R., "Corrugated Waveguide Monopulse Feed," International Symposium on Antennas and Propagation, Session 4, pp. 61-68, Los Angeles, CA, June 1981.

3. Chadwick, G.G., Shelton, J.P., "Two Channel Monopulse Techniuques — Theory and Practice," 9th Military Electronics Conference Proc., pp. 177-181, Washington, D.C., 1965.

4. Cooper, D.N., "A New Circularly Polarized Monopulse Feed System," Proc. IEEE, 53, No. 9, pp. 1242-1254, Sept. 1965.

5. Cook, J.S., Lowell, R., "The Autotrack System," Bell System Tech. J., 42, No. 4, Part II, pp. 1283-1307, July 1963.

6. Deschamps, G.A., Dyson, J.E., "The Logarithmic Spiral in a Single-Aperture Multimode Antenna System," IEEE Trans. on Ant. Prop., Vol. AP-19, No. 1, pp. 90-96, Jan. 1971.

7. Jensen, P.A., Ajioka, S.J., "Feed Design for Large Antennas," IEEE NEREM Record, pp. 62-63, 1962.

8. Potter, P.D., "Feasibility of Inertialess Conscan Utilizing Modified DSN Feed Systems," DSN Progress Report 42-51, JPL, pp. 85-93, March-April 1979.

9. Vu, T.B., "Corrugated Horn as High-Performance Monopulse Feed," Int. J. Electronics, Vol. 34, No. 4, pp. 433-444, 1973.

10. Yodokawa, T., Hamada, S.J., "An X-Band Single Horn Autotrack Antenna Feed System," International Symposium on Antennas and Propagation, Session 4, pp. 86-89, Los Angeles, CA, June 1981.

III.12 (SS.4)

INTERPRETATION OF HIGH-RESOLUTION POLARIMETRIC RADAR TARGET DOWN-RANGE SIGNATURES USING KENNAUGH'S AND HUYNEN'S TARGET CHARACTERISTIC OPERATOR THEORIES

Anthony C. Manson and Wolfgang-M. Boerner

Communications Laboratory, Electromagnetic Imaging Division, Department of Electrical Engineering & Computer Science, M/C 154, 851 S. Morgan St., P.O. Box 4348, 1141-SEO, 4210/11-SEL, Chicago, IL 60680, USA

ABSTRACT

Basic polarimetric backscattering characteristics of elongated, truncated, cylindrical targets with attachments, such as, a spherical-capped nose, a conical-capped aft, and side-protruding fins and wings are analyzed and interpreted by adhering strictly to Kennaugh's optimal polarization target characteristic operator theory derived from the 2x2 polarization radar scattering matrix [S], and Huynen's N-target decomposition theory of the 4x4 Mueller matrix [M]. Use is made of the polarimetric target-downrange signatures collected by Lee A. Morgan and Stephen Weisbrod on the Teledyne-Micronetics out-door broadband polarization scattering matrix measurement range. It is demonstrated here that the above thoeries are correct and are applicable to the electromagnetic inverse problem of deriving target classification operators from such complete polarimetric high resolution target down-range signatures.

1. Basic Theory

It is well known (Kennaugh, 1952; Huynen, 1970; Boerner, 1981; and Cloude, 1983) that the received voltage V(t) from a radar target, with scattering matrix [S(t)], obtained by transmitting an elliptically polarized wave $\underline{h}_T$ and using a receiver antenna with receiver-polarization $\underline{h}_R$ is given by

$$V(t) = [S(t)]\underline{h}_T \cdot \underline{h}_R \qquad (1)$$

Equation (1) clearly distinguishes between antenna polarizations, $\underline{h}_T$ and $\underline{h}_R$, and target observables, represented by scattering ma-

W.-M. Boerner et al. (eds.), Inverse Methods in Electromagnetic Imaging, Part 2, 695–720.

trix [S(t)]. Hence, one often hears a target being described by its "polarization properties". Almost all practical targets produce a different voltage when illuminated by horizontal polarization (H) or with vertical (V). This is due to the vector nature of interaction of the electromagnetic wave and the three-dimensional target structure. In the phenomenological approach (Huynen 1970), we have to consider the question: How does this information relate to the target as a physical object as is discussed in detail in the companion paper by Huynen (1984).

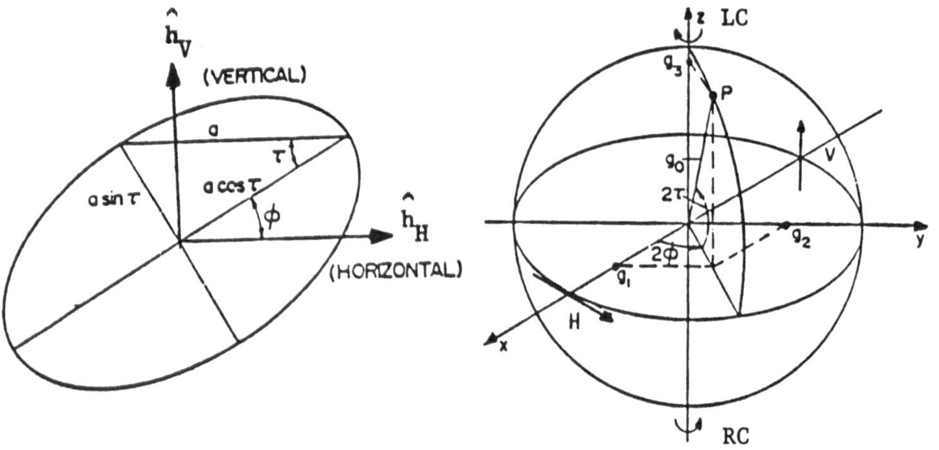

Fig. 1 Polarization Ellipse & Sphere

We have stressed the significance of target orientation ψ because it relates directly to antenna orientation ϕ. For the transmit wave we can write in terms of geometrical parameters:

$$\underline{h}(a,\alpha,\phi,\tau) = ae^{i\alpha}\begin{bmatrix}\cos\phi & -\sin\phi\\ \sin\phi & \cos\phi\end{bmatrix}\begin{bmatrix}\cos\tau\\ i\sin\tau\end{bmatrix} \quad (2)$$

where α is an absolute phase factor of the polarization vector $\underline{h}$ and τ is the ellipticity angle.

We will show shortly that every target at a given exposure and frequency has its own natural orientation ψ. For a roll-symmetric object, this angle is simply related to the target roll axis, but any odd-shaped target has its unique orientation ψ. As the target moves, each exposure produces a different orientation, which may depend on frequency. The target orientation angle ψ is easily calculated from the scattering matrix [S]; it is the orientation of the ellipse associated with the maximum polarization or x-pol nulls, which is characteristic for the target. It should be clear by now that the only relevant orientation, as far as the target's illumination is concerned, is the relative orientation Φ = ϕ - ψ.

This observation, however, does not infer that information on the relative orientation angle ψ is of no use: in fact, it may represent one of the most important radar characteristic parameters in polarimetric target downrange mapping, namely providing orientation information on the attachment of fins, wings, ducts, rods, etc. along downrange extension of a cylindrical target body as will be demonstrated in Section 6.

2. The Target Scattering Matrix

If we start with the most general polarization basis (AB), then the target matrix is simply the collection of four complex numbers:

$$[S(AB)] = \begin{bmatrix} S_{AA} & S_{AB} \\ S_{BA} & S_{BB} \end{bmatrix} \tag{3}$$

where in this study we strictly assume to be dealing with a monostatic radar system placed in reciprocal propagation space for which $S_{AB} = S_{BA}$.

A unitary transformation $[U(AB;A'B')]$ applied to (3) results in the matrix [S] presented in a new basic frame (A'B'):

$$[S(A'B')] = [U]^T[S(AB)][U] \tag{4}$$

Matrix [U] can be given a particularly simple form (Kennaugh, 1950 and 1954):

$$[U] = \frac{1}{\sqrt{1 + |\rho|^2}} \begin{bmatrix} 1 & -\rho^* \\ \rho & 1 \end{bmatrix} \tag{5}$$

where ρ is the complex polarization ratio.

The coefficients of S(A'B'), based on (4) are found as (Boerner, 1981)

$$\begin{aligned} S_{A'A'} &= (1 + |\rho|^2)^{-1}(S_{AA} + \rho^2 S_{BB} + 2\rho S_{AB}) \\ S_{A'B'} &= (1 + |\rho|^2)^{-1}(-\rho^* S_{AA} + \rho S_{BB} + [1-|\rho|^2)S_{AB}] \\ S_{B'B'} &= (1 + |\rho|^2)^{-1}(\rho^{*2} S_{AA} + S_{BB} - 2\rho^* S_{AB}) \end{aligned} \tag{6}$$

with the span and the determinant being transformation invariants (Boerner, 1980):

$$\text{Span}\ \{[S]\} = |S_{AA}|^2 + 2|S_{AB}|^2 + |S_{BB}|^2 = \text{Span}\ \{[S']\} = |S_{A'A'}|^2 + + 2|S_{A'B'}|^2 + |S_{B'B'}|^2 = \text{invariant} \tag{7}$$

$$|\det\{[S(AB)]| = |\det\{[SA'B')]\}| = \text{invariant}. \tag{8}$$

We may use for [U] the orthonormal basis vectors $[U] = [\underline{m},\underline{m}\perp]$, where $\underline{m} = \underline{h}_m$, the so-called maximum polarization or x-pol null is the eigenvector solution of the eigenvalue problem for [S(AB)]:

$$[S(AB)]\underline{x} = s\underline{x}^* \tag{9}$$

Because of the orthonormal properties of $\underline{m}$ and $\underline{m}\perp$, which satisfy (9), we find the condition that the off-diagonal terms $[S_{A'B'}]$ in (7) become zero, and this fact can be used in turn to solve for ρ and hence for $\underline{m}$ and $\underline{m}\perp$, without solving the eigenvalue problem (9) directly.

We now write for $\underline{m} = \underline{h}_m$ in geometrical variables:

$$\underline{m}(\phi, \tau_m) = \begin{bmatrix} \cos\phi & -\sin\phi \\ \sin\phi & \cos\phi \end{bmatrix} \begin{bmatrix} \cos\tau_m \\ i\sin\tau_m \end{bmatrix} \tag{10}$$

For the complex valued eigenvalues s_1 and s_2, which satisfy (9), we write:

$$s_1 = me^{2i(\nu+\beta)} \quad ; \quad s_2 = m\tan^2\gamma e^{-2i(\nu-\beta)} \tag{11}$$

We have now a complete description of the scattering matrix [S(AB)] in terms of geometrical target parameters. From (4) we obtain:

$$[S(AB)] = [U^*(\underline{m},\underline{m}\perp)] \begin{bmatrix} me^{2i(\nu+\beta)} & 0 \\ 0 & m\tan^2\gamma e^{-2i(\nu-\beta)} \end{bmatrix} [U^*(\underline{m},\underline{m}\perp)]^T \tag{12}$$

The geometric parameters are m, γ, φ, τ_m, ν and β. The positive quantity m denotes target magnitude. It may be viewed as an overall measure for target size and can be related to the span of the matrix as defined in (7). The angle γ is the characteristic angle; it determines separation of the targets copol-nulls on the polarization sphere. The angle φ is the celebrated target orientation angle.

As soon as the angle φ is found, it can be separated from all other target parameters and hence these target parameters are orientation independent (but are still dependent on target exposure and frequency, etc.) and can be used to characterize target structure. Without the basic mathematical framework (12) in which to express

[S(AB)], it would not have been possible to compute ψ and at once to eliminate it from the other target parameters.

Also, it is easily seen, that the combination of equations (2), (10), and (12) into (1) results in only the relative orientation $\Phi = \phi - \psi$ having significance in the target return, as required by common sense. The three angles ψ, τ_m, and ν are simply the Eulerian rotation angles about three orthogonal axes (see Fig. 2, which is the geometrical equivalent to equation (12)). Finally, β is the absolute phase of the target; it disappears with power measurements and provides information only in an interferometric (holographic) sense (Boerner, 1980).

Aside from its geometrical significance, the Eulerian angles are also powerful indicators of target structure: ν is called the skip-angle because it relates to double bounce or in general to even versus odd bounce scattering, τ_m is the helicity-angle and is a powerful indicator of target symmetry or non-symmetry (for symmetric targets, $\tau_m = 0$). As a note of caution, it must be empha-

$0 < \gamma < 45°$
$-45° < \nu < 45°$
$-45° < \tau_m < 45°$
$-90° < \psi < 90°$

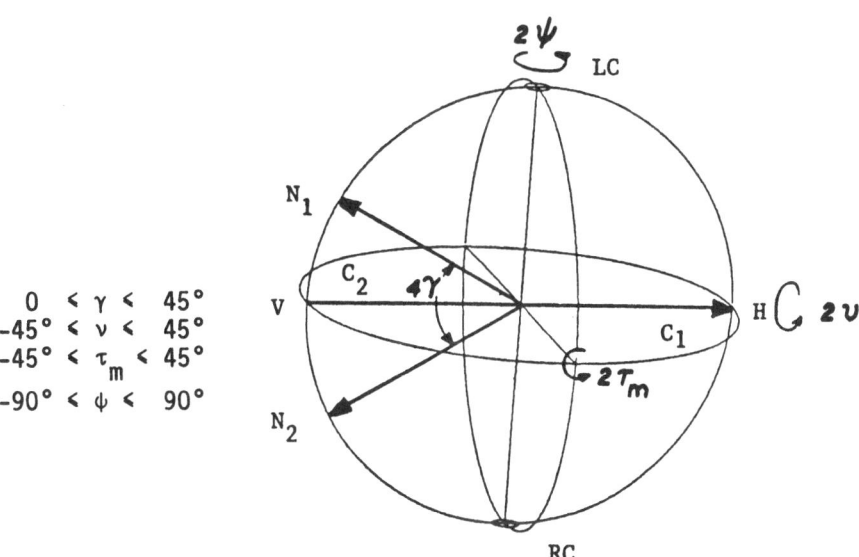

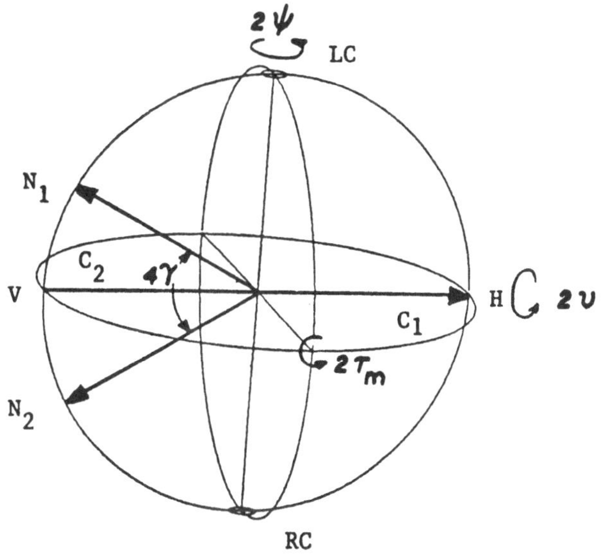

Fig. 2 Kennaugh's Target Characteristic Operator: Huynen's Polarization Fork (Illustrated for ψ=90°, τ_m=0, ν=0)

sized here that in the monostatic, nonreciprocal propagation case for which $S_{AB} \neq S_{BA}$; in addition to the absolute phase β and the five independent target characteristic parameters, m, ψ, τ_m, ν, δ, two additional characteristic angles need to be introduced as was shown in detail in (Davidovitz and Boerner, 1983) and as it was derived in the M.Sc. thesis of Davidovitz (1983). In those monostatic, nonreciprocal or in the bistatic cases for which $S_{AB} \neq S_{BA}$ the target characteristic angle concept of Kennaugh (1952) and Huynen (1970) does not hold and requires extension.

All this information, and much more to come, would have been impossible to extract if we blindly followed some general sort of data processing scheme without keeping a mental focus on data relevance to target structure. The polarization-null plot technique was developed by the late Professor Edward M. Kennaugh and we refer to his illuminating reports [Kennaugh, 1949-1954] as summarized in Moffatt [1983] and in Kennaugh and Moffatt [1984].

3. The Received Power (Huynen, 1970)

The derivation of received backscattered power, based upon equation (1), is tedious but straightforward. It turns out that to the voltage equation, $V = [S]\underline{h}_T \cdot \underline{h}_R$, there corresponds a similar linear relationship for power received. Let $P = |V|^2$, then

$$P(t) = [M(t)]\underline{g}_T \cdot \underline{g}_R \qquad (13)$$

where [M] is the 4x4 real-valued Stokes matrix (also called Mueller matrix in optics) which represents the target in terms of power measurements, and $\underline{g}_T$, $\underline{g}_R$ are the Stokes vectors which correspond to the elliptically polarized antennas in power. Most often (in optics) the Stokes vector $\underline{g}$ is given in geometrical parameters as follows:

$$\underline{g}(g_0, \phi, \tau) = \begin{bmatrix} g_0 \\ g_0 \cos 2\phi \cos 2\tau \\ g_0 \sin 2\phi \cos 2\tau \\ g_0 \sin 2\tau \end{bmatrix} = \begin{bmatrix} I \\ Q \\ U \\ V \end{bmatrix} \qquad (14)$$

The nomenclature given by (14) supplies the same information as in (2), with $g_0 = a^2$, except that the absolute phase α disppears with power measurements. Now the Stokes matrix [M] in (13) can be made equally analogous to the scattering matrix [S(AB)] in (12). We thus find:

$$[M] = \begin{bmatrix} A_0+B_0 & C & H=0 & F \\ C & A_0+B & E & G \\ H=0 & E & A_0-B & D \\ F & G & D & -A_0+B \end{bmatrix} \tag{15}$$

where

$$\begin{aligned} A_0 &= Q\ f\ \cos^2 2\tau_m \\ B_0 &= Q(1 + \cos^2 2\gamma - f \cos^2 2\tau_m) \\ B &= Q[1 + \cos^2 2\gamma - f(1 + \sin^2 2\tau_m)] \\ C &= 2Q \cos 2\gamma \cos 2\tau_m \\ D &= Q \sin^2 2\gamma \sin 4\nu \cos 2\tau_m \\ E &= -Q \sin^2 2\gamma \sin 4\nu \sin 2\tau_m \\ F &= 2Q \cos 2\gamma \sin 2\tau_m \\ G &= Q\ f \sin 4\tau_m \end{aligned} \tag{16}$$

$$Q = \frac{P_A^{\ 2}\ P_B^{\ 2}\ m^2}{8\cos^4\gamma} \tag{17}$$

$$f = 1 - \sin^2 2\gamma \sin^2 2\nu \tag{18}$$

Notice that the target orientation angle ψ does not appear in the defining equations (15, 16, 17). Instead it has been incorporated with the antennas $\underline{g}_T$ and $\underline{g}_R$, where it appears as $\Phi = \phi - \psi$ instead of ϕ in (14). This was an essential requirement for our approach to orientation-independent target discriminants. Hence all target parameters in (15) are orientation independent. It would be an easy matter to transform ψ back into the Stokes matrix (19). We would then obtain

$$[M_\psi(t)] = \begin{bmatrix} A_0+B_0 & C_\psi & H_\psi & F \\ C_\psi & A_0+B & E_\psi & G_\psi \\ H_\psi & E_\psi & A_0-B & D_\psi \\ F & G_\psi & B_\psi & -A_0+B_0 \end{bmatrix} \tag{19}$$

where:

$$H_\phi = C \sin 2\phi$$
$$C_\phi = C \cos 2\phi \tag{20}$$
$$G_\phi = G \cos 2\phi - D \sin 2\phi$$
$$D_\phi = G \sin 2\phi + D \cos 2\phi \tag{21}$$
$$E_\phi = E \cos 4\phi + B \sin 4\phi$$
$$B_\phi = -E \sin 4\phi + B \cos 4\phi \tag{22}$$

4. Target Structure Diagram

The target diagram is a pictorial representation of a single target, with orientation-bias ϕ removed. The diagram is a result of the following three equations which are defined in detail in Huynen (1970, 1978, 1982a, 1982b):

$$\begin{aligned} Q_1 &= 2A_0(B_0 + B) - (C^2 + D^2) > 0 \\ Q_2 &= 2A_0(B_0 - B) - (G^2 + H^2) > 0 \\ Q_3 &= (B_0{}^2 - B^2) - (E^2 + F^2) > 0 \end{aligned} \tag{23}$$

For a single object (at given exposure-look, pulse shape, frequency, etc.) the right-hand sides of Q_1 to Q_2 are zero. This poses an interesting set of conditions. For example, let us assume that $A_0=0$ in Q_1 and Q_2, then it must follow (because $Q_1-Q_2=0$) that $C=D=G=H=0$! Also, from these same equations we find that B_0-B and B_0+B are non-negative. Hence we find if $B_0-B=0$ or $B_0=B$, then from Q_2 and Q_3 it follows (since $Q_2=Q_3=0$) that: $E=F=G=H=0$. Finally, if $B_0+B=0$, then by Q_3 and Q_1: $E=F=C=D=0$. For these reasons the diagonal elements A_0, B_0+B, and B_0-B are called the generators of the off-diagonal Stokes parameters. A_0 is the generator of target symmetry. $B_1=(B_0-B)/2$ is, in general, the generator of target non-symmetry (if $B_1=0$ or $B_0=B$ we have a symmetrical target), while $B_2=(B_0+B)/2$ is, in general, the generator of target irregularity (if $B_2=0$, the target is regular). From these two definitions we have:

$$B_0 = B_1 + B_2$$

which again emphasizes that B_0 is the sum total of non-symmetrical and irregular target components.

We are now ready to assemble all pieces of information obtained thus far on single target parameters into a complete structure diagram (Figure 3). The diagram shows a threefold symmetry between target parameters. The three structure generators are: A_0, the generator of target symmetry; B_1, the generator of target non-symmetry; and B_2, the generator of target irregularity.

Each generator is responsible for generating two pairs of adjacent off-diagonal parameters: thus, A_0 generates the pair C&D and G&H. We already mentioned C&D. The pair G&H are coupling terms. H is a measure of coupling due to target orientation ψ. We found that if $\psi=0$, then H=0; whereas G couples the symmetric and nonsymmetric parts of the target: if G=0 (with $\psi=0$), then either the target is purely symmetric or nonsymmetric.

If $A_0=0$, then C=D=G=H=0 and we have the class of non-symmetric N-targets. N-targets play an important role in the theory of distributed targets. There they represent "residue" or "target noise" at the higher frequencies. N-targets produce the most asymmetric type of scattering (large helicity, $\tau_m = \pm 45°$), such as produced by troughs, edged interacting surfaces, helices, etc. The single N-target is given by four parameters: $B_0 > 0$, B, E, and F, for

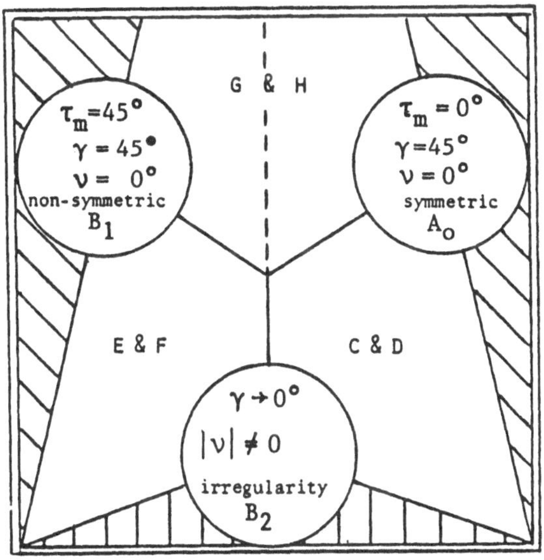

Figure 3: Target Structure Diagram for Single Radar Target

which $B_0{}^2 = B^2 + E^2 + F^2$ (or $Q_3^N = 0$) as was discussed in greater detail in (Huynen 1978, 1982 a, b).

For a target 'on-axis', with $\psi = 0$, the three generators have the following interesting relationships:

$$A_0 = Q\, f \cos^2 2\tau_m$$
$$B_1 = Q\, f \sin^2 2\tau_m$$
$$B_2 = Q(\cos^2 2\gamma + \sin^2 2\gamma \sin^2 2\nu)$$

with parameters defined previously. The first two equations are indicative for target symmetry or non-symmetry. The expression for B_2 shows that $B_2=0$, when the target is regular only if $\gamma = 45°$ and $\nu = 0°$, which defines convex or specular type of scattering.

5. Discussion of Target Parameters

The nine target parameters are of importance because they basically define a correlation matrix (after averaging). Secondly, they are model free, hence applicable to any radar target. And, thirdly, they exhibit genuine physical significance related to the target. The three basic parameters are the 'generators' $2A_0$, B_0-B, and B_0+B. Notice that the physical significance attached to these parameters is true for either beam-filled (coarse resolution) or hi-resolution targets. The former relates to averaged (integrated) global target characteristics, whereas the hi-resolution case focuses on target detail. Very fine detail is not always a desirable characteristic for target sorting, but will help for target identification purposes.

$2A_0$: Perhaps the most important parameter, and the generator of target symmetry, if $2A_0 = 0$, the target is non-symmetric or an N-target. It is associated with regular, smooth, convex types of surface scattering, which often contribute to specular returns. Since $2A_0$ is orientation independent (see Fig. 1), it serves as a good overall measure of target symmetry (for man-made targets). Incidentally, $2A_0$ is measured directly and simply by the (RC,LC) circularly polarized returns.

B_0-B: is a measure of target non-symmetry. This parameter can generally be expected to be small for man-made objects (if B_0=B we have a symmetrical target on axis). It must be noted that B_0 - B is a parameter that has been derived after the orientation effect ψ has been removed. Since B_0 - B is given by:

$B_0 - B = 2Qf\sin^2 2\tau_m$, where $f = 1 - \sin^2 2\gamma \sin^2 2\nu$

one can see that for a linear scattering matrix (Huynen, 1970: LSM) if $\tau_m \simeq 0$, $B_0 - B$ will be zero and for $\tau_m \simeq 45°$; $B_0 - B \to 2Qf$. Thus, $B_0 - B$ shows the intrinsic non-symmetry of the object. It tells us that there is no intrinsic plane of symmetry dividing the object and, therefore, the cross polarization basis will be elliptical or circular.

B_0+B: is a measure of target irregularity. This is because it is the generator of C, D, E and F, i.e. of off-diagonal terms of symmetrical as well as non-symmetrical parts of the target. Irregularity in conjunction with these other four parameters gives us important clues about the physical nature of the targets. For in stance, for a line-target: $2A_0 = C = B_0+B$, and we can expect high returns for these three plots in which case $(B_0 + B)$ is given by:

$$B_0 + B = 2Q(\cos^2 2\gamma + \sin^2 2\gamma \sin^2 2\nu)$$

To get a better idea of what irregularity means, we must first define what regularity means in terms of a scattering matrix. If we take scattering measurements in a linear polarization base, and remove the Eulerian rotations ψ and τ_m we are left with a diagonal scattering matrix. If this matrix can be represented in the form:

$$[S(H,V)] = m \begin{bmatrix} 1 & 0 \\ 0 & 1 \end{bmatrix} \quad \psi \text{ and } \tau_m \text{ removed.}$$

where m is a constant, then the target is said to be regular (sphere, flat plate), if the two diagonal terms differ in phase, or amplitude, then the target contains irregularity. Thus, we can see from the equation for $(B_0 + B)$ that it is γ and ν that produce irregularity. If $\gamma \to 0°$ or $\nu \to 45°$ then irregularity becomes apparent. Gamma (γ), the target characteristic angle, makes the amplitudes between two co-polarized amplitudes (null locations) differ and ν makes the 2 co-polarized phases differ. Depending on the value of τ_m we see that irregularity has meaning for symmetrical ($\tau_m = 0$) or non-symmetrical ($\tau_m \neq 0$) targets. The target skip angle ν indicates the phase shift of the co-polarized returns and will indicate curvature variations at the specular point (Foo, 1982; Foo, Chaudhuri, Boerner, 1983) for convex targets or multibounce effects for targets like corner reflectors derived by Foo (1982) as $\frac{k_u - k_v}{2k} = -\tan \frac{\phi_{11} - \phi_{22}}{2}$, where k_u, k_v represent the principal curvatures at the specular point and ϕ_{11}, ϕ_{22} the polarization phases of the S_{11}, S_{22} elements of [S(1,2)], respectively.

C: is a measure of global shapes. For a sphere C=0, while for a line target $C = 2A_0 = B_0 + B$ as indicated above. Since C is given by

$$C = 2Q\cos 2\gamma \cos 2\tau_m$$

we see that C measures global shape for predominantly symmetric targets. Examples of targets rendering high C values are long symmetric targets like dipoles that have linear scattering matrices of the form:

$$[S(H,V)] = \begin{bmatrix} 1 & 0 \\ 0 & 0 \end{bmatrix}$$

D: is a measure of local shape for convex surfaces and it relates to the local radii of curvature of the specular point on the surface, i.e. D=0 if the local radii at the specular point are equal, but D≠0 if they differ. In general, D is small for man-made targets, but it can be large for a cylindrical surface at broadside (see M.Sc. thesis of Bing-Yuen Foo, 1982).

E: is a parameter related to target non-symmetry (torsion); it is usually small for man-made objects. The parameter E is analogous to parameter D except that E is most sensitive when the target has circular cross-polarization nulls, whereas, D is most sensitive when the target has linear cross-polarization nulls. E is given by:

$$E = -Q\sin^2 2\gamma \sin 4\nu \ \sin 2\tau_m$$

It is the sine opposed to co-sine relation that makes this parameter significant in a circular polarization base. E is sensitive to phase difference of the two co-polarized terms of a scattering matrix in a circular basis. E will be maximized if the relative scattering matrix in the circular basis is:

$$[S(LC,RC)] = \begin{bmatrix} +j/2 & 0 \\ 0 & -j/2 \end{bmatrix}$$

E, therefore, measures phase irregularity for targets that are non-symmetric.

F: is the target helicity parameter. It is measured directly through the difference in (RC-RC) and (LC-LC) polarized returns from the target (hence the term helicity, which relates to a preference for sense of circular polarization). F is analogous to the parameter C but is sensitive when one has circular cross polarization nulls. F responds to targets with circular x-pol nulls and with gamma type irregularity, i.e., with a scattering matrix of the form:

$$[S(LC,RC)] = \begin{bmatrix} 1 & 0 \\ 0 & 0 \end{bmatrix}$$

Note, the similarity to the scattering matrix for a dipole in a linear base: "Targets giving rise to F may be imagined to be dipoles wrapped around in a helix".

G: is called the coupling parameter. It couples the symmetric and non-symmetric parts of the target. Usually G is small for man-made targets because the non-symmetric part usually is small. G is given by:

$$G = Qf\sin 4\tau_m = Q(1 - \sin^2 2\gamma \sin^2 2\nu)\sin 4\tau_m$$

This shows that G is maximized when $\gamma = 0$, $\nu = 0$ and $\tau_m = 22.5°$.

This occurs at an ellipticity halfway between circular and linear polarization and shows that it is a transition or "coupling" parameter as one goes from a symmetric to a non-symmetric type of target or vice-versa. In the intermediate stages of distorting a linear dipole into a helix one can expect to see G rise and fall. A flat type of target with no plane of symmetry will produce a large G because there are no phase irregularities so that $\nu = 0$. This very specific property should enable us to see dynamic changes between oblate-to-spherical-to-prolate-spheroidal shape states and this must be a very important parameter in further developing radar meteorology (see M.Sc. thesis, Jerald D. Nespor, July 1983).

H_ψ: is the coupling due to target orientation ψ. Hence H_ψ can be made zero by properly aligning the roll of the target along the radar line of sight.

6. Radar Target Classification Using Polarimetric Target Slant Range Signatures

In the following basic polarimetric backscattering characteristics of simple to increasingly more complex shaped missile-type targets are analyzed and interpreted by strictly adhering to Kennaugh's target characteristic operator concept and Huynen's target Mueller matrix decomposition theories which were introduced above. Scatterer model data was computer-generated on the DEC-VAX 11/750-780 Research Computer Processing System of the Communications Laboratory, Electrical Engineering & Computer Science Department, Uni-

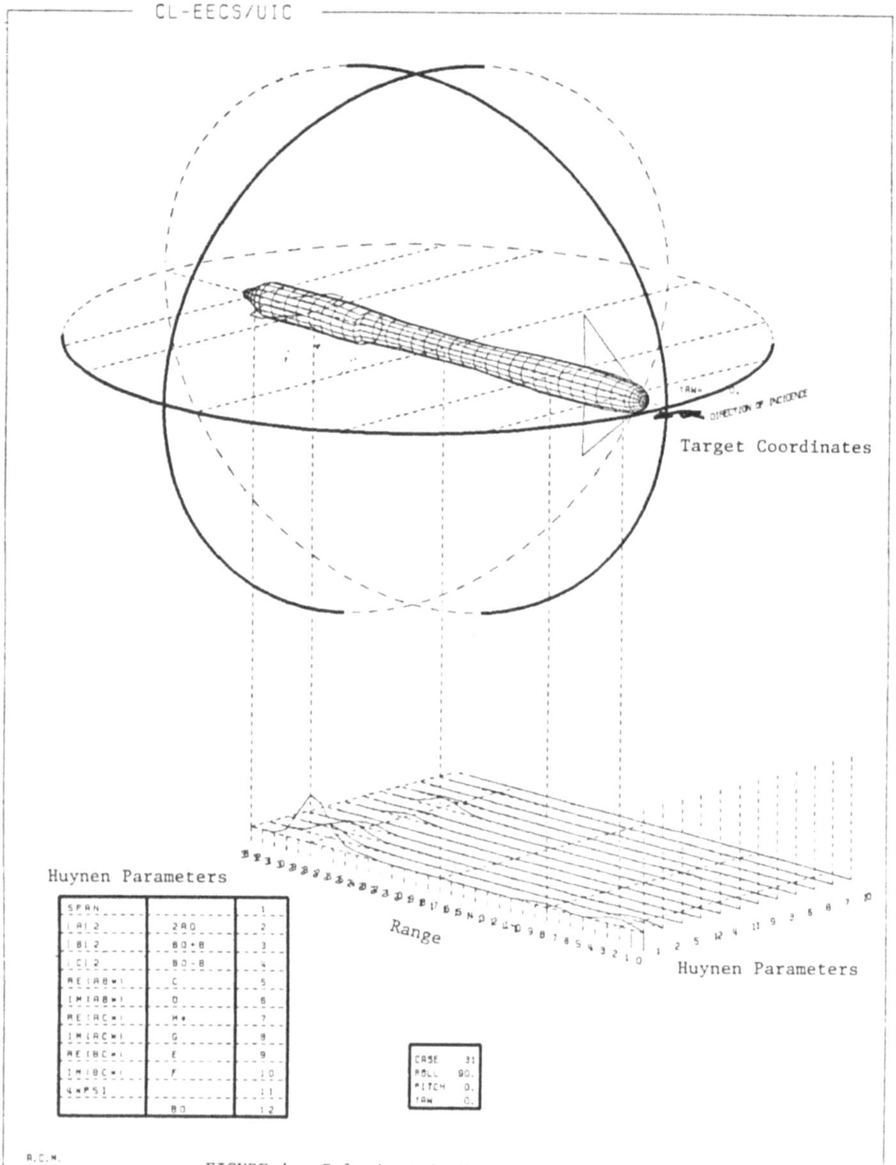

FIGURE 4: Polarimetric Target Downrange Signatures

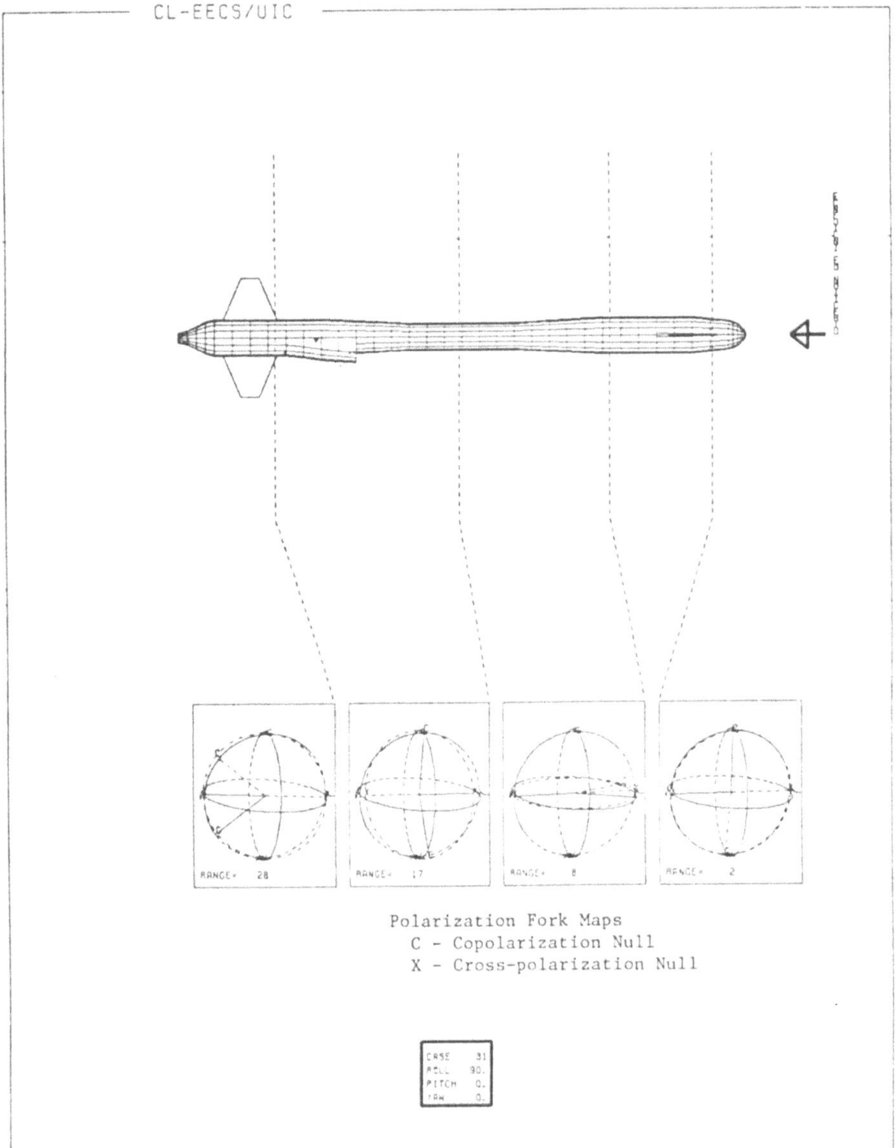

FIGURE 5a: Kennaugh's Target Characteristic Operators

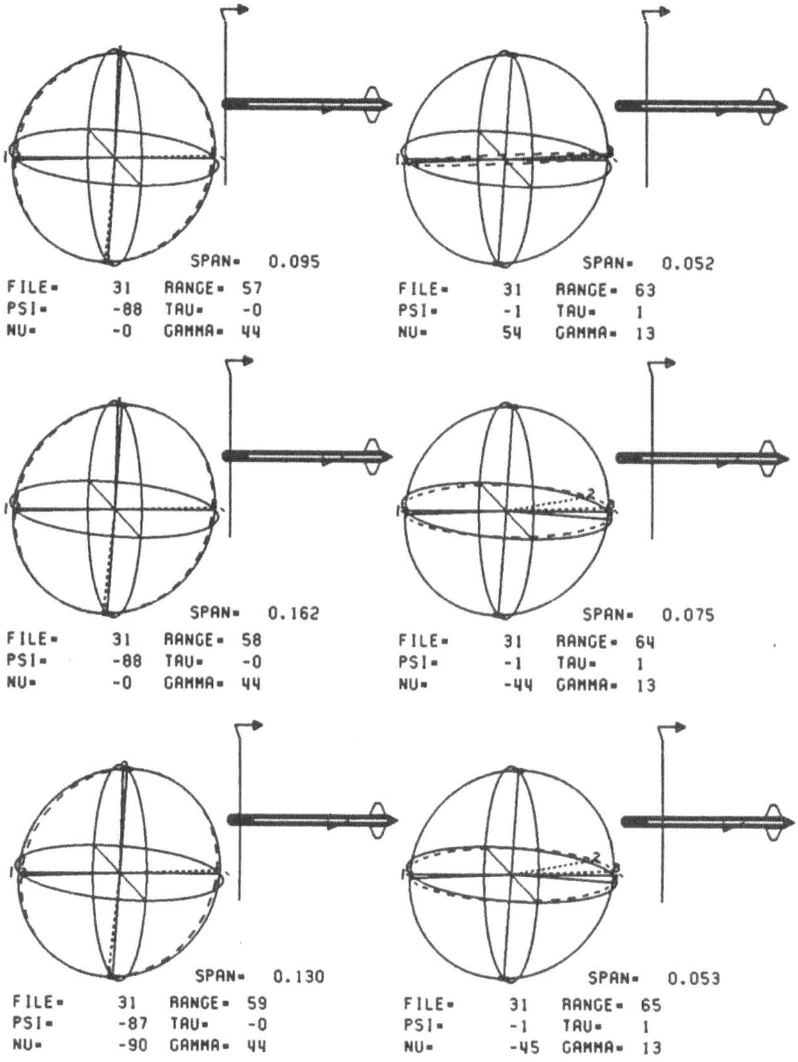

FIGURE 5b: Kennaugh's Characteristic Operators continued

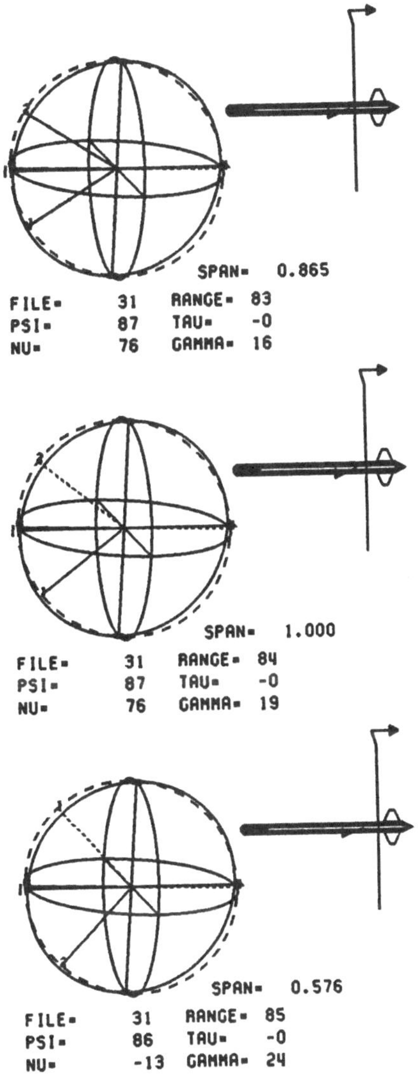

FIGURE 5c: Kennaugh's Characteristic Operators continued

versity of Illinois at Chicago, and compared with measurement data for missile-type composite targets by Teledyne Micronetics, collected on their S-band (2.4 to 3.975 GHz in steps of 25 MHz) and X-band (9.0 to 10.575 GHz) outdoor facilities for vertical (V) and horizontal (H) antenna polarization bases as described in L.A. Morgan, S. Weisbrod, High Resolution RCS Matrix Studies of Simple Targets, Report Number NADC 83016, June 30, 1982,

Using a broadband, stepped frequency dual (H, V) polarization radar measurement set up, the stepped frequency data was transformed into the time domain in order to "range-gate" undesired and/or desirable features of the order of 10 cm. Thus for each target down range swath of about 10 cm width an effective scattering matrix is recovered and those for about 128 swaths along downrange (as is explained in detail in Boerner, Manson, Huynen, 1983). The specific targets analyzed are truncated cylinders of about 2 meters lengths with wings and fins attached. In the following only one out of 482 available data sets made available to us is evaluated to verify the theories introduced here. The specific case chosen corresponds to the missile target in its full configuration, i.e. with the wings, fins and duct being present. The roll angle of the missile is 90°, meaning that the wings (front) are vertically oriented, the fins (aft) are horizontally oriented and the duct will be on the left side of the missile. The yaw angle is 0° so that we will be looking at the missile nose-on (Figure 4). The downrange features of the target can be divided into three well-defined scattering regions. The first region corresponds to the frontal sphere-cap of the missile, the second to the wings following close behind the frontal cap; and the third to the fin assembly at the rear of the missile. Each of the regions demonstrates easily identifiable scattering center characteristics and the following polarimetric signature plots are presented.

Region 1; Span:

The span plot for this region (ranges 57-59) shows a significant response due to the direct backscatter from the frontal hemispherical cap. This response holds for 3 range cells or approximately 6 inches. This agrees well with the fact that this scattering center is at a length on the order of the radius of the cap. However, due to the spreading of the simulated short pulse, the actual envelope will be larger (see Boerner, Manson, Huynen, 1983). When the wings, fins and duct are removed, this frontal specular area will produce the largest return, but for this case, the largest returns come from the fins and wings (Figure 4)

Mueller Matrix Parameters:

The parameters pertinent to Region 1 are $2A_0$, γ, ν, and τ_m. This region exhibits sphere-type target characteristics. Since a sphere has rotational symmetry, it will have an infinite number of symmetry planes intersecting the radar line of sight and hence have $\tau_m=0°$. This we observe for the polarization spheres corresponding to ranges 57-59 (Figure 5a). If B_0+B and B_0-B are zero, and $2A_0$ is the only generator present, the target will demonstrate this sphere-like characteristic. Since B_0+B produces contractions in γ and deviations of ν from 0°, we would expect that these values will be $\gamma = 45°$ and $\nu = 0°$ (Figure 5a).

Region 2; Span:

The span plot for this region shows a peak over ranges 63-65. This comes 4 range cells or about 8 inches after the cap region. This is the length of the wing downrange along the fusalage and shows that this return is from indirect backscatter from the back of the wings.

Mueller Matrix Parameters:

The pertinent parameters for interpreting this region are $2A_0$, B_0+B, C, γ, ν, ψ, and τ_m. This region, spanning ranges 63-65, differs significantly from Region 1. Here B_0+B is significant and of the same order of magnitude as $2A_0$, and can be expressed as:

$$B_0+B = 2Q(\cos^2 2\gamma + \sin^2 2\gamma \sin^2 2\nu)$$

with the orientation ψ taken out.

From this behavior we see the relationship of the polarization fork parameters to Huynen's Mueller matrix parameters. To under stand why $B_0 + B$ is large, it is helpful to look at the extremes of γ and ν which produce it. The parameter γ has the domain $0° < \gamma < 45°$ and ν has the domain $-45° < \nu < 45°$, or $|\nu| < 45°$. If $\gamma = 0$, the first term > 1 and the second drops out. When $\tau_m=0$, this situation gives rise to a scattering matrix of the form:

$$\underline{S} = \begin{bmatrix} 1 & 0 \\ 0 & 0 \end{bmatrix}$$

which is the form for a dipole target. Now the other extreme occurs when $\gamma > 45°$ and $|\nu| > 45°$. If $\tau_m=0$, this situation will be described by the scattering matrix:

$$\underline{S} = \begin{bmatrix} 1 & 0 \\ 0 & -1 \end{bmatrix}$$

which is the form for a corner reflector. The corner reflector produces a 180° inversion in phase because of the double bounce at the corner. In summary, $B_0 + B$ can indicate double bounce or dipole characteristics.

In region 63-65 we see a hybrid of dipole characteristic and multibounce scattering. This is apparent because $\gamma=13°$ and $|\nu| \approx 45°$ (Figure 2/3). Since the wing is a long, straight-wire-type target at this yaw angle, we expect a small γ, but the effect of ν needs study. The back edge of the wing most closely resembles a wire dipole, so we can expect it to produce these returns. But when one examines the ranges 63-65, it is evident that the backscatter corresponds to the back side of the wing and not the front (Figure 5b). This explains why $\nu \approx 45°$: because the returning wave is "actually creeping" over the back of the wing. This "creeping wave effect" will cause the phase of the horizontal channel to invert 180°, thus giving $|\nu| = 45°$ (Figure 5b). The parameter C given by:

$$C = 2Q\cos 2\gamma \, \cos 2\tau_m \qquad (\phi \text{ rotation removed})$$

shows that C will increase if the target possesses a plane of symmetry $\tau_m = 0$ and if $\gamma > 0$. Thus C is a shape indicator for symmetrical targets. Shape can be understood by considering the two extremes, namely a sphere and a dipole. Since both possess a symmetry plane, $\cos 2\tau_m = 1$. For a sphere $\gamma > 45°$, $C > 0$ and for a dipole $\gamma > 0$, $C > 2Q$. Clearly in our case $\gamma > 0$. The parameter D given by

$$D = Q\sin^2 2\gamma \, \sin 4\nu \, \cos 2\tau_m$$

also shows an important property, namely the phase difference between the component scattering matrices

$$[A] = e^{j\phi_A}\begin{bmatrix} 1 & 0 \\ 0 & 1 \end{bmatrix} \quad \text{and} \quad [B] = e^{j\phi_B}\begin{bmatrix} 1 & 0 \\ 0 & -1 \end{bmatrix}$$

if $D \approx 0$, this shows that the two components add in phase (for direct dipole scattering) or 180° out of phase (for indirect or backsurface wing scattering) to produce the final scattering matrix. This is the case for a target with dipole characteristics (see plot of D vs. range, Figure 6) as explained in the M.Sc. thesis of Bing-Yuen Foo (1982).

Even though the ϕ rotation is removed from the normalized Mueller matrix parameters, it is by no means useless. Here ϕ gives us important information about whether the wings are oriented horizontally or vertically, or in general at any other angle. From the sphere plots (Figure 2/3 & 4/5) one can see $\psi \approx 0$, which tells us that the wings are vertically oriented. Further downrange we will see the fins will have $\psi \approx 90°$, indicating horizontal orientation (Figure 1).

Region 3; Span:

In this region (ranges 83-85) we see the maximum return for the whole downrange. This can be attributed to the direct backscatter produced by the leading edge of the horizontally oriented fins.

Mueller Matrix Parameters:

This region, ranges 83-85, (Figure 4) contains the horizontal fins at the back of the target. We may expect the fins to behave somewhat like a dipole target but with some minor adjustments. Here the parameters of interest are ν, ψ, τ_m, γ, $2A_0$, $B_0 + B$, C, and D. The parameters $B_0 - B$, E, F, G, and H have not been ignored, but as these components are generated by $B_0 - B$ (the indicator of non-symmetry) which is small relative to the other generators, $2A_0$ and $B_0 + B$; E, F, G, and H are insignificant.

In this region the span (Figure 4) takes on its maximum value for all ranges. This can be attributed to the direct specular reflections produced by the incident wave from the edges of the fins. In this region we observe features very similar to the wing region. One similarity is that $2A_0$, $B_0 + B$, and C are of the same order of magnitude which shows the dipole characteristics of the fin edges. But whereas $D \approx 0$ and $|\nu| = 45°$ for the previous region, D demonstrates a persistent spike over the ranges, and has a significant magnitude. Since D is given by:

$$D = Q\sin^2 2\gamma \ \sin 4\nu \ \cos 2\tau_m$$

we find that D is produced by ν since $\tau_m \approx 0$ and γ is nonzero. Nonzero ν in this instance is not produced by a multibounce effect as for the wings, but could be generated by the tapering back of the fins. This tapering seems to produce a slight phase difference in the returns measured in the horizontal HH and vertical VV channels (Figure 4). The effect is similar to that observed when looking at a specular surface with two different principal curvatures, e.g. a spheroid as compared to a sphere (see M.Sc. thesis of Bing-Yuen Foo, 1982).

From the sphere plots we see that as we move further downrange in the fin region, γ starts to increase. This may be caused by the coupling of the scattering centers on the fin with those on the missile body (Figure 4). The angle $\psi \approx 0$ (Figure 2/3) is a direct indicator that the fins are horizontally orientated. This should be compared with ranges 63-65, (Figure 2/3 & 4/5) where $\phi \approx 0$ and the wings are oriented vertically.

Since the target extremities, wings and fins, produce the largest returns for this case, the return from the conical cap at the back has been thresholded out. But if one looks carefully at the $2A_0$ plot (Figure 4) approximately at range 88, one can detect its presence. This return combined with the frontal specular return can be used to provide an estimate of the targets total length.

Summary:

In summary, the target tells us it has a front specular region that is sphere-like. Shortly thereafter follows a symmetric oblong center showing a double bounce creeping wave behavior and a preferred vertical orientation. Finally, the target has a symmetric oblong region of great magnitude with a small phase difference and a preferred horizontal orientation.

Insensitive model analysis of random clutter showed that clutter nulls would move all over the polarization sphere which he mapped onto elliptical coordinates using an extended Lambertian mapping of Aitoff (Deetz and Adams, 1945) of the polarization sphere. However, analyses now initiated here (using a similar extended Lambertian mapping of Mollweide) shows that his model needs to be improved and must be made wavelength dependent.

As was also suggested in (Mieras et. al., 1983) a broadband high resolution polarimetric target signature mapping in terms of the polarization nulls may provide a target charactertic classification algorithm. Precisely such data, required to investigate this assumption, have been provided in the form of measurement data collected by Weisbrod and Morgan (1982). Namely, the broadband stepped frequency complete relative phase scattering matrix measurement data can be re-expressed by using a suitable Fourier transform of the stepped frequency data such that target downrange (slant) polarimetric signatures are obtained which can be expressed in terms of an equivalent or effective relative phase scattering matrix for each downrange resolution swath (cell) of about two (2) to eight (8) inches resolution at a center frequency of about 3 GHz, 2GHz bandwidth and 25 MHz stepped frequency increment (see Chapter 3). Applying Kennaugh's optimal polarization null theory as formulated by Huynen (1970, 1978, 1982, 1983) in terms of the 2x2 scattering matrix characteristic parameters (m, ϕ, τ, ν, γ), and its extension in terms of the "single target" Mueller matrix target characteristic decomposition parameters (A_0, B_0, B, C, D, E, F, G, H), we have found that all characteristic features that the interrogating wave observers can be explained.

In fact, for missile target shapes which are of relative simple structure, our results indicate strongly the feasibility of developing high-resolution polarimetric target signatures which could be used as input for generating target characteristic, identification algorithms. However, we require to further extend our investigations by supplementing the interpretation of the measurement data used with analytical complete relative phase scattering matrix target descriptions based on GTD algorithms and their extensions. In addition, we require similar broadband complete scattering matrix measurement data on specific types of clutter and for more complex shaped targets in isolation as well as embedded in clutter so that we are able to interpret all important target features observable by the interrogating wave in terms of the 3-D polarization fork graphical displays, target characteristic angles and Huynen's single target Mueller matrix parameter presentation.

The next step is to develop theoretical algorithms to derive the complete scattering matrix for multiple frequency and verify results of the data, further develop Huynen's N-target decomposition theory and related syntatic polarimetric target signature analyses. Of course, this analysis is preliminary and far from complete. A complete analysis would cover a class of targets of similar type and not just one object-set as was the case here. Also, clutter results would have to be analyzed, not only separately from the target but with the target imbedded in the clutter.

One approach to the target classification/identification problem was suggested by A. C. Manson from our lab: 'The target speaks' to us, as radar signatures are received. The clusterings of range data are the 'words', while one complete range profile produces a 'sentence'. The syllables used (phonemes) are the characteristic signatures which were observed. All this can be formulated within a linguistic framework with a target-eigen syntax and semantics.

Polarization Fork:
The polarization fork in this region is oriented so that the co-pol prongs point to the left ($\psi \approx 90°$) confirming a horizontal orientation. The roll of the fork ($\nu = -14°$) could possibly be attributed to the tapering back of the wings downrange. The location of the fork handle in the equation confirms that we are viewing a symmetric target and the contraction of the co-pol prongs shows the oblong characteristic of the fins.

Most essential for such a development is that the data can be analyzed from a phenomenological viewpoint, i.e. independently of a particular target model. The parameter approach above may be used for any type of radar target, at any frequency, either beam filled or for high resolution. Of course the results will depend on these conditions and we will be able to select the optimal conditions for each system at hand.

These preliminary results show the potential usefulness of such an approach and, thus justifies continued generous funding of our research efforts for advancing high resolution polarimetric target imaging.

7. Acknowledgements

This research was supported, in part, by the US Naval Air Systems Command, Contract No. N00019-82-C-0306, the US Naval Sea System Command, Contract No. N62269-83-R-0317 and the US Naval Air Development Center. We are grateful to Mr. James W. Willis, Mr. Charles Jeddrey for their support and particularly to Mr. Ray Dalton and to Dr. Otto Kessler for their advice and continued interest in our research. Spcecifically, We wish to express our extreme respect to Mr. Lee Morgan and Dr. Steven Weisbrod, who collected and preprocessed these excellent high resolution polarimetric measurement data. The skillful typing of the manuscript and the preparation of the drawings, done by Mr. Richard W. Foster, Mrs. Deborah A. Foster and Mr. Bill Huang, are gratefully acknowledged.

8. References

W-M. Boerner, "Polarization Microwave Holography: An Extension of Scalar to Vector Holography (INVITED), 1980 International Optics computing conference, SPIE's Techn. Symposium East, Washington, DC, April 9, 1980, Session 3B, paper No. 231-23, pp. 188-198, 1980.

W-M. Boerner, "Use of Polarization in Electromagnetic Inverse Scattering", Radio Science, Vol. 16, No. 6, (Special Issue including papers: 1980 Munich Symposium on Electromagnetic Waves), pp. 1037-1045, Nov/Dec, 1981.

W-M. Boerner, A.C. Manson and J.R. Huynen, "Radar Target Classification Using Polarimetric Target Slant Range Signatures", NASC Final Rept. #UIC-EMID-CL-83-06-15, Contract No. N00019-82-C-0306, June 15, 1983.

S.R. Cloude, "Polarimetric Techniques in Radar Signal Processing", Microwave Journal, Technical Feature, July 1983, pp. 119-127.

M. Davidovitz, "Analysis of Certain Characteristic Properties of the Bistatic, Asymmetric Scattering Matrix, M.Sc. Thesis, Communications Laboratory No. 83-04-15, April 15, 1983.

M. Davidovitz and W-M. Boerner, "Extension of Kennaugh's Optimal Polarization Null Theory of the Monostatic Reciprocal Scattering Matrix to the Bistatic Non-Symmetrical and/or Non-Reciprocal Monostatic Scattering Matrix Cases", Proceedings of 1983 Joint International URSI/IEEE-APS Symposium, Vol. 2: 83CH1860-6, May 23-26, 1983, Houston, TX. pp. 484-487.

C.H. Deetz and O.S. Adams, "Elements of Map Projection with Applications to Map and Chart Construction", U.S. Dept. of Commerce, Coast & Geodetic Survey, Special Publ. No. 68, 5th Ed., Revised 1944, U.S. Gov. Printing Office, Washington, 1945.

B-Y. Foo, "A High Frequency Inverse Scattering Model to Recover the Specular Point Curvature from Polarimetric Scattering Data", M.Sc. Thesis, Electr. Engr. & Comp. Sci. Dept., Univ. of Ill. at Chicago, IL, Communications Lab Report No. 82-05-21, May 21, 1982.

B-Y. Foo, S.K. Chaudhuri and W-M. Boerner, "A High Frequency Inverse Scattering Model to Recover the Specular Point Curvature from Polarimetric Scattering Data", in print, IEEE Trans. A&P, 1984.

J.R. Huynen, Phenomenological Theory of Radar Targets, Ph.D. Dissertation, Drukkerij Bronder-Offset, N.V. Rotterdam, 1970.

J.R. Huynen, "Phenomenological theory of Radar Targets, Chapter 11 in Electromagnetic Scattering, Academic Press, NY, edited by P.L.E. Uslenghi, 1978.

J.R. Huynen, 1982a, "A Revisitation of the Phenomenological Approach with Applications to Radar Target Decomposition", Dept. of Electr. Engr. & Comp. Sci., Univ. of Ill. at Chgo., Communications Lab Rept. No. EMID-CL-82-05-81-01, Contract NAVAIR Grant #N00019-80-C-0620, May 1982.

J.R. Huynen, 1982b, "Polarization Discrimination with Applications to Target Classification", Proceedings of the Second Workshop on Polarimetric Radar Technology, Vol. 1, US Army Missile Command, Redstone Arsenal, AL., May 3-5, 1983, pp. 197-216.

J.R. Huynen, 1985, "Towards a Theory of Perception for Radar Targets With Application to the Analysis of Their Data", Proc. NATO-ARW-IMEI-1983, Bad Windsheim, FR Germany, Ed. by Dr. Wolfgang-M. Boerner, D. Reidel Publishing Co., Paper No. IV.2 (RP.4).

E.M. Kennaugh, "Effects of Type of Polarization on Echo Characteristics", Ohio State University (Antenna Laboratory), Contract No. AF 28(099)-90, Report Nos. 389-1 to 389-15, and 389-17 to 389-24 (No. 389-16 nonexistent), Sept. 1949 to Oct. 1954.

E.M. Kennaugh, "Polarization Properties of Radar Reflections", Proceedings of the R&D Board Symposium on Radar Reflection Studies, Sept. 1950.

E.M. Kennaugh, "Polarization Properties of Radar Reflections", OSU (Antenna Lab), M.Sc. Thesis, Report No. 389-12, March 1952.

E.M. Kennaugh and D.L. Moffatt, "Transient Current Density Waveforms on a Perfectly Conducting Sphere", Proceedings of the NATO-ARW-1983 on Inverse Methods in Electromagnetic Imaging, Part 1, pp. 1-31, Sept. 18-24, 1983, D. Reidel Publishing Co.,

H. Mieras, R.M. Barnes, G.M. Vachula, J.N. Buchnam, C.L. Bennett and W-M. Boerner, "Polarization Null Characteristics of Simple Targets", Rome Air Development Center, Air Force Systems Command, Griffis Air Force Base, NY, Rept. No. RADC-TR-82-335, Final Technical Report, 1983.

D.L. Moffatt and T.C. Lee, "Time-Dependent Radar Target Signatures, Synthesis and Detection of Electromagnetic Authenticity Features", Proceedings of the NATO-ARW on Inverse Methods in Electromagnetic Imaging", Part 1, pp. 432-451, Sept. 18-24, 1983.

L.A. Morgan and S. Weisbrod, "High Resolution RCS Matrix Studies of Simple Targets", Teledyne Micronetics Final Report No. NADC 83016-30, Contract No. N62269-81-C-0802, Prepared for Naval Sea Systems Command, June 30, 1982.

J.D. Nespor, "Theory and Design of a Dual-Polarization Radar for Meteorological Studies", M.Sc. Thesis, Electr. Engr. & Comp. Sci. Dept., Univ. of Ill. at Chgo., Communications Lab Rept. No. 83-07-26, July, 1983.

III.13 (MM.5)

DEMANDS ON POLARIZATION PURITY IN THE MEASUREMENT AND IMAGING OF DISTRIBUTED CLUTTER

Andrew J. Blanchard, University of Texas at Arlington, Arlington, TX 76019 and Richard W. Newton, Remote Sensing Center, Texas A&M University, College Station, TX 77843

ABSTRACT

In recent years the use of polarization information contained in radar data has become increasingly important. A great deal of dual polarized backscatter measurements have been acquired using ground based scatterometers, airborne scatterometers and imaging systems. It was obvious from a cursory analysis of the data that polarization had something to offer. The problem then became a matter of unraveling the information contained in the polarization aspect of the radar data.

A great deal of effort was spent on the theoretical modeling of depolarized scatter from a variety of targets. The assumption was made that the data acquired by radar systems was uncontaminated. Recent work has indicated that system effects may have seriously contaminated the measured depolarized scatter. This paper addresses several major system parameters that affect the quality of polarization data.

The most serious parameter is that of inadequate isolation between the orthogonal polarization states of the antenna system. This is especially important when measuring backscatter from distributed targets such as clutter. In this case the isolation parameter must be maintained across the entire null-to-null beamwidth.

Another effect that contributes to error in the polarization measurement is a mismatch between the polarization state of the antenna and the target. These effects are most severe near the nadir measurement direction. They can be minimized by proper

W.-M. Boerner et al. (eds.), Inverse Methods in Electromagnetic Imaging, Part 2, 721–737.

orientation of the transmit/receive antenna.

This paper presents the criteria necessary to reduce errors in recovering polarized scattering parameters from point and distributed targets.

INTRODUCTION

The capability of target acquisition algorithms for seeker systems is directly dependent upon the information used to develop those algorithms. There appears to be considerable effort in characterizing return from target structures, but there appears to have been little work done in characterizing the backscatter (clutter) resulting from nontarget scene constituents that fall within the field of view of seeker systems. Development of target acquisition algorithms based only on the knowledge of backscatter characteristics of target structures might provide acceptable results for specific and selected target/clutter scenarios. However, the performance for arbitrary target/clutter scenarios cannot be predicted; it can only be measured experimentally. The difficulty of this approach is that all possible scenarios cannot be measured and if the algorithm does not provide acceptable performance, there is no base of knowledge or understanding upon which to develop a solution.

The only viable procedure in developing target acquisition and classification algorithms is to utilize knowledge of the target backscatter and clutter characteristics. The purpose of this paper is not to describe how to implement these algorithms, but to illustrate selected clutter and measurement characteristics that must be taken into account during the algorithm development procedure. It is also important to appreciate the fact that the backscatter from nontarget constituents (which constitute clutter) is dependent upon many physical parameters through nontrival relationships, and that analytic descriptions of the electromagnetic scattering process are difficult and imperfect. For these reasons, a full understanding of these processes can be obtained only through a combination of experimental measurements and model development.

Scene constituents that produce clutter can be placed in several generic classifications such as vegetation, snow, water, bare soil, urban, etc. However, the clutter signal resulting from any of these general clutter classes is not necessarily simple and distinct. The clutter signal resulting from any of these scene constituents is dependent upon many physical parameters that may be highly variable in time and in location. For instance, vegetation can be subclassed into agricultural, range, including both grasses and woody plants, forest, and wetlands. Within each of these subclasses, the clutter signal can be highly variable based on the structure and orientation of the vegetation and the environmental and meteorological effects. There are also two other factors that affect a backscatter measurement that should be taken into account during algorithm development. One factor relates to the sensor system, primarily the antenna, while the other relates to coherency effects in the electromagnetic backscattering process, and is thus termed a measurement effect.

SYSTEM INDUCED MEASUREMENT IMPURITIES

Recently, the use of backscatter due to so-called clutter or distributed targets in remote measurements has seen successful applications in agriculture and geology. Early analysis using dual polarized airborne measurements (1-3) showed unique polarization signatures of scene constituents of interest. However, the interaction processes controlling this phenomena are complex and were not well understood. Considerable effort has been put forth to develop an understanding of this phenomenon both theoretically and experimentally, using airborne and truck mounted measurement systems. It was determined that not only must a proper understanding of the interaction processes that govern the scattering process be developed, but that effects of the sensor system and measurement configuration on data quality must also be understood.

For example, theoreticians (6,7,8) generally model the scattering process with a plane wave confined to an infinitely small beamwidth. Is this an adequate model for computing the polarized backscatter measurements in view of the manner in which measurements are actually made? Real systems have finite antenna beamwidth, antennas with imperfect polarization isolation, and use measurement configurations in which the antenna polarization reference frame does not match the polarization reference frame that is assumed in analytical models. Several questions immediately arise. How important are these differences? Do they impact the quality of the acquired scattering cross sections? How do the inherent characteristics of imaging vs. non-imaging systems affect the quality of acquired scattering cross sections?

The attempt of this section is to assess the inherent error induced in the measurement of radar backscatter cross section by realistic radar systems. It will specifically address the following two points:

1) Effects of antenna systems with non-ideal polarization characteristics, and

2) Errors introduced by practical measurement configurations using antenna systems with ideal polarization characteristics.

Measurements by Non-Ideal Antennas

Ideally, linearly polarized antennas have no response to energy in the orthogonal polarization. Realistically, however, these systems do exhibit some response. This is illustrated schematically in Figure 1. Both the transmit and receive antennas have some residual polarization response in channels which are orthogonal to their nominal linear polarization. The receiver will therefore receive signals which are functions of all four backscattering cross sections (σ_{HH}, σ_{HV}, σ_{VH}, σ_{VV}). The effect of this nonperfect isolation has been analyzed to assess the ability of a real system to make a depolarized cross section measurement (4). The polarization reference frame of the antennas is assumed to match that of the surface for this analysis. The results indicate that antenna polarization isolation must be adequate across the entire beamwidth, not just along the antenna boresight, in order to retrieve an accurate measure of cross polarized radar cross section for distributed scene constituents. This result specifies the isolation requirements.

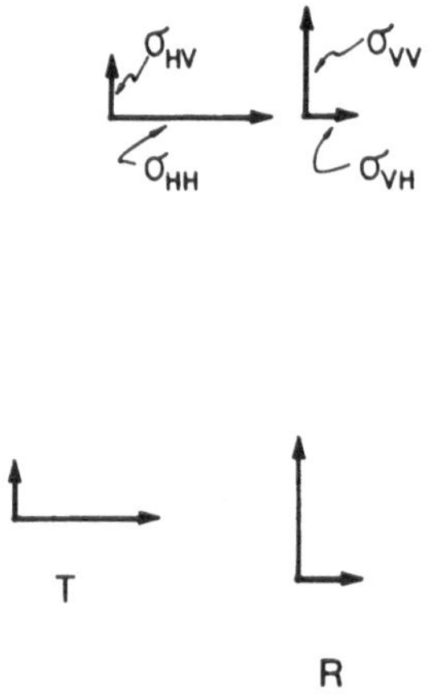

Figure 1. Schematic representation of normally linear polarized antenna with some individual response in the orthogonal polarization.

In this analysis the antennas are described in terms of a four component, normalized power radiation pattern. The four components correspond to four orthogonal polarization states identified as f_{HH}, f_{HV}, f_{VH} and f_{VV}, where H and V refer to horizontal and vertical polarization, respectively. An "ideal" antenna will have completely independent vertical and horizontal channels. For an ideal linearly polarized antenna, only one component, say f_{HH}, will be non-zero. The power pattern for a real antenna will contain all components and, consequently, can be written as

$$f(\theta,\phi) = \begin{bmatrix} f_{HH} & f_{HV} \\ f_{VH} & f_{VV} \end{bmatrix} \qquad (1)$$

where

f_{HH} = horizontal channel response to a horizontally polarized signal
f_{HV} = horizontal channel response to a vertically polarized signal
f_{VH} = vertical channel response to a horizontally polarized signal
f_{VV} = vertical channel response to a vertically polarized signal

All components are, of course, functions of the angular coordinates θ and ϕ, as emphasized by the notation $f(\theta,\phi)$. Figure 2 presents measurements of f_{HH} and f_{HV} for a real antenna system. The thin line is a measure of the response of the nominally horizontally polarized antenna to horizontal polarization (f_{HH}) as a function of angle. The heavy line is the response of the same antenna to a vertically polarized input wave (f_{HV}).

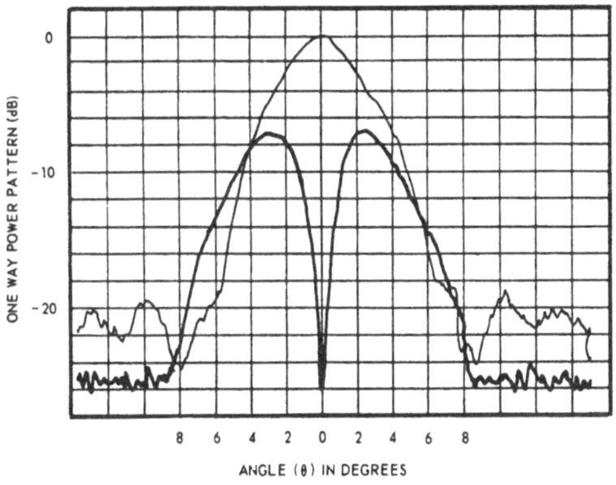

Figure 2. One way antenna pattern (H planar). Narrow line, like polarized pattern. Heavy line, cross polarized feed-through pattern.

Estimates of the normalized cross section σ_{HH}^0 and σ_{HV}^0 are computed in terms of the received power P_{RHH} and P_{RHV} and the normalized power pattern $f(\theta,\phi)$. The estimated values, denoted

σ_{HH} and σ_{HV}, are actually weighted averages of the true normalized cross sections and may be computed simply as:

$$\sigma^0_{HH} = \left(\frac{P_{RHH}}{P_T}\right) \frac{(4\pi)^3 R^2 \Omega_T}{G_T G_R \lambda^2 \int_{\Omega_T} f_{THH} f_{RHH} d\Omega} \tag{2}$$

and

$$\sigma^0_{HV} = \left(\frac{P_{RHV}}{P_T}\right) \frac{(4\pi)^3 R^2 \Omega_T}{G_T G_R \lambda^2 \int_{\Omega_T} f_{THH} f_{RVV}\, d\Omega} \tag{3}$$

If we define a set of "true" cross section parameters

$$\sigma^0_{ijT} = \left(\frac{P_{Rij}}{P_T}\right) \frac{(4\pi)^3 R^2}{G_T G_R \lambda^2}$$

where i represents the transmitted polarization and j is the received polarization, then the estimates σ_{HH} and σ_{HV} may be written in terms of these "true" values as

$$\begin{aligned} \sigma^0_{HH} = \sigma^0_{HHT} & \frac{1}{\Omega_T} \int_{\Omega_T} f_{THH} f_{RHH}\, d\Omega \\ + \sigma^0_{HVT} & \frac{1}{\Omega_T} \int_{\Omega_T} f_{THH} f_{RHV}\, d\Omega \\ + \sigma^0_{VHT} & \frac{1}{\Omega_T} \int_{\Omega_T} f_{THV} f_{RHH}\, d\Omega \\ + \sigma^0_{VVT} & \frac{1}{\Omega_T} \int_{\Omega_T} f_{THV} f_{RHV}\, d\Omega \end{aligned} \tag{4}$$

and

$$\begin{aligned} \sigma^0_{HV} = \sigma^0_{HHT} & \frac{1}{\Omega_T} \int_{\Omega_T} f_{THH} f_{RVH}\, d\Omega \\ + \sigma^0_{HVT} & \frac{1}{\Omega_T} \int_{\Omega_T} f_{THH} f_{RVV}\, d\Omega \\ + \sigma^0_{VHT} & \frac{1}{\Omega_T} \int_{\Omega_T} f_{THV} f_{RVH}\, d\Omega \\ + \sigma^0_{VVT} & \frac{1}{\Omega_T} \int_{\Omega_T} f_{THV} f_{RVV}\, d\Omega \end{aligned} \tag{5}$$

By examining (4) and (5) one observes that the like polarized measurement contains only a small error, since each error term involves the product of two cross polarized parameters. Applying this same criterion to the measurement of $\sigma_{HV}{}^{0}$, however, shows that the first, second, and fourth terms are of comparable order in the sense that each contains one factor which represents a cross polarized quantity. The third term is insignificant since it represents cross polarized transmission, cross polarized scattering, and cross polarized reception. The impact of the error terms on the cross polarized measurement is now evaluated.

Calculations have been presented by Blanchard and Jean (4) which simulate the results that are obtained in measuring the cross polarized radar cross section of extended targets with a specific set of non-ideal antennas. Figure 3 represents the like- and cross-polarized backscatter predicted by the model for an inhomogeneous volume bounded by a rough surface. A simulated measurement system response was calculated based upon the predicted values of cross section and the antenna characteristics shown in Figure 2. The results of these calculations are plotted in Figure 4.

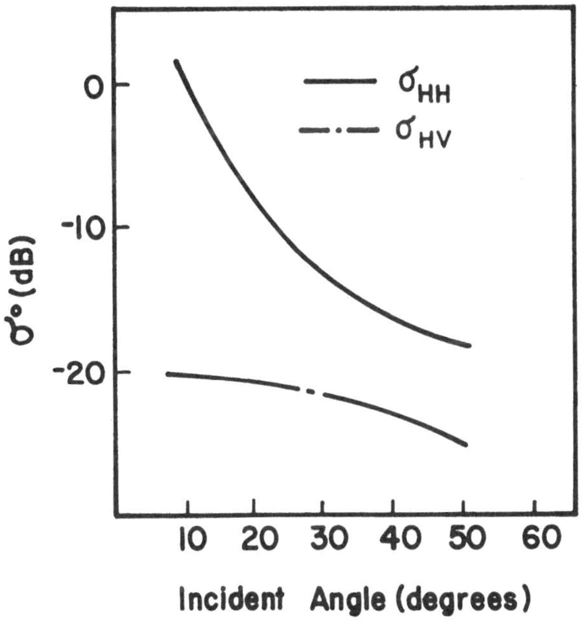

Figure 3. Theoretical backscatter cross section versus incident angle (From Blanchard [5]).

From the results of the simulation, one must conclude that the polarization properties of the antennas are inadequate for

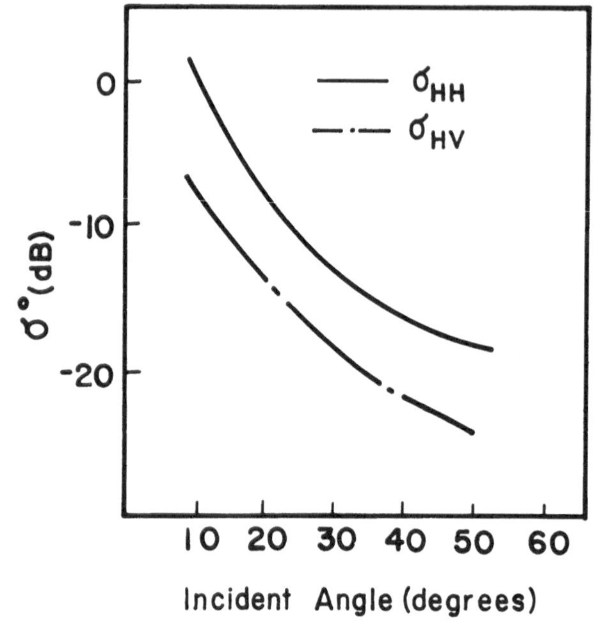

Figure 4. Simulated measurements of the backscatter cross section shown in Figure 3 using non-ideal antennas (isolation 12.5 dB).

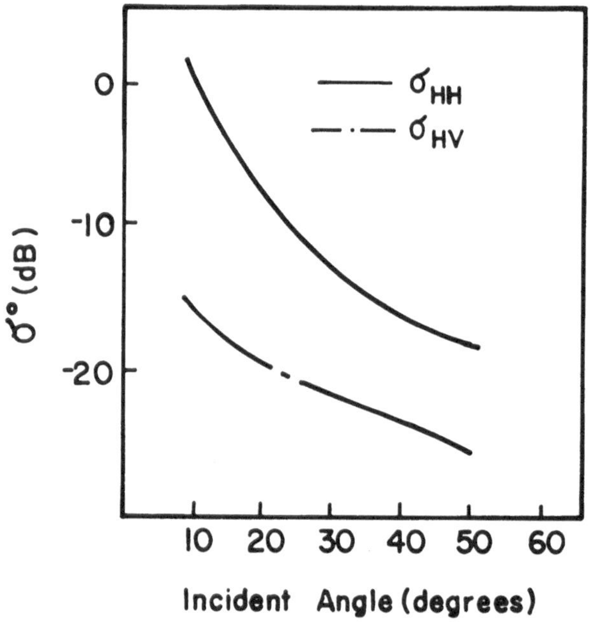

Figure 5. Simulated measurements of the backscatter cross section shown in Figure 3 using non-ideal antennas (isolation 20 dB.)

making the desired measurement. Had these data been actual measurements, evaluated without properly accounting for the antenna effect, the only conclusion that could have been drawn is that the model predictions and measured data do not match.

The simulation was repeated with the assumption that the polarization isolation ratio, computed over the 3-dB beamwidth, was increased to a value corresponding to an antenna isolation equal to 20 dB. Figure 5 illustrates that an antenna of such quality could provide an accurate measurement for incident angles of about 20 deg or greater. The simulated response near nadir still indicates a sharp increase in the measurement of $\sigma_{HV}{}^0$ that is not predicted by the model without taking the antenna effects into account. These results show that the boresight polarization isolation is not adequate to characterize antenna isolation performance in the measurement of distributed targets which produce the clutter return. The isolation figure of merit must be based on the integrated isolation over the processed beamwidth. If this integrated isolation is not adequate, the cross polarized scattering coefficients measured by such systems will be a function of system characteristics, thereby defeating the purpose of defining a system independent scattering cross section.

Also note that the feed-through effect due to imperfect isolation is most prominent at nadir. The ability to acquire valid polarization information is enhanced by making measurements at larger incident angles. This will be appropriate since the characteristic behavior of cross polarized backscatter is relatively insensitive to incident angle variations.

Errors Due to Measurement System Configuration

The following discussion analyzes the situation where the antenna polarization coordinate frame does not match the surface polarization coordinate frame. This assumption is incorporated into the radar backscatter equation to predict any errors that might be induced due to polarization coordinate system mismatch between the antenna and the surface. The results of this section indicate that this mismatch can produce significant differences between measured cross polarized backscatter and theoretical predictions.

Historically, the system engineer defines the polarization reference frame of the antenna with respect to the antenna coordinate frame. The theoretician specifies the polarization reference frame of an incident wave on a surface with respect to the surface coordinate frame. In general, these two coordinate frames do not coincide. One may transform the polarization vectors specified by the antenna polarization frame into the polarization vectors specified by the surface polarization frame. The formulation introduced by Claassen and Fung (5) is used to define the transformation.

In the antenna (primed) coordinate system shown in Figure 6, two orthogonal linear polarizations are specified in spherical coordinates (i.e., horizontal polarization will be defined by the unit vector $\bar{a}'_\theta$ and vertical antenna polarization will be defined by unit vector $\bar{a}'_\phi$). The polarization reference frame of the surface is specified by the unprimed coordinate system. Similarly, vertical polarization is defined by unit vector $\bar{a}_\theta$ and horizontal polarization by unit vector $\bar{a}_\phi$.) If the origin of the two frames is joined, a simple rotation will transform the polarization states of the antenna reference frame to the polarization states of surface reference frame. The transformation is as follows:

$$\begin{bmatrix} \bar{a}_\theta \\ \bar{a}_\phi \end{bmatrix} = [T] \begin{bmatrix} \bar{a}_{\theta'} \\ \bar{a}_{\phi'} \end{bmatrix} \qquad (6)$$

$$\text{where } T = \begin{bmatrix} \cos\Psi & \sin\Psi \\ -\sin\Psi & \cos\Psi \end{bmatrix}$$

Define Ψ so that

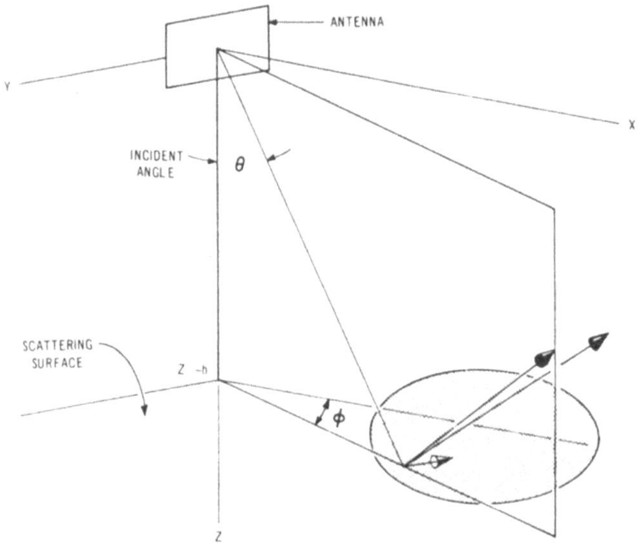

Figure 6. Coordinate reference frame.

$$\bar{a}_\theta \cdot \bar{a}_\theta{}' = \bar{a}_\phi \cdot \bar{a}_\phi{}' = \cos \Psi$$

$$\bar{a}_\theta \cdot \bar{a}_\phi = -\bar{a}_\phi \cdot \bar{a}_\theta{}' = \sin \Psi$$

These definitions can be used to calculate the distribution of energy in the two polarization states on the surface. Suppose a radar system transmits a horizontally polarized wave E_ϕ when pointed in the direction θ_0. Assuming the polarization transmitted by the antenna is pure, then the total power incident on the surface is given by

$$P_S = \frac{1}{2Z_0} \int \left|E_\phi\right|^2 d\Omega \qquad (7)$$

where $d\Omega = \sin\theta d\theta d\phi$. Using the coordinate transform, this can be decomposed into orthogonal surface components as:

$$P_\phi = \frac{1}{2Z_0} \int \left|E_\phi{}'\right|^2 \left[\left|a_\theta \cdot a_\phi{}'\right|^2 + \left|a_{\phi'} \cdot a_\phi\right|^2\right] d\Omega \qquad (8)$$

where Z_0 is the free space impedance.

Equation (8) indicates that although the antenna transmits pure horizontal polarization, both horizontal and vertical polarization exist when referenced to the surface coordinate frame. This is a result of the dot product $\bar{a}_\theta \cdot \bar{a}_\phi{}'$. The dot products in equation (8) show that the percent power in the vertical polarization on the surface increases (even though the antenna transmitted only horizontally polarized energy) with increasing beamwidth and decreasing angle θ_0 (which in this case also corresponds to the incident angle). Therefore, it becomes more difficult to acquire true polarized scatter near nadir for systems with large beamwidth. Stated another way, one must use narrower beamwidth to accurately measure scattering coefficients near nadir.

The results previously presented are based only on the geometry problem associated with the mismatch between the antenna and surface coordinate systems. The more significant source of error is due to the coupling between the like and cross backscattering cross sections that occur at the surface.

Although the results expressed in equation (8) are helpful in defining the degradation in polarization mismatch between the surface and antenna reference frames, the influence of this polarization mixing in the measured scattering cross section is of more importance. The power returned to the receiver will be calculated using the previous example of transmitting a horizontally polarized wave. The following assumptions are made: 1) the receiver antenna can detect both horizontal and vertical polarization without mutual coupling; 2) the surface scattering coefficients are random and $\sigma_{HV} = \sigma_{VH} = 0$ ($-\infty$ dB). The power detected by the receiver antenna can then be written as:

$$P_{R\phi'} = \frac{P_T G_T G_R \lambda^2}{(4\pi)^3} \int_{\Omega T} A^2 \sigma_{HH} \left|\bar{a}_\phi \cdot \bar{a}_\phi'\right|^2 + B^2 \sigma_{VV} \left|a_\theta \cdot a_{\phi'}\right|^2$$

$$+ 2\mathrm{Re}(A\sqrt{\sigma_{HH}}(\bar{a}_\phi \cdot \bar{a}_\phi')\ B\sqrt{\sigma_{VV}}(\bar{a}_\phi' \cdot \bar{a}_\theta))d\Omega \qquad (9)$$

$$P_{R\theta'} = \frac{P_T G_T G_R \lambda^2}{(4\pi)^3} \int_{\Omega T} A^2 \sigma_{HH} \left|a_\phi \cdot a_\phi'\right|^2 + B^2 \sigma_{VV} \left|a_\theta \cdot a_\phi'\right|^2$$

$$- 2\mathrm{Re}(A\sqrt{\sigma_{HH}}\ B\sqrt{\sigma_{VV}}\left|\bar{a}_\phi \cdot \bar{a}_\theta'\right| \left|\bar{a}_\theta \cdot \bar{a}_\theta'\right|)d\Omega \qquad (10)$$

where $A = \left|\bar{a}_\theta \cdot \bar{a}_\theta'\right|/R^2$

and $B = \left|\bar{a}_\phi \cdot \bar{a}_{\theta'}\right|/R^2$

Equation (9) represents the power returned to the receiver whose polarization is oriented the same as the transmitter antenna. This is the power one would normally use to calculate the σ_{HH} scattering cross section under the assumption that the transmit antenna was horizontally polarized. We find however that the returned power is made up of both σ_{HH} and σ_{VV} which are generally very close to each other in magnitude. This effect does not drastically influence the overall accuracy of the measurement.

The results presented in equation (10) are more interesting. This represents the power returned to the receiver in a polarization orthogonal to the original transmit polarization. Recall that it is assumed that $\sigma_{HV} = \sigma_{VH} = 0$ ($-\infty$ dB) so there is no target induced depolarization. Yet equation (10) predicts that the antenna will receive energy that will be interpreted as depolarized backscatter. This energy is a function of the difference in the like polarized backscatter coefficients. It is interesting to note that if $\sigma_{VV} = \sigma_{HH}$ and the phase difference between them is zero, the power predicted by equation (10) is zero. If, however, there is the slightest difference in either the phase or magnitude of the like polarized scattering coefficients, then significant power is introduced into the cross polarized receiver. This is not due to a depolarization effect on the surface, but to an artifact introduced by a mismatch between antenna and surface polarization. This effect becomes more pronounced for increasing beamwidths and measurements made closer to nadir.

The behavior of (10) is illustrated in Figures 7 and 8. Figure 7 presents the amount of induced depolarization measured by the receiver as a function of incident angle for various values of σ_{HH} and σ_{VV}. Recall in this analysis that it is assumed that the surface does not depolarize so that the measured depolarized cross section should be zero. Therefore the curves represented here portray the amount of induced depolarization because the antenna coordinate frame and the surface coordinate frame do not match. We plot this response for a number of σ_{HH} and σ_{VV} values. In general very slight differences in like-polarized scattering coefficients induced depolarization that is significant in the range of angles less than approximately 20°. If the difference in like-polarized scattering cross sections is greatly different, such as 3 dB, then the induced-depolarization effects become significant. The greater the difference between σ_{HH} and σ_{VV} the larger the induced-depolarization effect. What is significant here is that the error induced in the cross-polarized measurement is a strong function of very slight differences in the like-polarized measurement.

The data in Figure 7 represent the induced-polarization effects as a function of slight changes in cross section magnitude. In Figure 8 we present the induced-depolarization effects, not as a function of a change in magnitude, but as a change in phase angle between σ_{HH} and σ_{VV}. This graph presents induced-depolarization effect as a function of incidence angle for magnitude $\sigma_{HH} = \sigma_{VV} = 3$ dB and a variation of the phase difference between σ_{HH} and σ_{VV}. As one might expect there is a maximum depolarization effect when the phase angle between the two reaches 90°. What is important here is that this effect is not due to a depolarization effect on the surface, but to an artifact introduced by a mismatch between antennas and surface polarization.

In general, very slight differences in like polarized scattering coefficients will induce depolarization that is significant in the range of angles less than approximately 20°. If the difference in like polarized scattering cross sections is greatly different, such as 3dB, then the induced depolarization effects become significant. The greater the difference between σ_{HH} and σ_{VV} the larger the induced depolarization effect. It is significant that the error induced in the cross polarized measurement is a strong function of very slight differences in the like polarized measurement.

It has been well known that polarization states of the antenna and surface did not agree; however, this effect is now viewed as a significant influence on the measurement of depolar-

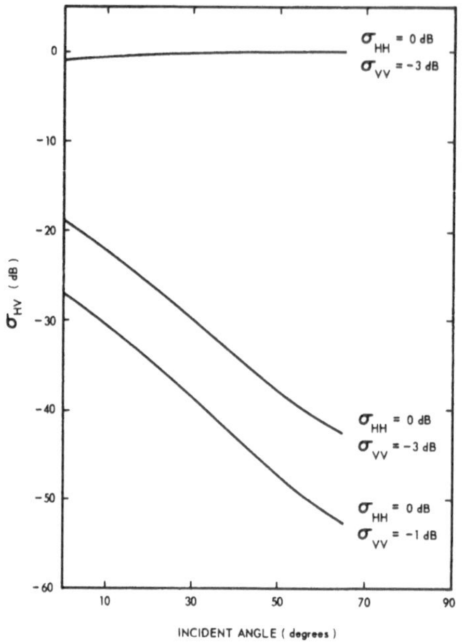

Figure 7. Measured cross-polarized scattering cross section versus incident angle as a function of differences in the true like-polarized scattering coefficients (σ_{HH}, σ_{VV}). In this case σ_{HV}=0.

ized cross sections. The effect becomes more pronounced for increasing beamwidths and for measurements made close to nadir. This analysis indicates that some effects that have been seen in depolarization measurements, such as the peak in σ_{HV} and σ_{VH} near nadir, were functions of system parameters, in some cases antenna feedthrough, and in some cases the effects just described. In the future, it is a requirement that the accuracy of the cross polarized measurement be assessed to ascertain whether the measurement is indeed contaminated by measurement system artifacts or if it represents a true measure of the depolarization caused by the surface.

MEASUREMENT UNCERTAINTIES DUE TO FADING

The phenomena of fading occurs whenever a coherent signal scatters from targets with nonuniform geometrical properties (i.e. in situations where the scattering surface does not conform to the phase front of the incoming wave). In dealing with scattering from statistically rough surfaces, the fading introduces uncertainty in the measurement being made. In the case of clutter returns, the fading due to the statistical nature of the target may require a large number of uncorrelated measurements to

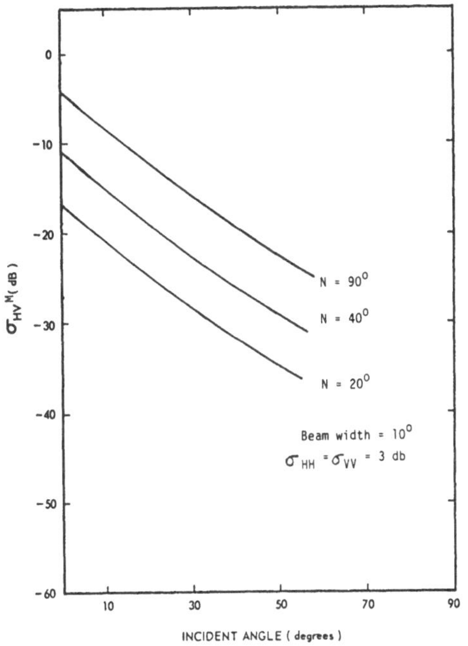

Figure 8. Measured cross-polarized scattering cross section versus incident angle as a function of phase differences in the true like-polarized scattering coefficients (σ_{HH}, σ_{HH}). In this case $\sigma_{HV}=0$, $\sigma_{HH} = \sigma_{VV} = +3dB$.

provide an adequate estimate of the mean value of the signal. For incoherent scattering, the amplitude of the backscatter is Rayleigh distributed. Therefore, the variance of the scattering return will be proportional to the return signal amplitude. To obtain a significant estimate of the mean of the signal, independent samples of the scattering amplitudes must be measured. Generally, as one scans out in incident angle, the mean of the scattering signal is much easier to obtain.

It is more difficult to acquire a suitable number of independent sample as one makes measurements closer to the nadir angle. At an incident angle of 0°, both coherent and incoherent scattered fields will be present at the receiver. The probability density function for the amplitude is no longer Rayleigh distributed. More independent samples are required than are required for larger incident angles to obtain an adequate estimate of the mean. Physically, when both coherent and incoherent scatters are present the fluctuation in signal amplitudes is much greater than when only incoherent scatter is present. Therefore, to minimize the fading problem one must obtain the scattering measurement at high incident angles where noncoherent scatter is dominant.

SUMMARY

There are two major results of the analysis presented. One is that the antenna polarization isolation criteria must be defined over the processed beamwidth in order to insure that accurate cross polarized backscatter measurements are made. Also, the isolation between antenna polarizations should be at least 16 dB greater than the difference between the true like and cross polarized backscatter coefficients. The second major result is that the cross polarized backscattering coefficient is a function of the difference between like polarized scattering cross section amplitudes and the phase difference between like polarized scattering cross sections (i.e. σ_{HH}, σ_{VV}). It was shown that the induced depolarized cross section is nonexistent if the scattering coefficients are identical in both amplitude and phase. In addition, the distortion disappears if the antenna is oriented such that the mismatch between the antenna coordinate frame and the surface coordinate frame is eliminated. It was found that the overall error in the cross polarized measurement increases as the antenna beamwidth increases, and that the error induced in the like polarized cross section is minor. Also, the induced error is greatest near the nadir look direction.

REFERENCES

1 Fung, A.K., 1977, "On depolarization of electromagnetic waves backscattered from a rough surface," Planetary Space Sci. 14, pp. 563-568.

2 Leader, J.C., 1971, "Bidirectional scattering of electromagnetic waves from rough surfaces," J. Appl. Phys., 42, pp. 4808-4816.

3 Rouse, J.W., "The effect of the subsurface on the depolarization of rough-surface backscatter," 1972, Radio Sci., 7:10, pp. 889-895.

4 Blanchard, A.J., and Jean, B.R., 1983, "Antenna effects in depolarization measurements," IEEE Trans. on Geosci. and Remote Sensing, GE-21, 1, pp. 113-117.

5 Claassen, J.P. and Fung, A.K., 1974, "The recovery of microwave scattering parameters from scatterometric measurements with special application to the sea," NASA Contract No. NAS1-10048, Tech. Rep. 195, Univ. of Kansas Center for Research, Inc.

6 Blanchard, A.J., and Rouse, J.W., Jr., 1980, "Depolarization of electromagnetic waves scattered from an inhomogeneous half space bounded by a rough surface," Radio Science, 15:4, pp. 773-779.

7 Blanchard, A.J., Newton, R.W., Tsang, L., and Jean. B.R., 1982, "Volumetric effects in cross polarized airborne radar data," IEEE Trans. on Geosci. and Remote Sensing, GE-20:1, pp. 36-41.

8 Benton, A.R., Blanchard, A.J., and Newton, R.W., 1983, "Determination of mobility and trafficability indicators using multi-frequency imaging radar," International Optical Engineering Symposium, Optical Soc. of Am., Crystal City, Virginia.

III.14 (SP.2)

POLARIZATION VECTOR SIGNAL PROCESSING FOR RADAR CLUTTER SUPPRESSION

VINCENT C. VANNICOLA and STANLEY LIS

Surveillance Division
Rome Air Development Center
Griffiss Air Force Base, New York 13441, U.S.A.

ABSTRACT

A model is presented for a radar scatterer which simultaneously incorporates range, doppler, and polarization characteristics in the statistical sense. Any scatterer, whether it be a target or clutter, can be represented by such a model. It is shown how this environmental model contains discriminants used to design an optimal waveform and an optimum receiver system for doubly spread (doppler and range) scatterers.

This paper will provide the reader with five conceptual issues relating to polarization processing, all of which are in analytical form. These are:

a. General model for the polarization scattering matrix with doppler, range, polarization sense and statistical properties simultaneously. Such a model is an aid to the systems engineer in providing properties to look for in a clutter/target model. It also provides parameters which should be measured in a clutter/target site survey.

b. A method for satisfying the conditions for an optimal polarization vector waveform.

c. A generalization of the ambiguity function; the ambiguity tensor which results when the waveform is a vector from a multiple-channel system.

W.-M. Boerner et al. (eds.), Inverse Methods in Electromagnetic Imaging, Part 2, 739–770.

d. The analytical procedure for designing the conventional and optimum vector (multi-channel) receiver.

e. Performance expressions for the clutter/target model and subsequent transmitter/receiver.

The discussion starts with a review of modeling and system optimization for the single-channel (scalar) case. After this brief review the paper extends the subject to the multiple-channel (polarization) case resulting in the manipulation of vectors, matrices, and higher order tensors.

INTRODUCTION

A backscatter radar environment is a stochastic phenomena having properties associated with doppler, range, and electromagnetic scattering (polarization scattering matrix). The environment consists of undesired (clutter) scatterers and desired (target) scatterers, each having its own characteristic doppler, range, and polarization properties. In optimization theory analytical approaches are used [1] to solve a set of integral equations for the analytical design of a radar system optimized over a set of radar environmental factors. The factors include the clutter scattering function on the transmit waveform. Such a scattering function is directly related to an autocovariance function of the noise resulting from the backscattered clutter with the transmitted incident waveform factored and deconvolved out. Such a function for a single-channel system incorporates all the statistics one needs to know about the doppler and range spread characteristics of the environment in order to optimize the radar. For a multiple-channel system the transmit waveform is a vector and the sattering function is a tensor of rank four. This has been indicated by several investigators [2] ,[3] where the parameters within the tensor were time averaged but doppler and range spread correlation were not considered. In the latter [3] the author indicates that benefits can be obtained with respect to target discrimination by controlling the transmitted polarization waveform and receiving in such a manner as to take advantage of the degree of coherence of the signals in the receiving channels.

There have been recent papers on such subjects. Stochastic models [4] describing the spectral and polarization characteristics of radar echoes from chaff clouds consisting of rotating dipoles with completely random or preferred orientations have been used.

Other papers [5], [6] include modeling and polarization processing with optimum transmit waveform vectors for the enhancement of target detection in clutter. The approaches utilize radar scattering matrix properties in the design of a two-channel vector receiver which discriminates between the scattering matrix-doppler-range spread characteristics of the target and clutter. Computer simulations were used to varify some simple models.

This paper is an analytical development of such a subject providing the basic mathematical expressions and their justification derived from stochastic processes and optimum filter theory. Its goal is the design of the optimum system making use of the two channel polarization-doppler-range spread scatterer properties with reference to a single channel development presented by Van Trees [1]. The two channel case treated in this paper is an extension of the single channel case.

This paper provides the following:

a. The statistical model for the doubly spread scatterer for a two-channel (polarization) radar.

b. An approach for optimal waveform design using this model.

c. The model for the ambiguity tensor function for a two-channel system.

d. Optimum receiver for the two-channel system.

e. System performance and evaluation.

A radar backscatter environment can be represented by a statistically time and range varying polarization scattering matrix. The model is general and can be used for clutter as well as the target. Once the targets are specified, the radar design, i.e., transmitter waveform and receiver, can be optimized. Performance of the system can also be evaluated. The procedures for system design and performance evaluation are merely extensions of present day optimum filter theory [1], the primary difference being the target model to which the procedures are applied is a scattering matrix random process. In addition to serving as design criteria for the radar system, such a model also designates the required field measurements for completely specifying such a scatterer. As a consequence of this model an approach for the generalized ambiguity function of multiple-channel systems is introduced. This paper extends the basic concept of the doubly spread doppler-range target to include the interrelation between scattering matrix elements. Hopefully these discriminants between the target and the clutter will enable a new dimension for improving target detection in a real world clutter environment.

In the body of the paper we will first review target modeling and system optimization for the single-channel radar system (1). Using the review as a basis we will extend the treatment to the multiple channel system specialized for the polarization case.

SINGLE-CHANNEL OPTIMUM RECEIVER

First, we would like to briefly review the procedures for deriving the target model, the optimal waveform and the optimum receiver for the single-channel case, that is when only one polarization sense is implemented. These procedures can be found in reference [1]. The signal received by the radar (1) from a slowly fluctuating point target in a clutter environment is

$$r(t) = bf_d(t) + n_c(t) + w(t) \qquad : \quad H_1 \tag{1}$$

$$r(t) = n_c(t) + w(t) \qquad : \quad H_0 \tag{2}$$

NOTE: All signals and noise are complex low pass equivalents.

- H_1 hypothesizes target present, H_0 target absent
- $f_d(t)$ is the complex range delayed doppler shifted replica of the transmitter waveform f(t) (see Figure 1),
- b is a complex random variable and represents the target backscatter, propagation losses, and antenna responses,
- $n_c(t)$ is the received signal from the clutter and is a complex random process, and
- w(t) is the additive white noise in the receiver.

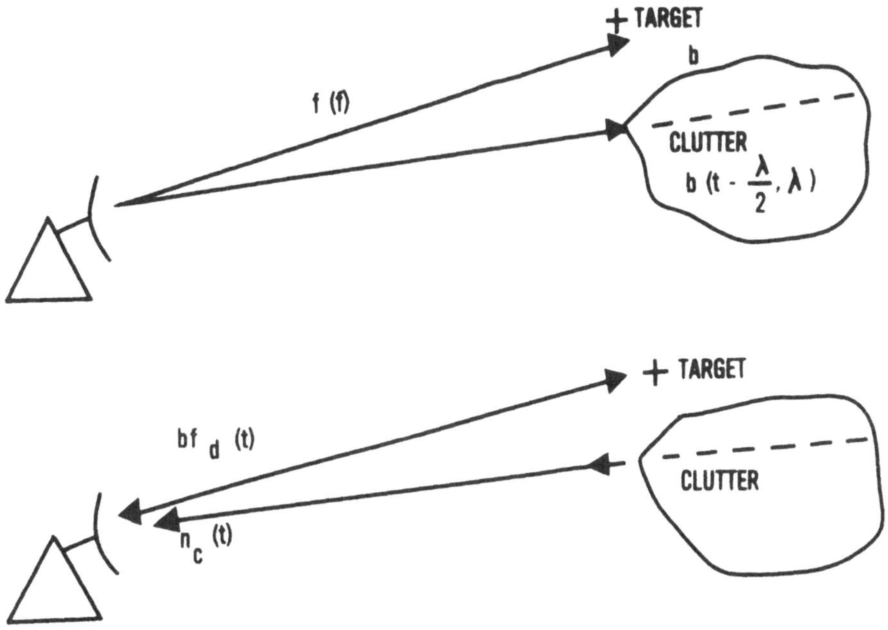

FIGURE 1. TRANSMIT AND RECEIVED SCALAR WAVEFORM

CLUTTER MODELING

The received signal from the clutter is the convolution and product of the clutter with the transmit waveform. Convolution is with respect to the range variable λ and the product is with respect to the time variable t.

$$n_c(t) = \sqrt{E_t} \int_{-\infty}^{\infty} f(t-\lambda)\, b(t-\tfrac{\lambda}{2}, \lambda)\, d\lambda \qquad (3)$$

where E_t is the energy in the transmit waveform.

This is a zero-mean complex Gaussian random process with the covariance function [1]

$$K_{n_c}(t,u) = E_t \int_{-\infty}^{\infty} f(t-\lambda)\, K_{DR}[t-u,\lambda]\, f^*(u-\lambda)\, d\lambda \tag{4}$$

or alternatively

$$K_{n_c}(t,u) = E_t \int_{-\infty}^{\infty} f(t-u)\, S_{DR}[f,\lambda]\, f\,(u-\lambda)\, e^{j2\pi f(u-u)} df\, d\lambda \tag{5}$$

where the correlation function, $K_{DR}(\rho,\lambda)$, is a two-variable function that depends on the reflective properties of the target and $S_{DR}(f,\lambda)$ represents the spectrum of the process $b(t,\lambda)$ and is called the scattering function of the clutter. The two functions K_{DR} and S_{DR} are a Fourier Transform pairs.

The white noise $w(t)$ is likewise a complex Gaussian random process but with covariance function $K_w(t,u) = N_0\delta(t-u)$.

CONVENTIONAL RECEIVER

The conventional receiver is designed for detecting a target in white noise only. We compute

$$\ell \triangleq \int_{-\infty}^{\infty} r(t)\, f^*(t-\rho_d)\, e^{-j2\pi f_d t}\, dt \tag{6}$$

and make the threshold comparison test

$$:\ell:^2 \underset{H_0}{\overset{H_1}{\gtrless}} \beta \tag{7}$$

The performance, i.e., output (S/N_0) is

$$\Delta_{wo} = \frac{\bar{E}_r / N_0}{1 + \rho_r}$$

$$\Delta_{wo} = \frac{\bar{E}_r/N_0}{1 + (E_t/N_0) \int_{-\infty}^{\infty} S_{DR}[f,\lambda]\, \Omega[\rho_d-\lambda,\ f-f_d]\, df\, d\lambda} \tag{8}$$

ρ_r represents the degradation due to the clutter

$f(t)$ is the transmit waveform which is used here as a modulation or filter function as shown in Figure 2.

ρ_d is the target round trip delay

f_d is the target doppler frequency

β is a predetermined threshold

$\bar{E}_r$ is the average energy received from the target

E_t is the transmit energy

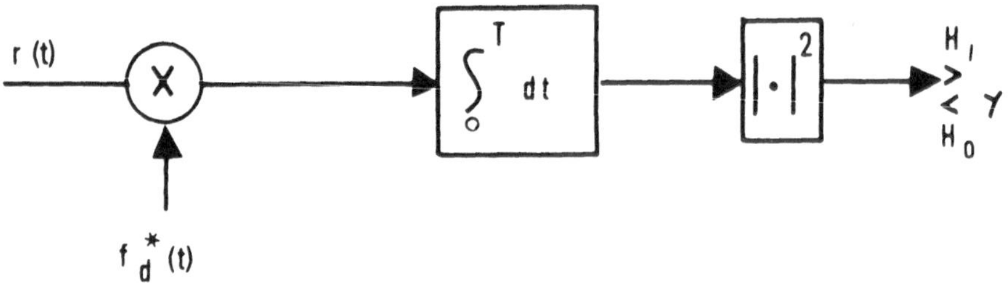

FIGURE 2a. MODULATION RECEIVER

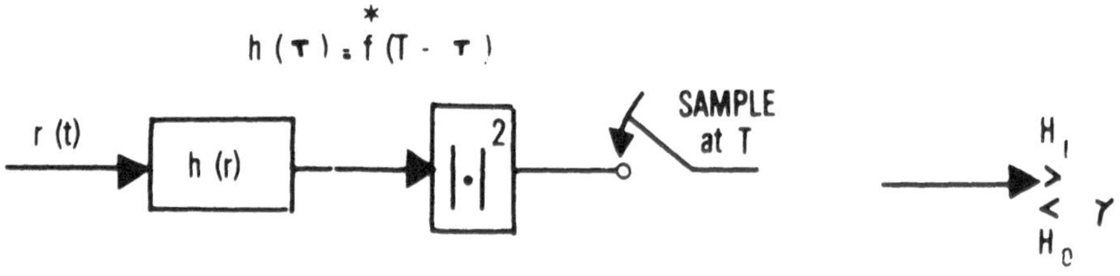

FIGURE 2b. MATCHED FILTER RECEIVER

N_o is the white noise energy (kT)

$\Omega(\lambda,f)$ is the waveform ambiguity function.

Next we shall review the optimum receiver, i.e., the single-channel receiver optimized for the clutter as well as the white component of the noise environment.

OPTIMUM RECEIVER

The optimum receiver computes

$$\ell \underline{\Delta} \int_{-\infty}^{\infty} r(t)\, g^{*}(t)\, dt \tag{9}$$

where g(t) satisfies the integral equation

$$f(t-\rho_d)\, e^{j2\pi f_d t} = E_t \int_{-\infty}^{\infty}\int_{-\infty}^{\infty} f(t-\lambda) K_{DR}(t-u,\lambda) f^{*}(u-\lambda) g(u)\, du\, d\lambda + N_o g(t) \tag{10}$$

The function, g(t) is a modulation or filter function and is used in place of f(t) in Figure 2 when colored noise is present.

The performance is obtained by evaluating

$$\Delta = \bar{E}_r \int_{T_i}^{T_f} f(t)\, Q_n^{*}(t,u)\, f^{*}(u)\, dt\, du \tag{11}$$

$$\text{or} \quad \Delta = \bar{E}_r \int_{T_i}^{T_f} f(t)\, g^{*}(t)\, dt \tag{12}$$

where the inverse Kernal $Q_n(t,u)$ satisfies

$$\int_{T_i}^{T_f} [\ K_{n_c}(t,z) + N_0\ d(t-z)\ ]\ Q_n(z,u)\ dz = d(t-u) \tag{13}$$

Then false alarm and detection probability are related by

$$P_F = (P_D)^{1+\Delta} \tag{14}$$

For certain special conditions of the scattering function we can simplify equations (9) and (10) further.

SPECIAL CASE - UNIFORM DOPPLER-RANGE INVARIANT CLUTTER

When the clutter has constant Doppler profile over infinite range, then ρ from equation (8) for the conventional receiver becomes

$$\rho_r = \frac{E_t}{N_0} \int_{-\infty}^{\infty} S_{n_c}(f)\ S_f(f-f_d)\ df \tag{15}$$

where

$$S_{n_c}(f) = S_{Du}(f) * S_f(f) \tag{16}$$

$S_{Du}(f)$ is the range invariant scattering function of the clutter

$$S_f(f) \triangleq | F(f) |^2 \tag{17}$$

$F(f)$ is the Fourier Transform of the transmit waveform $f(t)$ and f is frequency.

The symbol $*$ denotes convolution.

For the optimum receiver, $g(t)$ is given by the spectrum

$$G(f) = \frac{F(f-f_d)}{N_o + E_t\, S_{Du}(f) * S_f(f)} \tag{18}$$

and the performance is obtained from

$$\Delta_o = E_r \int_{-\infty}^{\infty} \frac{| F(f-f_d) |^2}{N_o + E_t\, S_{Du}(f) * S_f(f)} \tag{19}$$

OPTIMAL WAVEFORM DESIGN

The optimal waveform $f_o(t)$ must satisfy [1] the following integral equation

$$\int_{T_i}^{T_f} h_{ou}(t,u)\, f_o(u)\, du + \lambda_E\, f_o^*(t) - \lambda_B\, \ddot{f}(t) = 0 \tag{20}$$

where

$h_{ou}(t,u)$ is the optimum unrealizable filter satisfying

$$N_o h_{ou}(t,u) + \int_{T_i}^{T_f} h_{ou}(x,u)\, K_{n_c}(t,x)\, dx = K_{n_c}(t,u) \quad : \quad T_i \le t,u \le T_f \tag{21}$$

$K_{n_c}(t,u)$ satisfies equation (4) and (5). λ_E and λ_B are LaGrange multipliers with an energy and bandwidth constraint.

$$\int_{T_i}^{T_f} \left| f(t) \right|^2 dt = 1 \tag{22a}$$

$$\int_{T_i}^{T_f} \left| \frac{df(t)}{dt} \right|^2 dt = B^2 \tag{22b}$$

where $f_0(T_i) = f_0(T_f) = 0$

This concludes the one-channel receiver and waveform review.

Now that we have reviewed the optimum system for the single-channel case, let us extend these concepts to the two-channel case, i.e. a polarimetric system.

DUAL POLARIZATION CHANNELS

When we transmit and receive over two channels as we do in the polarization case, the received signals given in equations (1) and (2) are represented by two-element vector waveforms

$$\mathbf{r}(t) = b\mathbf{f}_d(t) + \mathbf{n}_c(t) + \mathbf{w}(t) \quad : \quad H_1 \tag{23}$$

$$\mathbf{r}(t) = \mathbf{n}_c(t) + \mathbf{w}(t) \quad : \quad H_0 \tag{24}$$

where boldface denotes vectors defined as follows:

$$\mathbf{r}(t) \triangleq \begin{bmatrix} r_V(t) \\ r_H(t) \end{bmatrix} \quad \text{the received signal, a vector function of time having a vertical and a horizontal component} \tag{25a}$$

$$b \triangleq \begin{bmatrix} b_{VV} & b_{VH} \\ b_{HV} & b_{HH} \end{bmatrix} \quad \text{a scattering matrix for an assumed slowly fluctuating point target with zero mean complex Gaussian random elements} \tag{25b}$$

$$f_d(t) \triangleq \sqrt{E_t}\, f(t-\rho_d)\, e^{j2\pi f_d t} \quad \text{the time delayed Doppler shifted replica of the transmit waveform vector} \tag{25c}$$

$E \triangleq$ the average energy in the transmit waveform vector,

$$f(t) \triangleq \begin{bmatrix} f_V(t) \\ f_H(t) \end{bmatrix} \tag{25d}$$

$$w(t) \triangleq \begin{bmatrix} w_V(t) \\ w_H(t) \end{bmatrix} \quad \text{the receiver noise vector assumed white} \tag{25e}$$

$$n_c(t) \triangleq \begin{bmatrix} n_{cV} \\ n_{cH} \end{bmatrix} = \sqrt{E_t} \int_{-\infty}^{\infty} b(t - \tfrac{\lambda}{2}, \lambda)\, f(t - \lambda)\, d\lambda \triangleq \text{colored noise vector due to backscatter from clutter} \tag{25f}$$

where t is time and λ is range expressed in time.

CLUTTER MODELING

The received vector, n_c, due to the clutter is obtained by convolving the Doppler-range variant scattering matrix of the clutter process.

$$b(t-\tfrac{\lambda}{2},\lambda) \triangleq \begin{bmatrix} b_V^T(t-\tfrac{\lambda}{2},\lambda) \\ b_H^T(t-\tfrac{\lambda}{2},\lambda) \end{bmatrix} \triangleq \begin{bmatrix} b_{VV}(t-\tfrac{\lambda}{2},\lambda) & b_{VH}(t-\tfrac{\lambda}{2},\lambda) \\ b_{HV}(t-\tfrac{\lambda}{2},\lambda) & b_{HH}(t-\tfrac{\lambda}{2},\lambda) \end{bmatrix} \tag{26}$$

with the transmit waveform vector, f(t), which is normalized to unit energy (see Figure 3). The scattering matrix, b(t,λ), is a mean Gaussian random process.

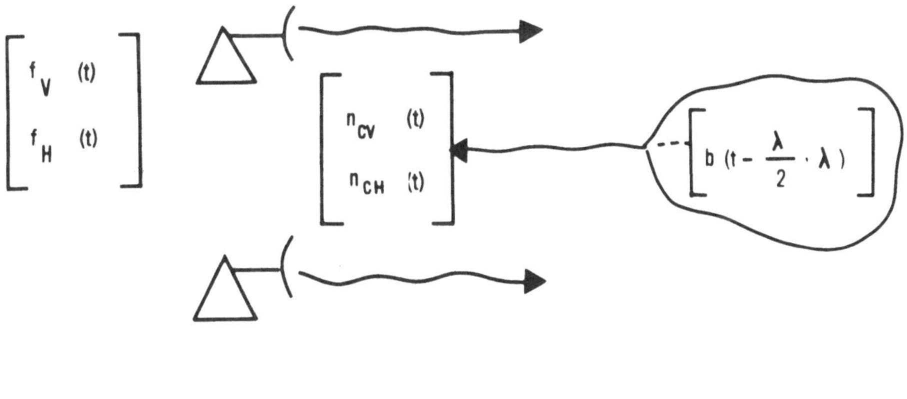

FIGURE 3. TRANSMIT AND BACKSCATTERED VECTOR FROM CLUTTER

The clutter vector covariance function is a matrix

$$K_{n_c}(t,u) \triangleq E[n_c(t)\ n_c^{\dagger}(u)] = \begin{bmatrix} K_{n_c VV}(t,u) & K_{n_c VH}(t,u) \\ K_{n_c HV}(t,u) & K_{n_c HH}(t,u) \end{bmatrix} \tag{27}$$

Substituting equation (25f) into equation (27) and bringing the expectation operator into the integral, the clutter covariance matrix becomes

$$K_{n_c}(t,u) = E_t \int_{-\infty}^{\infty} E \left\{ \left[b(t-\frac{\lambda}{2},\lambda)\ f(t-\lambda) \right] \right.$$

$$\left. \times \left[b(u-\frac{\lambda_1}{2}, \lambda_1)\ f(u-\lambda_1) \right]^{\dagger} \right\} d\lambda\ d\lambda_1 \tag{28}$$

With some matrix rearrangement, equation (28) becomes

$$K_{n_c}(t,u) = E_t \int_{-\infty}^{\infty} {}^{T}f\,(t-\lambda)\; E \left\{ \begin{bmatrix} b_V(t-\frac{\lambda}{2},\lambda) \\ b_H(t-\frac{\lambda}{2},\lambda) \end{bmatrix} \right.$$

$$\left. \times \left[b_V{}^{\dagger}(u-\frac{\lambda_1}{2},\,\lambda_1)\; b_H{}^{\dagger}(u-\frac{\lambda_1}{2},\,\lambda_1) \right] \right\} f^*(u-\lambda_1)\; d\lambda\; d\lambda_1 \qquad (29)$$

Assuming the returns from different range intervals are statistically independent and that the return from each interval is a sample vector function of a stationary zero-mean Gaussian random process, equation (29) reduces to

$$K_{n_c}(t,u) = E_t \int_{-\infty}^{\infty} f^T(t-\lambda)\; K_{DR}(t-u,\lambda)\; f^*(u-\lambda)\; d\lambda \qquad (30)$$

or alternatively

$$K_{n_c}(t,u) = E_t \int_{-\infty}^{\infty} f^T(t-\lambda)\; S_{DR}(f,\lambda)\; f^*(u-\lambda)\; e^{j2\pi f(t-u)} df\; d\lambda \quad (31)$$

THE TENSOR FUNCTIONS, $K_{DR}(\rho,\lambda)$ AND $S_{DR}(f,\lambda)$

The tensor correlation function $K_{DR}(\rho,\lambda)$ is generally a two-variable fourth-rank tensor that depends on the reflective properties of the scatterer. It is obtained from the expectation term in equation (29) above. It is written

$$E\left\{\begin{bmatrix} b_V(t-\frac{\lambda}{2},\lambda) \\ b_H(t-\frac{\lambda}{2},\lambda) \end{bmatrix}\begin{bmatrix} b_V{}^{\dagger}(u-\frac{\lambda_1}{2},\lambda_1) & b_H{}^{\dagger}(u-\frac{\lambda_1}{2},\lambda_1) \end{bmatrix}\right\}$$

Because of the statistical independence of the range intervals and the stationarity of each interval, the above expression with some further matrix manipulation reduces to

$$E\left\{\begin{matrix} b_V(t-\frac{\lambda}{2},\lambda)\, b_V{}^{\dagger}(u-\frac{\lambda_1}{2},\lambda_1) & b_V(t-\frac{\lambda}{2},\lambda)\, b_H{}^{\dagger}(u-\frac{\lambda_1}{2},\lambda_1) \\ b_H(t-\frac{\lambda}{2},\lambda)\, b_V{}^{\dagger}(u-\frac{\lambda_1}{2},\lambda_1) & b_H(t-\frac{\lambda}{2},\lambda)\, b_H{}^{\dagger}(u-\frac{\lambda_1}{2},\lambda_1) \end{matrix}\right\}$$

$$= \begin{bmatrix} K_{DRVV}(\rho,\lambda) & K_{DRVH}(\rho,\lambda) \\ K_{DRHV}(\rho,\lambda) & K_{DRHH}(\rho,\lambda) \end{bmatrix} \delta(\lambda-\lambda_1) \qquad (32)$$

which can be written

$$[\ K_{DR}(\rho,\lambda)\ ]\ \delta(\lambda-\lambda_1)$$

where

$$\rho = t-u$$

The subscripts DR denote that the clutter is doubly spread in Doppler and range. Substitution of this expression into equation (29) above and evaluation of the integral with respect to λ_1 gives the result in equation (30).

The function $S_{DR}(f,\lambda)$ is generally a two-variable fourth-rank tensor representing the spectrum of the process and is related to $K_{DR}(\rho,\lambda)$ by the Fourier transform

$$K_{DR}(\rho,\lambda) = \int_{-\infty}^{\infty} S_{DR}(f,\lambda)\ e^{j2\pi f\rho}df \tag{33}$$

This expression is denoted as the tensor scattering function of

$$b(t,\lambda)\ \underline{\Delta} \begin{bmatrix} b_{VV}(t,\lambda) & b_{VH}(t,\lambda) \\ b_{HV}(t,\lambda) & b_{HH}(t,\lambda) \end{bmatrix} \tag{34}$$

The elements of $K_{DR}(\rho,\lambda)$ may be found by carrying out the expectation operation indicated in equation (32) and recalling from equation (26a) that the vectors representing the clutter process scattering matrix are

$$b_V(t,\lambda) = \begin{bmatrix} b_{VV}(t,\lambda) \\ \\ b_{VH}(t,\lambda) \end{bmatrix} \tag{35a}$$

$$b_H(t,\lambda) = \begin{bmatrix} b_{HV}(t,\lambda) \\ \\ b_{HH}(t,\lambda) \end{bmatrix} \tag{35b}$$

Carrying out the expectation in equation (32) and omitting the variable ρ and λ (to conserve space) we have

$$K_{DR}(\rho,\lambda) = E\left\{\begin{bmatrix} \begin{bmatrix} b_{VV}\, b_{VV}^{*} & b_{VV}\, b_{VH}^{*} \\ b_{VH}\, b_{VV}^{*} & b_{VH}\, b_{VH}^{*} \end{bmatrix} & \begin{bmatrix} b_{VV}\, b_{HV}^{*} & b_{VV}\, b_{HH}^{*} \\ b_{VH}\, b_{HV}^{*} & b_{VH}\, b_{HH}^{*} \end{bmatrix} \\ \begin{bmatrix} b_{HV}\, b_{VV}^{*} & b_{HV}\, b_{VH}^{*} \\ b_{HH}\, b_{VV}^{*} & b_{HH}\, b_{VH}^{*} \end{bmatrix} & \begin{bmatrix} b_{HV}\, b_{HV}^{*} & b_{HV}\, b_{HH}^{*} \\ b_{HH}\, b_{HV}^{*} & b_{HH}\, b_{HH}^{*} \end{bmatrix} \end{bmatrix}\right\}$$

Taking the expectation inside the matrices we have

$$K_{DR}(\rho,\lambda) = \begin{bmatrix} \begin{bmatrix} K_{VV,VV} & K_{VV,VH} \\ K_{VH,VV} & K_{VH,VH} \end{bmatrix} & \begin{bmatrix} K_{VV,VV} & K_{VV,VH} \\ K_{VH,VV} & K_{VH,VH} \end{bmatrix} \\ \begin{bmatrix} K_{VV,VV} & K_{VV,VH} \\ K_{VH,VV} & K_{VH,VH} \end{bmatrix} & \begin{bmatrix} K_{VV,VV} & K_{VV,VH} \\ K_{VH,VV} & K_{VH,VH} \end{bmatrix} \end{bmatrix} \tag{36}$$

Equation (36) provides 16 different elements (discriminants) when one considers the statistical behavior of the polarization random process scattering matrix. Assuming the scattering matrix is a non-negative Hermitian process then the number of independent elements reduces to 10, still providing a wide selection of discriminants.

These tensor element functions completely describe the target and/or clutter irrespective of the transmit waveform and receiver design. They can be used in the waveform and receiver optimization as well as the performance equations in the same sense that the scalar correlation and scattering functions are used in the design equations of the single channel system. We will now consider the optimization of waveform and receiver for the polarization case.

CONVENTIONAL RECEIVER - POLARIZATION CASE

The conventional receiver is one which is optimized for a signal corrupted by white noise instead of the clutter colored noise. In this case the optimum filter is a vector matched filter. We compute

$$\ell \underline{\Delta} \int r^T(t)\, f^*(t-\rho_d)\, e^{-j2\pi f_d t}\, dt \tag{37}$$

and make the same threshold comparison as we did in equation (7).

The performance degradation is given by equation (8). For the vector case we leave the ambiguity function in tensor/integral form.

$$\rho_r = \frac{E_t}{N_0} \iiiint f^{\dagger}(t-\rho_d)\, f^{T}(t-\lambda)\, e^{j2\pi(f-f_d)t}\, S_{DR}(f,\lambda)\, f^{*}(u-\lambda)\, f(u-\rho_d)$$

$$\times\, e^{-j2\pi(f-f_d)u}\, dt\, du\, d\lambda\, df \qquad (38)$$

Equation (38) introduces another model, that of the ambiguity function for the multi-channel case. It takes on a form suggested by the integral of the outer product of all the transmit vectors f in Equation (38).

In equation (8) the integral contained the waveform ambiguity function

$$\Omega(\rho-\lambda,\ f-f_d)$$

The vector form for the performance degradation Equation (38) also has the appropriate functions and integrals which constitute the ambiguity function. However, it involves outer products with waveform vectors, a procedure which removes us from the more directly related subject matter. This mathematical model for the ambiguity function of a multiple-channel radar system should be studied further to determine what potential it may have with respect to better understanding and improved processing in ambiguity removal systems.

Next we present the optimum receiver for the multi-channel (polarization) system.

OPTIMUM RECEIVER - POLARIZATION CASE

The optimum receiver computes

$$\ell \triangleq \int_{-\infty}^{\infty} r^T(t)\, g^*(t)\, dt \tag{39}$$

for the vector case and makes the same threshold comparison as cited in equation (7), see figure 4. g(t) satisifies the matrix equation

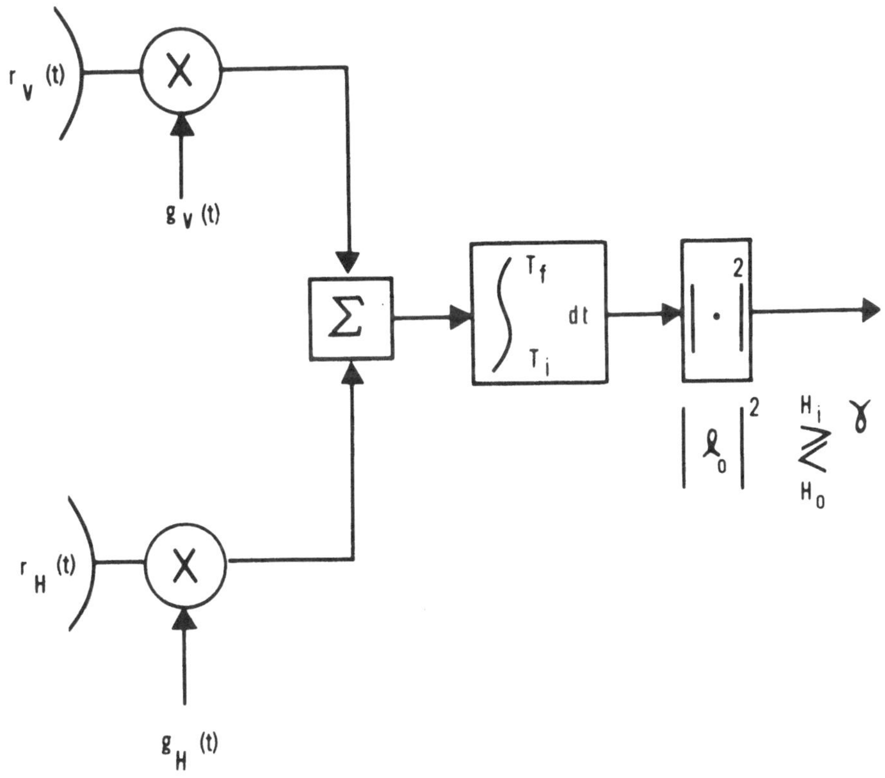

FIGURE 4. POLARIZATION MODULATION RECEIVER

$$f_d(t) = \int_{-\infty}^{\infty} K_{n_c}(t,u)\, g(u)\, du + N_0 g(t) \tag{40}$$

$$f_d(t) = \iint_{-\infty}^{\infty} f^T(t-\lambda)\, K_{DR}(t-u,\lambda)\, f^*(u-\lambda)\, g(u)\, du\, d\lambda + N_0 g(t) \tag{41}$$

and is defined as

$$g(t) \triangleq \int_{T_i}^{T_f} Q_{n_c}(t,u)\, f(u)\, du \qquad : \quad T_i \le t \le T_f \tag{42}$$

where $Q_{n_c}(t,u)$ is the inverse autocovariance matrix and satisifies the matrix integral equation

$$\int_{T_i}^{T_f} Q_{n_c}(t,x) \left[K_{n_c}(x,u) + N_0\, I\delta(x-u) \right] dx = I\delta(t-u) \tag{43}$$

$$: \quad T_i \le t,u \le T_f$$

The performance (output signal-to-noise ratio) of the optimum receiver is specified by

$$\Delta = \bar{E}_r \iint_{T_i}^{T_f} f_d^T(t)\, Q^*(t,u)\, f^*(u)\, dt\, du \tag{44}$$

or

$$\Delta = \bar{E}_r \int_{T_i}^{T_f} f_d^T(t)\, g^*(t)\, dt \tag{45}$$

The average energy received from the target is

$$\bar{E}_r = E \left\{ \int_{-\infty}^{\infty} s^T(t)\, s^*(t)\, dt \right\} E_t \tag{46}$$

where $s(t) = b\, f_d(t)$.

This evaluates to

$$\bar{E}_r = 2(\sigma^2_{VV} + \sigma^2_{HV})\, E_{t_V} + 2(\sigma^2_{HH} + \sigma^2_{VH}) E_{t_H}$$

$$+ 4\sqrt{E_{t_V} E_{t_H}}\; \mathrm{Re}(\rho_t[\sigma^2_{VV,VH} + \sigma^2_{HV,HH}]) \qquad (47)$$

where the σ^2_{VV} is the variance of the b_{VV} element in the target scattering matrix and $\sigma^2_{VV,VH}$ is the covariance between the b_{VV} andthe b_{VH} element of the target scattering matrix.

E_{t_V} and E_{t_H} is the transmit energy of the vertical channel and the horizontal channel respectively.

ρ_t is the correlation coefficient between the vertical and horizontal transmit waveform.

The quantity, Δ, for which expressions have been given throughout this discussion, is used to relate the detection and false alarm probability

$$P_F = (\ P_D)^{1+\Delta} \qquad (48)$$

The above optimization procedures involve integral and tensor equations which are very difficult to solve. In general even the scalar case is not trivial. The procedures become less difficult when certain special conditions arise.

SPECIAL CASES-UNIFORM DOPPLER-RANGE INVARIANT CLUTTER

CASE 1:

For example, consider the special case when the clutter has a range invariant tensor correlation or scattering function

$$K_{DR}(\rho,\lambda), \quad S_{DR}(f,\lambda)$$

and extends well beyond the range of a possible target. Furthermore, the signal is from a point target (no Doppler or range spreading).

Recall that the conventional receiver computes

$$\ell \triangleq \int_{-\infty}^{\infty} r^T(t)\, f^*(t-\rho_d)\, e^{-j2\pi f_d t}\, dt \tag{49}$$

and the optimum receiver computes

$$\ell_0 \triangleq \int_{-\infty}^{\infty} r^T(t)\, g^*(t)\, dt. \tag{50}$$

Let

$$f(t) \longleftrightarrow F(w) \tag{51}$$

denote a Fourier transform pair.

From equation (13-126) of reference [1] we may find $\mathbf{g}(t)$

$$g(t) \longleftrightarrow G(w) = [\, N_0\, I + S_{n_c}(w)\,]^{-1}\, F(w) \tag{52}$$

where

$$S_{n_c}(w) = \int_{-\infty}^{\infty} F^T(\alpha)\ S_{Du}(w-\alpha)\ F^*(\alpha)\ d\alpha \tag{53}$$

The performance is obtained from equation (13-92) of [1]

$$\rho_r = \frac{1}{N_0} \int_{-\infty}^{\infty} F^T(w-w_d)\ S_{n_c}(w)\ F^*(w-w_d)\ df \tag{54}$$

for the conventional receiver, i.e.,

$$\Delta_{W_0} = \frac{\bar{E}_r / N_0}{1 + \rho_r} \tag{55}$$

and

$$\Delta_0 = \bar{E}_r \int_{-\infty}^{\infty} F^T(w-w_d)\ [\ N_0 I + S_{n_c}(w)^{-1}]\ F^*(w-w_d)\ df \tag{56}$$

for the optimum receiver.

CASE 2: When the target is Doppler spread only and the noise is white, then we determine the vector/matrix equivalent expression

$$\mu(s) = \frac{T}{2} \left\{ (1-s) \int_{-\infty}^{\infty} \ell n\left[1 + \frac{S_s(w)}{N_0} \right] - \ell n\left[1 + \frac{(1-s)\, S_s(w)}{N_0} \right] \frac{dw}{2\pi} \right\} \quad (57)$$

where the parameter, $0 \leq s \leq 1$ and the functions, $S_s(w)$, is a scaler. Once $\mu(s)$ has been formulated for vector/matrix notation and evaluated, the Chernoff bounds and approximations may be determined. Another measure of performance is the mean square realizable filter error of the optimum filter output.

$$\Sigma_p = N_0 \int_{-\infty}^{\infty} \ell n\left[1 + \frac{S_s(w)}{N_0} \right] \frac{dw}{2\pi} \quad (58)$$

OPTIMAL WAVEFORM DESIGN - POLARIZATION CASE

The extension of the waveform design case to the polarization channels involves vectors, matrices and tensors in equations (20) through (22). The optimum unrealizable filter, h_{ou}, is a 2 X 2 matrix, the optimum waveform, f_o, is a vector, the autocovariance function, K_{n_c}, is a matrix, and the multiplications indicated in equations (22a) and (22b) become vector dot products. It must be remembered that the autocovariance matrix contains factors of the waveform vectors as defined in Equation (30). Consequently, the filter must satisfy the equation

$$N_o h_{ou}(t,u) = E_t \int_{T_i}^{T_f} \int_{-\infty}^{\infty} f_o^T(t-\lambda) K_{DR}(t-x,\lambda) f_o(x-\lambda)[Id(x-u) - h_{ou}(x,u)] \times d\lambda\, dx \tag{59}$$

and the optimal waveform vector must satisfy the integral equation

$$\int_{T_i}^{T_f} h_{ou}(t,u)\, f_o(u)\, du + \lambda_E\, f_o(t) - \lambda_B\, \ddot{f}_o(t) = 0 \tag{60}$$

The energy and bandwidth constraints become

$$\int_{T_i}^{T_f} f^T(t)\, f^*(t)\, dt = 1 \qquad (61)$$

and

$$\int_{T_i}^{T_f} \dot{f}^T(t)\, \dot{f}^*(t)\, dt = B^2 \qquad (62)$$

SUMMARY

We have extended the statistical model of a scatterer illuminated and observed on one channel to suit the two-channel polarization case. In that model we have accounted for the statistical range and doppler spread characteristics. The mathematical formulation for the polarization sensitive scatterer was written in the framework of a random process scattering matrix whose covariance properties took on a sixteen-element tensor of the fourth rank. Each element was expressed in terms of its correlation with respect to range and doppler. This tensor contained all the backscatter information necessary to determine the range/time/polarization dependent behavior of the received signal for any arbitrary transmit polarized waveform.

From such an analytical model we extended the conventional (matched filter) as well as the optimum receiver to include the polarization case. We also gave an expression for their performance. Special cases were covered wherein the clutter parameters were assumed range invariant so one can use spectral processing for the receiver. As we can see from the extended results, the receiver contains two channels, each having its own filter or modulating function determined from the time/range statistical properties of the transmit polarization waveform reacting with the clutter/target model discussed in the preceding paragraph. The transmit, receive and modulating waveforms are each two-element vectors.

We also gave an expression which an optimal transmit waveform must satisfy for the polarization case. This again is an extension of the scalar case given previously.

No attempt was made to work out an illustrative example. The nature of these equations renders their solution a subject in itself and will be treated in a subsequent paper.

Comments are invited which may provide further insight into the advantages and disadvantages of such a target model and the resulting processing.

REFERENCES

[1] Van Trees, H. L., "Detection, Estimation and Modulation Theory, Part III" John Wiley Sons, Inc., 1971.

[2] Peolman, A. J., "Cross Correlation of Orthogonally Polarized Backscatter," IEEE AES-12, No. 6, pp 674-682.

[3] Varshanchuk, M. L. and Kobak, V. O., "Cross Correlation of Orthogonally Polarized Components of Electromagnetic Field, Scattered by an Extended Object," Radio Eng. Electron Phys., Vol 16, pp 201-205, February 1971.

[4] Ioannidis, G. A., "Model for Spectral and Polarization Characteristics of Chaff," IEEE AES-15, No. 5, pp723-726, September 1979.

[5] Rome Air Development Center, "Polarization Processing Techniques Study", Final Report, RADC-TR-79-285, November 1979, AD No. A080 565, Contract No. F30602-78-C-0119.

[6] Fujita, M., and A. Klein, "Adaptive Polarization Processing," Workshop on Polarimetric Radar Technology, U. S. Army Missile Command, Redstone Arsenal, 25-26 June 1980.

IV.1 (RS.1)

THE RADIATIVE TRANSFER APPROACH IN ELECTROMAGNETIC IMAGING

Akira Ishimaru

Department of Electrical Engineering
University of Washington
Seattle, Washington 98195 USA

In the past, the radiative transfer theory has been largely concerned with the propagation of scalar intensities. In recent years, however, there has been an increasing interest in the propagation of complete arbitrarily polarized electromagnetic waves. We first present the general formulation of the complete radiative transfer equation using the Stokes' vectors, including the extinction matrix and the Mueller scattering matrix. The coherent and incoherent Stokes' vectors are defined corresponding to the coherent and incoherent field. The coherent Stokes' vector is shown to have a depolarization effect. Solutions of the vector radiative transfer equation require Fourier decomposition to account for the polarization effects. Solutions for linearly polarized and circularly polarized waves in spherical particles are presented, showing the angular dependence and the degree of polarization. It is indicated that even in spherical particles, depolarization occurs due to multiple scattering. Discussions are also presented on the propagation of Stokes' vectors in nonspherical particles. Depolarization effects and the deterioration of images are discussed in terms of the solutions of the radiative transfer equation. The limitations of the radiative transfer equation and its relation to multiple scattering theory are also discussed. It is shown that when the density is high, the radiative transfer theory becomes increasingly approximate.

W.-M. Boerner et al. (eds.), Inverse Methods in Electromagnetic Imaging, Part 2, 771–795.

1. INTRODUCTION

In recent years, there has been an increased interest in the use of polarization effects for radar and electromagnetic imaging problems (References 1, 2, and 3). The problem of electromagnetic imaging can be divided into the following areas: (1) Propagation of the Stokes' vector from the transmitter to the target region through various atmospheric conditions (rain, dust, fog, clouds, turbulence, etc.). (2) Scattering of the Stokes' vector from the object. (3) Scattering of the Stokes' vector from the rough surface, terrain, and the volume scattering. (4) Propagation of the Stokes' vector from the target region to the receiver. (5) The characteristics of the receiver relating the Stokes' vector to the output.

The propagation characteristics of the Stokes' vector through various media can be described by the equation of transfer. Even though the scalar equation of transfer has been studied extensively in the past, the vector equation of transfer has not received as much attention. In recent years, however, a need for further study of the vector radiative transfer theory has become increasingly evident and several important studies have been reported. This paper presents a general formulation of the vector theory of radiative transfer under general anisotropic scattering conditions. Some useful solutions are also presented for several practical situations.[4-8]

2. GENERAL FORMULATION OF VECTOR RADIATIVE TRANSFER THEORY

Let us consider the plane-parallel problem shown in Figure 1. We use E_v, the vertical, and E_h, the horizontal components of the electromagnetic wave propagating in the direction $\hat{s}$. We define the modified Stokes' vector $[I] = [I(\bar{r},\hat{s})]$[4,9,10,11]

$$[I] = \begin{bmatrix} I_v \\ I_h \\ U \\ V \end{bmatrix} = \begin{bmatrix} \langle E_v E_v^* \rangle \\ \langle E_h E_h^* \rangle \\ 2\mathrm{Re}\langle E_v E_h^* \rangle \\ 2\mathrm{Im}\langle E_v E_h^* \rangle \end{bmatrix} , \tag{2.1}$$

where Re and Im denote the real part and the imaginary part, the angle bracket < > indicates the ensemble average, and $\hat{s}$ is the unit vector in the direction (μ,ϕ), $\mu = \cos\theta$.

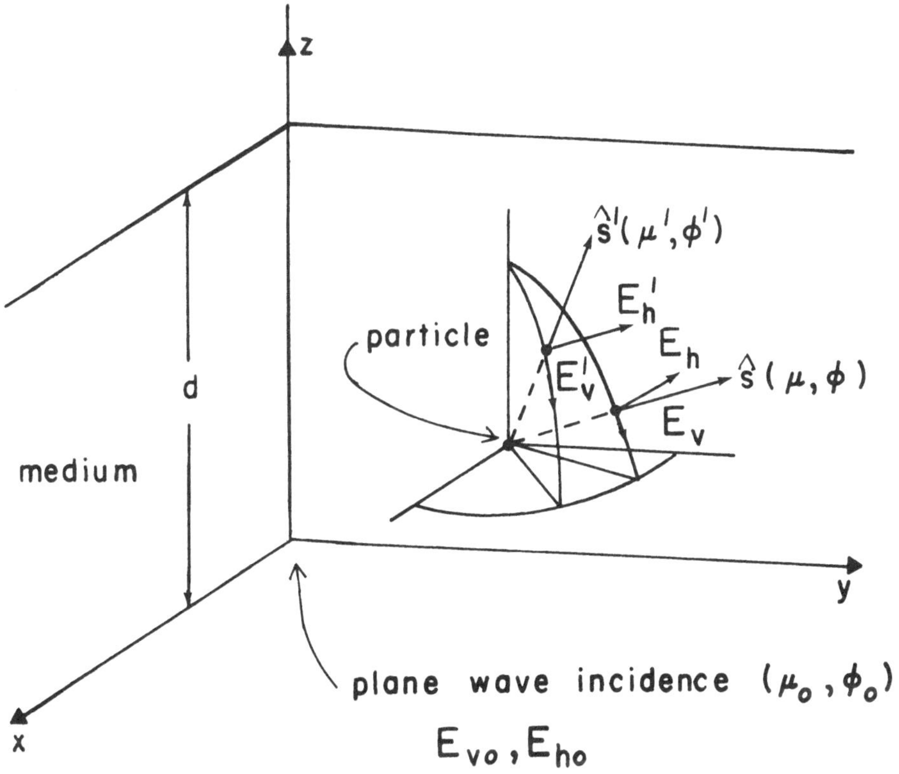

Figure 1. Scattering amplitude matrix.

The equation of transfer can be written as follows:

$$\frac{d}{ds}[I] = [\Lambda][I] + \int [S][I']d\omega' \quad , \tag{2.2}$$

where $[\Lambda] = [\Lambda_{ij}]$ = 4 x 4 extinction matrix and

$[S] = [S_{ij}]$ = 4 x 4 scattering (Mueller) matrix.

These two matrices $[\Lambda]$ and $[S]$ will be shown in detail in the following sections. $[I'] = [I(\bar{r},\hat{s}')]$ is the Stokes' vector pointed in the direction s', $d\omega'$ is the elementary solid angle in the direction of $\hat{s}'$, and ds is the differential distance along $\hat{s}$.

Let us write the Stokes' vector [I] as a sum of the coherent Stokes' vector $[I_c]$ and the incoherent (diffuse) Stokes' vector $[I_d]$

$$[I] = [I_c] + [I_d] \quad . \tag{2.3}$$

The coherent intensity $[I_c]$ satisfies the equation

$$\frac{d}{ds}[I_c] = [\Lambda][I_c] \quad . \tag{2.4}$$

Therefore, the equation of transfer (2.2) can be rewritten as

$$\frac{d}{ds}[I_d] = [\Lambda][I_d] + \int [S][I_d']d\omega' + [I_i] \quad , \tag{2.5}$$

where $[I_i] = \int [S][I_c']d\omega'$ = incident intensity matrix (reduced intensity), $ds = \frac{dz}{\mu}$, $\mu = \cos\theta$. In the following sections, we will examine the matrices $[\Lambda]$, $[S]$, and $[I_i]$.

3. SCATTERING AMPLITUDE MATRIX [F]

In order to study the extinction matrix [E], the Mueller (scattering) matrix [S], and the incident intensity matrix $[I_i]$, it is first necessary to express the scattering amplitude matrix [F] for an arbitrarily shaped particle.[11-15]

Consider the wave [E'] incident on a particle in the direction of $\hat{s}'$ and the scattered wave [E] in the direction of $\hat{s}$ at a distance R in the far field zone of the particle (see Figure 1). They are related through [F]

$$[E] = \frac{e^{ikR}}{R}[F][E'] \quad , \tag{3.1}$$

where

$$[E] = \begin{bmatrix} E_v \\ E_h \end{bmatrix} , \quad [E'] = \begin{bmatrix} E_v' \\ E_h' \end{bmatrix} , \quad [F] = \begin{bmatrix} F_{11} & F_{12} \\ F_{21} & F_{22} \end{bmatrix} = [F_{ij}] \quad .$$

The scattering amplitude matrix [F] is a function of $\hat{s}$ and $\hat{s}'$ and depends on the particle characteristics. In the following sections, we will discuss several representations of [F].

3.1 Scattering amplitude matrix [F] for an arbitrarily shaped particle

The scattering amplitude matrix [F] is expressed in the coordinate system (x-y-z) of the medium (Figure 1). However, the scattering characteristics of a particle are often most conveniently described in the coordinate system (x_b-y_b-z_b) appropriate to the particle. The relationship between these two coordinate systems is given by Euler's transformation[16]

$$\begin{bmatrix} \hat{x}_b \\ \hat{y}_b \\ \hat{z}_b \end{bmatrix} = [A_e] \begin{bmatrix} \hat{x} \\ \hat{y} \\ \hat{z} \end{bmatrix} . \tag{3.2}$$

$$[A_e] = [a_{ij}] .$$

There are many representations of the Euler transformation, and they are given in Reference 16.

Suppose that the scattering amplitude matrix $[F_b]$ is known in the coordinate system (x_b-y_b-z_b) (see Figure 2). The scattered field $[E_b]$ and the incident field $[E_b']$ are related by

$$[E_b] = \frac{e^{ikR}}{R} [F_b][E_b'] , \tag{3.3}$$

where
$$[E_b] = \begin{bmatrix} E_{bv} \\ E_{bh} \end{bmatrix} , \quad [E_b'] = \begin{bmatrix} E_{bv}' \\ E_{bh}' \end{bmatrix} .$$

Note that for the given choice of the coordinate transformation between (x-y-z) and (x_b-y_b-z_b), the Euler matrix $[A_e]$ can be determined and, therefore, is well known. The scattering amplitude matrix $[F_b]$ is also assumed to be known as a function of $\hat{s}(\mu_b,\phi_b)$ and $\hat{s}'(\mu_b,\phi_b')$ (see Figure 3).

Equation (3.3) is given in the x_b-y_b-z_b system. However, we need to express (3.3) in the x-y-z system (3.1). This is done by finding the matrix [B] and [B'] where

$$[E] = [B][E_b] \text{ and } [E'] = [B'][E_b'] .$$

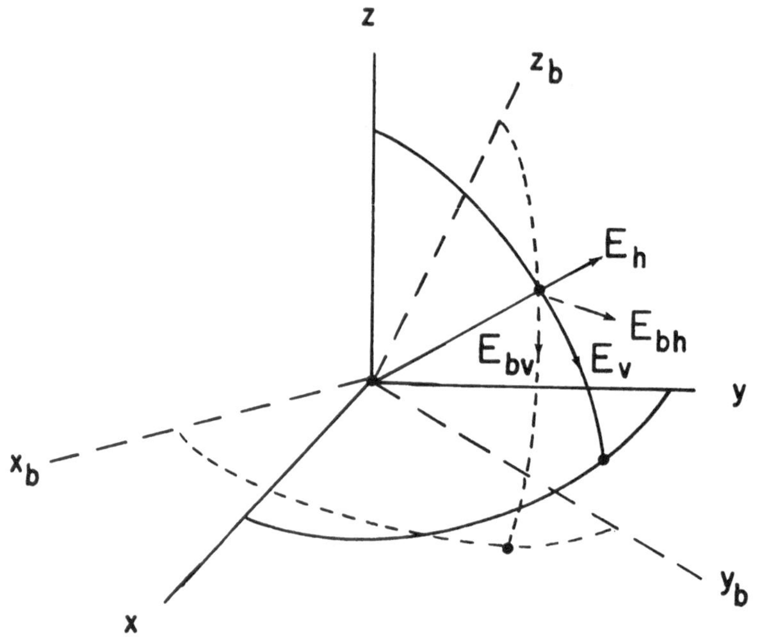

Figure 2. Medium coordinate system x-y-z and the coordinate system x_b-y_b-z_b attached to the particle.

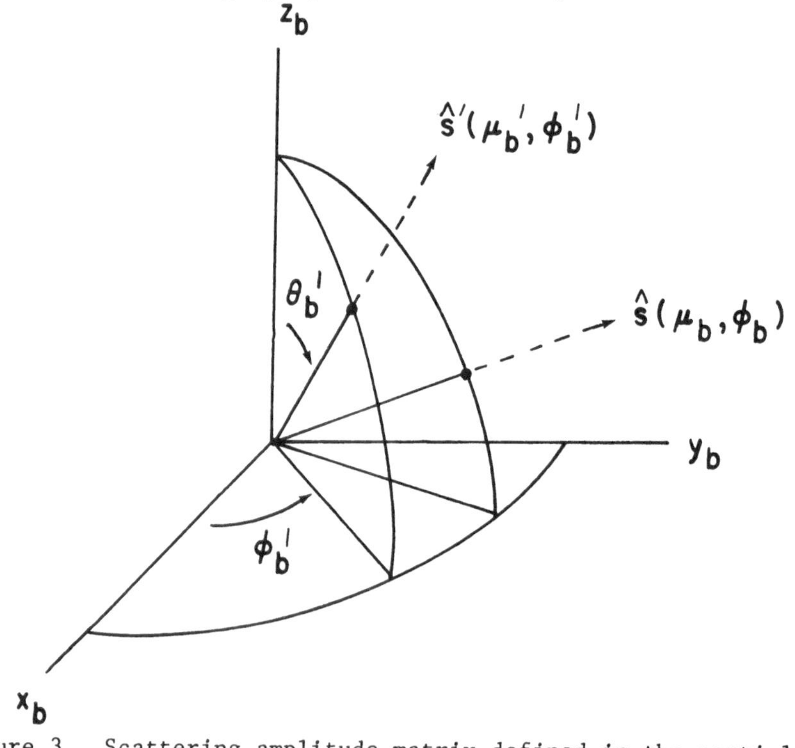

Figure 3. Scattering amplitude matrix defined in the particle coordinate x_b-y_b-z_b.

$$[B] = \begin{bmatrix} \hat{v}\cdot\hat{v}_b & \hat{v}\cdot\hat{h}_b \\ \hat{h}\cdot\hat{v}_b & \hat{h}\cdot\hat{h}_b \end{bmatrix} , \quad [B'] = \begin{bmatrix} \hat{v}'\cdot\hat{v}_b' & \hat{v}'\cdot\hat{h}_b' \\ \hat{h}'\cdot\hat{v}_b' & \hat{h}'\cdot\hat{h}_b' \end{bmatrix} . \tag{3.4}$$

Using (3.3) and (3.4), (3.1) is given by

$$[F] = [B][F_b][B']^{-1} . \tag{3.5}$$

Now it is necessary to express (3.5) in terms of (μ,ϕ), (μ',ϕ') and Euler's angle a_{ij}. Since $[F_b]$ is given only in terms of (μ_b,ϕ_b) and (μ_b',ϕ_b'), we need to express (μ_b,ϕ_b) and (μ_b',ϕ_b') in terms of (μ,ϕ), (μ',ϕ'), and a_{ij}. To do this, consider the Eulerian transformation of the components of the unit vector $\hat{s}$ in the x-y-z system and the x_b-y_b-z_b system.

$$\begin{bmatrix} \sin\theta_b \cos\phi_b \\ \sin\theta_b \sin\phi_b \\ \cos\theta_b \end{bmatrix} = [A_e] \begin{bmatrix} \sin\theta \cos\phi \\ \sin\theta \sin\phi \\ \cos\theta \end{bmatrix} . \tag{3.6}$$

From this, we get

$$\cos\theta_b = a_{31} \sin\theta \cos\phi + a_{32} \sin\theta \sin\phi + a_{33} \cos\theta ,$$

$$\cos\phi_b = \frac{a_{11} \sin\theta \cos\phi + a_{12} \sin\theta \sin\phi + a_{13} \cos\theta}{\sin\theta_b} , \tag{3.7}$$

which gives θ_b and ϕ_b in terms of θ, ϕ, and a_{ij}. Similarly, θ_b' and ϕ_b' can be given in terms of θ', ϕ', and a_{ij}.

$$\cos\theta_b' = a_{31} \sin\theta' \cos\phi' + a_{32} \sin\theta' \sin\phi' + a_{33} \cos\theta' ,$$

$$\cos\phi_b' = \frac{a_{11} \sin\theta' \cos\phi' + a_{12} \sin\theta' \sin\phi' + a_{13} \cos\theta'}{\sin\theta_b'} . \tag{3.8}$$

Using (3.7) and (3.8), we obtain θ_b, ϕ_b, θ_b', and ϕ_b' from θ, ϕ, θ', ϕ', and Euler's transformation a_{ij}. We can then use these θ_b, ϕ_b, θ_b', and ϕ_b' to obtain $[F_b]$.

Next consider $[B] = [B_{ij}]$ in (3.4). In order to express B_{ij} in terms of μ, ϕ, and a_{ij}, we need to express $\hat{v}_b$ and $\hat{h}_b$. This can be done by noting

$$\hat{h}_b = \frac{\hat{z}_b x \hat{s}}{|\hat{z}_b x \hat{s}|} \quad , \qquad \hat{v}_b = \hat{h}_b x \hat{s} \quad , \tag{3.9}$$

and expressing $\hat{z}_b$ and $\hat{s}$ in terms of μ, ϕ, and a_{ij}.

To do this, we use the Euler transformation (3.2) and the representation of $\hat{s}$, $\hat{v}$, and $\hat{h}$ in terms of $\hat{x}$, $\hat{y}$, and $\hat{z}$.

$$\begin{bmatrix} \hat{s} \\ \hat{v} \\ \hat{h} \end{bmatrix} = [b] \begin{bmatrix} \hat{x} \\ \hat{y} \\ \hat{z} \end{bmatrix} \quad , \tag{3.10}$$

where

$$[b] = \begin{bmatrix} \sin\theta\cos\phi & \sin\theta\sin\phi & \cos\theta \\ \cos\theta\cos\phi & \cos\theta\sin\phi & -\sin\theta \\ -\sin\theta & \cos\phi & 0 \end{bmatrix} \quad .$$

Combining (3.2) and (3.10), we get

$$\begin{bmatrix} \hat{x}_b \\ \hat{y}_b \\ \hat{z}_b \end{bmatrix} = [A_e][b]^{-1} \begin{bmatrix} \hat{s} \\ \hat{v} \\ \hat{h} \end{bmatrix} = [d] \begin{bmatrix} \hat{s} \\ \hat{v} \\ \hat{h} \end{bmatrix} \quad , \tag{3.11}$$

$$[d] = [A_e][b]^{-1} = [d_{ij}] \quad .$$

Making use of (3.9) and (3.11), and noting $\hat{z}_b = d_{31}\hat{s} + d_{32}\hat{v} + d_{33}\hat{h}$, we get

$$\hat{h}_b = \frac{d_{33}\hat{v} - d_{32}\hat{h}}{\sqrt{d_{33}^2 + d_{32}^2}} \quad , \qquad \hat{v}_b = \frac{-d_{32}\hat{v} - d_{33}\hat{h}}{\sqrt{d_{33}^2 + d_{32}^2}} \quad . \tag{3.12}$$

Therefore, we get

$$[B'] = [B_{ij}] = \frac{1}{(d_{33}^2 + d_{32}^2)^{\frac{1}{2}}} \begin{bmatrix} -d_{32} & d_{33} \\ -d_{33} & -d_{32} \end{bmatrix} \quad . \tag{3.13}$$

Similarly, we get

$$[B'] = [B_{ij}'] = \frac{1}{(d_{33}'^2 + d_{32}'^2)^{\frac{1}{2}}} \begin{bmatrix} -d_{32}' & d_{33}' \\ -d_{33}' & -d_{32}' \end{bmatrix} , \qquad (3.14)$$

where

$$[d'] = [d_{ij}'] = [A_e][b']^{-1} ,$$

$$[b'] = \begin{bmatrix} \sin\theta' \cos\phi' & \sin\theta' \sin\phi' & \cos\theta' \\ \cos\theta' \cos\phi' & \cos\theta' \sin\phi' & -\sin\theta' \\ -\sin\theta' & \cos\phi' & 0 \end{bmatrix} .$$

In summary, this section gives the scattering amplitude matrix [F] in (3.5) in the medium coordinate in terms of the scattering amplitude matrix $[F_b]$ given in the coordinate system attached to the particle. For given $\hat{s}(\mu,\phi)$, $\hat{s}'(\mu',\phi')$, and Euler's transformation $[A_e]$, we calculate $[F_b]$ using (3.7) and (3.8), [B] using (3.13) and [B'] using (3.14), yielding $[F] = [B][F_b][B']^{-1}$.

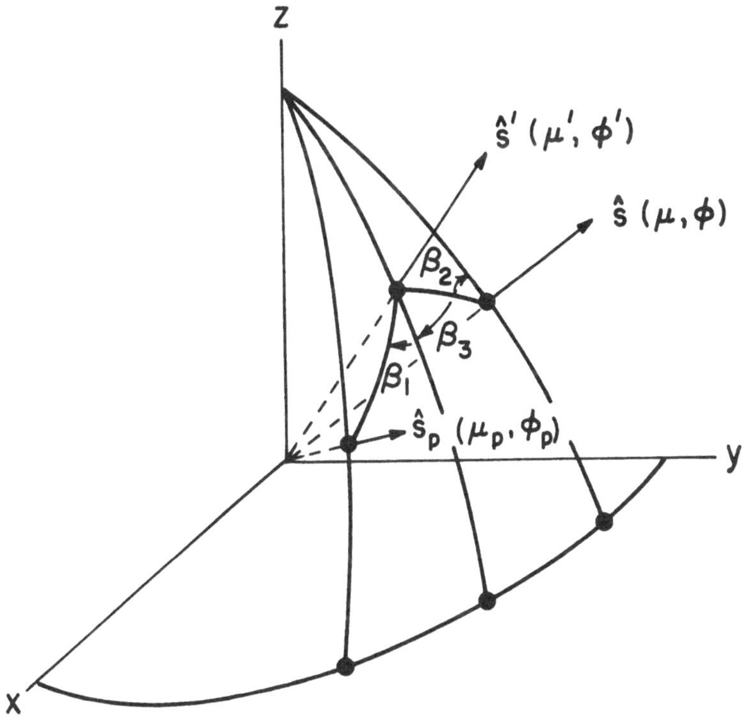

Figure 4. Transformation parameters used for axially symmetric particles.

3.2 Scattering amplitude matrix [F] for an axially symmetric particle

If the particle is axially symmetric, certain simplifications can be made. For example, in the previous section, the coordinate system x_b-y_b-z_b can be oriented so that the z_b axis coincides with the axis of symmetry. Euler's transformation is then given by two angles of rotation and the scattering amplitude matrix $[F_b]$ is a function of μ_b, μ_b', and $\phi_b-\phi_b'$, rather than ϕ_b and ϕ_b'. In this section, however, we present a simpler alternative method suited for axially symmetric bodies.[22,23]

Consider an axially symmetric particle oriented in the direction $\hat{s}_p$. We wish to express the scattering amplitude matrix [F] in terms of $\hat{s}(\mu,\phi)$, $\hat{s}'(\mu',\phi')$, and $\hat{s}_p(\mu_p,\phi_p)$ (Figure 4).

Let us assume that the scattering amplitude matrix $[F_p]$ is given in the coordinate system x_p-y_p-z_p shown in Figure 5.

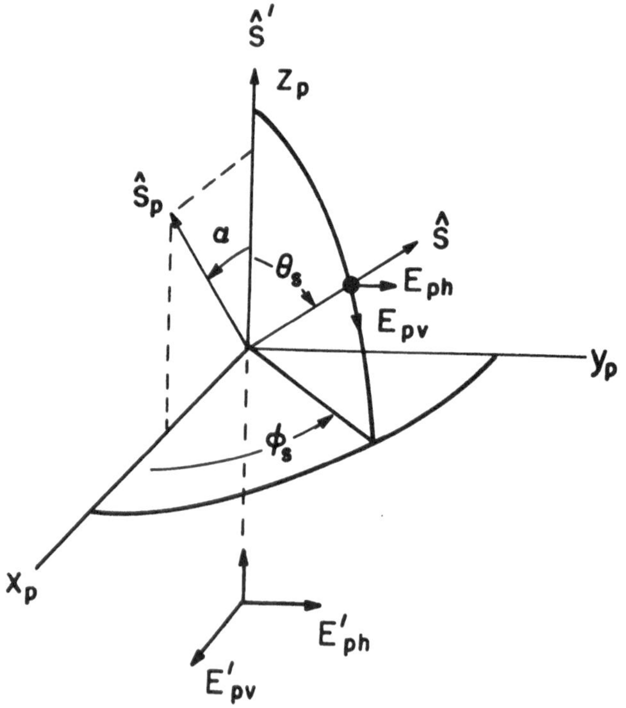

Figure 5. Scattering amplitude matrix for an axially symmetric particle defined in the particle coordinate x_b-y_b-z_b.

$$[E_p] = \frac{e^{ikR}}{R} [F_p][E_p'] \quad , \tag{3.15}$$

$$[E_p] = \begin{bmatrix} E_{pv} \\ E_{ph} \end{bmatrix} \quad , \qquad [E_p'] = \begin{bmatrix} E_{pv}' \\ E_{ph}' \end{bmatrix} \quad .$$

The symmetry axis $\hat{s}_p$ is in the x_p-z_p plane and is inclined from the z_p axis by the angle α. The scattering amplitude matrix $[F_p]$ is, therefore, assumed to be known as a function of α, θ_s, and ϕ_s.

Now as can be seen from Figure 5, [E'] and $[E_p']$ are simply related by the rotation matrix $[R(\beta_1)]$.

$$[E_p'] = [R(\beta_1)][E'] \quad , \tag{3.16}$$

where $$[R(\beta_1)] = \begin{bmatrix} \cos\beta_1 & -\sin\beta_1 \\ \sin\beta_1 & \cos\beta_1 \end{bmatrix} \quad .$$

Similarly, [E] and $[E_p]$ are related by

$$[E] = [R(\beta_2)][E_p] \quad . \tag{3.17}$$

Therefore, we get

$$[F] = [R(\beta_2)][F_p][R(\beta_1)] \quad . \tag{3.18}$$

Now if we can express θ_s, ϕ_s, α, β_1, and β_2 in terms of θ, ϕ, θ', ϕ', θ_p, and ϕ_p, then the scattering amplitude function [F] can be expressed by $\hat{s}$, $\hat{s}'$, $\hat{s}_p$, and $[F_p]$. θ_s and α can be given by

$$\cos\theta_s = \hat{s}\cdot\hat{s}' = \cos\theta\cos\theta' + \sin\theta\sin\theta'\cos(\phi-\phi') \quad ,$$

$$\cos\alpha = \hat{s}_p\cdot\hat{s}' = \cos\theta_p\cos\theta' + \sin\theta_p\sin\theta'\cos(\phi_p-\phi') \quad . \tag{3.19}$$

Also, noting

$$\cos\theta_p = \cos\alpha\cos\theta' - \sin\alpha\sin\theta'\cos\beta_1 \quad , \tag{3.20}$$

and using (3.19), we get

$$\cos\beta_1 = \frac{\sin\theta_p \cos\theta' \cos(\phi'-\phi_p) - \sin\theta' \cos\theta_p}{\sin\alpha} . \tag{3.21}$$

Note that when $\phi' \neq \phi_p$, $0 < \beta_1 < \pi$ if $0 < \phi' - \phi_p < \pi$ and $-\pi < \beta_1 < 0$ if $-\pi < \phi' - \phi_p < 0$. Also when $\phi' = \phi_p$, $\beta_1 = 0$ if $\theta' \leq \theta_p$ and $\beta_1 = \pi$ if $\theta_p < \theta'$. In particular, when $\theta_p = 0$, $\beta_1 = 0$ if $\theta' = 0$ and $\beta_1 = \pi$ if $0 < \theta'$.

To obtain β_2, note

$$\cos\theta' = \cos\theta_s \cos\theta + \sin\theta_s \sin\theta \cos\beta_2 . \tag{3.22}$$

We then get

$$\cos\beta_2 = \frac{\cos\theta' \sin\theta - \sin\theta' \cos\theta \cos(\phi-\phi')}{\sin\theta_s} . \tag{3.23}$$

Note that when $\phi \neq \phi'$, $0 < \beta_2 < \pi$ if $0 < \phi - \phi' < \pi$ and $-\pi < \beta_2 < 0$ if $-\pi < \phi - \phi' < 0$. Also when $\phi = \phi'$, $\beta_2 = 0$ if $\theta' \leq \theta$ and $\beta_2 = \pi$ if $\theta < \theta'$. In particular, when $\theta = 0$, $\beta_2 = 0$ if $\theta' = 0$ and $\beta_2 = \pi$ if $0 < \theta'$.

Next consider $\phi_s = \beta_1 + \beta_3$ where

$$\cos\beta_3 = \frac{\cos\theta' \cos\theta_s - \cos\theta}{\sin\theta' \sin\theta_s} , \tag{3.24}$$

when $\phi \neq \phi'$, $0 < \beta_3 < \pi$ if $0 < \phi - \phi' < \pi$ and $-\pi < \beta_3 < 0$ if $-\pi < \phi - \phi' < 0$. When $\phi = \phi'$, $\beta_3 = 0$ if $\theta' \leq \theta$ and $\beta_3 = \pi$ if $\theta < \theta'$.

Equations (3.19), (3.21), (3.23), and (3.24) give θ_s, ϕ_s, α, β_1, and β_2 in terms of $\hat{s}$, $\hat{s}'$, and $\hat{s}_p$. Using these in (3.18), [F] is now expressed in terms of $\hat{s}$, $\hat{s}'$, and $\hat{s}_p$.

3.3 Scattering amplitude matrix [F] for a small ellipsoidal particle[17,18]

If the particle sizes are much smaller than a wavelength, the scattering amplitude $\bar{f}(\hat{o},\hat{i})$ is given in terms of the dipole moment $\bar{p}$ of the particle.

$$\bar{f}(\hat{o},\hat{i}) = \frac{k}{4\pi\varepsilon_o} V[\bar{p} - \hat{o}(\hat{o}\cdot\bar{p})] , \tag{3.25}$$

where V is the total volume of the particle and $\hat{i}$ and $\hat{o}$ are the unit vectors in the direction of propagation for the incident and the scattered waves, respectively.

Consider a small ellipsoidal particle with the principal axes oriented in the coordinate system x_b-y_b-z_b, and the surface given by

$$\frac{x_b^2}{a^2} + \frac{y_b^{\,2}}{b^2} + \frac{z_b^{\,2}}{c^2} = 1 \quad . \tag{3.26}$$

Euler's transformation $[A_e]$ given in (3.2) relates the x-y-z system to the x_b-y_b-z_b system.

Let us express (3.25) in matrix form

$$[F][E'] = \frac{k^2 V}{4\pi\varepsilon_0} [C_1][P] \quad , \tag{3.27}$$

where

$$[C_1] = \begin{bmatrix} \hat{v}\cdot\hat{x}_b & \hat{v}\cdot\hat{y}_b & \hat{v}\cdot\hat{z}_b \\ \hat{h}\cdot\hat{x}_b & \hat{h}\cdot\hat{y}_b & \hat{h}\cdot\hat{z}_b \end{bmatrix} .$$

$$[P] = \begin{bmatrix} P_1 \\ P_2 \\ P_3 \end{bmatrix} \quad , \quad \bar{p} = P_1\hat{x}_b + P_2\hat{y}_b + P_3\hat{z}_b \quad .$$

The polarization [P] of the dipole moment is given by the polarizability tensor $[\alpha]$ and the component $[E_b]$ of the incident wave in the direction of the principal axes.

$$[P] = \varepsilon_0[\alpha][E_b] \quad , \tag{3.28}$$

where

$$[\alpha] = \begin{bmatrix} \alpha_1 & 0 & 0 \\ 0 & \alpha_2 & 0 \\ 0 & 0 & \alpha_3 \end{bmatrix} \quad , \quad \alpha_i = \frac{\varepsilon_r - 1}{1 + (\varepsilon_r - 1)L_i} \quad ,$$

$$L_1 = \frac{abc}{2} \int_0^\infty (s+a^2)^{-1}[(s-a^2)(s+b^2)(s+c^2)]^{-\frac{1}{2}}\, ds \quad ,$$

with appropriate exchanges of a, b, and c for L_2 and L_3.

We can also express $[E_b]$ in terms of the incident wave [E'].

$$[E_b] = [C_2][E'] \quad ,$$

$$[E_b] = \begin{bmatrix} E_{bx} \\ E_{by} \\ E_{bz} \end{bmatrix} , \quad [C_2] = \begin{bmatrix} \hat{x}_b \cdot \hat{v}' & \hat{x}_b \cdot \hat{h}' \\ \hat{y}_b \cdot \hat{v}' & \hat{y}_b \; \hat{h}' \\ \hat{z}_b \cdot \hat{v}' & \hat{z}_b \cdot \hat{h}' \end{bmatrix} . \tag{3.29}$$

Finally, we obtain

$$[F] = \frac{k^2 V}{4\pi}[C_1][\alpha][C_2] , \tag{3.30}$$

where $[C_1]$, $[\alpha]$, and $[C_2]$ are given in (3.27), (3.28), and (3.29), respectively.

3.4 Scattering amplitude matrix [F] for a spherical particle[13,8]

For spherical particles, it is convenient to express f_{ij} using the following notations developed by Chandrasekhar and Sekera.

$$f_{11} = (\ell,\ell)T_1 + (r,r)T_2 , \quad f_{12} = -(r,\ell)T_1 + (\ell,r)T_2 ,$$

$$f_{21} = -(\ell,r)T_1 + (r,\ell)T_2 , \quad f_{22} = (r,r)T_1 + (\ell,\ell)T_2 , \tag{3.31}$$

$$(\ell,\ell) = [(1=\mu^2)(1-\mu'^2)]^{1/2} + \mu\mu' \cos(\phi'-\phi) ,$$

$$(\ell,r) = -\mu' \sin(\phi'-\phi) , \quad (r,\ell) = \mu \sin(\phi'-\phi) ,$$

$$(r,r) = \cos(\phi'-\phi) ,$$

and

$$T_1(\chi) = \frac{A_{rr}(\chi) - \chi A_{\ell\ell}(\chi)}{1-\chi^2} , \quad T_2(\chi) = \frac{A_{\ell\ell}(\chi) - \chi A_{rr}(\chi)}{1-\chi^2} ,$$

where

$$\chi = \cos \textcircled{H} = [(1-\mu^2)(1-\mu'^2)]^{1/2} \cos(\phi'+\phi) + \mu\mu' ,$$

$$\mu = \cos\theta , \quad \mu' = \cos\theta' .$$

(θ',ϕ') and (θ,ϕ) correspond to the incident and scattered wave directions, respectively, and Ⓗ is the angle between the incident and scattered waves. $A_{\ell\ell}$ and A_{rr} are functions of Ⓗ, and they are related to the scattering functions S_2 and S_1, respectively, used by van de Hulst for the Mie solution.

$$A_{\ell\ell} = \frac{i}{K} S_2^* \quad \text{and} \quad A_{rr} = \frac{i}{K} S_1^* . \tag{3.32}$$

Here the conjugates of S_1 and S_2 are taken because van de Hulst and Kerker used exp(jωt) dependence while this paper uses exp(-iωt) dependence.

4. COHERENT FIELD[12]

In section 3, we presented various representations of the scattering amplitude matrix [F]. We will now go back to Section 2 and examine the coherent field in terms of the scattering amplitude matrix [F].

Let us consider a coherent wave propagating in the direction $\hat{s}_i(\theta_i,\phi_i)$. Let us write the coherent wave as $[E_c]\exp(iks)$ where s is the distance along $\hat{s}_i$. Then $[E_c]$ satisfies the equation

$$\frac{d}{ds}[E_c] = [M][E_c] \quad , \tag{4.1}$$

where $[M] = [M_{ij}]$, $M_{ij} = \frac{i2\pi\rho}{k} f_{ij}(\hat{s}=\hat{s}'=\hat{s}_i)$. ρ is the number density and f_{ij} is evaluated in the forward direction $\hat{s} = \hat{s}'=\hat{s}_i$.

The solution of (4.1) can be obtained by using the eigenvalue technique. We let $[E]\sim\exp(\gamma s)$ where γ is the eigenvalue and is given by

$$\begin{vmatrix} M_{11}-\gamma & M_{12} \\ M_{21} & M_{22}-\gamma \end{vmatrix} = 0 \quad . \tag{4.2}$$

From this, we get two eigenvalues γ_1 and γ_2

$$\left.\begin{matrix}\gamma_1\\ \gamma_2\end{matrix}\right\} = \frac{1}{2}\left\{(M_{11}+M_{22}) \pm [(M_{11}-M_{22})^2 + 4M_{12}M_{21}]^{\frac{1}{2}}\right\} \quad , \tag{4.3}$$

and the eigenvectors

$$\begin{bmatrix} A_{11} \\ A_{21} \end{bmatrix} = \begin{bmatrix} M_{12} \\ \gamma_1 - M_{11} \end{bmatrix} \qquad \text{for } \gamma_1 \quad ,$$

$$\begin{bmatrix} A_{21} \\ A_{22} \end{bmatrix} = \begin{bmatrix} M_{12} \\ \gamma_2 - M_{11} \end{bmatrix} \qquad \text{for } \gamma_2 \quad . \tag{4.4}$$

The general solution is then given with two arbitrary constants C_1 and C_2.

$$\begin{bmatrix} E_v \\ E_h \end{bmatrix} = C_1 e^{\gamma_1 s} \begin{bmatrix} A_{11} \\ A_{21} \end{bmatrix} + C_2 e^{\gamma_2 s} \begin{bmatrix} A_{12} \\ A_{22} \end{bmatrix}$$

$$= [A] \begin{bmatrix} e^{\gamma_1 s} & 0 \\ 0 & e^{\gamma_2 s} \end{bmatrix} \begin{bmatrix} C_1 \\ C_2 \end{bmatrix} , \quad (4.5)$$

where

$$[A] = \begin{bmatrix} A_{11} & A_{12} \\ A_{21} & A_{22} \end{bmatrix} .$$

Now we assume the boundary condition at $s = 0$

$$[E(s=0)] = \begin{bmatrix} E_v(s=0) \\ E_h(s=0) \end{bmatrix} . \quad (4.6)$$

Using (4.6) and (4.5), we get the solution

$$[E(s)] = [A] \begin{bmatrix} e^{\gamma_1 s} & 0 \\ 0 & e^{\gamma_2 s} \end{bmatrix} [A]^- \; [E(s=0)] = [T][E(s=0)]$$

$$= \left\{ [T_1]e^{\gamma_1 s} + [T_2]e^{\gamma_2 s} \right\} [E(s=0)] , \quad (4.7)$$

where

$$[T_1] = [A] \begin{bmatrix} 1 & 0 \\ 0 & 0 \end{bmatrix} [A]^{-1} , \quad [T_2] = [A] \begin{bmatrix} 0 & 0 \\ 0 & 1 \end{bmatrix} [A]^{-1} .$$

5. PROPAGATION OF COHERENCY MATRIX AND STOKES' VECTOR

In the last section, we expressed the coherent field $[E(s)]$ at s in terms of the field $[E(o)]$ at $s = 0$.

$$[E(s)] = [T][E(o)] \quad (5.1)$$

or

$$\begin{bmatrix} E_v \\ E_h \end{bmatrix} = \begin{bmatrix} T_{11} & T_{12} \\ T_{21} & T_{22} \end{bmatrix} \begin{bmatrix} E_v' \\ E_h' \end{bmatrix} .$$

We now wish to express the relationship (5.1) in terms of the Stokes' vectors [I] and [I'].

$$[I] = [Q][I'] \ , \tag{5.2}$$

where

$$[I] = \begin{bmatrix} \langle E_v E_v^* \rangle \\ \langle E_h E_h^* \rangle \\ 2\mathrm{Re}\langle E_v E_h^* \rangle \\ 2\mathrm{Im}\langle E_v E_h^* \rangle \end{bmatrix} ,$$

and [I'] is expressed similarly with E_v and E_h replaced by E_v' and E_h'. [Q] is the 4 x 4 matrix.

To find [Q], it is convenient to consider the coherency matrix [J] and the matrix [D].

$$[J] = [J_{ij}] = \begin{bmatrix} \langle E_v E_v^* \rangle & \langle E_v E_h^* \rangle \\ \langle E_h E_v^* \rangle & \langle E_h E_h^* \rangle \end{bmatrix} , \quad [D] = \begin{bmatrix} J_{11} \\ J_{12} \\ J_{21} \\ J_{22} \end{bmatrix} . \tag{5.3}$$

It is easy to see that [D] and [I] are related simply by

$$[I] = [P][D] \quad \text{and} \quad [I'] = [P][D'] \quad , \tag{5.4}$$

where

$$[P] = \begin{bmatrix} 1 & 0 & 0 & 0 \\ 0 & 0 & 0 & 1 \\ 0 & 1 & 1 & 0 \\ 0 & -i & i & 0 \end{bmatrix} .$$

Next we relate [D] to [D']. From (5.1), we get

$$[D] = [T] \otimes [T^*][D'] \quad , \tag{5.5}$$

where $[A] \otimes [B]$ is the "direct product" of [A] and [B] defined by

$$[A] \otimes [B] = \begin{bmatrix} A_{11}[B] & A_{12}[B] \\ A_{21}[B] & A_{22}[B] \end{bmatrix} \tag{5.6}$$

$$= \begin{bmatrix} A_{11}B_{11} & A_{11}B_{12} & A_{12}B_{11} & A_{12}B_{12} \\ A_{11}B_{21} & A_{11}B_{22} & A_{12}B_{21} & A_{12}B_{22} \\ A_{21}B_{11} & A_{21}B_{12} & A_{22}B_{11} & A_{22}B_{12} \\ A_{21}B_{21} & A_{21}B_{22} & A_{22}B_{21} & A_{22}B_{22} \end{bmatrix} .$$

Combining (5.4) and (5.5), we get

$$[I] = [Q][I'] \quad , \quad [Q] = [P][T] \otimes [T^*][P]^{-1} \quad ,$$

$$[P]^{-1} = \begin{bmatrix} 1 & 0 & 0 & 0 \\ 0 & 0 & 1/2 & i/2 \\ 0 & 0 & 1/2 & -i/2 \\ 0 & 1 & 0 & 0 \end{bmatrix} . \tag{5.7}$$

Equation (5.7) gives the expression relating [I] to [I'] when [E] and [E'] are related by [T] given in (5.1).

We can also express the propagation of the coherency matrix [J] by noting

$$[J] = <[E][\tilde{E}^*]> \quad , \quad [J'] = <[E'][\tilde{E}^*]> \quad . \tag{5.8}$$

Then since [E] = [T][E'], we get

$$[J] = [T][J'[[\tilde{T}^*] \quad . \tag{5.9}$$

Therefore (5.7) and (5.9) are equivalent.

6. PROPAGATION OF COHERENT INTENSITY

From (4.7) and (5.7), we get the coherent Stokes' vector $[I_c]$ at s when the incident Stokes' vector $[I_o]$ is given at s = 0. Noting

$$[T] = [T_1]e^{\gamma_1 s} + [T_2]e^{\gamma_2 s} \quad ,$$

we get

$$[I_c] = [Q][I_o] \quad , \tag{6.1}$$

$$[Q] = [P][T_1] \otimes [T_1^*][P]^{-1} e^{(\gamma_1+\gamma_1^*)s}$$
$$+ [P][T_1] \otimes [T_2^*][P]^{-1} e^{(\gamma_1+\gamma_2^*)s}$$
$$+ [P][T_2] \otimes [T_1^*][P]^{-1} e^{(\gamma_2+\gamma_1^*)s}$$
$$+ [P][T_2] \otimes [T_2^*][P]^{-1} e^{(\gamma_2+\gamma_2^*)s} \quad .$$

7. EXTINCTION MATRIX [Λ]

Let us now find the extinction matrix [Λ]. We note

$$\frac{d}{ds}[J] = \langle \frac{d}{ds}[E][\tilde{E}^*]\rangle + \langle [E]\frac{d}{ds}[\tilde{E}^*]\rangle$$
$$= [M]\langle [E][\tilde{E}^*]\rangle + \langle [E][\tilde{E}^*]\rangle [\tilde{M}^*]$$
$$= [M][J][U] + [U][J][\tilde{M}^*] \quad , \tag{7.1}$$

where [U] is a unit matrix.

Noting (5.9), we get

$$\frac{d}{ds}[I_c] = [\Lambda][I_c]$$
$$[\Lambda] = [P]\{[M] \otimes [U] + [U] \otimes [\tilde{M}^*]\} \quad . \tag{7.2}$$

Rewriting [Λ], we get

$$[\Lambda] = \begin{bmatrix} 2\mathrm{Re}M_{11} & 0 & \mathrm{Re}M_{12} & \mathrm{Im}M_{12} \\ 0 & 2\mathrm{Re}M_{22} & \mathrm{Re}M_{21} & -\mathrm{Im}M_{21} \\ 2\mathrm{Re}M_{21} & 2\mathrm{Re}M_{12} & \mathrm{Re}(M_{11}+M_{22}^*) & -\mathrm{Im}(M_{11}+M_{22}^*) \\ -2\mathrm{Im}M_{21} & 2\mathrm{Im}M_{12} & \mathrm{Im}(M_{11}+M_{22}^*) & \mathrm{Re}(M_{11}+M_{22}^*) \end{bmatrix} \quad . \tag{7.3}$$

8. SCATTERING (MUELLER) MATRIX

The Mueller matrix [S] represents the scattered Stokes' vector [I] in the direction of (θ,ϕ) when the Stokes' vector [I'] is incident on the volume dV in the direction of (θ',ϕ').

$$[I] \sim \frac{1}{R^2} [S][I']dV \quad . \tag{8.1}$$

Noting $dV = R^2 d\omega' ds$, we get the second term on the right side of (2.2)

$$\frac{d[I]}{ds} \sim \int [S][I']d\omega' \quad . \tag{8.2}$$

In terms of the scattering amplitude matrix $[F] = [F_{ij}(\theta,\phi;\theta',\phi')]$, we get, following Section 5,

$$[S] = \rho[P][F] \otimes [F^*][P]^{-1}$$

$$= \begin{bmatrix} \rho|f_{11}|^2 & \rho|f_{12}|^2 & \rho\mathrm{Re}(f_{11}f_{12}^*) & -\rho\mathrm{Im}(f_{11}f_{12}^*) \\ \rho|f_{21}|^2 & \rho|f_{22}|^2 & \rho\mathrm{Re}(f_{21}f_{22}^*) & -\rho\mathrm{Im}(f_{21}f_{22}^*) \\ \rho 2\mathrm{Re}(f_{11}f_{21}^*) & \rho 2\mathrm{Re}(f_{12}f_{22}^*) & \rho\mathrm{Re}(f_{11}f_{22}^*+f_{12}f_{21}^*) & -\rho\mathrm{Im}(f_{11}f_{22}^*-f_{12}f_{21}^*) \\ \rho 2\mathrm{Im}(f_{11}f_{21}^*) & \rho 2\mathrm{Im}(f_{12}f_{22}^*) & \rho\mathrm{Im}(f_{11}f_{22}^*+f_{12}f_{21}^*) & \rho\mathrm{Re}(f_{11}f_{22}^*-f_{12}f_{21}^*) \end{bmatrix}$$

9. INCIDENT INTENSITY MATRIX $[I_i]$

When the coherent intensity $[I_c]$ propagates in the direction $\hat{s}_i(\theta_i,\phi_i)$, $[I_i]$ is given by

$$\begin{aligned} [I_i] &= [I_i(\theta,\phi;\theta_i,\phi_i)] \\ &= \int [S][I_c']d\omega' \\ &= [S(\theta,\phi;\theta_i,\phi_i)][Q(\theta_i,\phi_i)][I_o] \quad , \end{aligned} \tag{9.1}$$

where $[Q]$ and $[I_o]$ are given in Section 6.

10. CP REPRESENTATION

Instead of the representation (E_v,E_h), the circular polarization (CP) representation (E_+,E_-) is often used

$$E_+ = \frac{1}{\sqrt{2}}(E_v - iE_h) \quad , \quad E_- = \frac{1}{\sqrt{2}}(E_v + iE_h) \quad , \tag{10.1}$$

where E_+ and E_- are the right-handed (RHC) and left-handed (LHC) circular polarization. The corresponding Stokes' vector for CP representation is

$$[I_{cp}] = \begin{bmatrix} I_c \\ I_+ \\ I_- \\ I_c^* \end{bmatrix} = \begin{bmatrix} \langle E_+E_-^*\rangle \\ \langle E_+E_+^*\rangle \\ \langle E_-E_-^*\rangle \\ \langle E_+^*E_-\rangle \end{bmatrix} . \tag{10.2}$$

$[I_{cp}]$ is related to $[I]$ by

$$[I_{cp}] = [Q_{cp}][I] \tag{10.3}$$

$$[Q_{cp}] = \frac{1}{2}\begin{bmatrix} 1 & -1 & -i & 0 \\ 1 & 1 & 0 & -1 \\ 1 & 1 & 0 & 1 \\ 1 & -1 & i & 0 \end{bmatrix} .$$

11. FOURIER EXPANSION OF RADIATIVE TRANSFER EQUATION

Consider the equation of transfer (2.5). We expand $[I_d]$, $[\Lambda]$, $[S]$, and $[I_i]$ in Fourier series in ϕ.

$$[I_d] = \sum_m [I]_m e^{-jm\phi} \quad , \quad [\Lambda] = \sum_m [\Lambda]_m e^{-jm\phi} \quad ,$$

$$[S] = \sum_m \sum_{m'} [S]_{m,m'} e^{-jm\phi+jm'\phi'} \quad , \quad [I_i] = \sum_m [I_i]_m e^{-jm\phi} \quad , \tag{11.1}$$

where $[I]_m = [I]^*_{-m}$, $[\Lambda]_m = [\Lambda]^*_{-m}$,

$$[I_i]_m = [I]^*_{-m} \; , \quad [S]_{m,m'} = [S]^*_{-m,-m'} \; .$$

We then obtain

$$\mu \frac{d}{dz} [I]_m = \sum_{m'} [\Lambda]_{m-m'} [I]_{m'} + \sum_{m'} \int [S]_{m,m'} [I']_{m'} d\mu' + [I_i]_m \; . \tag{11.2}$$

The integral in (11.2) can be approximated by a series using quadrature formula and (11.2) can be expressed in a matrix form, which can be solved by the eigenvalue-eigenvector technique.

12. PROPAGATION AND SCATTERING OF STOKES' VECTOR IN SPHERICAL PARTICLES

When the particles are spherical, considerable simplifications are possible. If a wave is normally incident on spherical particles, the Fourier components are all independent and there are only two terms m = 0 and m = 2. It is shown that in the first-order scattering, the cross polarization is proportional to $\sin^2 2\phi$, and therefore, the cross polarization disappears at $\phi = 0$, ±90°, and 180°. However, the complete multiple scattering yields the cross polarization for all ϕ angles. It is also shown that the degree of polarization decreases with the optical depth. This has been discussed in detail in Reference 8.

If the incident wave is obliquely incident, all Fourier components are independently excited. This has been discussed in Reference 7.

13. PROPAGATION AND SCATTERING OF STOKES' VECTOR IN NONSPHERICAL PARTICLES

The first-order solution, sometimes called the distorted Born approximation, has been presented in References 17 and 18. A more complete solution must be based on (11.2).

14. LIMITATION OF THE RADIATIVE TRANSFER THEORY

The radiative transfer theory is applicable when the particle density is approximately less than 1% in volume. Recent studies indicate that the effective cross sections are different at high densities from the low density value and they also depend on the particle sizes in wavelength. When the particle sizes are small, the effective cross sections are reduced while for large particles, they are slightly higher. These were discussed recently in References 19, 20, and 21.

15. FIRST-ORDER SCATTERING APPROXIMATIONS FOR ELECTROMAGNETIC IMAGING

Let us now consider the electromagnetic imaging of an object in the presence of particulate matter in the atmosphere. In the first-order scattering approximations, we assume that the incident Stokes' vector $[I_0]$ propagates through the medium with the propagation constant of the coherent Stokes' vector represented by $[Q_1]$ in (6.1) resulting in the Stokes' vector $[Q_1][I_0]$ which is incident on the target. The target has scattering characteristics represented by the Mueller matrix $[S_0]$.

The scattered Stokes' vector is, therefore, given by $[S_o][Q][I_o]$. This is then propagated back to the transmitter through the medium with $[Q_2]$. Therefore, the received Stokes' vector $[I_r]$ is given by (Figure 6)

$$[I_r] = [Q_2][S_o][Q_1][I_o] \quad , \tag{15.1}$$

where [Q] is given in (6.1) with s = propgation distance, and $[S_o]$ has the same form as [S] in (8.2) with f_{ij} of the target and $\rho = 1$.

The polarization loss factor K is then given by

$$K = \frac{I_{o1}I_{r1} + I_{o2}I_{r2} + U_oU_r - V_oV_r}{(I_{o1} + I_{o2})(I_{r1} + I_{r2})} \quad , \tag{15.2}$$

where

$$[I_o] = \begin{bmatrix} I_{o1} \\ I_{o2} \\ U_o \\ V_o \end{bmatrix} , \quad \text{and} \quad [I_r] = \begin{bmatrix} I_{r1} \\ I_{r2} \\ U_r \\ V_r \end{bmatrix} .$$

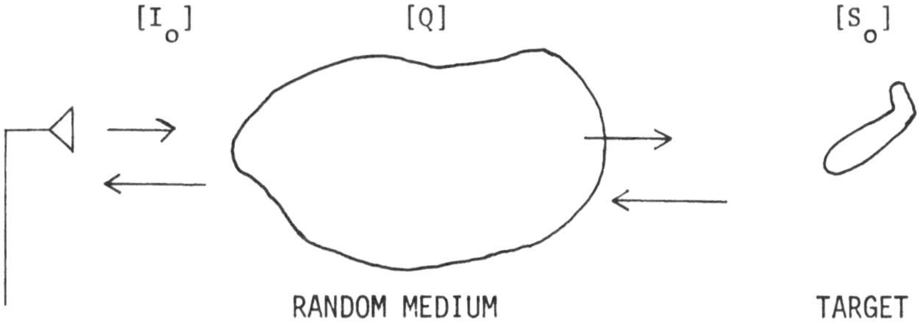

Figure 6. Electromagnetic imaging of target in random medium.

REFERENCES

1. W.-M. Boerner, A. K. Jordan, and I. W. Kay, "Introduction to the special issue on inverse methods in electromagnetics," IEEE Trans. Antennas Propagat., 29(2), pp. 185-191, 1981(a).

2. W.-M. Boerner, M. B. El-Arini, C.-Y. Chan, and P.M. Mastoris, "Polarization dependence in electromagnetic inverse problems," IEEE Trans. Antennas Propagat., 29(2), pp. 262-271, 1981(b).

3. J. R. Huynen, "Phenomenological theory of radar targets," Ph.D. Thesis, Techn. Univ. Delft, Drukkerij Bronder-Offset N.V., Rotterdam, Netherlands, 1970.

4. A. Ishimaru, *Wave Propagation and Scattering in Random Media*, Vol. 1, Academic Press, New York, 1978.

5. A. Ishimaru and R. L.-T. Cheung, "Multiple scattering effects on wave propagation due to rain," Ann. Télécommun., 35, pp. 373-379, 1980(a).

6. A. Ishimaru and R. L.-T. Cheung, "Multiple-scattering effect on radiometric determination of rain attenuation at millimeter wavelengths," Radio Science, 15(3), pp. 507-516, 1980(b).

7. A. Ishimaru, R. Woo, J. A. Armstrong, and D. Blackman, "Multiple scattering calculations of rain effects," Radio Science, (special issue on NASA Propagation Studies) 17, pp. 1425-1433, 1982.

8. R. L.-T. Cheung and A. Ishimaru, "Transmission, backscattering and depolarization of waves in randomly distributed spherical particles," Applied Optics, 21(20), pp. 3792-3798, October 1982.

9. S. Chandrasekhar, *Radiative Transfer*, Clarendon, Oxford, 1950.

10. D. Deirmendjian, *Electromagnetic Scattering on Spherical Polydispersions*, Elsevier, New York, 1969.

11. H. C. van de Hulst, *Multiple Light Scattering*, Vol. 2, Academic Press, New York, 1980.

12. T. Oguchi, "Scattering from hydrometeors: A survey," Radio Science, 16(5), pp. 691-730, 1981.

13. Z. Sekera, "Scattering matrices and reciprocity relationships for various representations of the state of polarization," J. Opt. Soc. Am., 56(12), pp. 1732-1740, 1966.

14. H. C. van de Hulst, *Light Scattering by Small Particles*, John Wiley, New York, 1957.

15. M. Kerker, *The Scattering of Light*, Academic, New York, 1969.

16. H. Goldstein, *Classical Mechanics*, Addison-Wesley, Reading, Massachusetts, 1981.

17. L. Tsang, M. C. Kubacsi, and J. A. Kong, "Radiative transfer theory for active remote sensing of a layer of small ellipsoidal scatterers," Radio Science, 16(3), pp. 321-329, 1981.

18. R. H. Lang, "Electromagnetic backscattering from a sparse distribution of lossy dielectric scatterers," Radio Science, 16(1), pp. 15-30, 1981.

19. A. Ishimaru and Y. Kuga, "Attenuation constant of a coherent field in a dense distribution of particles," J. Opt. Soc. Am., 72(10), pp. 1317-1320, 1982.

20. V. N. Bringi, V. V. Varadan, and V. K. Varadan, "Coherent wave attenuation by a random distribution of particles," Radio Science, 17(5), pp. 946-952, 1982.

21. L. Tsang, J. A. Kong, and T. Habashy, "Multiple scattering of acoustic waves by random distribution of discrete spherical scatterers with the quasicrystalline and Percus-Yevick approximation," J. Acoust. Soc. Am., 71, pp. 552-558, 1982.

22. C. Yeh, R. Woo, J. W. Armstrong, and A. Ishimaru, "Scattering by Pruppacher-Pitter raindrops at 30 GHz," Radio Science, 17(4), pp. 757-765, 1982(a).

23. C. Yeh, R. Woo, J. W. Armstrong, and A. Ishimaru, "Scatter-by single ice needles and plates at 30 GHz," Radio Science, 17(6), pp. 1503-1510, 1982(b).

ACKNOWLEDGMENT

This research was supported by the U.S. Army Research Office and the Office of Naval Research.

IV.2 (RP.4)

TOWARDS A THEORY OF PERCEPTION FOR RADAR TARGETS

With application to the analysis of their data base structures and presentations

J. Richard Huynen

Dept. of Electrical Engineering and Computer Science
University of Illinois at Chicago
Box 4348 851 S. Morgan St.
Chicago, Illinois 60680

1. Abstract

Visual perception depends on observation of three-dimensional objects, based upon a set of given exposures of the objects relative to the observer. Hence there is in the data base a built-in observer bias which has to be eliminated if the target is to be recognized as a physical object, independently of the observer.

Radar observation is built on the same principle, except that due to polarization dependence of the target a "polarizationbias" is introduced, which corrupts the target data base. Elimination of the polarization bias can be achieved if the target scattering matrix is known for the monostatic radar case. This approach was called phenomenological in previous work by this author.

The report reviews the methods of data presentation of objects based upon fields and upon power. For the polarized scattered return this amounts to a coherent wave-field or a completely polarized (cp) stokes vector power-presentation. For targets this amounts to a presentation based upon the scattering matrix or the stokes-matrix (Mueller-matrix). A novel vector formulation is introduced which relates to the cognitive requirements for a string of target features.

The power-presentations are useful to consider object mixtures which lead to a new class of partially polarized (pp) waves and of distributed radar targets. From the object mixtures we wish to formulate the concept of 'the average object'.

This can be accomplished through the so-called 'N-target

W.-M. Boerner et al. (eds.), Inverse Methods in Electromagnetic Imaging, Part 2, 797–822.

decomposition theorem', introduced by the author in 1970.

Concept formulation of general object structures is shown to be diagrammically related to hierarchical object tree structures, widely used in AI work. The work closes with detailed accounts for mixtures of the polarized wave object (n = 2), the radar-target (n = 3) and for the general object of index n. The generally 'useful' attributes are linearly related to the index: (2n - 1), while the 'N-target residue' which is put aside, is of order $(n-1)^2$. Hence there is great economy in preserving those meaningful parameters by which the idealized object is recognized.

These novel perceptually related data presentations should be useful as a guide to the radar-analyst in his search for meaningful target designators.

2. Introduction

The study of nature through controlled, coherent, electromagnetic illumination became possible only with the development of radar systems during and after World War II. The primary objective in those early days was the detection of targets through rain, dense fog, and generally adverse meteorological conditions, and the ability to observe at night, when ordinary visual observation became impaired. A short transmitted pulse provided a means for measuring distance towards the target, by measuring the time it takes to bounce off and return to the receiver. The coherent source enables one to focus on a desirable frequency band, and obtain doppler information. The phase coherency also provides for control of polarization.

Since the emphasis was on target detection, very little use was made of the information contained in the scattered return which relates to polarization. This information is expressed mathematically by the target scattering matrix S. More recently, emphasis has moved to investigations related to radar target discrimination, sorting, classification, and ultimately identification.

The thrust for this development has come from improvements in radar technology and data processing techniques which allow for greater resolution, faster data gathering (below clutter-target decorrelation times), and data handling processes [1]. This emphasis on technological improvement and data processing tends to obscure a basic objective of what is ultimately aimed for. We will call this the perception of radar targets as physical objects. The final purpose of the technological development is to establish existence for these objects (detection), and to determine their physical identity through discrimination and identification. What has been lacking thus far is an emphasis on the pre-processing stage of data analysis towards these aims.

3. Visual and radar perception of targets

Perception is a general requirement for all biological species. It enables an observer to make contact with objects in an already existing world, (although some logical positivists will disagree with this interpretation, but it is easy to show their fallacy), and to assess possible beneficial use or threat, for a particular life situation at hand. The methods used to achieve these goals are of course multivaried and we will restrict ourselves in this report to visual perception.

With visual observation of an object, a two-dimensional image is projected onto a sensitive aperture by means of an optical lens system, much like the principles used in photography. Radar imaging systems have been developed that reproduce radar images of land surface structures in much the same fashion, although the optical processing may be quite involved.

The essential characteristic of visual perception is the projection of a physically three-dimensional object structure onto a two-dimensional image which depends on the direction from which the object is being observed. This direction determines the target's exposure to the observer. For this moment we will exclude the possibility of holographic presentation and observation.

A single projection of a three-dimensional physical object usually does not give sufficient information to fix its identity. Experience has given us decision thresholds by which from a set of multiple observations an object may be identified with sufficient confidence.

The same basic principles apply to the perception of radar targets. However, differences occur, which are due to the polarization properties of the coherent radar, which do not appear as prominent in vision, because the light source is usually not coherent, and in any case the eye is not sufficiently polarization sensitive.

These differences in perception appear with monostatic radar, where transmitter and receiver are at the same location, and even more so in the bistatic case. The differences are due to the sensitivity of targets to the direction of linear polarization of the incoming wave. Usually the RCS patterns are labeled by HH, or VV, or on occasion the HV cross polarized pattern, is measured. The difference in RCS for HH and VV is due to target orientation angle bias, which is a dynamic motion variable. It would be like a human observer looking at an object in a fixed direction, who would see a different object, simply by moving his head 90° sideways!

This certainly would create havoc in ordinary perception of objects, as this phenomenon has created confusion among early radar researchers. This 'polarization effect' at best was considered a nuisance to be ignored, if at all possible, for radar detection purposes. However in the monostatic case, with the advent of complete polarimetric radar technology, the target-orientation angle bias due to dynamic motion at fixed exposure can be removed completely by a simple orthogonal transformation of the measured scattering matrix. The reason why this is possible for monostatic radar is that target rotation along the line of sight direction is equivalent to rotation of the antenna polarization reference axis along the same direction. However no such correspondence exists with single targets for the so called bistatic mode of operation.

4. Monostatic motion parameters

The monostatic radar is illustrated in figure 1. The radar antennas for both transmit and receive mode of operation are located at a single site. The dynamic coordinates of the target are given by the roll-angle about some defined target axis and the aspect direction which points to the radar site location. The aspect direction fixes those parts of the target which become exposed to the incoming plane wave radiation. Hence the aspect direction defines target exposure.

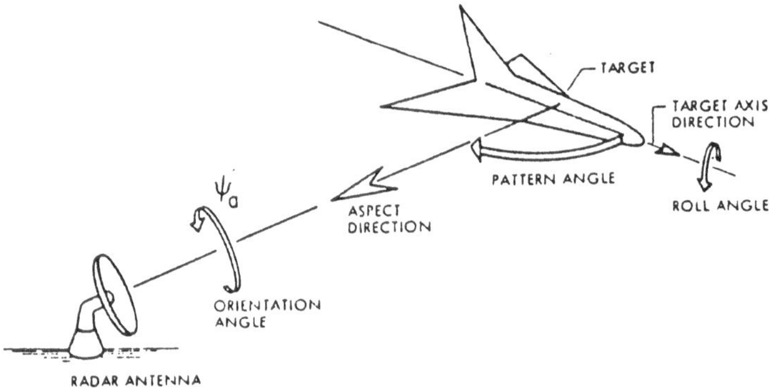

Fig.(1) Target Aspect Direction and Orientation Angle

However, we notice a third motion possibility, which is rotation of the target along the fixed line of sight direction. This rotation is shown in the figure by the angle ψ_a. The three types of motion are usually referred to as: roll, pitch, and yaw, of the airplane, but care should be exercised to relate these coordinates to the ones given in the figure because the definitions depend on the order in which rotation operations are given (rotations in 3dimensional space are not commutative). Other nomenclatures often used are: canting-angle, banking-angle, target attitude, etc.

The rotation motion along the line of sight direction leaves target exposure otherwise unaltered. If we use the visual perception analogy: looking at an object, say a glass in one's hand with outstretched arm, and pitching the glass by rotating the arm, does not alter our perception of the glass, nor does it expose different parts of the glass to our eyes, although the image formed on the retina will rotate with the rotation movement of the glass. Apparently our brain has learned at an early stage of evolution to compensate for this type of orthogonal transformation of images on the retina.

Now the same principle can be applied in the monostatic case to radar reception. The effect of change of orientation of the target can be compensated for, by simply rotating the coordinate frame (HV) of the antenna at the site location. In order to effect this change, an orthogonal transformation representing a physical rotation has to be applied to the measured scattering matrix S of the target at that particular exposure.

The reader may at this point be somewhat puzzled, why such a long, careful, and perhaps boring detailed account is sought to be necessary to explain such an almost obvious and self-evident phenomenon? The reason is that the process being described becomes self-evident only if we link the perception of objects by radar to our natural visual perception!

To the average 'data processing engineer' or scientist, the relevance of incoming data is not challenged, until after data processing is completed, and data can be shown to have significance and relevance or not. This approach works well if object structures are relatively simple and characteristic parameters are relatively few. For example a search for IQ distribution factors among various social groups. However, one interesting aspect of radar target feature research is that we are dealing with an enormous variety of possible physical objects, each of which may have its own complicated physical shape.

A radar-system may best be considered as an extension of ordinary human observation, and hence perception of radar targets can best be mirrored along the lines of visual perception. As we have seen visual perception is able to identify objects from observations based on a set of given target exposures. A minimum set of exposures is necessary for target identification. Each exposure is defined by the target aspect direction (see fig. 1) and hence is a function of only two dynamic motion parameters. The third motion parameter is orientation angle ψ_d, which can be compensated for.

What we have tried to emphasize is the need for preprocessing of target data before classification/identification algorithms can be applied successfully [2].

Another way of looking at the difficult art of target identification is to realize that physical objects are given as existing objects which have physical structure and attributes, and which display themselves in time and space to an observer (the 'existence' part will be discussed later). The display sets up a subjective bias which is due to the various geometrical position parameters, the target has relative to the observer, and which biases information related to the basic target structure. The identification in final analysis will be based on identifying basic target structure parameters after eliminating the miscellaneous geometrical motion parameter biases. This conceptual point is missed by many data processors, who treat all incoming data as unbiased inputs.

We shall discuss later-on in this report how these same principles can be extended to select other parameters derivable from the data base which relate to objects as irreducible physical constructs. This is an important result, which follows from the scattering matrix analyses. These ideas and subsequent mathematical techniques in turn will have important application to general perception, where the concept of an object is the irreducible atom, on which our understanding of the world around us is built.

5. Some processes which lead to concept formulation

The use of radar has opened up a new method for viewing the world. A plane wave is transmitted and the wave reflected from the target is observed and analysed. The information contained in the reflected wave is biased by the parameters of the incoming wave (frequency, pulse-shape, polarization etc.) and the target exposure to the incoming wave. Our ultimate aim is to characterize the target as a physical object, which as it were is cleansed from all the observation biases ('das Ding an Sich', as Kant would have preferred to call it [3]).

In a real-world situation there is always the problem of noise due to the environment, or clutter due to unwanted interference from other targets. The target itself may contain elements which corrupt, rather than enhance, the information base related to target characterization. We all know that with ordinary perception a substantial amount of information which is received, is cast aside, as it were, in favor of certain desirable special types of feature characteristics which define the object. Observing a tree, for instance, initially gives us all the detailed information regarding its special shape, foilage clusterings, branchings etc., but finally all this detail structure is crystallized into the realization of a concept, which is 'tree' and perhaps a special variety of tree.

At first observation, it might seem like a hopeless task to even attempt to arrive at a meaningful concept, like that of a tree, because of the enormous variety of manifestations it presents to an observer. On the other hand every normal child is able to perform this task in relatively short time, which is an indication, that the processes involved which lead to concept formulation must have a basic common structure.

If this is the case, than a corresponding basic mathematical structure must also be present, which merely maps the physical process into an analytical representation. We will find shortly, that this is indeed the case. Another relevant point is that because concept formulation is such a basic aspect of all cognition, the same basic structure may be found in other areas of physics. We will show that indeed the same structure can be found in quantum mechanics. All this will be of interest in a wide variety of fields where cognition is important.

With radar targets, the emphasis is on characterization of physical objects through the analysis of the return signal. Because the signal might be corrupted by clutter and/or noise, our first task will be to find a representation of polarized return wave signal, which is 'pure', or uncorrupted by clutter and/or noise. This will be a topic for discussion in the following sections. The next task will then be to unmask the target feature characteristics from the various biases, due to the transmitted wave, and those due to the set of multiple exposures, through which the target is being observed.

The underlying mathematical structure is surprisingly exposed through an analysis of the properties of a stokes vector, which gives the structure of a polarized wave, and the extension to radar targets, which leads to the author's so-called 'N-target decomposition theorem'. Both structures can be shown to have a similar representation, which in fact can be generalized to any object structure.

The basic idea is that an object is given by a string of n feature parameters. This idea is well known and accepted in general pattern recognition, where the features are mapped as a point in an n-dimensional feature space. A set of observations then leads to a set of points in the feature space. What one hopes for, is that a clustering of points in the feature space can be associated with one type of target. Different targets can hence be identified by their clusters and optimal target separation techniques become available through analytic data processing schemes. All these schemes at present fail in one important aspect: to define the concept of the object in terms of its feature parameters.

In other words, I want to define the concept 'tree' if what is given are bunches of leaves, branches, and shape distributions of each of these. This task of defining the concept of the object, based upon given features may seem to lead to insurmountable difficulties, but in fact it is very simple.

One of the approaches is found from the stokes vector properties. As one knows, the stokes vector defines an elliptically polarized wave uniquely in terms of power related quantities. This is distinct from a representation in terms of fields. The underlying physical processes usually are analysed in terms of fields, whereas measurements are presented as measures of power. There is some similarity and also a vast difference in conceptual understanding between these two approaches. The similarity lies in the fact that initially the two methods of presentation should describe the same underlying physical phenomenon.

Having said this, there is also where the correspondence stops. The representation based upon fields describes one unique physical state, whereas the power formulation can be used to assess the results of a series of independent measurements. Therein lie the conceptual differences. The 'decomposition' of a given physical field in terms of other fields may lead to different mathematical representations of the same field, whereas the decomposition (summation) in terms of powers relates physically to independent measurements in general of different physical events.

This distinction is crucial in the case of a statistical evaluation of a physical process. Such processes invariably lead one to consider a mixture of physical states, whereas the field-representation always refers to a so-called single, or pure, physical state (the terminology mixture, and pure state is taken from quantum mechanics). Hence, if one considers 'measurables' in the everyday sense of practical experience, one almost always has in mind a statistical average of some measurable, observable quantity. Our conceptual frame work will be based on these three types of target descriptors.

First we establish the field-variables, which are descriptors of a single, or pure object state. Secondly we consider the corresponding target observables which are presented in terms of an equivalent set of power quantities. These two representations completely describe the same physical state. Next one performs a statistical averaging on a set of single target (power) measurements, which invariably leads to a consideration of target mixtures. The final, and crucial step will be to retrieve from the target mixture information the concept of the 'average single object', which corresponds to our everyday conceptual framework.

This will be our program for the remaining part of this report. Several, very interesting correspondences between and/or object defining tree structures, used extensively in artificial intelligence (AI) work will also emerge as a result of this investigation.

6. The 2-dimensional elliptically polarized wave

The plane wave, transmitted by a radar antenna, may be analysed by a two component complex valued vector $\underline{a}$:

$$\underline{a} = \begin{bmatrix} a_x \\ a_y \end{bmatrix} = pe^{i\alpha} \begin{bmatrix} \cos\phi & -\sin\phi \\ \sin\phi & \cos\phi \end{bmatrix} \begin{bmatrix} \cos\tau \\ i\sin\tau \end{bmatrix} \qquad (1)$$

Here, (x,y) gives the horizontal-vertical reference base at the radar site location (fig. 2).

The form on the right above shows the plane wave expressed in geometrical parameters: wave amplitude p, absolute phase α, orientation angle ϕ, and ellipticity τ. The propagation factor $(\omega t - kz)$ is omitted as usual. The geometrical representation has special significance, because it relates in a direct way to a basic requirement of cognition, that useful target parameters must be free of target orientation bias. This means that by viewing the same target at the same exposure, but by merely changing the direction of orientation ϕ, relative to the targets' own orientation ψ_a (see fig. 1) cannot alter the nature of target features. This is because these target features are inherent to the target's physical structure, and cannot depend on the method by which the target is being observed.

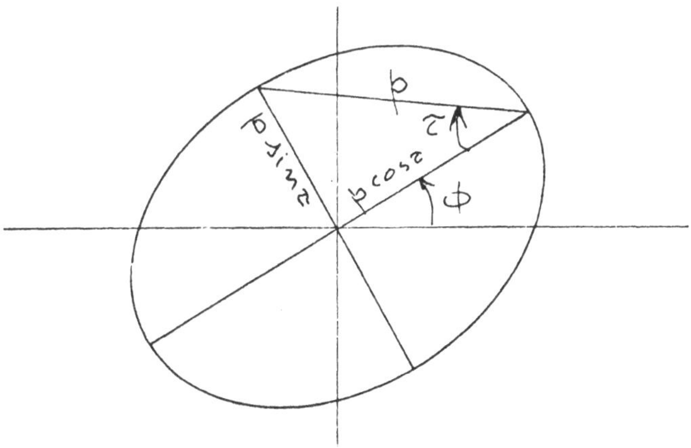

Fig. (2) Left-sensed polarization ellipse in fixed plane.

However, the target scattering characteristics in general depend on polarization illumination. Hence it is of foremost importance to eliminate the 'polarization-bias', introduced by the dynamical variable (is called a dynamical variable because the relative orientation (-) by which the target is viewed, depends on the dynamic motion of the target). Once the target scattering matrix is known, the polarization-orientation-bias can be removed from the target data base by a simple mathematical procedure [4,5].

Expression (1) gives the e.p. wave in terms of the field-components in x and y directions. However, there is another useful presentation of the polarized wave state, which relates to power-measurements. This technique consists of forming the outer product of vector $\underline{a}$:

$$\underline{a} \times \underline{a}^* = \begin{bmatrix} a_x a_x^* & a_x a_y^* \\ a_x^* a_y & a_y a_y^* \end{bmatrix} = \frac{1}{2} \begin{bmatrix} g_o + g_2 & g_3 - ig_i \\ g_3 + ig_1 & g_o - g_2 \end{bmatrix} \qquad (2)$$

$$\text{Here } g = (g_o;\ g_1, g_2, g_3) = (g_o; \underline{g}) = g(\underline{a}) \qquad (3)$$

is called a completely polarized (cp) stokes vector, which belongs to the coherent, monochromatic wave $\underline{a}$. The hermetian matrix (2), when averaged, is called the coherency matrix. It contains the four essential measurable parameters which define the polarized wave state. If we substitute the geometrical variables of (1) into (2) we find:

$$g = \begin{bmatrix} g_o \\ \hline g_1 \\ g_2 \\ g_3 \end{bmatrix} = \begin{bmatrix} I \\ \hline V \\ Q \\ U \end{bmatrix} = \begin{bmatrix} p^2 \\ \hline p^2 \sin 2\tau \\ p^2 \cos 2\tau \cos 2\phi \\ p^2 \cos 2\tau \sin 2\phi \end{bmatrix} \qquad (4)$$

This definition agrees with the author's thesis convention (1970). In optics the convention g = (I, Q, U, V) is often used. The reason for preferring (4) will become clear later on.

The stokes vector, as we observe equation (4) has a very peculiar and very important structure. First of all, it is not a vector, in the ordinary sense, because for the coherent (cp) case the first component g_o is already determined by the three-vector $\underline{g}$

components because:

$$g_o^2 = g_1^2 + g_2^2 + g_3^2 \tag{5}$$

It thus seems, that the extra value, g_o and the extra dimension, associated with g_o are redundant pieces of information. The full significance of the stokes vector structure (i.e. of the extra dimensional component) becomes clear after we prove the following important property:

"THE STOKES VECTOR IS IRREDUCIBLE"

Next, we have to explain what 'irreducbility' means. It means, that unlike the field notation (1), the (cp) stokes vector (4) cannot be decomposed, or subdivided into a sum of other completely polarized stokes vectors. We illustrate this property by simply adding two (cp) stokes vectors g and h, and observing the properties of the sum s. Consider s = g + h, or

$$(s_o;\ \underline{s}) = (g_o;\ \underline{g}) + (h_o;\ \underline{h})$$

then $\quad s_o = g_o + h_o \quad$ and $\quad \underline{s} = \underline{g} + \underline{h} \qquad (6)$

also, we know for the two stokes vectors:

$$g_o = |\underline{g}|, \quad \text{and} \quad h_o = |\underline{h}|.$$

Next we compute s_o and $|\underline{s}|$:

$$s_o^2 = g_o^2 + h_o^2 + 2g_oh_o$$

$$|\underline{s}|^2 = |\underline{g}|^2 + |\underline{h}|^2 + 2\underline{g}\cdot\underline{h} = g_o^2 + h_o^2 + 2g_oh_o\cos\delta$$

From these two relations, we find that:

$$s_o^2 \geq |\underline{s}|^2 = s_1^2 + s_2^2 + s_3^2 \tag{7}$$

Hence the sum vector s in general does not present a <u>pure</u>, or coherent monochromatic completely polarized wave, instead it represents a <u>mixture</u> or a partially polarized (pp) wave. In order to prove the irreducibility property, all we have to do, is observe that equality in (7), when s is a cp wave, can result only if $\cos\delta = 1$; in which case the two vectors $\underline{g}$ and $\underline{h}$ are parallel. In the latter case the stokes vectors g and h into which s is decomposed are proportional, such that s, g and h are the same, except for proportionality factors.

The irreducibility property for the 'pure' case of a cp (coherent) wave has important conceptual consequences. It shows that the pure or monochromatic case has an 'atomic' character: it cannot be split up in smaller components, each of which are cp waves. This remarkable property contrasts sharply with the decomposition in fields, based on equation (1), which always can be done. This will be discussed next.

7. Significance of addition with fields and with power

We are thus led to examine more closely, what the function of 'addition' means in the two cases of fields and power. Addition of fields is a coherent, vector type of addition, and there are no restrictions whatever placed upon how vectors are to be added or subtracted. This decision is left to the convenience of the analyst who wishes to initiate some type of orthonormal expansion of fields, for example. What is even more relevant is the fact that the physical state always remains in the pure form, i.e. the physics is describable by a single field entity (given by ψ in quantum mechanics).

On the other hand, the description of the physical state (cp waves) in terms of power relates to incoherent addition. To illustrate this, let the addition of fields be given by $\underline{c} = \underline{a} + \underline{b}$, then the formula for power:

$$C = |\underline{c}|^2, \quad A = |\underline{a}|^2, \quad B = |\underline{b}|^2 \text{ becomes:}$$

$$|\underline{c}|^2 = (\underline{a}+\underline{b})\cdot(\underline{a}^*+\underline{b}^*) = |\underline{a}|^2 + |\underline{b}|^2 + 2\mathrm{Re}(\underline{a}\cdot\underline{b}^*)$$

or

$$C = A + B + 2\mathrm{RE}(\underline{a}\cdot\underline{b}^*) \tag{8}$$

Hence $C = A + B$ has significance if the fields $\underline{a}$ and $\underline{b}$ are treated as being uncorrelated.

Physically this means that addition of powers representing two physical states ignores the phase relationships the two systems might otherwise have. Instead of a single 'pure' state we have now a mixture or incoherent addition of two pure states. In short: addition of power reflects the real world expediency of dealing with independent physical systems by performing an independent measurement on each system separately, and combining these results for analytical assessment.

Several important consequences follow from the preceding discussion. Whereas the field-description of a physical state always deals with a single coherent or pure state, the power-representation, in terms of stokes vectors, allows one to consider families, ensembles, or mixtures of pure states, which defines a larger class of physical entities. The power sum of two pletely polarized (cp) waves, as we have seen in equation (7), gives a stokes vector which no longer describes a cp wave. Instead a new physical entity emerges which is a partially polarized (pp) wave, for which the inequality $s_0 \geq |\underline{s}|$ holds, with equality applied only for the special case of the cp case. The stokes vector, power type, of addition becomes relevant in statistical sampling procedures. The processes of statistical sampling invariably leads one to constructions of pure object mixtures, and as we have seen, these lead to a wider class of physical entities, with additional parameters necessary for a description of these states. The extra parameter in the stokes vector for the pp case is s_0, which for the pure case is determined by the structure components of $\underline{s}$: $s_0 = |\underline{s}|$, but for the pp case s_0 becomes an independent wave descriptor.

The same phenomenon will be observed if we consider the class of 'pure' or coherent single targets, given in a field-description by a scattering matrix S (see next section). In this case there also exists a power formulation, which is given by the stokes reflection (Mueller) matrix M. Whereas the matrix S is given by five independent parameters, the matrix M can be shown to have nine parameters, which for the pure case are not all independent. However if we now consider a family or ensemble of single targets, the resulting mixture of targets (chaff, meteorological clouds, environmental clutter, etc) will have in general nine independent parameters. The statistical averaging process thus has introduced four new independent variables.

This surprising development has caused controversy and puzzlement among investigators who approach the statistical problem from a strictly single target, 5 parameter, point of view, (for example by considering a distribution of representative target null-polarizations on a Poincaré sphere), and one cannot reconcile the extra four parameters within this framework. The fact is that there is an important conceptual difference at stake. A statistical sampling of field-parameters (as is done by the null-polarization approach) is conceptually quite different from a sampling based on the stokes matrix (power) parameters. Based on our previous discussion the first approach could be termed 'coherent' sampling, whereas the second is 'incoherent' sampling; although one should be careful not to confuse this terminology with other already existing nomenclature.

By the coherent sampling method one wishes essentially to investigate how a single, coherent object behaves as it exposes different aspects of itself to an observer. Presumably these aspects are related to each other in a 'coherent' or relatively orderly progression, i.e. the object must be relatively simple in structure, with respect to the frequency of radar illumination used. This could then be called the low-frequency case. For the highfrequency case the different exposures become essentially uncorrelated and the incoherent sampling, based upon the stokes-matrix, is in order. More discussion on these topics will follow, after we introduce the target scattering matrix in the next section.

8. The target scattering- and stokes-matrices

In what follows, we will only summarize the most essential parts of the theory. Full details can be found in the author's thesis [4]: "Phenomenological Theory of Radar Targets: (1970, available upon request, from the author: J.R. HUYNEN, 10531 BLANDOR WAY, LOS ALTOS HILLS, CA 94022). Appropriate equation and section numbers will refer to this source.

The radar system formalization is done very simply. An electromagnetic wave $\underline{a} = (a_x, a_y)$ of the form equation (1) is transmitted to the target. The returned back scattered wave (for the monostatic case) is given by $\underline{c} = T\underline{a}$ where T is the 2x2 target scattering matrix:

$$T = \begin{bmatrix} t_{11} & t_{12} \\ t_{21} & t_{22} \end{bmatrix} \qquad (9)$$

Due to reciprocity $t_{12} = t_{21}$; the t_{ij} are complex numbers which give target scattering coefficients for the various modes of polarization used, i.e. HH, HV, and VV.

Let the antenna receiver be given by the field $\underline{b}$, then the equation for voltage V received at the radar site location is: $V = T\underline{a} \cdot \underline{b}$ ('70 - 9.3) [4]. The matrix T is given by three complex, or six real numbers. One of these is the absolute phase, which changes with every slight movement of target, or radar, along the line of sight direction. Hence the absolute phase cannot be used as a target parameter, and we are left with five useful target parameters. Notice that indeed the absolute phase disappears with power measurements, supporting our view above, that it can be put aside. Although these ideas seem to appear quite natural, there still exists in the literature some disagreement about how the absolute phase should be defined. We will stick to the convention introduced in ('70 - 10.22), which is invariant to unitary transformations (orthonormal coordinate-systems in which the matrix T may be expressed).

Because this will be needed shortly, we now give a simple expression for a unitary transformation which is due to Kennaugh [6,7]:

$$U = \frac{1}{(1+|\rho|^2)^{1/2}} \begin{bmatrix} 1 & -\rho^* \\ \rho & 1 \end{bmatrix} \qquad (10)$$

where ρ is a complex polarization ratio. It is easily checked that U is unitary: $U^T U^* = I$, where the T-index denotes matrix transposition and the star is complex conjugation.

There also exists a description of the radar operation which is based upon power-measurements. Let $P = |V|^2$ be the received power, then it can be shown ('70 - 11.11), that

$$P = M\, g(\underline{a}) \cdot h(\underline{b}) \qquad (11)$$

where $g(\underline{a})$ is the stokes vector for the transmitted wave and $h(\underline{b})$ that for the radar receiver polarization. The matrix M is called the stokes reflection matrix, or Mueller-matrix; it is the equivalent to the matrix T, for power measurements:

$$M = \begin{bmatrix} A_o+B_o & F & C & H \\ F & -A_o+B_o & G & D \\ C & G & A_o+B & E \\ H & D & E & A_o-B \end{bmatrix} \qquad (12)$$

We notice nine parameters in M: A_o, B_o, B, C, D, E, F, G, and H. However, for the 'pure' case of a single, coherent target, there must be a one to one correspondence between matrix M and T formulations, while the latter had only five independent parameters. This shows that not all nine parameters in M can be independent. There must be four dependency relationships. These dependencies and their relationship to scattering matrix parameters become clear after we introduce a second type of formulation of the scattering matrix and stokes matrix. The new target description is based on a vector-formulation! This will be discussed next.

9. Vector formulation of the target scattering matrix

The following material is introduced as an alternative and very useful novel method of target-representation, which appears to be very different from the usual scattering matrix or stokes matrix approaches. It may at first seem rather odd, that a matrix approach can be made to correspond to a vector-presentation of some kind. After all, we can normalize vectors and speak of orthogonal vectors and we can form scalar products, but no such operations appear naturally with matrices.

Nevertheless, what follows derives naturally from ideas in cognition, where our concept of a target is associated with a string of measurable target features. A string of parameters is more naturally given by a vector rather than by a matrix. These different points of view are reconciled, as we shall see, when we write the target matrix T of equation (9) in the following form:

$$T = \begin{bmatrix} a+b & c \\ c & a-b \end{bmatrix} \tag{13}$$

Here again, a, b and c are complex numbers, which are related to the t_{ij} by simple expressions:

$$a = \tfrac{1}{2}(t_{11} + t_{22}), \quad b = \tfrac{1}{2}(t_{11} - t_{22}) \text{ and } c = t_{12}$$

Associated with (13) we now introduce the following target-vector formulation:

$$t_3 = (a,\ b,\ c) \tag{14}$$

It is clear at once, that the information content of scattering matrix T is preserved by the target-vector t_3.

Next we have to verify what happens if we apply a unitary transformation (10) to the matrix T. Let $T' = U^T\ T\ U$ be the transformed matrix, then we find after simple algebra that the new coefficients (a', b', c') are related to the old ones by the following quasi-hermetian transformation:

$$\begin{bmatrix} a \\ b' \\ c' \end{bmatrix} = \frac{1}{2(1+|\rho|^2)} \begin{bmatrix} (2+\rho^2+\rho^{*2}) & \rho^{*2}-\rho^2 & 2(\rho-\rho^*) \\ \rho^2-\rho^* & (2-\rho^2-\rho^{*2}) & 2(\rho+\rho^*) \\ 2(\rho^*-\rho) & 2(\rho^*+\rho) & -2(1-|\rho|^2) \end{bmatrix} \begin{bmatrix} a \\ b \\ c \end{bmatrix} \tag{15}$$

Notice that although the matrix H in (15) is hermetian, the

unitary transformation also introduces a change of sign in c', of the transformed target vector, hence the term quasi-hermetian is used to indicate this fact. We also notice the interesting property: trace H = 1. The result (15) establishes a unique correspondence between the target scattering matrix T representation, and the target-vector t_3 formulation.

There are still some features, which may initially seem puzzling to an alert reader. Apparently there could be some difficulty with the introduction of transmit and receive antennas with the new vector formalism. The natural way to obtain the scalar received voltage from the target vector t_3 is through a dot product operation with another vector. But how can both transmit and receive polarizations be given by one vector? This problem can be resolved in an elegant manner, but as this development distracts form our present objectives, we will reserve the discussion for a future publication.

Next we have to show what consequences the new vector-presentation of the target may have. Because it is much simpler in general to manipulate vectors, and to attach conceptual significance to each of their components, rather than to work with matrices, we can expect many useful results to emerge. That this is indeed the case is shown in the next section.

10. The 3 x 3 equivalent stokes reflection matrix: M_3

If we write the target scattering matrix in the form (13), the coefficients of the stokes matrix, given by (12) are easily expressed in terms of the complex field quantities a, b and c; as follows:

$$A_o = \tfrac{1}{2}|a|^2 \quad (16)$$

$$B_o = \tfrac{1}{2}(|b|^2 + |c|^2) \quad (17)$$

$$B = \tfrac{1}{2}(|b|^2 - |c|^2) \quad (18)$$

$$\left.\begin{array}{l} C - iD = ab^* \\ E + iF = bc^* \\ H + iG = ac^* \end{array}\right) \quad (19)$$

We notice in particular, that all the stokes matrix parameters are thus obtained from (a, b, c) by simple dual product relationships. This suggests, that the formation of the outer product: $t_3 \times t_3^*$, will reproduce all the stokes matrix parameters.
Let us define:

$$M_3 = t_3 \times t_3^* = \begin{bmatrix} |a|^2 & ab^* & ac^* \\ a^*b & |b|^2 & bc^* \\ a^*c & b^*c & |c|^2 \end{bmatrix} \qquad (20)$$

then

$$M_3 = \begin{bmatrix} 2A_o & C-iD & H+iG \\ C+iD & B_o+B & E+iF \\ H-iG & E-iF & B_o-B \end{bmatrix} \qquad (21)$$

The 3 x 3 coherency-matrix M_3 contains exactly the same information as the 4 x 4 stokes matrix M. However, conceptually there are major differences. The process of averaging becomes more natural with M_3, because of its significance as a correlation matrix.

In that case $\langle M_3 \rangle$ represents a target mixture given by nine independent parameters, whereas the vector t_3 gives the field representation of the pure or coherent target state. The 'N-target decomposition' theorem ('70 - section 38) of the author [4,5] takes on a simpler and more natural form with M_3 as compared with M. Conceptually, everything seems to fall into place with the t_3 - M_3 formalism.

The case for n = 3 also is easily generalized: The single target (or object) is given in general by a string of n complex parameters, which defines the t_n target-vector in terms of a field representation. However, the mixture, obtained by averaging $\langle M_n \rangle = \langle t_n \times t_n^* \rangle$, results in a target mixture given by a string of n^2 real and measurable object parameters. The conceptual approach leads one to consider the feasibility of extracting from the n^2 parameter target mixture a 'single averaged object', given by (2n - 1) real parameters. This would be the conceptual object or the objectconcept. The remaining $(n - 1)^2$ parameters now determine a 'residue' which is a physically realizable object mixture of one order less than that of the original target mixture.

There is also the matter of hierarchical ordering of the target parameter data sets. We will introduce a method based on and/or object defining tree structures familiar to AI researchers, which however is a novel approach in radar target technology. These topics are dealt with in detail in the following sections.

11. Target decomposition theorems

The introduction of the coherency matrix M_3 for radar targets leads us to familiar grounds. For n = 2 we found a description of the stokes vector by M_2:

$$M_2 = \frac{1}{2}\left[\begin{array}{c|c} g_0+g_3 & g_2-ig_1 \\ \hline g_2+ig_1 & g_0-g_3 \end{array}\right] \quad (22)$$

↳ submatrix

For this case we already know that for the cp coherent pure case $\underline{g} = (g_1, g_2, g_3)$ can be chosen independently and then $g_0 = |\underline{g}|$ is determined.

In the general case of a partially polarized (pp) wave, $\langle M_2 \rangle$ produces $n^2 = 4$ independent parameters. Now $g_0 \geq |\underline{g}|$, and we know it is always possible to decompose the partially polarized (pp) wave into a completely polarized (cp) coherent part $g_c = (g_0'; \underline{g})$ for which $g_0' = |\underline{g}|$, plus a residue: $g_r = (g_0 - g_0'; 0)$. The residue obviously is physically realizable, and is chosen such that the vector-part is zero.

Conceptually this means that the three structure parameters (g_1, g_2, g_3) of the object mixture, suffice here to completely define the concept: 'average cp wave' from the pp mixture. We notice that the 2n - 1 = 3 parameters chosen for the 'average cp wave' appear in the first row of matrix M_2, above the dotted line. This will be a general trademark, by which to find those (2n - 1) parameters in a decomposition of an object mixture which are 'preserved' for the desirable part.

The submatrix of order n - 1 also shown in equation (22) consists here of the single term $g_0 - g_3$. This submatrix will generally contain the parts which are being split up, and which contribute exclusively to the residue part. Why did we single out g_0 to be split up, while g_3 was being preserved? Presumably, we could have preserved $g_0 + g_3$ and split up $g_0 - g_3$, or perhaps some other combination is called for. The answer is, that we want the residue part of the decomposition to be of the same type under a certain class of transformations. And obvously this residue must be shown to be physically realizable. Clearly the null-vector $\underline{g}_r = \phi$ satisfies these conditions for all rotations on the so-called Poincaré sphere. Note that trace $M_2 = g_0$, which supports this viewpoint. These properties are generally applicable also to other values of n.

The case n = 3 can be formulated in a similar way.
We repeat the expansion of (21):

$$M_3 = \left[\begin{array}{c|cc} 2A_o & C-iD & H+iG \\ \hline C+iD & B_o+B & E+iF \\ H+iG & E-iF & B_o-B \end{array}\right] \rightarrow \text{submatrix} \qquad (23)$$

As in the previous case, the 2n - 1 = 5 terms, which are 'preserved' in the decomposition theorem are the terms above the horizontal dotted line, i.e. A_0, C, D, G and H. The terms that are split up are contained by the submatrix of order (n - 1) = 2. There are $(n - 1)^2 = 4$ such terms.

Hence the 'N-target decomposition theorem', proved in 1970 by the author, splits up the nine parameters of the mixture into an 'average single target' (5 parameters) and an N-target-residue (4 parameters). What is not obvious at this point, is that the decomposition always is physically realizable, but this was shown to be the case ('70-section 38).

The N-target residue has the property $A_0^N = 0$, because the original term A_0 is preserved to contribute exclusively to the 'average single target'. This property guarantees that the N-target structure (with $A_0^N = 0$) is preserved under changes of orientation angle, by either observer, or the target (see figure 1). Hence the target decomposition satisfies all the properties of common perception and it is physically realizable and unique. We notice for the trace rule: trace $M_3 = 2(A_0 + B_0)$, which gives an expression equivalent to total object power for n = 3.

For general n, we start out with a string of n complex field components. One of these components is singled out to be the first component: a_0. Usually a_0 may be characterized as having invariance under some type of transformation, such as orientation-invariance. The other n - 1 complex variables may be combined into a vector $\underline{a}$, such that the target/object vector description is: $t_n = (a_0, \underline{a})$. The corresponding coherency-matrix is written in the usual form:

$$M_n = t_n \times t_n^* = \left[\begin{array}{c|c} |a_0|^2 & a_0\underline{a}^* \\ \hline a_0^*\underline{a} & \underline{a}\times\underline{a}^* \end{array}\right] \qquad (24)$$

$\hookrightarrow$ submatrix

The object-mixture decomposition may be read off from the matrix in (24). The (2n - 1) real terms in the first row above the horizontal dotted line are 'preserved' for the averaged conceptual object. The $(n - 1)^2$ parameters in the submatrix are split up, one part contributing to the averaged object, the other part being 'residue'. Notice that the target residue derives its terms exclusively from the submatrix. Of course, one has to prove the physical realizability for the object decomposition (certain terms representing power must be positive). This can be done along lines similar to the case n = 3.

We now address the question of hierarchical ordering of the target parameter data base.

12. Hierarchical object-concept tree structures

We are trying to put into analytic form the idea, which lies behind the common expression: "The whole is more than the sum of its parts". What comes to mind are organizational structures, which are presented by the familiar organization-charts or an inverted tree-structure, also widely used in AI work [8]. Of course the same structure is basic to many other areas such as linguistics, manufacturing processes, economy, social structures etc. Of special interest for this report is the application to radar targets and more generally to object-concepts. A simple example is the concept 'chair', which may be defined as an object which has legs, a back and a seat. The structural diagram is given in figure (3).

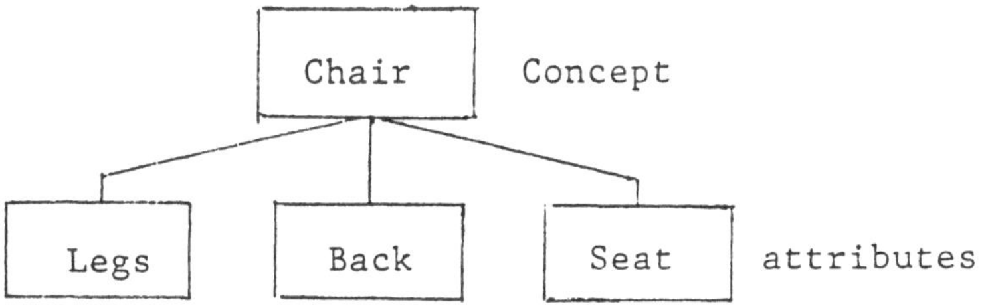

Figure (3) Object-concept tree for a chair.

What we have to distinguish at the very onset is the difference between one actual existing single object which is a chair given by its attributes, and what is obtained as a concept after we look at many different chairs. The latter is called the mixture of chairs. The concept actually defines existence for the object to the observer. From the object mixture one derives averages for legs (four legs of average length and shape), backing and seat. The process of mixing and averaging also requires that each time a new object is presented to the mix, the concept 'chair' which defines as it were its reality as an object, will have to be adjusted to the new information content. What finally emerges after many samples are taken, is a stable 'concept-chair' which is defined by its average attributes.

What we have described above is the process, which was developed in detail in previous sections, by which from a mixture of single objects the concept 'average single object' was derived. In particular the associated analysis will be immediately applicable, and this could be useful in artificial intelligence (AI) work. Conversely, the object-concept tree structure of AI work will be useful for radar target investigations.

Without further ado we now present the object tree structure for the case n = 2:

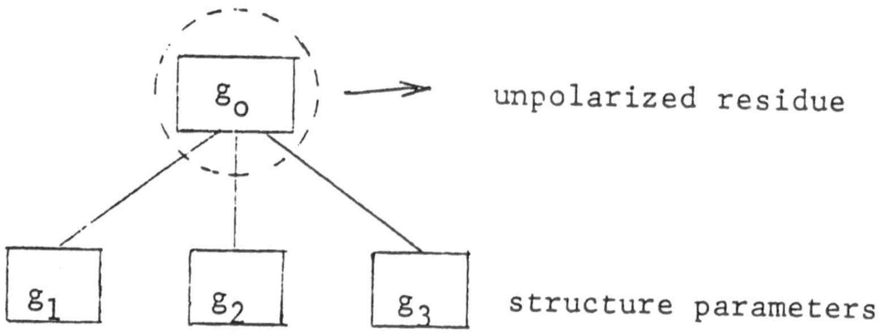

Figure (4) Tree structure for the polarized wave object.

The electromagnetic polarized wave, which is the 'object' in figure (4), is given as we have seen by the three stokes vector attributes (g_1, g_2, g_3). These parameters are 'preserved' after object mixing and contribute exclusively to the average object attributes. The 'concept' of the object is given by the block g_0. However the average of changing concepts $\langle g_0 \rangle$ is not the same as the concept 'average object' which would be based on average attribute information.

The difference between the two is the object-residue which in this case takes the form of an unpolarized (up) wave. The dotted outline in the figure indicates the part of the object which is split up after mixing, and which contributes exclusively to the residue part. The important rule, which 'holds together' the object structure is the idea of normed-addition of attributes.

$$g_0^2 = g_1^2 + g_2^2 + g_3^2 \qquad (25)$$

This idea may have major impact on other AI work, where this process currently is not used. It follows that the stokes vector rule (25) will have many other applications; it is related to the Hamiltonian equation in classical relativistic - and quantum mechanics.

Other properties, which we have not yet mentioned, can be shown to have fundamental significance in general cognition [9]. For single objects: $M_n^2 = M_n$ (with proper normalization). This follows immediately from the outer product definition. The same idempotent rule applies in quantum theory, where M_n is called the Von Neumann density matrix.

We next proceed to the case n = 3. The object for this case is the radar target, already a truly universal type of object structure, given by (2n - 1) = 5 real and measurable parameters. The object-tree structure is shown in figure (5).

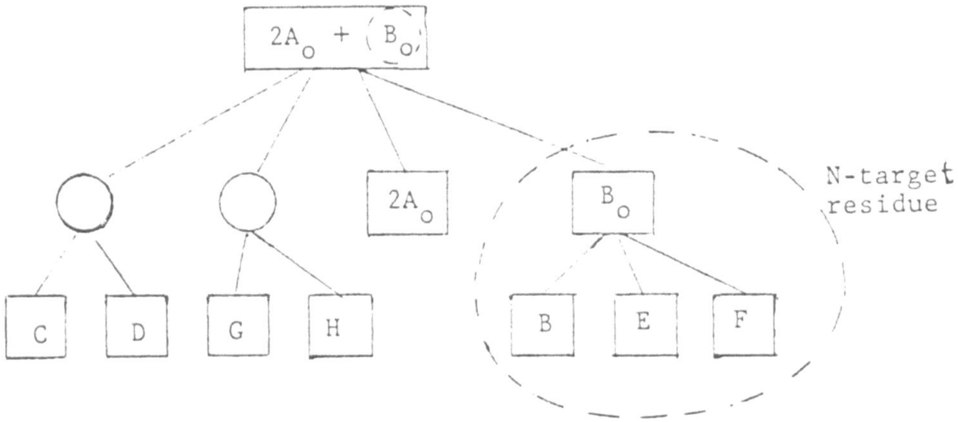

Figure (5) Tree structure for radar target.

Notice that two types of boxes are shown. The square boxes contain target attributes. The circled boxes do not contain attributes, although <u>values</u> might be inserted with them. The circled boxes are called <u>junction boxes</u>, their values are derived from the target attributes which they <u>generate</u>. For example, the foremost left junction box has value $\sqrt{C^2 + D^2}$, while the other circled box has value $\sqrt{G^2 + H^2}$. For target mixtures, parameters in the square boxes are averaged over, but values in circled boxes are <u>not</u> averaged.

In this way a consistent methodology for analysing target trees is obtained. The single, pure, coherent target is given by five independent variables: A_o, C, D, G, and H. Notice that these were also the parameters which appear in the first row in the matrix of (23). The other parameters B_o, B, E and F appear in the dotted encirclement. They are not free for the single target, but are determined by the five given above. Notice that these parameters appear in the submatrix of (23).

For a mixture, the first five parameters, after averaging, contribute exclusively to the 'average single target' of the target decomposition theorem. The parameters in the dotted circle are split up, one part contributing to the 'single average target', the other part produces the N-target residue.

Throughout the tree structure, the normed-addition rule (25) prevails. The reader is asked to verify that the top level 'concept' box must have value $2A_o + B_o$ as shown. The B_o part of this term is encircled by a dotted line, because it is the same B_o which contributes to the N-target residue.

These concepts can be generalized to any value of n. The cases n = 2 and n = 3 contain all the basic structure from which the general case is generated. For general n we have the following tree structure, based on (24):

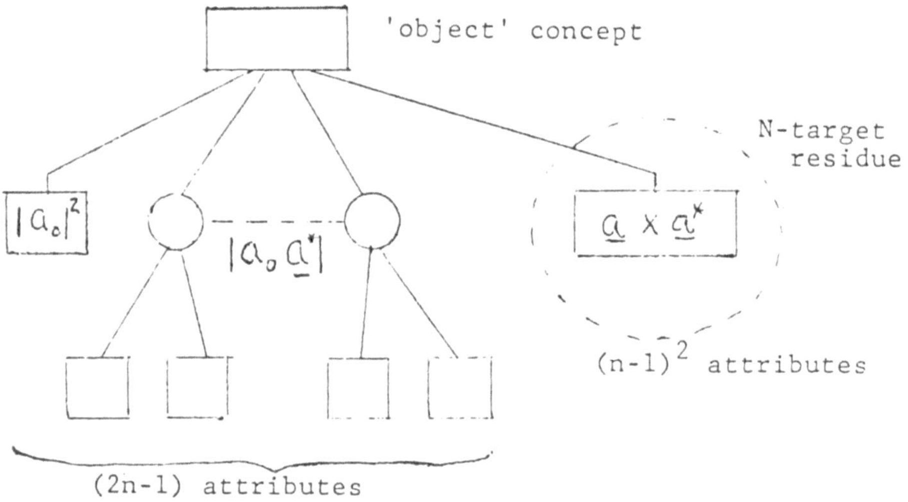

Figure (6) Tree-structure for general object

The (2n -1) attributes which determine a 'pure' object are shown on the left hand side of the figure. The parameters which are split up, and which contribute to the N-target (object) residue are shown in the dotted encirclement on the right. The concept 'object' which denotes object-existence is shown at the top of the hierarchical chain.

We observe, that if n increases, the object attributes for the 'pure' case, which fixes the idealized object, increases linearly with n, while the 'noise' aspect, generated by a target mixture, increases with n^2. Hence there is great economy in preserving those meaningful parameters by which the concept-object, or the idealized object is recognized. The greatest part of the n^2 available parameters of an object-mixture can thus be done away with, as presenting N-target residue.

This completes our discussion of object-tree structures. We showed that the radar target case (n = 3) forms the prototype for general object parametrization. The 'pure' object-concept idealization from a class of similar given objects is shown to derive in a simple systematic manner. The demonstration of physical realizability of the general decomposition theorem, is one of the key factors which makes idealization into a pure object structure (the Platonic 'idea') possible.

13. Acknowledgements

The author wishes to thank Dr. W. -M. Boerner for his continued support and encouragement, which made this work possible. We also wish to thank Mr. James Willis and Dr. Richard G. Brandt for their interest in and support for this research.

The task of transcribing my notes into a word-processed document was ably executed by Ms. Jeanne Martin. Final draft was provided by Ms. Kathleen Slius, for which I express my gratitude.

14. References

[1] Boerner, W. M., Use of polarization in electromagnetic inverse scattering, Radio Science, 16(6), Nov./Dec. 1981, 1037-1045.

[2] David Reade, private communication.

[3] Immanuel Kant: 'Kritik der Reinen Vernunft' (Critique of Pure Reason) Verlag Von Felix Meiner, Leipzig,1919.

[4] J. R. Huynen, Phenomenological theory of radar targets, Doctoral Thesis, Technical University, Delft, The Netherlands (1970) (obtainable from the author).

[5] J. R. Huynen, Phenomenological theory of radar targets, Chapter 11 in "Electromagnetic Scattering", Academic Press, N.Y., 1978, edited by P. L. E. Uslenghi.

[6] Kennaugh, E. M., Polarization properties of radar reflections, M.Sc. Thesis, Dept. of Elect. Eng., The Ohio State Univ., Columbus, Ohio 43212, 1952.

[7] Kennaugh, E. M., Effects of type polarization on echo characteristics monostatic case, Rpts. 389-1 (Sept. 16, 1949) and 389-4 (June 16, 1950), The Ohio State Univ., Antenna Laboratory, Columbus, Ohio 43212.

[8] Avron Barr and Edward A. Feigenbaum, The Handbook of Artificial Intelligence, Vol I, Heuris Tech Press, Stanford, California, 1981.

[9] J. R. Huynen, Theory of cognitive systems, July 1982 Dept. of Information Engineering, University of Illinois at Chicago, Box 4348, 851 S. Morgan St. Chicago, IL 60680. (Also available from the author: 10531 BLANDOR WAY, LOS ALTOS HILLS, CA 94022. Tel (415)-941-2374)

IV.3 (SI.4)

RADAR IMAGERY

Torleiv Orhaug
National Defence Research Institute
P.O. Box 1165
S-581 11 LINKÖPING SWEDEN

ABSTRACT

This tutorial paper discusses radar imagery from several viewpoints. Various processing tasks of radar data and imagery data are briefly discussed. Some of the main properties of radar images are pointed out and the differences between optical and radar imagery are discussed. The problem of image quality is reviewed with reference to works in optical and radar imagery. Computer processing of radar imagery is considered and various tasks like image encoding and enhancement, rectification and feature processing are briefly commented, using examples from radar. Finally, computer analysis of imagery in general and radar imagery in particular is discussed and the limitation of pattern recognition methodology for image analysis is pointed out.

1. INTRODUCTION

Gathering of information from the physical environment is classically carried out by the visual system of man. The photographic camera was the first fully technical remote sensor, producing (imagery) data which could be stored on and visualized by the film plate. The sensor was in this manner acting as a "substitute" for the eye, making vision independent of time and location. In order to be useful information, the imagery data must be read and interpreted. In a similar way as direct seeing must be acquired so also image reading, and image interpretation must be learned.

W.-M. Boerner et al. (eds.), Inverse Methods in Electromagnetic Imaging, Part 2, 823–839.

The evolution of science and technology has produced a wide repertoire of imaging instruments and man is nowadays dealing with a very large amount of images of various kinds and with varying characteristics and properties. An important characteristic of imagery data is the relation between the data and the physical parameters of the scene being sensed. Other significant problems are the relation between the imagery data and the sensing instrument and the question of how the physical characteristics of the image affect the interpretability or the possibility of information extraction.

Images result from interaction between scene illumination and physical objects in the scene. Active sensors provide their own illumination which can both be controlled and changed; this provides interesting flexibility and features to the imaging sensor. One question of particular interest is the amount of new information available through active technique, and in which way such information can be used for scene analysis and interpretation/understanding.

Understanding the environment through gathering and interpretation of imagery data is of great importance. Since many years computers are being used to help man deal with imagery data. Considerable progress has been made in this respect; still, "automatic" image interpretation is difficult to achieve but for rather simple cases. Therefore, human (visual) interpretation will be important also for the future. The question of image quality is of significance since the image either will be used as input data for machine processing/analysis or be used for visual interpretation. This paper focuses on the discussion of radar images, in particular SAR imagery data. The discussion of several topics like image quality will be based upon a general discussion not necessarily limited to radar imagery only.

The paper is organized in the following way. After a brief introductory remark of various processing/conversion tasks related to imagery data (chapter 2) the main characteristics of radar images are introduced in chapter 3. A discussion of image quality problems in general is presented in chapter 4 together with a brief review of results of quality assessment of radar imagery. Some problems related to computer processing of radar images like image enhancement, rectification and speckle processing are dealt with in chapter 5. Finally, in chapter 6 problems of image analysis are discussed. Two different approaches to image analysis, pattern recognition and image understanding are briefly considered.

2. CONVERSION AND PROCESSING OF IMAGERY DATA

Imagery data can be processed and/or converted by technical means and the following tasks may be relevant (see Orhaug (1)):

- Data to image transformation:

  Such tasks may be of various types ranging from simple format conversion to more elaborate processing/conversion tasks found in tomography and SAR.

- Image to image transformation:

  This covers various tasks of machine processing and human interpretation and is defined as those transformations that result in interpretation/description of the information content of the imagery data.

- Data to description transformation:

  This is of particular interest to SAR-data since this transformation covers those processing tasks for which raw data are transformed into description without using data in image form. An example of this is the derivation of ocean spectra from SAR-data.

3. RADAR IMAGERY

The observable quantity in radar imagery is governed by the wellknown equation (see Bennett and McConnell (2)):

$$I = S/N = \frac{P_T A_r^2 L}{4\pi \lambda^2 R^4 k T_o B F_N} \cdot \sigma_0 \cdot G_{re} \cdot G_{al} \cdot M_p \cdot N_p \qquad (1)$$

where

S/N	:	received signal-to-noise-ratio
P_T	:	peak transmitter power
A_r	:	physical aperture size (receiver and transmitter)
L	:	system loss factor
λ	:	wavelength
R	:	slant range
k	:	Boltzmann's constant
B	:	system bandwidth
F_N	:	receiver noise factor

σ_0 : target radar-cross-section
G_{re} : single look range processor gain
G_{al} : single subaperture azimuth processor gain
M_p : no. of statistically individual looks in range
N_p : no. of statistically individual looks in azimuth.

The measurable quantity may be observed both as a function of deception angle (Θ), polarization (φ), and wavelength, or of

$$I = D \circ G_0(\Theta, \varphi, \lambda) \tag{2}$$

where D is a quantity given by eq. (1).

The characteristics of the radar image may be summarized as follows:

* the use of microwaves gives a sensor independent of weather and the use of active sensing makes the sensor independent of (outside) illumination,

* the SAR sensor provides an image resolution in principle independent of range (and of wavelength) while conventional radars (and microwave radiometers) suffer from their fixed (and often poor) angular resolution tied to the real aperture size of the microwave sensor with a spatial resolution comparable to what optical sensors can provide (in particular for spaceborne altitudes),

* the observable quantity is proportional to the backscatter coefficient of the material in the line of sight and at microwaves this reveals phenomena different from those observed at visual and IR-wavelengths,

* at optical wavelengths the spectral properties depend strongly on chemical compositions of the material while at microwaves the backscatter depends both on the physical properties of the surface (slope angle, roughness) and on the electrical properties (related to porosity of soil and water content),

* due to the design of the sensor and the geometry of sensing, the radar image shows various kinds of radiometric and geometric artifacts,

* the major radiometric artifact is the appearance of speckle phenomena due to the coherent, monochromatic illumination of the sensor; the speckle adds to the thermal random noise of the image signal and degrades image visual quality,

* the major geometric artifacts are layover, range compression (foreshortening) and shadowing,

* these effects can, however, also be used for interpretation of terrain images since they enhance certain topographic features,

* the tonal (textural) variation of radar imagery is of two distinct types:

  - large scale variations (i.e. low spatial frequencies) dominated by surface backscatter variations due to vegetation, rock-type etc,

  - small scale variations (high spatial frequencies) dominated by topographic slope effects (aspect-angle effect),

* the dynamic range of radar imagery is typically 50-80 db which is far larger than the corresponding range of optical imagery (10-20 db) (see Mitchell, Marder (3)),

* due to the large dynamic range, the characteristics of the effective impulse response function affect the imagery data differently than in optical imagery:

  - near sidelobes may cause a strong scatterer to produce confusing returns in competition with neighboring weaker scatterers,

  - far sidelobe peaks may cause a large radar cross-section to produce signals mistaken as a return from a real object.

4. IMAGE QUALITY

General

The problem of how image parameters like contrast, resolution, noise etc affect subjective ("cosmetic") image quality and image interpretability has been the target of numerous investigations since a long time, particularly for optical (visual) images. There are several ways of addressing this problem. One method is to study how certain image parameters influence image interpretability by actually performing interpretation experiments (observing numbers of detected, recognized and identified targets) using imagery of various values for the image parameters studied. Another method involves subjective judgement of image quality while varying image parameters. This method rests upon observations of high correlation between subjective image quality and image interpretability.

Another line of approach is the search for an image quality factor and its relation to image (physical) parameters. Since images (often) are thought to be interpreted by man, the properties of the of the visual system are often incorporated into the quality factor. Granger and Cupery (4) introduced a measure called subjective quality factor (SQF) using the sensor system MTF (Modulation Transfer Function) and the resolution properties of the visual system. The SQF does not take noise into account. Another quality measure (MTFA) introduced by Charman and Olin (5) is based upon the area defined by the MTF of the imaging sensor and a threshold curve related to the visual system. The use of the MTFA measure for a raster scan sytem has been studied by Snyder (6).

An alternative method of investigating perception of imagery data is that of multidimensional scaling. In these experiments observers are presented images pair-wise and asked to rate similarity. Subjective (perceptual) distances are then derived from the similarities by assuming that the images can be represented as points in an n-dimensional space where each dimension corresponds to a separate image quality component. This wellknown method in psychophysics was first applied to pictorial information by Marmolin and Nyberg (7). They investigated the number of "perceptual dimensions" and found for the image data being used that the number of dimensions is about 4-5. Another interesting question concerns the relation between the perceptual dimensions and the physical parameters of the image; the method of investigation used does not, however, directly give this relation.

The frequent used of digital image data has lately spurred the investigation of consequences of quantization on image perception. Although the effects from a data/signal point of view are analyzed by methods in communication theory (Nyquist sampling, aliasing, quantization noise) the corresponding perceptual effects have to be separately evaluated. Examples of sampled imagery can be found in (8). More detailed investigations of the perceptual effects of sampling using various combinations of prefiltering and smapling frequency have recently been conducted by Linde et al. (9), and by Arguello et al. (10).

Quality of Radar Imagery

The discussion so far has concerned image quality in general and the imagery used for quality studies has often been optical imagery. We shall now make a short review of some of the recent work made to make quality studies using radar imagery (most often SAR imagery).

In an earlier study, Mitchell (11) investigated the tradeoff between speckle noise and resolution of extended targets (i e target dimensions much larger than the size of the resolution cell). Such targets can in particular be found among terrain features but may also be man-made objects. For extended targets having few strong scatterers which are individually resolved, interpretation is normally made with no difficulty since the association of a particular response to a given target is relatively easy. For many scatterers per resolution cell, so called "coherent breakup" caused by mutual interference occurs. Using sets of simulated images having different resolution/speckle properties the tradeoff could be investigated. The findings indicated that about 100 resolution cells were needed for target recognition (and less if noncoherent averaging is performed).

In a recent study, Ford (12) investigated the tradeoff between resolution and speckle for real SAR images used for geological applications. Using Seasat SAR data from 3 test sites, 13 different resolution/look combinations were genereated by processing. Resolution combinations (range resolution meter/azimuth resolution meter) were produced in the range 50/25 - 300/300. Multiple looks were generated both by noncoherent averaging and by spatial processing technique. For almost all conditions investigated, interpreters preferred multiple look images. For most targets, two-look images with best resolution were rated as best. For constant data volume a higher resolution is more important for target discrimination than is a greater number of looks. One exception to this finding was for sanddunes which required more looks than other geological targets investigated. Also in this study, multiple look images were preferred over the corresponding single-look image at all resolutions.

An extended study of the tradeoff relation between pixel size (and form) and noncoherent averaging has been made by Moore (13). Both synthetic and real aperture images were used and from original images degraded ones were produced (using optical method) and the resulting images were rated (using a scale 0-4) by trained interpreters with regard to their ability to interpret various specific features on the images. The interpretability (P) was found to be related to a "spatial-grey-level-volume" (SGL-volume) defined by

$$V = r_a \circ r_R \circ r_g(N) \tag{3}$$

where r_a is the azimuth (along-track) resolution, r_R is the range (across-track) resolution and $r_g(N)$ is a grey-level resolution, defined by the ratio of the 90-percent level to the 10-percent level on the fading distribution (for N samples independently averaged foreach pixel). The experimentally found interpretability data could be described by the following simple analytic expression

$$P = P_o \exp[-V/V_c] \tag{4}$$

where P_o is a scaling constant and where V_c is an application dependent parameter. Since rectangular pixels could be simulated it was possible to investigate the effect of not only pixel size (area) but also pixel rectangularity. The results clearly indicated that (at least for the application tested) pixel size (area) is more important than individual pixel dimension (one-dimensional resolution); length-to-width ratios between 1 and 10 were investigated.

The simple expression for interpretability means that for quadratic pixels, the relation between interpretability and (one-dimensional) pixel size is approximately a Gaussian function. It could be interesting to investigate some of the results from earlier experimental studies of photographic imageries to see if a similar functional relationship may be obtained. Also, the earlier results relating interpretability to the MTFA-measure might intuitively be expressed in similar terms.

In a recent paper, Stiles et al. (14) investigated the interpretability of extended targets, focusing upon flat terrain and terrain with moderate relief. Using simulated imgery (incoroporating different terrain types through microwave scattering models) they quantitatively investigated the effect of hilly ground on pictorial presentation of the simulated imagery data.

5. RADAR IMAGE PROCESSING

General

In this chapter we shall address the problem of image processing limited to image-image transformation (using the terminology introuduced in chapter 2). Such processing may have several purposes:

* compensation for sensor imperfections (causing radiometric and geometric image artifacts)

* image quality improvement or image enhancement

* encoding of image features for visual presentation

* features extraction for visual or machine processing

and we shall give some examples of various processing methods.

Image Compensation

In radar imagery, two special types of radiometric artifacts are common: speckle variation and uneven scene illumination caused by uncorrected antenna pattern effect. The latter is caused by (uncompensated) sensor platform motion and the result is that the image grey level partly depends upon position and not only on the backscatter properties of the scene. Blom and Daily (15) discuss methods for compensating this effect by equalizing the image grey level mean and variance. Similar methods ("histogram processing") are often used for image enhancement and other image-image transformation tasks. The limitation of this technique is that signal-to-noise-ratios in partly illuminated regions are decreased. Also, this method is not very useful for terrain with natural backscatter monotonically varying over the image. This technique must therefore be applied with care and in close cooperation with trained observers.

Image Encoding and Enhancement

A multitude of methods have been developed in the field of image processing for image encoding and image enhancement and several of these methods have also been applied in the area of radar imagery. An interesting method of applying color coding to geological interpretation has been discussed by Blom and Daily (15). Using the fact that tonal (textural) variation in radar imagery depends upon different physical mechanisms causing different spatial scales of variation, they separated tonal variations into two spatial spectral regions (low frequencies corresponding to vegetation and rock types, high frequencies corresponding to local slope effects). They used the lower and higher frequencies to modulate hue and intensity, respectively. (Saturation was interactively chosen to produce visually pleasing imagery).

Another problem area for computer processing is to determine areas of shadows in radar imagery. This can only be done through map elevation data and precise rectification and will be commented upon later.

Rectification and Map-Matching

Image rectification is needed for several reasons: compensation for sensor effects, merging of radar imagery with maps or other imagery data (e g Landsat multispectral imagery), navigation purposes to name a few. Due to the special features of a radar sensor, the accurate map rectification of SAR imagery requires more complex procedures than needed for passive sensors (photography and MSS sensors). Two methods are currently used for radar rectification

* polynomial approximation algorithms using either a high order global polynomial or a series or regional low order polynomials

* use of light path model.

The drawback of the first method is the need for a large number of ground control points. The second method only requires a sparse grid of control points (see Guindon et al. (16)). They showed that for a mountainous area in Canada, the average topographic error for an airborne SAR (at 6400 m altitude) was as much as 75 meters (compared to the SAR resolution of 3 meters): Topography therefore constitutes a major problem in accurate map rectification.

Another problem which also must be solved with the use of elevation data is the identification of shadow regions. For the same mountainous area Guindon et al. estimated that for Seasat coverage approximately 28 % would be lost due to shadowing (with the 20 degree Seasat SAR look angle). Identification of pixels in shadow is important e g for later machine analysis (segmentation), in particular for acquiring training data.

Since one (of many) important application of radar imagery is navigation, scene matching has been much studied. Two main problems may be identified in this area: what image data (or features) to be used, and calculation of measures describing the degree of mismatch. Normally, shape features (lines, edges) are used since the contrast of mismatch measure is superior to using (unprocessed) image data. Also, such features are often stable over time and easily predictable. The vast literature in image processing gives many methods which can be used for feature processing from simple high-pass filtering, edge and line-detection algorithms using local detectors followed by various linking/search schemes (see Wong (17)).

Feature Processing

We have earlier mentioned processing of radar imagery for extraction features like shape features to be used e g for further machine processing. An important type of feature in radar imagery is texture. Texture is often used in visual interpretation of aerial photography (and other visual and nonvisual imagery) and since long one has attempted to extract such features by machine processing. Various methods are being used for features processing, the main being variance analysis, Fourier analysis and computation of grey level co-occurrance matrices (GLCM). An example of the application of the first two methods is discussed by Blom and Daily (15). The last method has been discussed by Shanmugan et al. (18). Most of the work using GLCM has been based on the use of second order matrices. These are defined as counts of the number of times two neighboring resolution cells separated by a given amount occur in the image (or in the window used for calculating the GLCM). Application of texture features will be commented upon in the next chapter.

Speckle Processing

It was indicated before (chapter 3) that radar imagery (like laser imagery) shows large intensity variations (speckle) due to the coherent nature and coherent processing of the returned signal. The speckle noise is multiplicative with respect to the signal term, in contrast to receiver thermal noise and radiometric quantization noise which are additive. The speckle noise is therefore proportional to the average grey level (or intensity) of local areas. The effect of speckle noise with respect to image quality and (visual) image interpretation was discussed in chapter 4. In this section we shall address the problem of processing radar (SAR) data for speckle reduction. Two principally different methods have been used for this purpose: averaging of several frames and spatial smoothing technique (post-image-formation-processing), respectively.

The multiframe technique for speckle reduction in SAR imagery uses multiple images generated by partitioning the available signal bandwidth in range (multiple looks) and/or in azimuth (subapertures). Each individual frame is processed independently and multiple frames are added (averaged) pixelwise to generate a speckle-reduced end product.

Normally, the individual multiframes are assumed to be statistically independent. The intensity of single pixel is exponentially distributed and for incoherent addition of several frames of the same area, the speckle variations are chi-square distributed. Bennett and McConnell (2) have investigared the case when the individual frames are partly correlated. Such correlation may be caused by overlapping subaperture filters. The effect of overlap is that the effective number of frames decreases. Hence, the ratio of the signal mean to the standard deviation does not any longer increase linearly with the square root of the number of multiframes (as is the cases with independent frames). For 50 % overlap, for example, 4 processed subapertures give the signal-to-noise enhancement of 1.8 (= $\sqrt{3.3}$) instead of 2 (= $\sqrt{4}$). For reasonable small overlap, however, the decrease of integration efficiency is comparatively small.

The technique of smoothing after image formation (spatial filtering) may be implemented by several different methods like linear filtering (frequency domain filtering), Bayesian estimation, median filtering, and adaptive filtering like the use of local statistics or the use of sigma-probability. As with multiframe averaging, the main problem is that of resolution detorioration which has the (unwanted) effect of deteriorating small signal details. The methods of Kalman filtering and Bayesian estimation are implemented using recursive filters. These methods are based upon statistical models for signal and noise and their efficiency depends upon the accuracy of these models. This is a limitation since accurate models for the signal (scene) characteristics are very difficult to build.

Instead of using a (general) signal model, the processing scheme may be adapted to the signal by introducing local signal knowledge (local signal statistics or features) extracted from the imagery data. One such method has been used by Frost et al. (19). They developed an adaptive Wiener-type filter being optimum in the mean squared sense. The filter paramters are estimated from local image data in a small window. The result from this filtering process is that for large image variations the filter impulse response is wide and hence its integration effect is large. If the window is located on a boundary (large signal variation associated with an edge) the corresponding filtering effect is small. The adaptive algorithm thus has a tendancy to preserve edges. A similar method of edge preserving filter has also been described by Lee (20). These methods are improvements over the simple linear frequency filter method and is also an extension of the Kalman-filtering method. Another method of edge preservation is the use of the nonlinear median filter. For a discussion of the properties of this filter, see Justusson (21). This filter has also been applied to speckle filtering (Lee (22)).

A further improvement of the adaptive filter is the nonlinear sigma-filter. By this method only pixel intensity values falling within a certain region (2-sigma) are used in the averaging process. Significantly different pixel values are thus excluded from the averaging process. The result is that pixel values of spot noise are excluded. Furthermore, a majority decision is made within the filter window to replace noise spots. Details are given in a paper by Lee (22).

6. RADAR IMAGE ANALYSIS

General

The purpose of imaging systems is to gather information of a scene (target) and in most cases information means scene description in some form. Imagery data can only be (visually) interpreted if the resolution (measured relative to the actual target dimension) is sufficient (see Jasani (23) for the resolution requirements for various visual interpretation tasks for different types of targets using optical images). The history of radar imagery is partly a history of finding means of achieving good resolution by unconventional means since the large wavelength (compared to optics) in most cases prohibits the use of conventional (physical aperture) methods.

Recently, other methods than conventional imaging for information acquisition have been discussed and developed. In inverse scattering one is faced with the problem of determining the size, shape and electromagnetic properties of an (isolated) target knowing the incident filed and observing the scattered field (Boerner (24)). Special problem areas are cases with sparse data, aperture-limitation and solution-stability. Conceptually, image interpretation can also be considered as a form of "inverse problem" and the scene has to be interpreted (understood and/or described) with ambiguous (and noisy) data being collected. This can only be done with knowledge of different kinds (sensor knowledge, illumination and scattering models, world models, etc).

Other systems being implemented are based on other principle tailoring the system parameters to the problem (target types being of interest) in a manner which may be described as multidimensional correlation (Gjessing (25)).

We shall, however, concentrate our discussion to the field of image analysis and we shall discuss two methods which come into use for machine analysis.

Pattern Recognition Approach

Visual image interpretation makes use of several image features like size, shape, shadow, tone/color, texture, pattern and context. In the pattern recognition paradigm of image (machine) analysis, one makes an attempt to analyze the image by using a few important features in a classification algorithm. The most wellknown application of this paradigm has been in multispectral image classification (segmentation). In radar imagery one has tried to utilize the same concept either for radar imagery alone or integrated with other data (Landsat and/or map data). Blom and Daily (15) used Seasat SAR data coregistered with Landsat MSS data for rock-type discrimination and found that SAR data increased accuracy by 7%. By incorporating SAR texture a significant increase of 21% was noticed. Shanmugan et al. (18) used texture data (GLCM) alone and showed that such data are efficient for classifying simple geological formations. Guindon and Goodenough (16) used map type (topographic) data in addition to (airborne) SAR data for forestry classification and found that for integration of topography (elevation, slope and aspect) and SAR data (two frequencies and two polarizations) classification accuracy was 72%. An example of using a similar technique for target identification is given by Ezquerra and Harkness (26).

The pattern recognition approach can be used for restricted cases (recognition of objects known a priori through training). As a general method for image interpretation it is seriously limited. One reason is the difficulty of defining and extracting features which are invariant with respect to illuminatiion and aspect angle. The image interpretation often becomes an unreasonable matching problem since little respect is paid to the various physical restrictions.

Image Understanding Approach

Due to the difficulties of using the pattern recognition approach as a more general tool in image analysis, the trend in computer image analysis (computer vision) is to design a machine system using some of the knowledge about (human) vision. Vision is considered as an active (reasoning) process using sensory data in terms of scene, illumination and sensor models to invert the image-formation process for the determination of configurations of objects and/or surfaces (Barrow, Tenenbaum (27)). An attempt to analyze and make explicit the reasoning process for a restricted interpretation task (interpretation of drainage pattern from Landsat imagery) has recently been made by Wang et al. (28). A feasibility study of applying the above methodology for target (ship) classification from ISAR data has been made by Drazowitch and Lanzinger (29).

ACKNOWLEDGEMENTS

The author greatfully acknowledges discussions with and suggestions from H Ottersten concerning the content and details of this paper. The help with typing and proofreading from G Börjeson , S Lundahl and A Lantz is also appreciated.

REFERENCES

(1) Orhaug, T. 1978: Pattern Recognition with special Emphasis on Image Processing, in Lund. T. (Editor): Surveillance of Environmental Pollution and Resources by Electromagnetic Waves, Proc. of NATO Adv. Study Institute, Spåtind, Norway, April 1978, Reidel Publishing Company, Dordrecht, Holland.

(2) Bennett, I.R., McConnell, P.R. 1981: Considerations in the design of optimal multilook processors for image quality, in SAR Image Quality, ESA SP-172, ESA 1981.

(3) Mitchell, R.M., Marder, S. 1981: Synthetic Aperture Radar (SAR) image quality considerations, SPIE vol. 310, Image Quality, pp 58-68.

(4) Granger, E., Cupery, K. 1972: An optical merit function (SQF) which correlates with subjective image judgements, Photographic Science and Engineering, Vol., 16, No. 23, pp 221-230.

(5) Charman, W.N., Olin, A. 1965: Image quality criteria for aerial camera systems, Photographic Science and Engineering, Vol. 9, No. 6, pp 385-397.

(6) Snyder, H.L. 1973: Image quality and observer performance, in Perception of displayed information, Editor: Biberman, L.M. Plenum Press.

(7) Marmolin, H., Nyberg, S. 1975: Multidimensional scaling of image quality, FOA report C 30039-H9 also in Proc. Second International Joint Conference on Pattern Recognition, August 1974, Copenhagen, Denmark, IEEE Catalog number 74CH0885-4C, pp 366-367.

(8) Scott, F., Hollanda, P.A., Harabedian, A. 1970: The Informative Value of Sampled Images as Function of the Number of Scans per Scene Object, Photographic Science and Engineering, Vol. 14, No. 1, pp 21-27.

(9) Linde, L., Marmolin, H., Nyberg, S. 1981: Visual effects of sampling in digital picture processing, IEEE Trans., Systems, Man and Cybernetics, Vol. SMC-11, No. 3, pp 201-207.

(10) Arguello, R.J., Kessler, H.B., Sellner, H.R. 1981: Effect of sampling, optical transfer function shape and anisotropy on subjective image quality, SPIE Vol. 310, Image Quality, pp 24-33.

(11) Mitchell, R.L. 1974: Models of Extended Targets and their Coherent Radar Images, Proc. IEEE, Vol. 62, No 6. pp 754-758.

(12) Ford, J.P. 1982: Resolution Versus Speckles Relative to Geological Interpretability of Spaceborne Radar Images: A Survey of User Preference, IEEE Trans. on Geoscience and Remote Sensing, Vol. GE-20, No. 4, pp 434-444.

(13) Moore, R.K. 1979: Tradeoff Between Picture Element Dimension and Noncoherent Averaging in Side-Looking Airborne Radar, IEEE Trans. Aero-Space and Electronic Systems, vol. AES-15, No. 5, pp 697-708.

(14) Stiles, J.A., Frost, V.S., Holtzman, J.C., Shanugan, K.S. 1981: The Recognition of Extended Targets: SAR Images for Level and Hilly Terrain, IEEE Trans. Geoscience and Remote Sensing, vol. GE-20, No. 2, pp 205-211.

(15) Blom, R.G., Daily, M. 1982: Radar Image Processing for Rock-Type Discrimination, IEEE Trans. on Geoscience and Remote Sensing, vol. GE-20, No. 3, pp 343-351.

(16) Guindon, B., Goodenough, D.G., Teillet, P.M. 1982: The Role of Digital Terrain Models in the Remote Sensing of Forests, Canadian Journal of Remote Sensing, vol. 8, No. 1, pp 5-16.

(17) Wong, R.Y. 1979: Radar to Optical Scene Matching, SPIE vol. 186, Digital Processing of Aerial Ima ges, pp 108-114.

(18) Shanugan, K.S., Narayanan, V., Frost, V.C., Stiles, J.A., Holtzman, J.C. 1981: Textural Features for Radar Image Analysis, IEEE Trans. on Geoscience and Remote Sensing, vol. GE-19, pp 153-156.

(19) Frost, V.S., Stiles, J.A., Shanmugan, K.S., Holtzman, J.C., Smith, S.A. 1981: An adaptive Filter for Smoothing Noisy Radar Images, Proc. IEEE, vol. 69, No. 1, pp 133-135.

(20) Lee, J.S. 1981: Speckle Analysis and Smoothing of Synthetic Aperture Radar Images, Computer Graphics and Image Processing, vol. 17, pp 24-32.

(21) Justusson, B. 1981: Median Filtering: Statistical Properties, Chapter 5 in Huang, T.S. (Editor): Two Dimensional Digital Signal Processing II, Topics in Applied Physics, vol. 43, Springer Verlag, Berlin.

(22) Lee, J-S. 1983: A Simple Speckle Smoothing Algorithm for Synthetic Aperture Radar Images, IEEE Trans. on Systems, Man and Cybernetics, vol. SMC-13, No. 1, pp 85-89.

(23) Jasani, B. 1982: Outer Space - A New Dimension of the Arms Race, chapter 4, p 47.

(24) Boerner, W-M., Ho, C-M., Foo, B-Y. 1981: Use of Radon's Projection Theory in Electromagnetic Inverse Scattering, IEEE Trans. on Antennas and Propagation, vol. AP-20, No. 2, pp 336-341.

(25) Gjessing, D.T., Hjelmstad, J., Lund, T. 1982: A Multi-frequency Adaptive Radar for Detection and Identification of Objects: Results on Preliminary Experiments on Aircraft against Sea-Clutter Background, IEEE Trans. on Antennas and Propagation, vol. AP-30, No. 3, pp 351-365.

(26) Ezquerra, N.F., Harkness, L.L. 1981: Recogntion Applications in Radar Data Processing, IEEE 0360-8913/81/0000-0024, pp 24-28.

(27) Barrow, H.G., Tenenbaum, J.T. 1981: Computational Vision, Proc. IEEE, vol. 69, No. 5, pp 572-595.

(28) Wang, S., Elliott, D.B., Campbell, J.B., Erich, R.N., Haralick, R.M. 1983: Spatial Reasoning in Remotely Sensed Data, IEEE Trans. on Geoscience and Remote Sensing, vol. GE-21, No. 1, pp 94-101.

(29) Drazowich, R.J., Lanzinger, T.X. 1981: Radar Target Classification, IEEE, CH 1595-8/81/0000/0496, pp 496-501.

IV.4 (SI.1)

INVERSE METHODS APPLIED TO MICROWAVE TARGET IMAGING

L.A. Cram

THORN EMI Electronics Ltd., Wells, Somerset, U.K.

ABSTRACT

Descriptions are given of many different methods currrently in use at a radar modelling facility where target images derived from microwave data are being investigated. These studies exploit measurements made at wavelengths between 0.3 m and 0.3 mm on both full-scale targets and suitably scaled models. Results of doppler processing are used to illustrate the equivalence of modelling and full-scale data. Both DFT methods and maximum entropy processing are applied to coherent-radar data measured in an arc-scan around the target. Stacked spectra, in particular those produced by the latter method, are very revealing. Examples are given of the processing of polarization scattering matrix data. Images obtained at millimetre wavelengths both by radiometry and by holographic methods are illustrated. The relative merits of the various imaging processes are discussed.

1.0 INTRODUCTION

The process of using a knowledge of the EM waves which have been reflected by a target in order to provide an image of that target may be described as inverse processing. It is in this wider sense that the term is used here. Some of the methods to be described incorporate the derivation of a mathematical filter or function which would remove all but noise-like information from the received signal. Then the reciprocal of the known characteristics of this filter provides

W.-M. Boerner et al. (eds.), Inverse Methods in Electromagnetic Imaging, Part 2, 841–870.

Fig. 1 General View of the Radar Modelling Facility

Fig. 2 Scale Model of a Tank

Fig. 3 Scale Model of an Aircraft

Fig. 4 Scale Model of a Ship

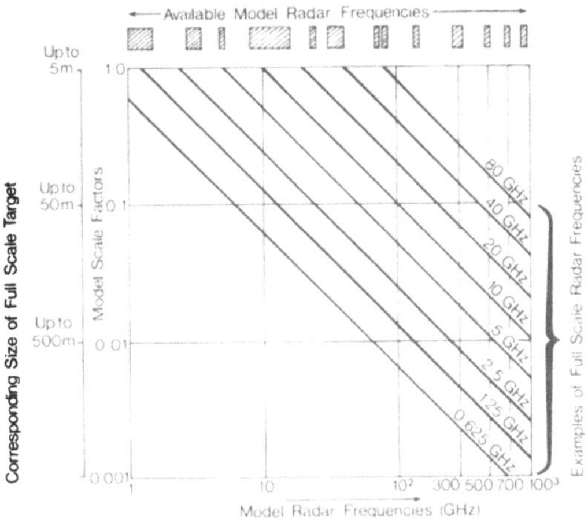

Fig. 5 Chart of Radar Modelling Facilities

Fig. 6 Model Aircraft Suspended from a Gantry

an image of the target which is far superior to that obtained by conventional Fourier transform methods. This approach is described as a maximum entropy method and comparative examples will be given of its effectiveness. Other imaging methods incorporate the Fourier transformation type of process, but not by digital computation. They either use the reception hardware (i.e. the aerial system) or, alternatively, they use analogue processing methods as in the reconstruction of holographic images with a laser. The various methods have been investigated using frequencies from about 1 GHz to 1 THz and the results will be described and discussed. The measurement and processing of ancillary target information, such as doppler side-band data and polarization transformation characteristics of a radar reflector, will also be discussed with examples.

Accurate and consistent measurement facilities have been developed for determining radar characteristics of targets. These use both field and laboratory measurements. Considerable use has been made of indoor measurements done under laboratory conditions by scale modelling techniques, with radars at millimetre wavelengths (and less), in order to investigate the signals scattered by scale model targets. A brief review of these measurement facilities will be given before presenting the results of applying a number of processing methods in various target imaging investigations.

2.0 MEASUREMENT FACILITIES

In testing or assessing an inverse processing method intended to produce an image of a microwave radar target, it is necessary to be confident as to how any characteristics in the image are related to the details of the target and of its aspect. Consequently, it is often highly desirable to be able to repeat a given series of measurements precisely or with only a single and known change to the target shape or aspect. However, a normal target for a microwave radar might be a large and complex object such as a ship, aircraft or land vehicle and, in any full-scale mobile trial, it is impossible to control, or even to know, the precise target aspect with adequate precision.

2.1 Scale Model Radars

By using millimetric (or sub-millimetric) radars rather than centimetric radars, and by using appropriately scaled models of the target vehicles, it becomes possible to analyse results with precision and to repeat them if necessary [1].

The targets are scaled to a size and weight which permits easy handling and accurate control of attitude. This can be achieved with supporting means whose radar reflections are themselves negligible. In order that the results can be a perfect analogue of those pertaining to the equivalent full-scale situation, it is a requirement that the same linear scaling ratio shall be used for wavelength as for target model size.

The measurement facility used by THORN EMI Electronics Ltd. is shown in Fig. 1. It has many radars and these are available, for frequencies from 0.8 GHz to 890 GHz, at about octave intervals, or less. Most of them use pulse range gating to avoid confusion by backscatter from the building in which the measurements are made. None of the radars is coherent from pulse to pulse, but all can use coherent reception so that phase and doppler measurements may be made.

2.2 Target Models

Target models of ships, aircraft, missiles or land vehicles are usually so scaled as to be less than 5 m in extent. Consequently, for small targets such as some missiles or land vehicles, it is practical to use the full-scale vehicles whereas, for aircraft or small boats, a model scaled about 1:3 or 1:10 is suitable. For still larger ships a scaling factor of 1:100 or 1:200 may be needed. Examples are illustrated in Figs. 2 to 4. The scaling is done with attention not only to the accuracy of the main contours, but also to the smaller detail of order of a few wavelengths and, where possible, the surface roughness is well represented. Also, any movable surfaces (including rotating or vibrating components) are correctly scaled. For example, the engine intakes and exhausts in jet aircraft are accurately scaled as far within as the third set of either the compressor or the rotor blades and these blades can be rotated by battery driven motors.

A summary of the facilities now available is shown in chart form in Fig. 5. This indicates the nominal frequencies at which model radars exist. It permits the choice of a suitable frequency and scale factor for modelling, so that any required full-scale radar/target engagement may be represented.

2.3 Target Support

Some target models may be viewed in a natural position on the ground; this is appropriate for models of land vehicles or of parked aircraft. Flying aircraft and missiles are usually supported, remote from other reflectors in the building, by means of dielectric strings whose lengths can be adjusted to change the attitude of the target. The suspension points are attached to a structure, as in Fig. 6, which can be rotated so as to cause different aspects of the target to be presented to the radar. Ships and some other models are usually supported from beneath by a tapered 'non-backscattering' pedestal which permits control of both attitude and aspect.

2.4 Scope of Measurement Facilities

Seven different model support facilities are available as shown in Fig. 7 and, for each of these, there are several radars from which any one can be chosen to view the targets. In several cases, the radar is stationary and the target is rotated in front of it. However, with other systems, the radar is mounted either on a smooth linear track or on a circular arc track centred at the target, and it is moved along this track during the measurement. Radars can be used for either monostatic or bistatic measurements without loss of the coherency of reception and, for the latter, the variation of the bistatic angle can be continuous.

The relative position of radar and target can be maintained stable enought to permit r.f. phase measurements with a long term accuracy of about 20 deg, even at frequencies of 140, 280 and 890 GHz. The short term accuracy is, of course, very much better than this.

A visual analogue display is provided for the operator to permit immediate knowledge of any potential faults in the system or of possible errors in the experimental set-up. The data are, at the same time, digitized and recorded so that they may subsequently be subjected to computer analysis by the mathematics and software teams.

2.5 Doppler Modulation Measurement

All the radars used with the seven measuring systems are capable of measuring the doppler sideband modulation components which arise, for example, from turbine fans on aircraft, rotor blades on helicopters, or even from scanning radars on ships. To display these doppler components, without folding of the spectrum, it is usual to use a second i.f. at a fairly low frequency of about twice the anticipated doppler frequency.

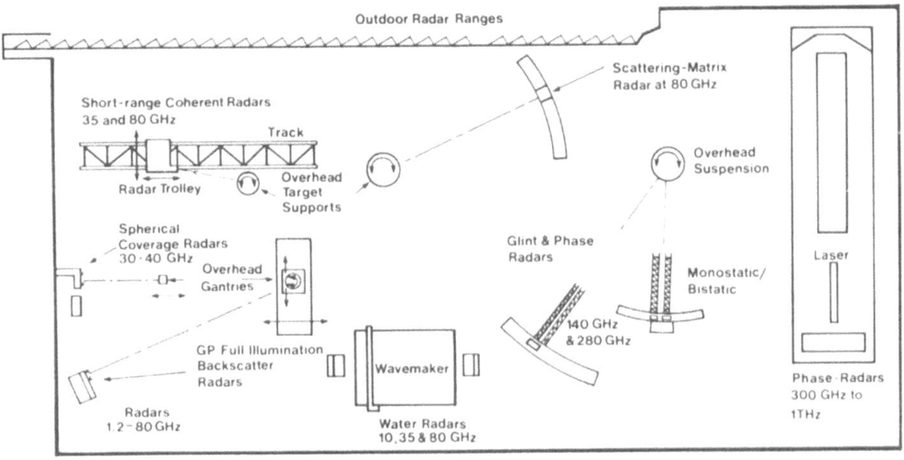

Fig. 7 Plan of Radar Modelling Facilities

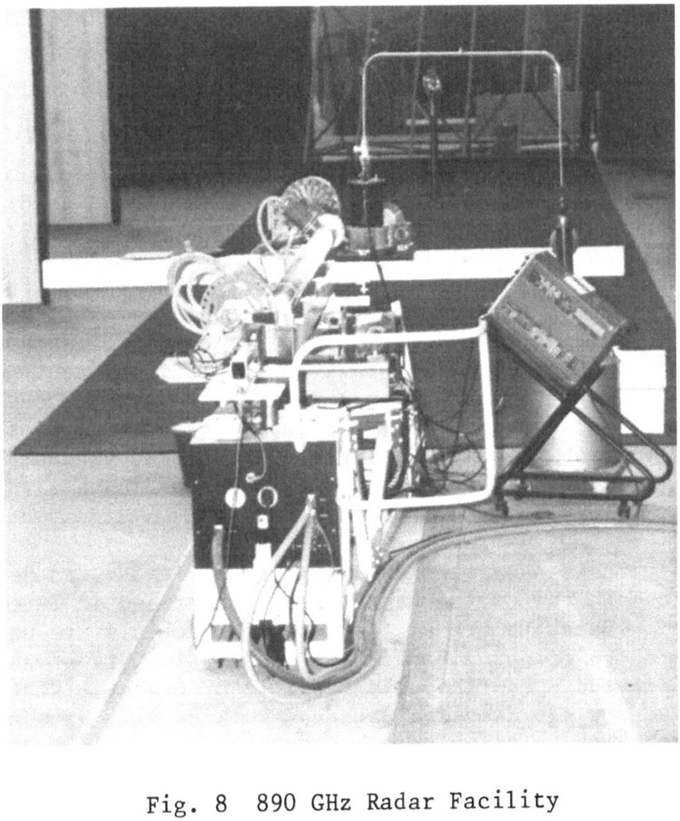

Fig. 8 890 GHz Radar Facility

2.6 Arc-Scan Synthetic Aperture Radars

Five of the measuring systems can be used for gathering synthetic aperture arc-scan data from a target. In three of these, the radar moves on a circular track centred at the target, while, in the others, the radar is stationary and the target is rotated. The three tracks with moving radars use millimetric wavelengths and operate at 80, 140 and 280 GHz respectively. Their targets are usually held stationary by suspension on thin dielectric strings with further strings to tether them. For some measurements, however, it is appropriate to support them on the ground.

To operate an arc-scan synthetic aperture measurement radar system at submillimetric wavelengths, the stability requirements are very high; for example at 890 GHz, a 1 deg phase change occurs for 0.5 μm relative movement. Consequently, for such work, the radar is kept stationary; the target is supported on a tapered pillar and is rotated after its attitude has been correctly adjusted. This radar and its target can be seen in Fig. 8. At longer wavelengths, a target can be rotated with adequate stability even when supported by strings and such a system is adopted at 16 GHz.

All of these radars use coherent reception. This is implemented by retaining and delaying a portion of the power from the transmitter source in order to use it as a reference in the receiver [2]. With the millimetre wave systems, at 80 to 280 GHz, the transmissions are pulsed and the delay of the pulse reference signal also provides range gating so as to eliminate interference by reflections from the walls of the room. The submillimetre wave system uses a CW transmitter and, in this case, interfering backscatter is eliminated by imposing on it a doppler frequency which lies outside the receiver bandwidth. This is achieved by placing a reciprocating screen behind the target. In all these systems, the coherent reference signal is square-wave modulated in phase between 0 and 90 deg. From the resulting time-multiplexed quadrature measurements, the r.f. phase and amplitude of the received signal can be determined.

These arc-scan systems are very versatile and can be adapted to make any of the measurements described in Sections 3, 4 and 5. Separate transmit and receive aerials are used and they can be either adjacent for monostatic data or set apart to provide a continuously variable or preset bistatic angle. The aerials used for arc-scan work usually have beams wide enough both to illuminate and to view the whole target. For some purposes, however, the beamwidths can be reduced.

2.7 Polarization Scattering Matrix Measurement

In principle, using any of the coherent radars, a time multiplex measurement of the full polarization scattering matrix could be made [3]. Three of the radars have now been adapted for this purpose. The linear polarization of the transmitted power is switched between two orthogonal planes and, for each condition, the plane of the receiver aerial polarization is switched in turn to be first parallel and then orthogonal to that of the transmission. The stability of the target relative to the radar is such as to permit the measurement, in each case, of the amplitude and phase of the received signal relative to the transmitted values. From these measurements the full scattering matrix may be determined. Such measurements may be done in an arc-scan around the target at 16 GHz, 80 GHz and 140 GHz.

2.8 Radio Hologram Recorder

Radio hologram recorders using several different methods have been investigated at frequencies from 10 GHz to 140 GHz [4]. In an early system, an analogue of the usual optical method was performed using a stationary and separate source of target illumination and a scanning receiver which performed a raster scan across the hologram window. The reference signal was radiated, to illuminate the hologram window directly, as is usual in optical holography. Other systems have coupled the reference by means of a transmission line, and for convenience, we have sometimes used a reciprocal arrangement in which the transmitter is scanned rather than the receiver. In some cases, the frequency of the reference signal has been side-stepped to provide, on reconstruction, increased angular separation between the target image and the incident optical beam. In a recent investigation, conducted with a near monostatic 140 GHz radar, both transmitter and receiver were scanned across the recording frame in a raster manner. This provided discrimination equivalent to a 280 GHz bistatic system and also overcame some practical problems related to coupling the reference at these wavelengths.

2.9 Spot Scanning

Large gain aerials can be fitted to a radar to provide a focused beam on either transmission or reception (or both) and the focused spot can be scanned across the target by rotating the aerial. Such facilities have been used at 890 GHz [5]. A different technique has been employed at 20 GHz, in which a focused spot radar was arranged to scan the target by moving the radar along a linear track past the target.

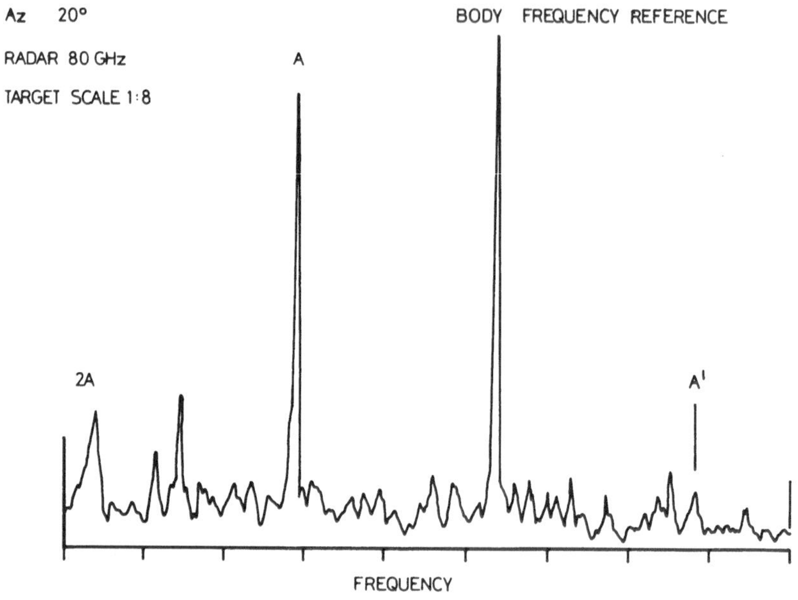

Fig. 9 Spectrum of Radar Signal from Scale Model Aircraft with Engines Rotating

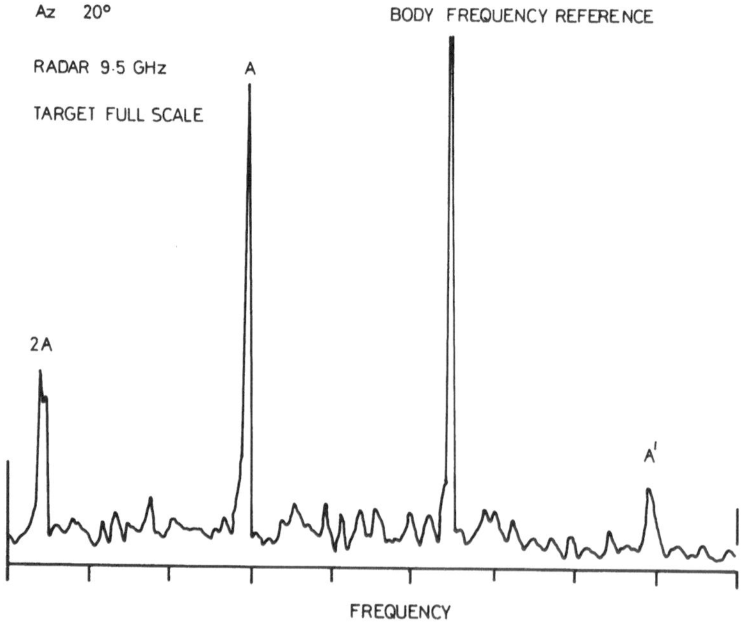

Fig. 10 Spectrum of Radar Signal from Full-Scale Aircraft Engine

2.10 Scanning Radiometer

Full-scale targets have been imaged in various weather conditions at 80 to 82 GHz by a radiometer incorporating Dicke switching. An aerial giving a focused spot was used and facilities were provided to rotate it on two orthogonal axes to permit a raster scan of the target. Linear polarization was employed and images were obtained with two orthogonal polarizations [6].

3.0 ROTATING REFLECTORS ON TARGETS - DOPPLER SPECTRAL ANALYSIS

Many radar targets incorporate components which must rotate in order to be functional - examples include the compressor and the rotor blades of turbine engines, the rotor blades of a helicopter or the antenna of a mechanically scanning radar system. One of the simplest techniques for processing coherent radar data so as to reveal the resultant modulations, is to perform a spectral analysis of the radar reflections from the target. A Fourier transform process applied to the coherently received radar waveform provides the skin doppler frequency, caused by the overall body motion of the target relative to the radar, together with any sidebands around this skin frequency, which have been caused by reflection from rotating elements in the target.

Doppler processing is a long established and familiar radar technique; nevertheless, the detailed and controlled investigation of the use of this processing method, made possible by the scale modelling facility, has been very rewarding. The model radars permit coherent processing whether they use CW or pulsed transmission. As mentioned above, the targets are made with precision and fidelity: the model turbine engines fitted in the jet aircraft models incorporate correctly shaped compressor and turbine blades as far in as the third stage. Furthermore, these can be rotated at a suitable controlled speed by an internal electric motor.

The turbine rotation speed is not required to be according to the scaling factor applied to the linear dimensions. The relevant bandwidths are such that the choice of a different scaling factor for speed is inconsequential. It is often convenient to rotate model turbine rotors very slowly and to have no relative body motion between the model radar and the target centre. In the absence of any corrective measure, this would produce a zero skin doppler frequency and, hence, a folding of the sideband spectrum so that upper and lower sidebands caused by the rotating element overlap and interfere. Such confusion may be overcome by creating a suitable pseudo skin doppler frequency by side-stepping the frequency of the reference r.f. signal.

Measurements have been made with models of turbines, as well as with tethered full-scale engines. The results have been related also to data obtained with flying aircraft targets. Fig. 9 shows an example of the spectral analysis of data from a 1:8 scale model aircraft engine obtained with a model radar operating at 80 GHz. The body frequency and engine sidebands can be seen.

A full-scale engine of the same type was examined at the equivalent aspect with a coherent radar at 9.5 GHz and spectral analysis of the results provided the power distribution shown in Fig. 10. The major features displayed by these doppler spectra from full-scale trials and 1:8 scale engine measurements are identical. In both cases, the sideband, A, to the lower frequency side of the skin doppler is much larger than its equivalent, A', on the other side. Also, the second harmonics at 2A are of lower level in both cases.

Much information can be obtained from such records and it is possible, by means of doppler sideband analysis, to resolve separately the echoes from two aircraft when they are both in the same range-azimuth slot.

Similar investigations of other rotating components can be conducted and the same methods can also be used to investigate vibrating components; if adequate signal-to-noise ratios can be achieved, such methods have great potential.

4.0 ARC-SCAN RADAR DATA

In order to devise and refine the inverse processing methods suitable for arc-scan radars, it has again been found advantageous to make use of data obtained via radar modelling. Work has been done with models of ships, aircraft, missiles and land vehicles and has usuallly been aimed at replicating full-scale operation in the band 8-12 GHz.

For the ship models, a scale factor of from 1:100 to 1:200 is used with a radar operating at 890 GHz. The relative angular movement is achieved with a stationary radar by rotating the target. For other models, a scale factor from 1:3 to 1:16 is used and the relative movement is achieved with the target stationary, but with the radar moving along an arc of a circle centred at the target.

Of course, both causes of relative angular movement may occur with operational radars and either may predominate. For example, an airborne operational radar may move past or round an essentiallly stationary target; alternatively, a ground based radar may see a target turning in its field of view.

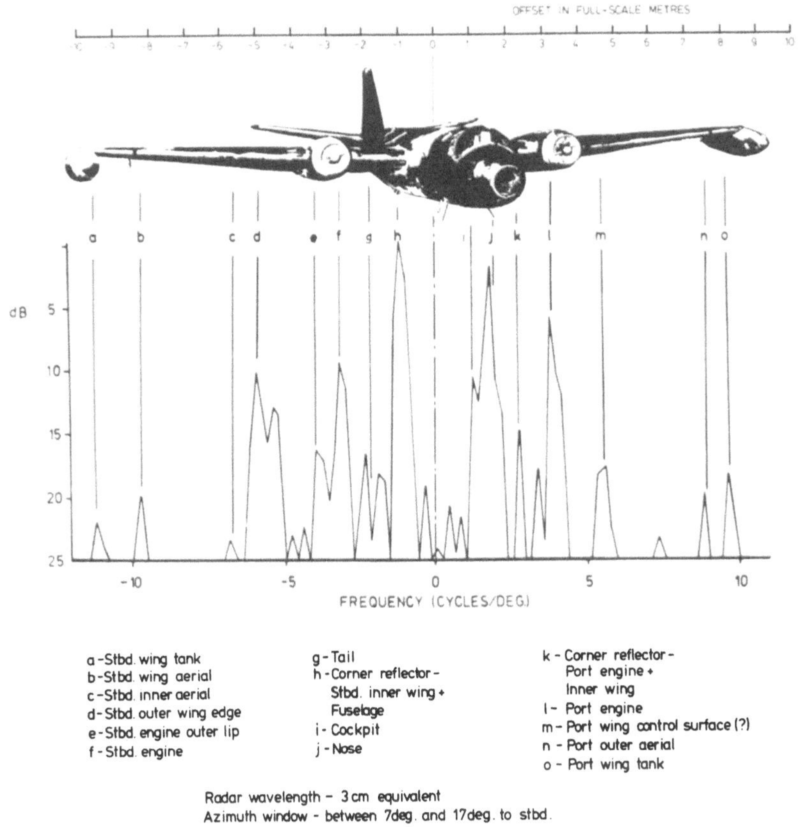

Fig. 11 Use of Angular Spectrum (by DFT) to Identify Reflection Sources

The recorded polar diagram (amplitude and phase) arises from the combination of many signals from different elementary reflectors on the target. The inverse process of decomposition of the composite signal may be achieved by subjecting the polar diagram to local spectral analysis in the angular domain. This provides, for each chosen aspect angle, a display of the significant target reflectors, showing their relative intensities and their lateral displacements from the centre of rotation. The size of the angular window used for the analysis determines the precision with which the elementary reflectors can be located, but the window must not be so wide as to permit either the projected angular position of any individual reflector or its polar diagram to change significantly across the window.

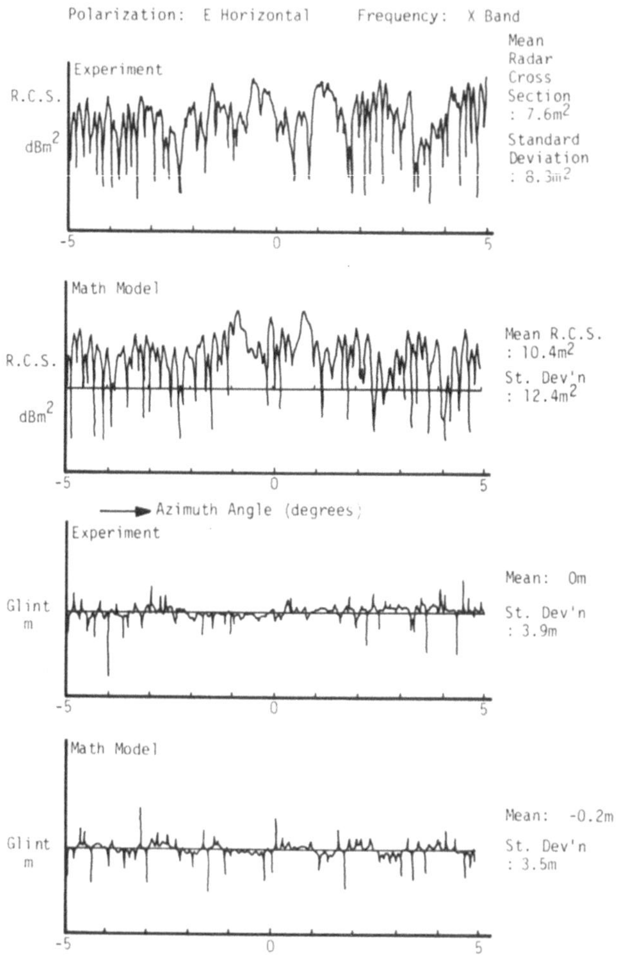

Fig. 12 Mathematical Model Results Compared with Experimental Plots

An analysis window with an aperture of n wavelengths can discriminate subreflectors separated angularly by 1/(2n) radians. Note, however, that, as a consequence, a large subreflector of greater angular subtense than 1/(2n) radians would have the details of its polar diagram smoothed by such a large window. Also, even a small reflector, displaced appreciably in range from the centre of rotation, could have its projected position change by more than 1/(2n) radians as the receiver moves from one edge of the window to the other. The optimization of the aspect angle window for the doppler analysis of the radar return from rotating targets has been discussed by Georg Graf [7].

Initially, the processing was done with digitized data using FFT algorithms. The window size and edge weightings were adjusted appropriately. In one case, amplitude and phase data were recorded from a model of an aircraft target within 12 $\pm$ 5 deg from head-on by using a correctly scaled model of an I band radar. By processing the data with an FFT algorithm, a plot of echoing area vs. deviation from target centre was obtained, as shown in the lower part of Fig. 11. The window size, in this case, was 10 deg and the weighting across it was uniform. Successive occlusion of various parts of the aircraft has confirmed that the several spectral peaks arose from the 15 components which are indicated on the photograph of the model. Similar plots from differently directed azimuthal windows permit the polar diagrams of the subreflectors to be ascertained as well as locating their approximate position in range within the target.

Such data for only the major 20 or 30 subreflectors are adequate to produce a relatively simple yet effective mathematical model of the target. It is possible to check the validity of this model by computing the expected radar signal properties and comparing these with similar data obtained more directly from the original measurements. The ability of the radar to measure both amplitude and phase permits the plotting of overall radar cross-section and the location of the effective phase centre of the reflections. (This location is not constant and its deviation from the target centre is known as the glint.) Fig. 12 shows a comparison of such plots obtained directly from the measured data with those obtained by the double inversion process through the mathematical model. The degree of similarity confirms that, for this target, the number of target elements included in the mathematical model is adequate to provide a good representation of such characteristics of the radar's performance.

4.1 Stacked-Spectra Displays

It has already been mentioned that, as the target rotates, a continuously changing aspect of it is presented to the window used for the analysis. The inverse processed data for adjacent windows show the variations in the polar diagrams of the individual subreflectors. One convenient way to display this is to stack the angular spectra from successive windows one above the other. Fig. 13 shows a set of stacked spectra derived by Discrete Fourier Transform (DFT) methods from data obtained with a model of a small aircraft target. The lowest of the set of graphs was obtained by analysing data from a 4 deg wide window centred 3 deg to starboard. The other graphs were obtained from similar windows successively 4 deg further to port until a window centred at 45 deg was reached.

The target is seen to have a radar width, when viewed head-on, of about 2 m. This increases to almost 4 m when viewed at 45 deg to head-on. Some of the reflection sources appear to persist from one angular window to another but, on the whole, discrimination of the reflection sources is too poor to permit much detailed comment. The 4 deg analysis window should give a discrimination of about 0.25 m, but there is, in fact, some smoothing occurring because of shortcomings in the DFT method.

4.2 Limitations of DFT Methods

The discrete Fourier transform analysis method has several shortcomings. These have been clearly stated, for example, by R. Benjamin [8]. Apart from the assumption that discrete, equally spaced, spectral lines will represent the signal within the window, there are several inconsistent and erroneous assumptions made about the signal outside the observation window. For example, it is incorrect to impose a sudden change at the window edge by assuming that the signal outside the window is zero, or that it consists of an infinite series of waveforms identical to that within. None of the assumptions used in DFT processing for the signals outside the window is correct or justified.

Fortunately, there are in many circumstances much better methods of doing spectral analysis. These were announced 25 years ago by J.P. Burg at an international meeting of geophysicists [9], but only much more recently have they come into wider use. The subject is discussed at length in [10].

4.3 Maximum Entropy Inverse Processing Methods

If the signal in an observation window is autoregressive

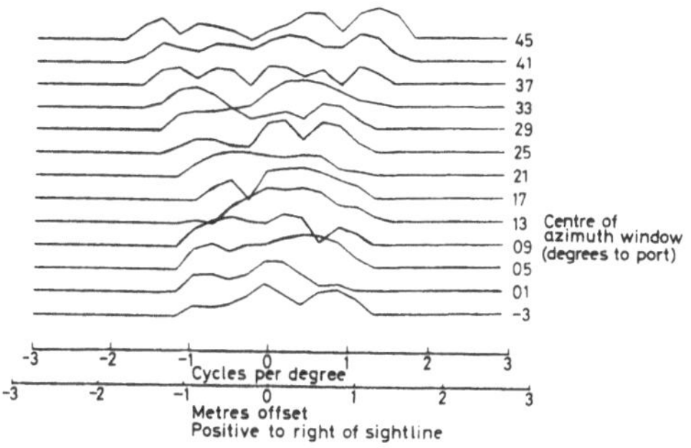

Fig. 13 Angular Spectra Derived from Scale Model by DFT Analysis

(that is, if it permits reasonable extrapolations for a limited range on either side of the observation window and if it is statistically representative of signals in other such windows), then Burg's maximum entropy methods can be used. The arc-scan radar data satisfy these criteria. The principle of the method is that, given a signal intercepted within the observation window, a specific optimum extrapolation function can be found such that the error in its predictions contains no structure, and is noise-like. Hence the method is to adjust the prediction procedure until the error is noise-like i.e. the information theory entropy of the error is maximized. Because the application of the prediction procedure to the signal leads to uniform noise, the spectrum of the signal must be the reciprocal of the frequency response imposed by that prediction procedure. Hence, once the frequency response is known, the signal is readily determined. We have extended the algorithms proposed by Burg to enable them to be used with the complex phase and amplitude data derived from the radar measurements.

The improvement achieved by applying maximum entropy processing to arc-scan data is displayed in Fig. 14. This illustrates the result of applying iterative processing to the same data as were used to produce Fig. 13. The subreflectors of the target are here far better characterized in their angular definition and location and in the intensity of their

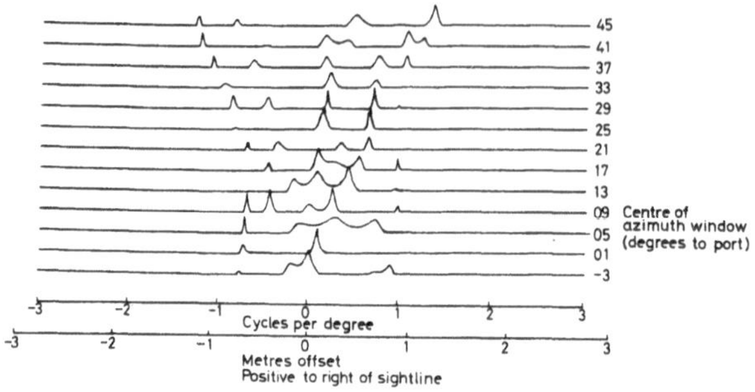

Fig. 14 Angular Spectra Derived from Scale Model by Maximum Entropy Analysis

reflected signal. Fig. 14 reveals several subreflectors within the target, which exhibit broad polar diagrams; they can be identified by the manner in which their lateral displacement from the central point varies as the aspect angle changes. The processing method not only provides an improved image of the target; it also permits, as a consequence, a much improved mathematical representation of that target in terms of a restricted number of elements. This, in turn, can be used to predict more effectively the signals that would be received from it by any radar at that frequency.

Such a mathematical model of the target reflection characteristics may be obtained entirely from scale model radar data, without the use of an actual target. It provides a succinct, yet accurate target description, suitable for storage in a computer and, together with similar models for other targets, may be used in the field to provide an instant target recognition aid.

5.0 POLARIZATION SCATTERING MATRIX

A radar target, in reflecting radar signals, acts as a polarization transformer. Incident radiation of either linear or circular polarization is, in general, transformed into elliptical polarization by reflection. An appreciation of the scatter polarization transformation characteristics of the target is necessary for the proper interpretation of signals received by even the simplest of radars. In the future, radars are likely to incorporate inverse processing methods and to derive increasing information and benefit from the measurement and understanding of the target polarization characteristics.

One can, for example, exploit a knowledge of the difference between the polarization characteristics for man-made targets and natural targets in order to enhance the target-signal-to-clutter ratio [11].

Polarization transfer characteristics of model targets can be determined by measurement of the full scattering matrix. These data can be obtained with a coherent scaled radar which makes rapid successive measurements of four amplitude and relative phase components, as described in Section 2.7.

A very discriminating test for the method is to use the linear polarization data to calculate both the co-polarized and the cross-polarized signals expected when the same target is illuminated with circular polarization. Such calculations can then be compared with further measurements actually made with circular polarization, as is illustrated in Fig. 15. The target, in this case, was a simple corner reflector consisting of two orthogonal planes. The measurements were made at 80 GHz. The target was deliberately observed from a non-symmetrical aspect to ensure that all the terms of the full scattering matrix were non-zero. This lack of symmetry is revealed by the azimuthal plot obtained by measurements made with circular polarization and is evident particularly in the nulls between the lobes of Fig. 15(b). Even these features are well replicated by the calculation made from scattering matrix measurements done with linear polarization as is shown in Fig. 15(a). Perhaps it is even more remarkable that the cross-polarized circular polarization signals (which are 30 - 40 dB smaller than the co-polarized signals) also display good comparison between the directly measured data and the values computed from the scattering matrix data as shown in Figs. 15(c) and 15(d).

A similar comparison of the measured circular polarization signals both cross- and co-polarized with those calculated from the scattering matrix data is shown in Fig. 16 for a more complex target. The target in this case was a metal step ladder: this was chosen because of the numerous orthogonal corner reflectors which it contains.

Data for a more typical radar target are shown in Fig. 17. For this, a 1:8 scale model of an HS 125 aircraft was measured at 16 GHz with a full scattering matrix radar. The measurements covered a 360 deg azimuthal scan in the wing plane. The resulting responses for VV, HH and VH, HV polarization combinations are shown in Fig. 17, where H signifies horizontal and V vertical polarization of electric vector and where the first symbol of a pair refers to transmission while the second refers to reception. The responses computed for co- and crossed-circular polarization are shown in Fig. 18, where R and L refer to right- and left-handed circular polarization. In this figure the RL plot shows single or triple bounce reflections while the RR plot shows double bounce reflections.

6.0 MILLIMETRE WAVE HOLOGRAPHY

A coherent radar system in which either the transmitter or the receiver is moved in a raster scan across a window region in front of a target can be used to record data which are equivalent to those recorded in an optical hologram. If these data are reproduced at a reduced scale as blackening on a photographic transparency, the information can be retrieved by using monochromatic light to provide a holographic optical reconstruction of the object as 'seen' by reflected millimetric radiation.

Data of this type have been recorded with a planar raster scan using frequencies of 10, 35 and 70 GHz as was described in [4]. In a later development of these methods at 140 GHz, both the transmitter and the receiver were jointly scanned across the window and this provided, as discussed in the same paper, a discrimination appropriate to a radar of twice the frequency, i.e. one operating at a wavelength of 1 mm.

The full impression of the value of this type of processing is only appreciated when the replay equipment is under manual control and can be focused at successive planes through the target object. Photographic reproductions of the replay image do not do justice to the method but, nevertheless, an illustration is given here in Fig. 19 of a reconstruction from the microwave hologram of a landrover vehicle, together with a photograph of the vehicle taken from the same location. This work was done at 140 GHz with scanning of both the transmitter and the receiver. It displays several interesting features including reflection of the vehicle in the ground, transparency of the canvas cover to this wavelength to display metal supports for the canopy, as well as an internally stored spare wheel, reflections from wheel arches and hubs and also from cavities around the edge of the door.

7.0 ACTIVE SPOT SCANNING TECHNIQUES

Another method of imaging a target, which has been investigated by scale modelling methods, is to focus a radar spot on to the target model and to raster scan this across the model. The scanning by the radar spot has been done in two

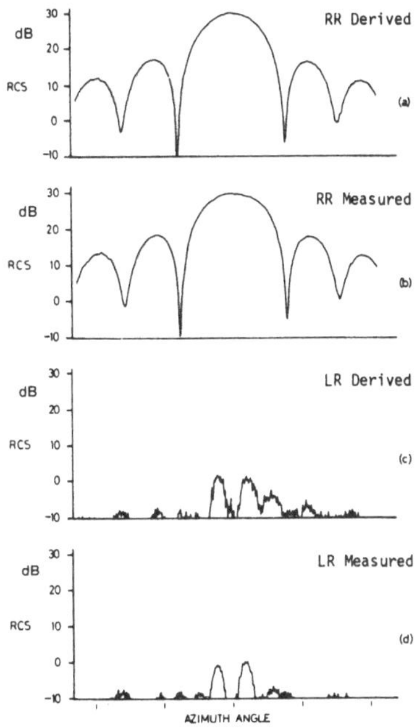

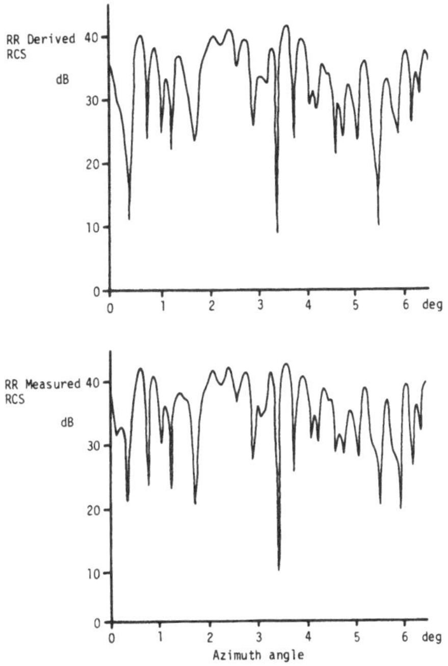

Fig. 15 *(LEFT)* Comparison of Derived and Measured Radar Cross Section of a Dihedral Reflector for Circularly Polarized Radiation

(a) Derived from Scattering Matrix Radar Data: Tx, Right Hand Circular (RHC); Rx, RHC

(b) Measured with Circularly Polarized Radar: Tx, RHC; Rx, RHC

(c) Derived: Tx, Left Hand Circular (LHC); Rx, RHC

(d) Measured: Tx, LHC; Rx, RHC

Fig. 16 *(RIGHT)* Comparison of Derived and Measured RCS of a Step Ladder for Circularly Polarized Radiation ways. The radar may be driven past the target on a straight track with the radar sightline orthogonal to the line of travel. Alternatively, the radar and target may both remain stationary while the radar aerial scans angularly in two orthogonal axes to produce the raster scan.

The latter modification was made [3] to a coherent 890 GHz radar to produce a focused spot of 15 cm diameter at the target. This system has been used with a 1:5 scale model of a tank and with 1:100 scale models of ships. Examples are given in Fig. 20 of the types of image which have been reconstructed from the received signal envelope, together with photographs of the corresponding model targets.

Any inverse processing involved in such a spot scanning method is incorporated in the aerial itself and is not in the same class as methods previously discussed where the processing is applied to the output data. Nevertheless, such a "hardware processing" system is in competition with the computer processing systems and should be considered in the same context. One limitation of spot scanning is the pre-focusing required. In the computer processing systems, such as may be used with arc-scan or holographic data, the focusing may be done at one or at several chosen ranges during the subsequent processing of the data. One important application is to compute results at a larger range than that scaled. On the other hand, if polarization matrix measurements are needed for the individual subreflector elements of a target, it may prove easier to do this with a spot scannning system.

8.0 PASSIVE SPOT SCANNING - RADIOMETRY

Electromagnetic imaging of targets can be done passively without any intentional illumination. As a result of its temperature, any object emits electromagnetic radiation over a band of wavelengths, peaking at a frequency dependent on the temperature. Even at normal temperatures, the radiation band includes the centimetric and millimetric bands and the energy emitted is a function of temperature and emissivity of the body. The body will also reflect some of the other thermal radiation which strikes it. To detect these small incoherent emanations needs a receiver with a large time-bandwidth product. Subsequent processing cannot be used to perform the focusing because the signals are fundamentally incoherent. Hence, as in Section 7, the body must be scanned by a pre-focused spot.

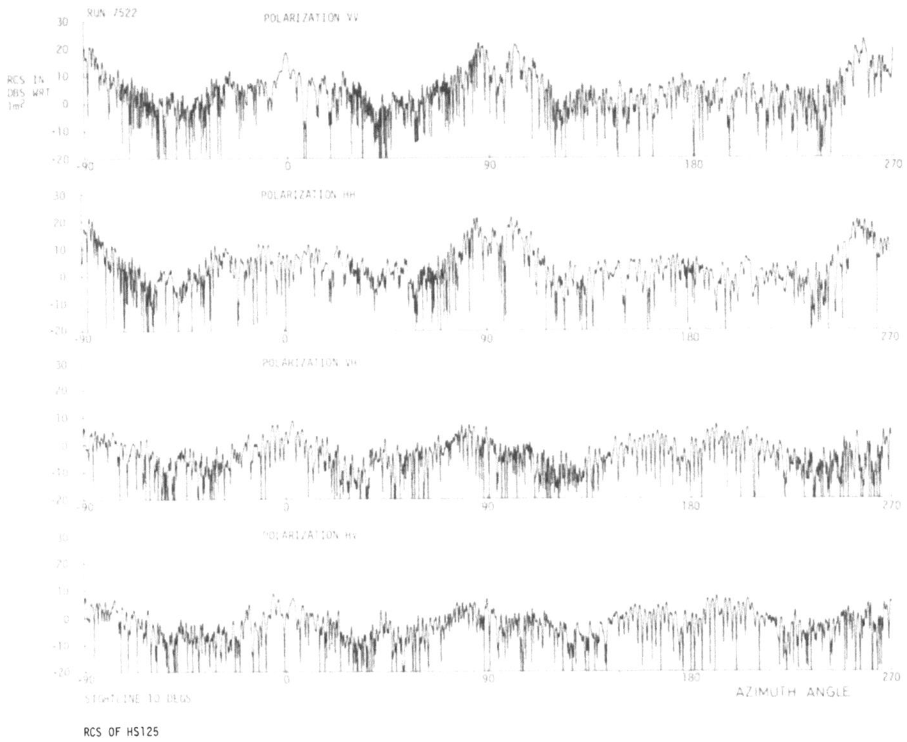

Fig. 17 Signal Amplitude Data from Full Scattering Matrix Measurements on a Model HS 125 Aircraft

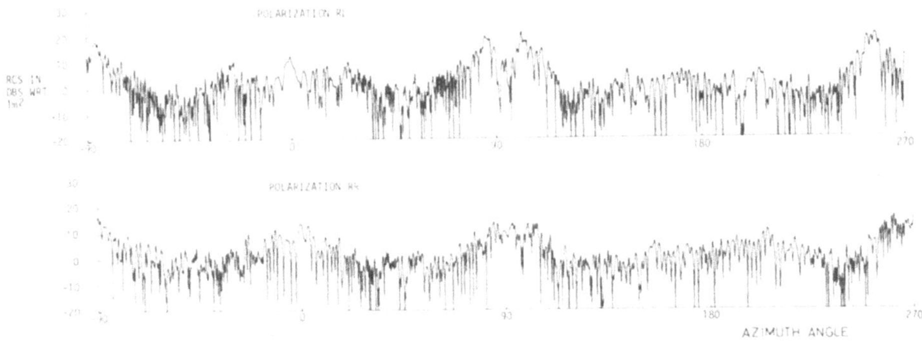

Fig. 18 RCS for Co- and Crossed- Circular Polarization Radars as Derived from Full Scattering Matrix Data on a Model HS 125 Aircraft

Experimental work has been done in the band 80 to 82 GHz to investigate the effects of sky temperature (which determines the illumination on the target) and of polarization which can have a critical effect on reflected power. This work was described in [6]. Fig. 21 shows images of a metal landrover obtained with two sky conditions. The field of view is only 20 x 10 spot sizes but, nevertheless, a recognizable image is obtained.

9.0 INTERRELATION OF COHERENT PROCESSING SYSTEMS

Most of the methods discussed have the common factor that they use coherent radar data. The holographic and arc-scan methods, in particular, are both concerned with analysing the phase and amplitude information to obtain target images. It is revealing to discuss the relative features of the methods used in these processes and also the influence played by polarization.

The holographic recording was made over a flat two-dimensional area while the arc-scan data involved only one-dimensional scanning; the scan being along a line at constant radius from a point within the target. The holographic recording, like that used in optical holography. records only "$a \cos \phi$" (that is, the real part of the complex amplitude $a \exp \{j\phi\}$). Consequently, the inversion process to regenerate the true image produces also a conjugate image. The two images may be confused if they fall in the same angular region. The data recorded in the arc-scan system included also the value of the imaginary part, "$a \sin \phi$", and consequently the inversion of the combined data does not generate any potentially confusing conjugate image.

The optical method of replaying and displaying the hologram is, of course, not an optimum method of using the data recorded. The analogue Fourier transform method suffers from the disadvantages discussed in Section 4 in that there is a presumed discontinuity at the window edge with, in this case, zero field assumed outside it. If the data be digitised for processing by computer, the entire dynamic range of the image may be readily utilised. However, the conjugate image problem still remains.

To produce a good two-dimensional image, it would be better to extend the arc-scan method to two dimensions either on the surface of a sphere or of a cylinder, Such an extension to a spherical cap has been discussed by R. Voles [12]. This method uses rotation of the target about two orthogonal axes

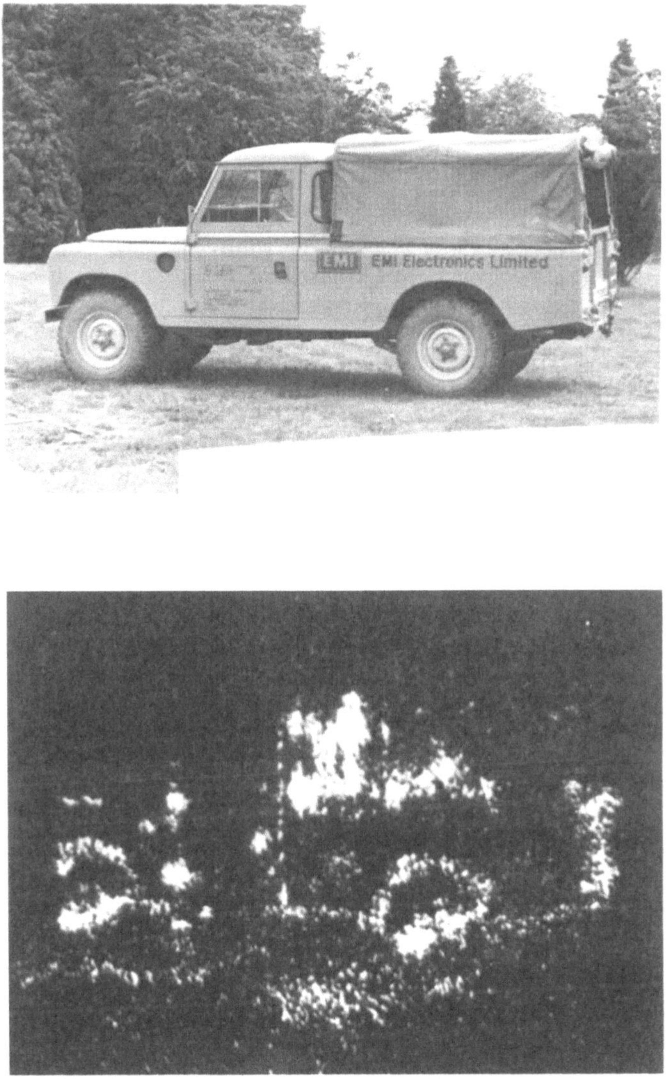

Fig. 19 Corresponding Optical and 140 GHz Holographic Images of a Landrover

to provide the necessary scanning of aspect angle by the radar. The limitation to the size of the window, whether it be planar, cylindrical, or spherical cap shaped, will arise from limitations on stability of the oscillator frequency and the mechnical stability of the radar and target mountings throughout the scanning period.

The target, of course, has a third linear dimension. Information about range to the individual target subreflectors may be obtained from both arc-scan and holographic data by observing the changes in the subtended directions to the subreflectors as the target aspect changes. Range information is also available from the focusing properties of either type of data.

The image of the target, whether it be obtained by holography or by arc-scan methods, will be a function of the receiver and transmitter polarizations which are used. If the data are gathered in an arc-scan by a radar measuring the full scattering matrix, then arc-scan data can be computed for any combination of polarizations as was shown in Fig. 18. From such data, it is possible by maximum entropy inverse processing to derive one-dimensional target images from any chosen aspect. Fig. 22 shows two stacks of such angular spectra for an HS 125 target model at a wavelength equivalent to a full-scale aircraft viewed at about 2 GHz. In this example, a 110 deg azimuthal scan around the target is shown. Both stacks are for circular polarization and Fig. 22(a) shows RL, (i.e. the sphere accept, or odd number of reflections, arrangement), while Fig. 22(b) shows RR (the double bounce, sphere signal reject arrangement). It is easy to identify a specular reflection sliding towards the nose region together with dihedral scattering located at the nose.

10.0 CONCLUSION

Controlled experiments to assess and develop inverse processing algorithms have been described. Many of these experiments employed scale modelling techniques.

Examples have been given of several methods of imaging radar targets. Two of these incorporate spot scanning techniques (active and passive) and the data require no subsequent processing to produce images. Two others, which record one- and two-dimensional phase coherent information respectively, require inverse processing methods to produce target images. In the two-dimensional millimetre wave holography, the subsequent processing used analogue optical methods. In the one-dimensional or arc-scan methods, the data are digitized and a comparison has been given of DFT methods and maximum entropy methods for producing the target image in one-dimension.

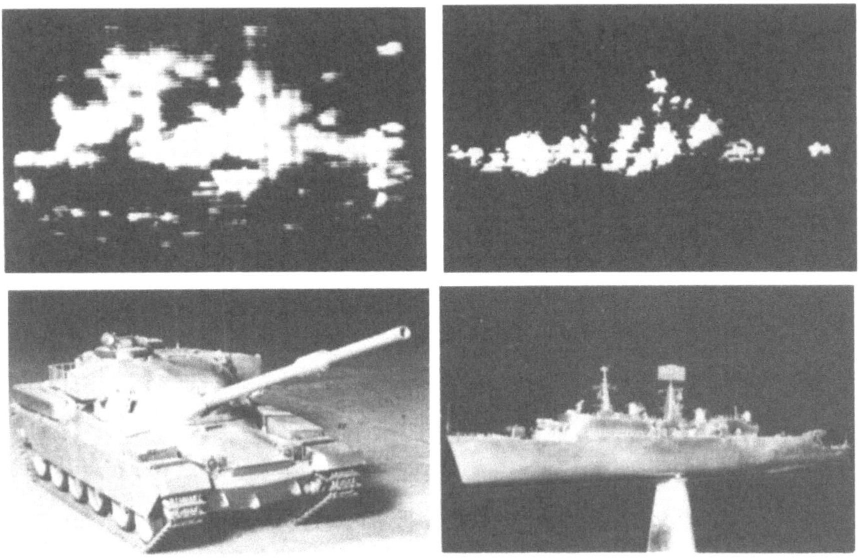

Fig. 20 Images Taken Using Spot Scanner

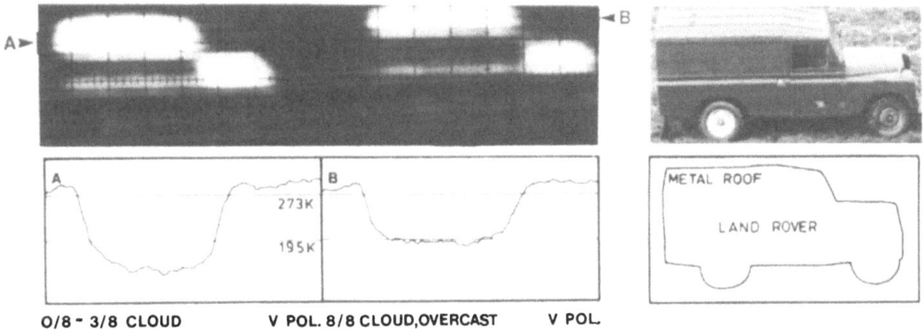

Fig. 21 Effect of Sky Temperature on Radiometric Image

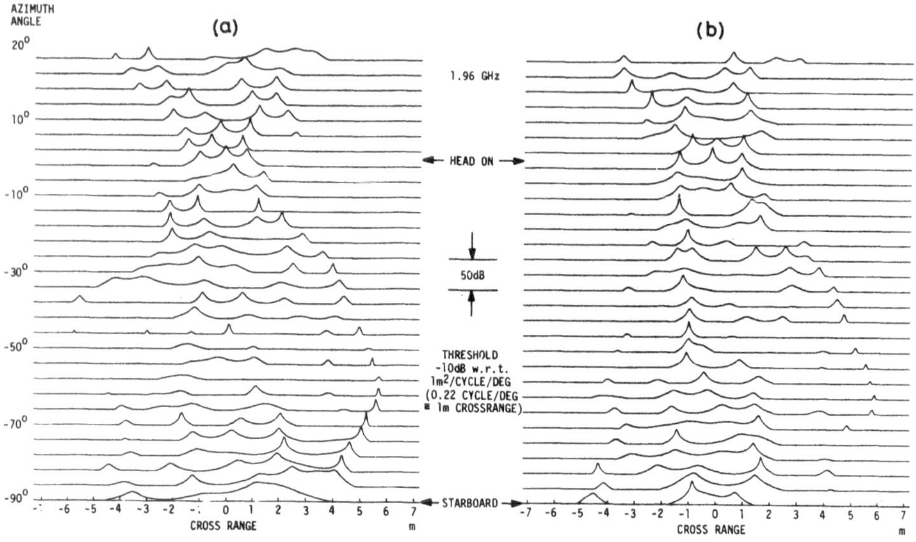

Fig. 22 Angular Spectra around an HS 125 Aircraft for (a) Circular Sphere Accept, (b) Circular Sphere Reject, Polarizations

Two other very significant methods of collecting and processing data are described: they do not provide optical type images of targets, but do contribute valuable ancillary information. One provides the doppler spectra of rotating or vibrating components in a target. The other is the measurement of the polarization scattering matrix for the target. An example has been given of some methods of displaying data from a combined arc-scan polarization scattering matrix measurement. This indicates the importance of paying attention to polarization aspects in any radar work.

Although natural and man-made objects tend to have little colour or frequency sensitivity in the reflection of radar wavelengths, there nevertheless exists a plethora of imaging information to be presented when polarization and time variations are taken into account during the construction of two- or three-dimensional images of a target.

11.0 ACKNOWLEDGEMENTS

The author wishes to acknowledge the support of MOD (PE) and that of his colleagues at THORN EMI Electronics Ltd., Wells, Somerset.

12.0 REFERENCES

[1] L.A. Cram and S.C. Woolcock, "Review of two decades of experience between 30 GHz and 900 GHz in the development of model radar systems", AGARD Conf. Proc. No. 245, pp. 6-1 to 6-15, Sept. 1978.

[2] E.G. Brown, M.W. Plaster and S.C. Woolcock, "Millimetre-wave measurement radars for 140 GHz and 280 GHz", Military Microwaves Conf. Proc., Microwave Exhibitors and Publishers, 1980, pp. 60-65.

[3] D. Bird, M.W. Plaster and C.G.C. Wilson, "Measurement of the depolarization of radar scattering using scale modelling techniques", Military Microwaves Conf. Proc., Microwave Exhibitors and Publishers, 1982, pp. 429-434.

[4] L.A. Cram, G.W. Newbery and K.O. Rossiter, "Microwave holography: a decade of development", AGARD Conf. Proc. No. 245, pp. 8-1 to 8-16, Sept. 1978.

[5] B.E. Prewer and H. Herman, "Scale model measurement facilities and their use for obtaining scattering data pertinent to millimetric wave radars", Military Microwaves Conf. Proc., Microwave Exhibitors and Publishers, 1982, pp. 414-422.

[6] S.E. Millard, K.O. Rossiter and G.R. Selby, "Radiometric measurement at 80 GHz", Military Microwaves Conf. Proc., Microwave Exhibitors and Publishers, 1980, pp. 492-497.

[7] G. Graf, "On the optimization of the aspect angle windows for the doppler analysis of the radar return of rotating targets", IEEE Trans. Antennas Propagat., Vol. AP-24, pp. 378-381, May 1976.

[8] R. Benjamin, "Generalisations of maximum entropy pattern analysis", IEE Proc. Vol. 127, Pt. F, No. 5, pp. 341-353, Oct. 1980.

[9] J.P. Burg, "Maximum entropy spectral analysis", presented at the 37th Annual Meeting Soc. Explor. Geophys. Oklahoma City, Okla., 1967.

[10] S.M. Kay and S.L. Marple Jr., "Spectrum analysis - A modern perspective", Proc. IEEE, Vol. 69, pp. 1380-1419, Nov. 1981.

[11] W-M Boerner, "Polarization control in radar meteorology", presented at URSI Radar Symposium, Bournemouth, U.K., Aug. 1982.

[12] R. Voles, "Radar target imaging by rotation about two axes", IEE Proc. Vol. 125, No. 10, pp. 919-921, Oct. 1978.

IV.5 (SR.1)(SP.4)

OPTIMUM DETECTION TECHNIQUES IN RELATION TO SHAPE AND SIZE OF OBJECTS, MOTION PATTERN AND MATERIAL COMPOSITION

Dag T. Gjessing, Jens Hjelmstad and Terje Lund

Royal Norwegian Council for Scientific and Industrial Research, Environmental Surveillance Technology Programme

ABSTRACT

Most of the existing detection/identification systems do not make optimum use of all the a priori information that one generally is in possession of with regard to the target of interest. Knowing the geometrical shape of the target of interest and its molecular surface structure (e.g., structure of paint), an illumination function can be structured (matched filter concept) to give optimum system sensitivity with respect to the target of interest relative to background objects (interferents). Theoretical and experimental results are given for a limited number of geometrical objects and for different molecular surface compositions. It is shown that the system sensitivity and specificity can be improved considerably using the suggested matched filter illumination technique.

1. INTRODUCTION

In introducing the topic of generalized inversion methods, attention should be focused on two aspects. The first is concerned with the opportunities which new technology in the field of physics and electronics offer; the second is concerned with the task of matching these new technological achievements to the particular detection problem.

Significant technological advances during the last few years in the fields of microwave techniques, electrooptics, and computer technology have opened new avenues with regard to detection and identification of objects and of chemical agents in our environment with previously unidentifiable properties. Some examples follow.

W.-M. Boerner et al. (eds.), Inverse Methods in Electromagnetic Imaging, Part 2, 871–905.

Wide-range tunable microwave sources make it possible to illuminate the object under investigation over a frequency band in a predescribed manner for the particular target "fingerprint" to be revealed.

Microwave receiver elements make it possible to determine the wavefront (amplitude and phase).

Similarly, in the electrooptics field, tunable lasers are available over a large range of wavelengths. Thus an illumination can be chosen so as to: (i) penetrate the medium (e.g., the atmosphere) between the observer and the object to be investigated, and (ii) give maximum information about the molecular structure (the chemical compounds) of the object surface and of the agents emitted from the object (exhaust gases, oil spills, fragrance, etc.).

Electrooptical receiving systems, specifically superheterodyne receiver techniques, give a dramatic improvement in sensitivity and bandwidth resolution.

Low-cost electronic data processors (microprocessors) make it possible to perform the intricate and previously time-consuming computations necessary in real time.

Methods in electromagnetic field theory are being developed from which the shape of the reflecting body (target in a background of interfering bodies) can be determined from measurements of the scattered EM field in space, time and frequency.

Finally, statistical estimation methods such as Kalman filtering and pattern recognition methods have been developed. From these it is possible to make optimum use of a redundant set of information (multisensor information about a target) so as to increase the overall sensitivity and the confidence level of a given set of observations.

In brief, the problem is the following: A method of optimum sensitivity for the detection and quantitative evaluation of an object or a chemical agent is sought. It is assumed that all the data pertaining to the target are known. These target "fingerprints" are typically: the shape, size, and texture of the object, i.e., the distribution of scattering elements in height, width, and depth (the macrostructure); the chemical composition of the surface; the chemical composition of the gases (or the liquids) which the object of interest may emit or leave behind it (scent from plants, odours from fermenting organic substances, gases from soil, exhaust gases from vehicles, etc.); the footprints, tracks, or traces which the object leaves behind.

The target will in general have fingerprints in various spectral ranges. Some of these are weak and some are pronounced. When it comes to determining which of these fingerprints the attention should be focused on, we shall also have to consider the fingerprints of the background, the noise, and we shall need information about the extent to which the medium through which the illuminating waves must pass in order to illuminate the target is transparent to the various frequency bands of potential use. Thus, consideration will have to be given to the following: characteristics (fingerprints) of the target (deterministic or statistical); characteristics (statistics) of the background (additive noise); characteristics (structure) of the propagation medium between the platform of observation and the target (distortion, attenuation, multiplicative noise source).

Since at least the two latter factors vary in time and space, it is not likely that there exists a unique frequency band within which optimum detection and identification capability are achieved at all times. One would therefore wish to have a redundant set of observations and leave it to a Kalman filtering method to sort out the best set of data under any given condition for the best assessment to be made.

We shall now proceed to discuss the two main aspects of the general complex of problems briefly mentioned above. These are the detection and identification of a given chemical agent, or a set of such, comprising the target surface, and the detection and identification of a given target shape.

2. OPTIMUM DETECTION AND IDENTIFICATION OF THE CHEMICAL COMPOSITION OF TARGET SURFACE

In order to illustrate the essential points and the potentials of the technique, let us define a challenging problem of practical importance, using it as an example. Our aim is to detect and identify a particular chemical agent, for example some sulfate deposited on vegetation in a certain geographical area. This could be an area contaminated from human activity or by deposits from general air pollution. A priori this agent may have its fingerprint (absorption lines) in any wavelength region from the ultraviolet region (λ = 0.2 - 0.4 μm) through the infrared region (2 - 20 μm) to the millimeter and microwave region (λ = 100 - 200 μm).

To reveal these fingerprints, we illuminate the ground on which this agent may be present with electromagnetic waves, changing the frequency over the frequency band of interest in some predescribed manner that is optimum with regard to the detection and identification of the particular chemical compound of interest. The situation is assumed to be as follows:

- We are looking solely for a specific chemical compound.

- We know the absorption spectrum or reflectance spectrum of this compound.

- We know nothing about any of the other agents (interferents) that may be present.

- We have very meager information about the vegetation (topography) on which the agents may be deposited, but we have some general idea about the roughness of the structure (grass, coniferous trees, rocky ground). Let us assume that the background is coniferous trees.

A hypothetical detection system will be discussed for the purpose of illustrating the essential points. Figure 2.1, which is subdivided into two sections, shows a remote probing system in symbolic form. Section A has a generator G providing the illumination. It illuminates an area on which contaminated trees are growing. The contaminating agent is assumed to be deposited over the area, which is viewed by two receivers having two separate not overlapping fields of view. When the frequency of the illuminator is changed in a linear manner (saw-tooth frequency-modulation) and the signals V_1 and V_2, which are received at the two receivers, are observed, (referring to the symbolic diagram of Figure 2.1), the signals

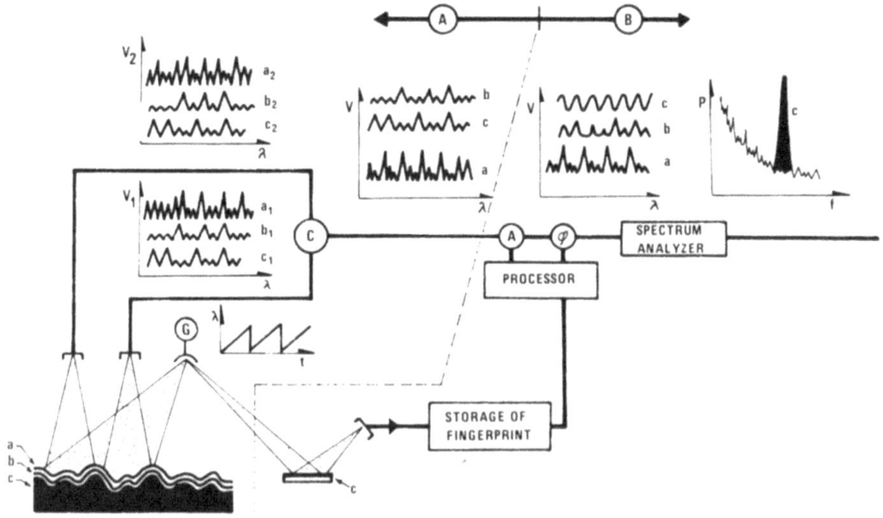

Figure 2.1 Symbolic diagram illustrating the optimum detection and identification principle.

have the following components:

- A random signal with large variance (Rayleigh distributed) resulting from the topography. The signal is a result of backscatter from scattering facets distributed in depth. They give rise to many interfering waves and result in large variations in field strength with frequency (signal trace marked "a").

- A deterministic signal component resulting from the complex surface chemistry (trace marked "b") appearing as modulation of the random topography signal.

- Deeply imbedded in the background chemistry is the component of interest, namely that stemming from the absorption spectrum of the particular chemical compound of interest (trace marked "c").

Our first task in the signal enrichment process is to reduce the dominating effect of the topography (traces a_1 and a_2). In order to achieve this, we shall make use of knowledge we already have about the signal (the deterministic molecular absorption spectrum) and about the noise (the random signal resulting from scattering from a random surface such as from a tree). The noise resulting from the scattering process will obviously be statistical in nature, and there will be no correlation between the signals at the two receivers if the vegetation within the two separate fields of view is not identical in form. Furthermore, if the scatterers (the branches of the trees) move only a fraction of a wavelength from one frequency scan to the next, the correlation from one frame to another at either receiver will be limited. Since the turbulence scale (i.e. the spatial correlation distance) of the surface wind causing the vegetation to move is small compared with the height of the vegetation, the above assumptions are justified.

Therefore our task is to find a process by which the uncorrelated noise component can be reduced relative to the correlated signal component. This is done in the correlator marked C in Figure 2.1. To ensure a thorough physical understanding of the statistical signal retrieval methods involved, this process will be discussed in some detail.

First let us consider the case where the two contributions are added together (additive noise). Figure 2.2 gives a plot in three dimensions of the data that form the basis for the correlation process. Consider the signals from receiver No.1 first. In our cartesian coordinate system the signal strength is plotted vertically, wavelength is plotted horizontally in the plane of the paper and the frame-number (or scan number) is plotted orthogonally to the V-λ plane. The signal resulting from the deterministic absorption spectrum is denoted "b" as in Figure 2.1 and the random signal stemming from the vegetation is denoted "a". To reduce the

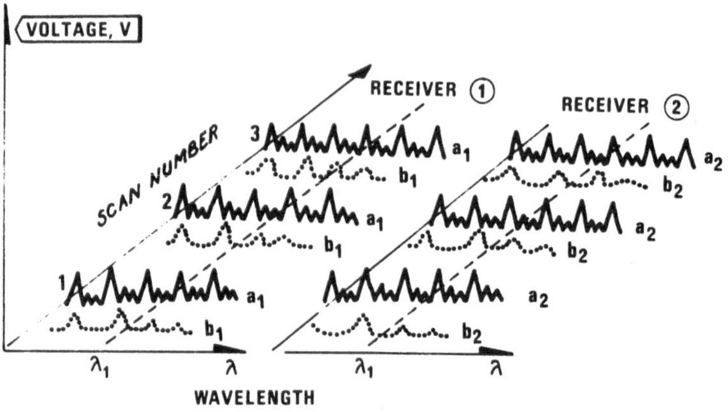

Figure 2.2 Data forming the basis for signal enhancement by a correlation process.

effect of the random uncorrelated contribution, we form an ensemble covariance function for each value of λ, by comparing (correlating) $V(\lambda_1)$ from each scan. Thus we will obtain a number of points on the covariance function determined by the number of frames (scans) and we will obtain a number of covariance functions determined by the wavelength resolution $\Delta\lambda$. If our ensemble then is N, our signal to noise ratio is enhanced by a factor $\sqrt{N}$.

However, the simple example with additive noise (addition of coherent and non-coherent components of the signal) is probably not the most realistic one. For example, if the coherent signal appears as an amplitude modulation on the incoherent signal, then it would be more meaningful to cross-correlate the spectral components of the signals from the two receivers by forming the coherence function instead of the covariance in the C processor.

In order to express this mathematically, we will use an approach based on work by Kjelaas, Nordal and Bjerkestrand (14). In the general case we may have both multiplicative and additive noise. If the spectral components of the noise are not overlapping those of the signal, a simple filtering technique can be employed. If the spectral components do overlap, we shall have to use methods that give us an optimum signal-to-noise ratio. First let us consider the case of additive noise.

If then $S(t)$ be the signal and $n(t)$ the noise, the resultant

$$F(t) = S(t) + n(t)$$

We then assume that there is no correlation between $S(t)$ and $n(t)$. The cross-covariance R_{fs} between the resultant signal f and the signal s is given by

$$R_{fs} = R_{(s+n)s}(t) = R_s(t) + R_{ns}(t) = R_s(t)$$

Similarly the autocovariance is

$$R_f(t) = R_{s+n}(t) = R_s(t) + R_n(t)$$

This can be written in the form

$$\left[R_s(t) + R_n(t)\right] \otimes h(t) = R_f(t)$$

where the symbol $\otimes$ denotes convolution and $h(t)$ is therefore the impulse response. The transfer function $H(\omega)$, which is the Fourier transform of the $h(t)$ function, is then given by

$$H(\omega) = \frac{P_s(\omega)}{P_s(\omega) + P_n(\omega)} = \frac{P_s(\omega)}{P_f(\omega)}$$

where $P(\omega)$ is the power spectrum so that $P(\omega)$ is the Fourier transform of the autocorrelation function $R(t)$.

Let us consider the case of a multiplicative noise source. The resultant signal may be written as

$$f(t) = S(t) \otimes h_n(t)$$

and

$$R_f(t) \otimes h_n(t) = R_{fs}(t)$$

which in terms of the transfer function gives

$$H(\omega) = \frac{R_{fs}(\omega)}{R_f(\omega)}$$

We can now express the transfer function $H(\omega)$ in terms of the coherence

$$coh_{sf}(\omega) = \frac{P_{sf}(\omega)}{P_s(\omega)\ P_f(\omega)}$$

where $P_{sf}(\omega)$ is the cross-spectrum and the transfer function is given by

$$H(\omega) = \frac{P_{sf}(\omega)}{P_f(\omega)}$$

$$= \frac{coh_{sf}(\omega)\ P_s(\omega)\ P_f(\omega)}{P_f(\omega)}$$

$$= P_s(\omega)\ coh_{sf}(\omega)$$

Thus, if we are dealing with additive uncorrelated noise, the signal-to-noise ratio is enhanced by forming the covariance function as implied in the discussion of Figure 2.2. The operator C in Figure 2.1 is a correlator giving us the ensemble covariance function. If, however, the noise is multiplicative, we shall have to cross-correlate the spectral components by forming the coherence function.

Concluding the discussion of section A in the symbolic Figure 2.1, we note that the uncorrelated effects of the rough scattering surface are suppressed during the correlation process. All the correlated factors, chemical compound of interest deeply imbedded in all the uninteresting molecular structures, are retained in unaltered proportions. Now let us proceed to section B of the diagram.

This section illustrates a method by which we can make use of the detailed information that we possess about the chemical compound of interest (molecular structure, reflectance spectrum) to perform a selective detection:

- The signal preprocessed in section A is "filtered" with the fingerprint (reflectance spectrum) of the chemical compound of interest. In the symbolic diagram of Figure 2.1 this is achieved by shining part of the energy from the illuminator onto a reflector containing a film of the chemical agent on which we are focusing our attention. The resulting reflectance spectrum contains the fingerprints that give us the basis for the second filtering operation.

- The operation involves a structuring of the original frequency sweep so as to limit the frequency content of the signal resulting from the chemical agent of interest to a delta function in the frequency domain (i.e., to a pure sinusoidal variation in time over the sweep period). In a practical system one would accomplish this by sweeping the transmitter frequency (the illuminator) in a particular manner (not with a saw-tooth as depicted in the schematic figure above) and at the same time and in synchronism with the frequency sweeping, amplitude modulate the illuminator (see Figures 2.3 and 2.4).

To illustrate this principle and its merits, let us assume that the ϕ and A operators of Figure 2.1 is a tape recorder. The signal is played into the tape recorder in real time and played off successively. The first operator, the "A function", involves an amplitude modulation of the signal so as to adjust the lines in the reflectance spectrum related to the agent of interest to the same level. This involves an amplitude modulation of the reflectance spectrum of interest. The second signal from the processor (the ϕ operator) adjusts the speed of the tape to obtain constant spac-

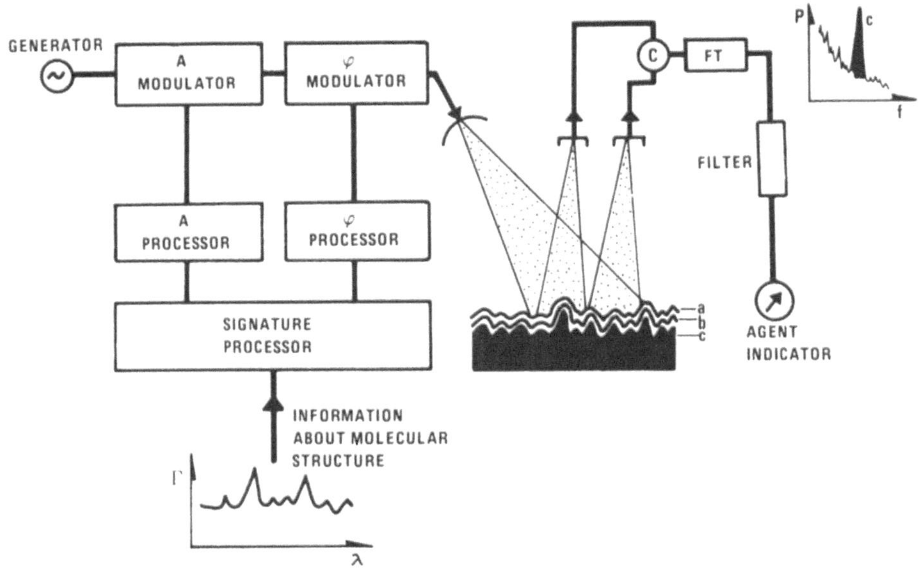

Figure 2.3 Schematic diagram of a practical system with matched illumination.

ing between the lines in the reflectance spectrum of interest. If the interval in the wavelength domain is large between two lines, the tape should be speeded up; conversely, if the distance is small the speed should be slowed. In this manner all the characteristic spectral lines will be equally spaced when originating from the chemical compound of interest and distributed in a disordered manner for all the interferents.

In principle, adjusting both the frequency and the amplitude of the generator illuminating the target of interest in great detail, can transform the signal component originating from our particular chemical compound to a pure sine wave requiring a scan-time limited bandwidth to be detected. In this manner the signal-to-noise level of our system can, in principle, be brought to a very high level. Limiting factors are determined by the available integration time and the degree to which it is practicable to structure the illumination function.

Finally, the basic Figure 2.1 shows the result of a final Fourier analysis of the signal. Our signal of interest is concentrated to a very narrow spectral region, whereas the effect of the interferents is distributed over the spectrum. Before we proceed to consider in some greater detail the character of the noise to be suppressed, we shall give a more realistic example of a selective

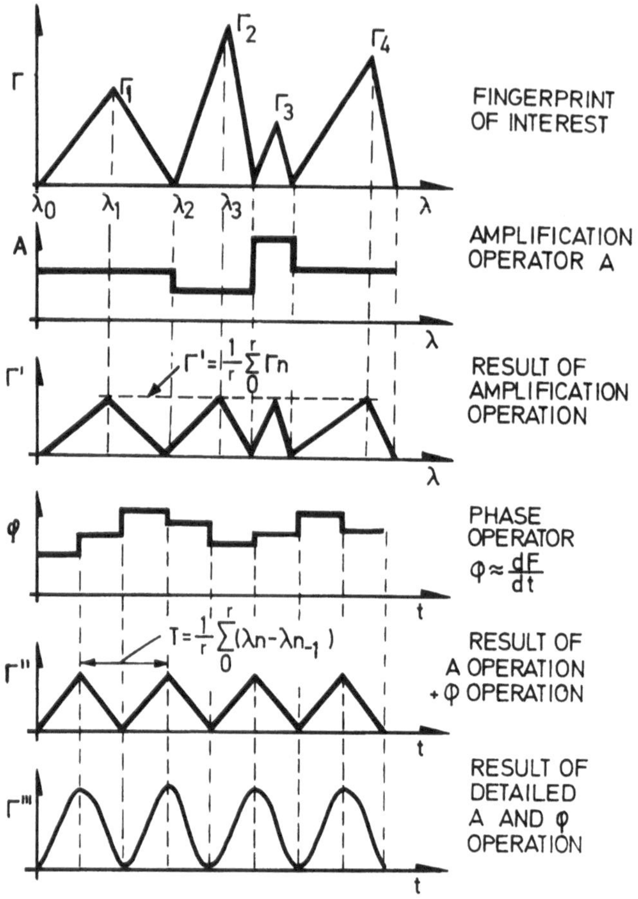

Figure 2.4 Example of an optimum structuring of the illumination function for the purpose of detecting/identifying a particular molecular structure.

detection method. Such a method is illustrated schematically in Figure 2.3.

The illumination (microwaves, IR, visible light, UV) is amplitude and frequency modulated by particular waveforms (see Figure 2.4). These waveforms are the result of a detailed processing of the molecular signatures (see Figure 2.4). The processed illumination is then transmitted to the object of interest. Since the illumination is matched to the molecular structure of interest, the signal appearing at the receivers has minimum information bandwidth and the bandwidth of the entire receiving system can be minimized. Minimum bandwidth gives minimum noise contribution.

Figure 2.4 illustrates the A and ϕ operation process. In the top curve of Figure 2.4 an idealized molecular spectrum is shown (spectrum of absorption, emission or reflectance). The reflectance Γ varies in a triangular manner with wavelength. If the illuminator is linearly frequency modulated, this is what the signal would look like. Amplitude modulating the illuminator in a manner illustrated in Figure 2.4, second curve from above, seeking the same strength of all the lines in the reflectance spectrum, gives the results shown.

Referring now to the fourth curve from above, we change the relative position of the maxima and of the minima of the spectrum so as to obtain a periodic function shown in the fifth graphical representation. This is achieved by changing the rate dF/dt at which the frequency is changed. It is then obvious that if a more detailed A and ϕ operation is applied, the result is a sinusoidal variation, shown at the bottom of Figure 2.4. To detect this requires only a very small bandwidth.

For the purpose of emphasizing the merits of the technique, some illustrative examples are given in Figure 2.5. Two different absorption spectra are considered. Type "A" is characterized by eight absorption lines, whereas type "B" has ten absorption lines. Each absorption line is Lorentz shaped.

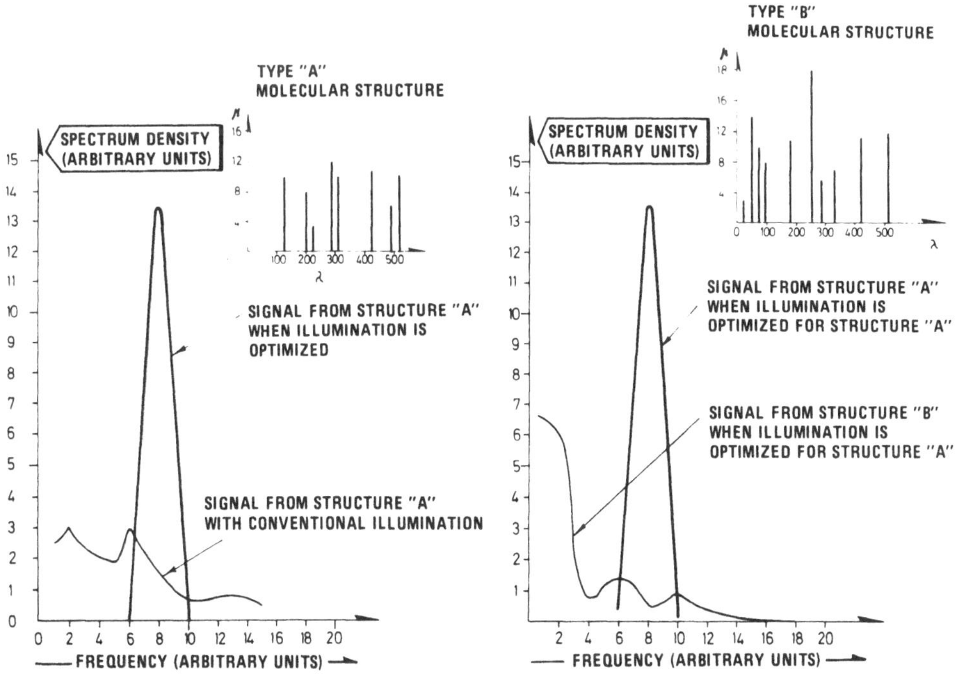

Figure 2.5 The effect of optimized illumination: theoretical results.

Adopting the amplitude- and frequency- modulation scheme illustrated in Figure 2.4 above, the illumination is structured for type "A" molecules. Figure 2.5 shows the result of this optimization process. Note that the spectrum density function associated with optimized illumination is narrow, whereas the spectrum resulting from conventional illumination is wide. Also, if an extensive structuring of the illumination had been accomplished, the resulting signal spectrum would have been a delta function. Figure 2.5 also shows the discriminating power of the technique (right-hand figure). Here an illumination structured for type "A" molecules is applied to structure "B". Note the marked difference in maximum spectral intensity and the shift in frequency of the peaks.

If a computer simulation program implementing this crude technique that involves "first-order matching" only for three frequently encountered compounds, namely epoxy, vinyl and biphinyl, is used, the result is Figure 2.6. Note that by the use of narrow filters

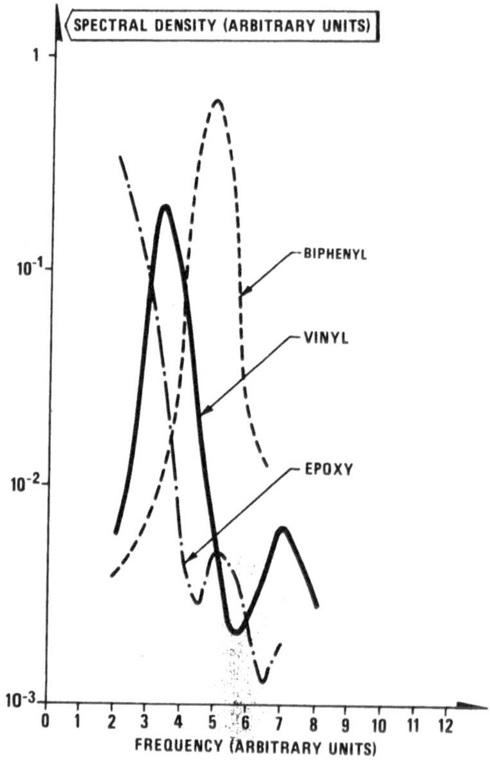

Figure 2.6 The effect of matched illumination for epoxy, vinyl and biphenyl. The illumination that has been structured for vinyl is applied to epoxy. Similarly, the curve for biphenyl is obtained by structuring the illumination for biphenyl. Computer simulations.

as depicted in Figure 2.3 above, the particular agent of interest can be selected when its absorption spectrum is known.

Let us now express this mathematically. On the basis of information about the absorption spectrum, we search the expression for the manner in which the frequency should be varied so as to give a sinusoidal variation of the received (reflected) signal. This we refer to as matched illumination.

For a linear frequency scan, the frequency as a function of time is given by

$$\omega = \omega_1 t' + \omega_0$$

where ω_1 and ω_0 are constants.

If $f(\omega)$ is the reflection/absorption spectrum, applying linear scanning we have

$$f(\omega) = f(\omega_1 t' + \omega_0)$$

We now want a modified sweep function giving us a periodic output function. Hence, we wish to obtain

$$f(\omega_1 t' + \omega_0) = \hat{f}(t') = g(t)$$

where $g(t)$ is our periodic function. This can be achieved by inverting the $f(t')$ function. Thus, the function should satisfy the condition

$$f^{-1}\hat{f}(t') = t' = f^{-1}[g(t)]$$

Therefore, in order to match the illumination to the agent of interest, a frequency sweep function $f^{-1}(\omega)$, which is the inverse of the reflection/absorption spectrum $f(\omega)$, should be used.

An example will clarify this. Let the absorption spectrum be of an exponential form

$$f(\omega) = e^{-a\omega} = g(t)$$

$$f^{-1}[f(\omega)] \rightarrow \omega = -\frac{1}{a} \ln g(t)$$

$$\omega(t) = -\omega_1 \frac{1}{a} \ln g(t) + \omega_0$$

Having given the general expression for the matched illumination function in terms of functions describing the material of interest, we shall go on to consider the effect of the topography (the surface roughness) in relation to that of surface chemistry. The question we now ask is: how should the transmitting/receiving system be structured in order to give maximum information about the geometrical shape of an object? Or, how should the system be structured to have optimum sensitivity in relation to a given object of known size and shape.

3. INVERSION TECHNIQUES FOR DETERMINATION OF OBJECT SIZE AND SHAPE

We are now faced with the consideration of three filter functions. One is constituted by the transmission medium between the observation platform and the target (this intervening medium gives rise to multiplicative noise and distortion). The second filter we shall have to consider is determined by the terrestrial background against which the target is viewed. Thirdly, we shall consider the target itself. The more detailed information we require about the particular target per unit time, the more widebanded must our radar illuminator be, and the larger must the bandwidth be of the propagation medium between this scene to be investigated and the observation platform. We must, therefore, tailor the illuminating waveform, so as to obtain maximum information about the object of interest and at the same time ensure minimum adverse influence of the intervening transmission medium (1,2).

In order to provide "matched illumination" in relation to a given target, we shall have to structure the illumination both in space and in time. If we have at our disposal a radar system which can be amplitude modulated (a pulsed radar), then we shall have to shape the radar pulse in such a way as to obtain maximum influence on the returned radar pulse by the particular target of interest. If we have at our disposal a radar illuminator which can be structured in the frequency domain, then we should compose an illuminating frequency spectrum in such a way as to obtain constructive interference by all the reflecting facets of the target.

In this brief contribution we shall concentrate on the multifrequency radar system. As we shall see, this system lends itself directly to simple computer control in a manner which is very familiar to the computer scientist.

Having structured the illumination in the time domain for optimum coupling to the target, it remains to shape the phasefront in space so as to obtain maximum coupling to the particular reflecting structure of interest. By making use of a matrix antenna (two-dimensional broadside array) as the radar receiver, the phase and amplitude at each receiver element can be controlled by a computer system so as to provide an antenna system which is matched to the phasefront of the wave system which is reflected back from the target of interest whereas the waves originating from the terrestrial background are suppressed.

Finally, we can manipulate the polarization properties of our transmit/receive system so as to investigate the polarization characteristics (the symmetry properties) of the target.

If the four signature domains which we have at our disposal (space, frequency, polarization and motion pattern) are statistically orthogonal as regards target, propagation medium and background we have a simple situation from an information retrieval point of view. We can treat each domain separately and the information processing system (multi-sensor data fusion) becomes comparatively simple. There are well established algorithms for such data handling. Examples are Kalman filtering, maximum entropy methods, optimum parameter estimation methods, etc. If the degree to which the signature domains are orthogonal vary with the conditions prevailing, and if also the signatures themselves (target, background, transmission medium) vary rapidly with time, the adaption process becomes increasingly complicated.

In this brief presentation we shall lean heavily on earlier contributions from the authors' laboratory (1,2,3,4,5), highlight these and describe a multi-frequency polarimetric radar system which presently is being developed by the authors. We shall present results from simple mathematical models and offer some preliminary experimental verifications.

3.1. Basic physical concepts. Formulation of the problem.

As introduced above, we shall be considering four signature domains:

a) By measuring the correlation properties $R(\Delta F)$ in the frequency domain of the waves scattered back from the illuminated area (target against background) we obtain information about the longitudinal distribution of the scatterers. Specifically it can be shown (1,2,3,4,5) that if we describe the distribution in range (longitudinally) of the scatterers by the <u>delay function</u> $f(z)$ which has dimension field-strength and is the square root of the scattering cross-section $\sigma(z)$, then the correlation function in the frequency domain $R(\Delta F)$ is the Fourier transform of the autocorrelation function of $f(z)$. A measure of $R(\Delta F)$ is obtained, as we know, by multiplying the scattered field-strength of frequency F by the complex conjugate of the field-strength at frequency $F+\Delta F$.

Thus we have

$$E(F)\,E^{*}(F+\Delta F) \sim R(\Delta F) \qquad (2.1)$$
$$\sim \mathcal{F}\tau\{R(\Delta z)\}$$

Note that if the target is illuminated with two frequencies spaced ΔF apart, then irregularities in the target with scale size $\Delta z = \frac{c}{2\Delta F}$ contribute to the scattered field.

b) By measuring the spatial correlation properties of the field scattered back from the target in a plane normal to the direction of propagation, i.e. transversely, we obtain information about the transverse distribution of the scatterers. If then the x and y directions are orthogonal to the direction of propagation z (direction from radar to target), then we measure the field-strength at the points x and $x+\Delta x$ in exactly the same way as above where we were dealing with different frequencies.

It can here readily be shown (page 18 of reference 2) that

$$E(x)\, E^{*}(x+\Delta x) \sim R(\Delta x)$$

$$\sim \mathcal{F}\tau\{\sigma(\tfrac{x}{R})\} \tag{2.2}$$

where $\sigma(x)$ is the transverse distribution of the scatterers over the scattering body in the x direction and R is the distance to the target. This, of course, is the same as saying that the spatial autocorrelation of the transverse field-strength is the Fourier transform of the angular power spectrum of the scattered wave (angle of arrival spectrum).

c) By measuring the temporal distribution of the scattered field (the power spectrum) information about the motion pattern of the target is obtained through the well-known Doppler relationship.

$$f = \frac{1}{2\pi}\vec{K}\cdot\vec{V} \tag{2.3}$$

where f is the Doppler frequency, $\vec{K}$ is the vector difference $\vec{k}_1 - \vec{k}_s$ between the wavenumber $\vec{k}_1$ of the illuminating (incident) wave and $\vec{k}_s$ the wavenumber of the scattered wave. $\vec{V}$ is the velocity of the scattering element.

Thus if we are dealing with a target (such as the sea surface) composed of many scattering centers or facets which have different velocity, we obtain information about the velocity distribution of scale size Δz by illuminating the target with two frequencies with frequency difference $\Delta F = \frac{c}{2\Delta z}$ and by measuring the temporal variation (power spectrum) of the quantity

$$W(\omega) \sim V(F,t)\, V^{*}(F+\Delta F,t) \tag{2.4}$$

Note that the velocity of the scattering element of scale size Δz is obtained from equation (2.3) by noting that it is the wavenumber $K = \frac{2\pi}{\Delta z}$ of the difference frequency ΔF that enters into the Doppler equation in this case.

Hence the Doppler frequency is given by

$$f = \frac{2\Delta F}{c} \cos \phi$$

For details the reader is referred to ref (5) page 15.

d) By measuring the distribution of the scattering centers (the $\sigma(z)$, $\sigma(x)$ and $\sigma(y)$ functions) for each element of the scattering matrix, we obtain information about the symmetry characteristics of the target.

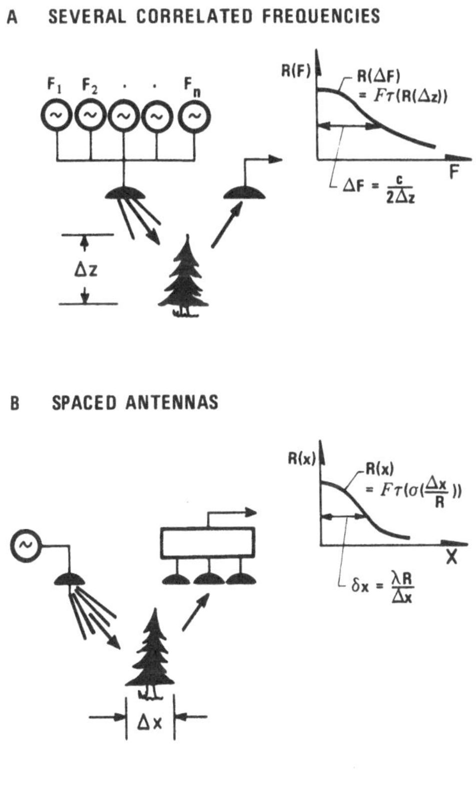

Figure 3.1 The basic concepts of the general adaptive (inverse scattering) radar system

Figure 3.1 finally sums up the basic concepts of the general adaptive (inverse scattering) radar system, whereas Figure 3.2 shows the radar signature (correlation function in the frequency domain) of some idealized scattering objects. Note that the spatial correlation function is obtained by changing the abscissa of Figure 3.2 from ΔF to $R\lambda/\Delta x$.

Before we proceed to give experimental verifications of the simple mathematical models based on first principle physics, let us consider the basics of polarimetry in relation to a multi-frequency radar system. Single frequency pulsed polarimetric radars have recently received considerable attention (6-9); in this very space limited presentation we shall confine ourselves to introducing the basic concepts involved in an experimental investigation which is in progress at the authors' laboratory aiming at increasing the target identi-

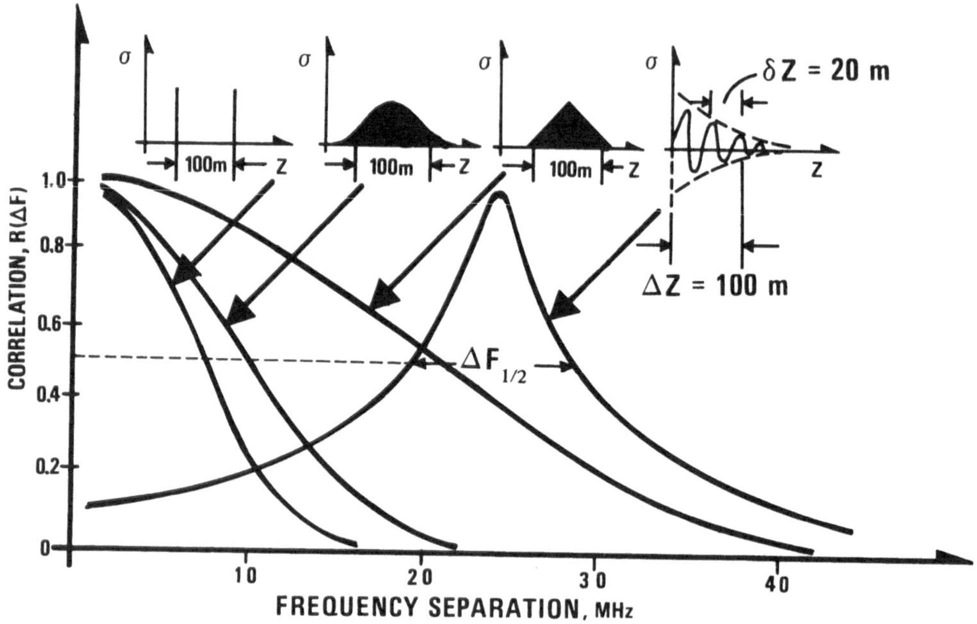

Figure 3.2 The radar signature (correlation function) of some simple scattering objects

fication potential of our multi-frequency radar system. Note that the present radar (see Figure 3.7) makes use of six correlated computer controlled frequency synthesizers which will give 15 different frequency spacings (can couple to 15 different target scales). The radar has two transmitters and two receivers with polarization control so as to enable us to determine the frequency covariance function

$$R(\Delta F) \sim V(F)\, V^{*}(F+\Delta F)$$

simultaneously for 15 frequency separations and for the three polarization combinations (horizontal/horizontal (S_{11}), horizontal/vertical (S_{12}), and vertical/vertical (S_{22}). The radar can also operate with an adaptive polarization basis, that is, the radar optimizes its detection and identification capability by transmitting the optimal elliptic polarizations.

After having completed the preliminary tests with this system, the program will be expanded to include six spaced receivers so as to allow us to determine the spatial correlation properties of scattered field with 15 spatial separations.

In order to introduce this concept, Figures 3.3 and 3.4 are presented. Here we have "modelled" a target in the form of an air-

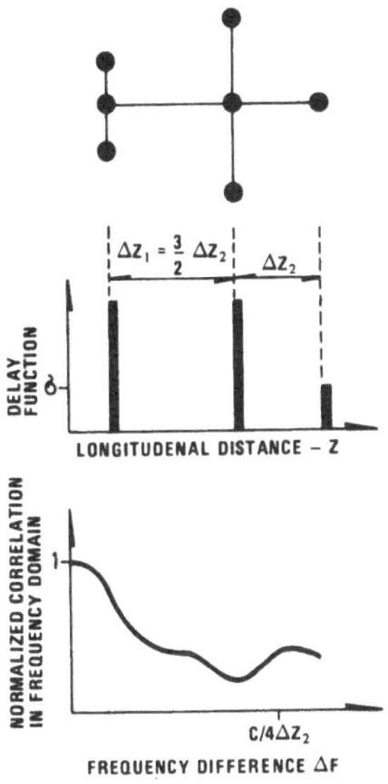

Figure 3.3

An idealized target consisting of isotropic polarization in variant scattering centers and the corresponding multi-frequency radar signature. The correlation function R(ΔF) is measured.

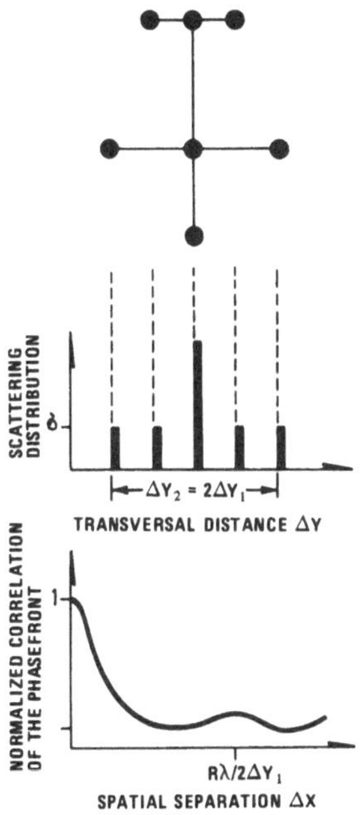

Figure 3.4

An idealized target viewed head on with a radar system having transversally spaced receiving antennas measuring the spatial correlation function F(Δx).

plane making use of seven isotropic and polarization invariant scattering centers.

Note that the scattering matrix elements merely are represented by their moduli. The actual system, however, measures their relative phase, thus allowing detailed analysis of the target's symmetry properties within each time/space resolution cell in an arbitrary polarization basis.

Figure 3.3 refers to the longitudinal case where the target is viewed head on with 15 coherent frequency spacings. The idealized delay function is shown and the corresponding frequency covariance function R(ΔF).

In exactly the same way, Figure 3.4 shows the spatial correlation of the scattered field when the target is viewed head on.

Having introduced the multi-domain adaptive radar designed primarily for environmental surveillance applications, we shall present a somewhat more general target configuration and give the $R(\Delta F)$ and $R(\Delta x)$ signatures for the three pertinent polarization configurations.

Figure 3.5 shows the simplified airplane target viewed head on with our 15 frequency spacings and three polarization combinations. Note that as in Figures 2.3 and 2.4 the target has seven scattering centers. However, as a means of illustrating the polarization issue, we have selected four different classes of scattering in Figures 3.5 and 3.6 with the following notations:

-- scatterers for horizontal/horizontal S_{11}

¦ " " vertical/vertical S_{22}

| odd-bounce scatterers

< even-bounce scatterers

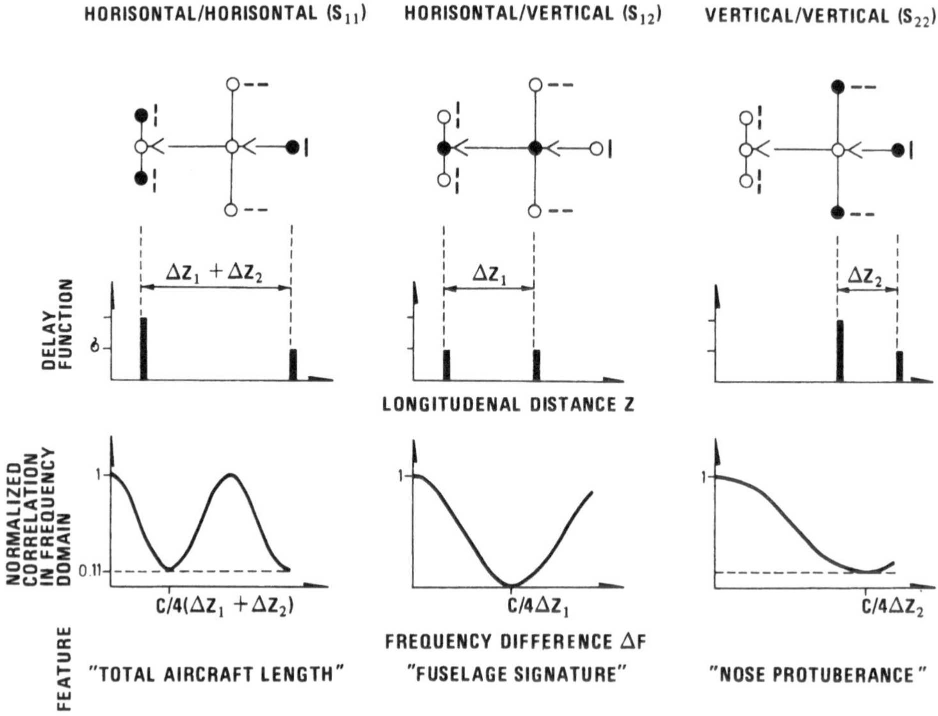

Figure 3.5 Measuring the frequency covariance function $R(\Delta F)$ for the three appropriate polarization combinations, information can in principle be obtained about the detailed longitudinal target dimensions.

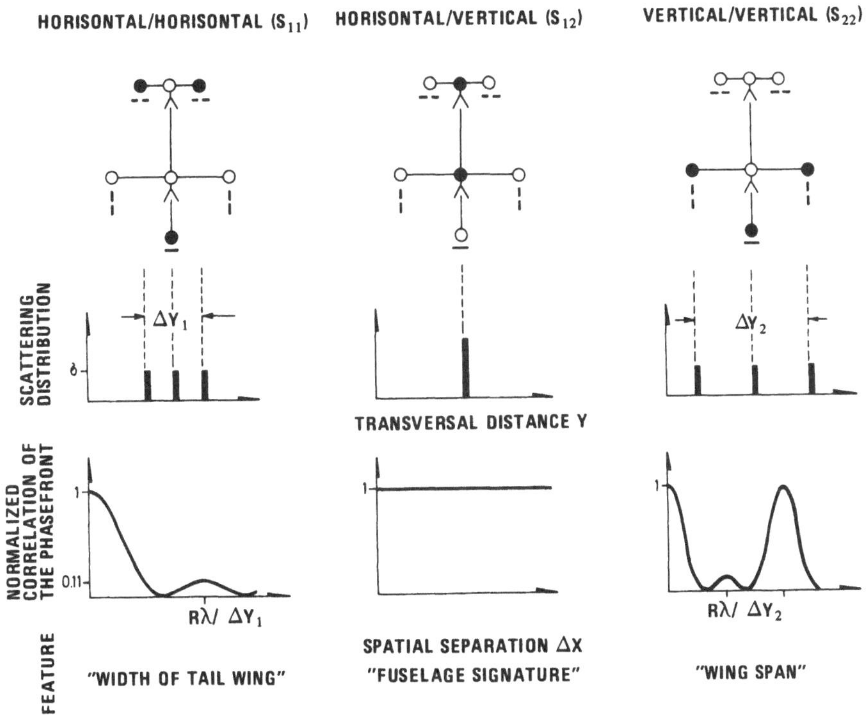

Figure 3.6 Measuring the spatial correlation properties of the scattered field for the three appropriate polarization combinations, information can in principle be obtained about the detailed transverse dimensions of the target.

Note that to the extent that the target has polarization sensitive scattering centers, we can draw conclusions regarding detailed aircraft dimensions, not merely overall length as in the case of the simplest scheme illustrated in Figure 3.2.

3.2. A brief description of the radar system

The adaptive polarimetric multi-frequency radar system shown in Figure 3.7 is made from a conventional 960 telephony microwave communications link. The system structure is, as indicated in the figure, very simple. The six frequencies in the 50 - 90 MHz band delivered from a common crystal oscillator are up-converted to the 6 GHz band giving a total output power of 200 mW (33 mW per frequency) to the transmitting antenna. Two different antennas were used for the various applications: a horn antenna with an aperture of 30 x 30 cm and a paraboloid with diameter 1.20 m.

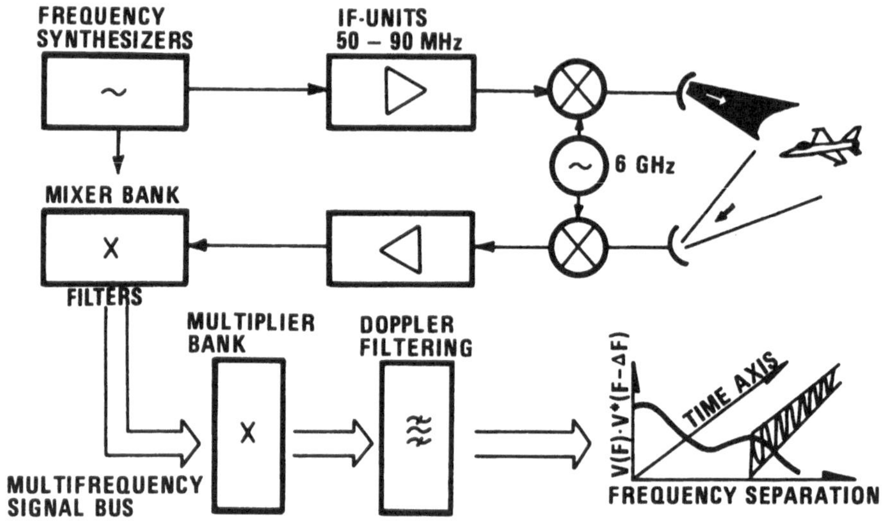

Figure 3.7 The multi-frequency radar system is based on a set of frequency synthesizers in the 50 - 90 MHz band which is up-converted by a 6 GHz source which also serves at the local oscillator for the receiver. Upon amplification the receiver IF signal is mixed with the output from the transmitting synthesizers so as to obtain a set of coherent signals which in turn are subjected to a correlation analysis to give the target signature in the time/frequency domain.

The receiving antenna was positioned adjacent to the transmitter (110 dB isolation). Transmitting and receiving antennas are identical. The backscattered signal is down-converted to the IF band (50 - 90 MHz) by means of a 6 GHz source. This source is x-tal-controlled and common to both the up- and down- convertors. Upon amplification and filtering, the six separate VHF receiver frequencies are mixed with the corresponding six frequencies from the transmitting synthesizers. The resulting six voltages giving 15 different frequency pairs are then multiplied and the products (the 15 covariances $V(F,t)\ V^*(F+\Delta F,t)$ resulting) are subjected to 15 sets of Fast Fourier Transform filters thus producing 15 power spectra (Doppler spectra).

The essence of these processes is illustrated in Figure 3.8.

Note that since the six receiver filters are tuned separately to the same frequency as that of the continuous wave transmitter allowing for Doppler broadening only, the system can be made very sensitive. Specifically, when the receiver bandwidth was limited

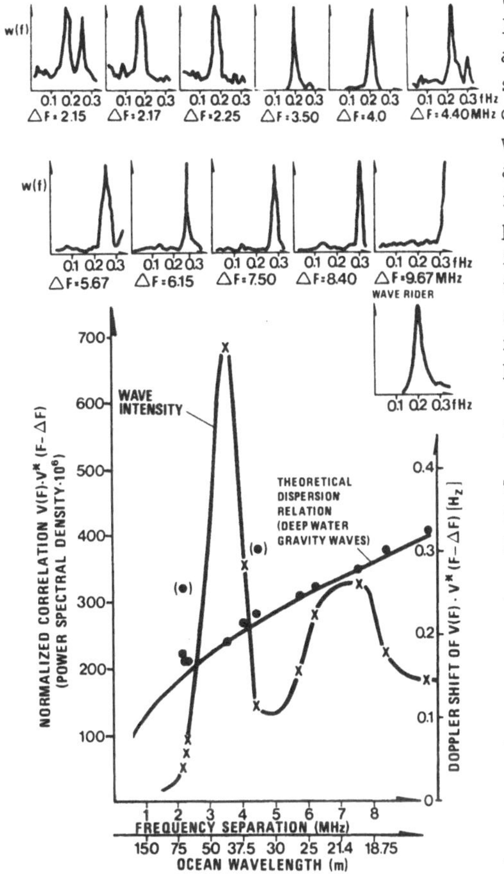

Figure 3.8

Wave intensity and wave velocity plotted to the basis of ocean wavelength (frequency separation). Note that the upper set of curves gives the Doppler spectra for various ocean wavelengths (frequency separation ΔF). For details see ref (4).

to some 10 Hz (integration time 1/10 sec) a small Cessna 173 aircraft was seen with a substantial signal to noise ratio out to ranges in excess of 10 km with 30 x 30 cm antenna aperture and 33 mW power transmitted per frequency line (total radiated power 200 mW). The more a priori information one has about the target of interest (size and shape, velocity) the more narrow-banded can our system be (see ref (1), page 85).

3.3. Experimental verifications

The multi-frequency radar system was used for three different investigations:

- Measurement of directional ocean wave spectra (wave intensity, velocity and direction).
- Investigation of ship signatures against a sea background.
- Classification of air targets (F-16 aircraft).

A brief highlighting of the results of these investigations will now be given.

Directional ocean wave spectra. Illuminating the sea surface from a cliff 50 m above the sea, the ocean wave spectra (wave intensity and wave velocity) was determined for various azimuth directions and for 15 different ocean wavelengths in the interval from some 5 m (couples to ΔF = 27 MHz) to 150 m (corresponding to ΔF = 1 MHz). For each frequency separation ΔF the power spectrum of the frequency covariance function V(F,t) V^*(F+ F,t) was computed.

Examples of such power spectra (Doppler spectra) are shown in the upper part of Figure 3.8. In the lower part of the figure, two curves are plotted: the curve marked with crosses gives the wave intensity (wave height) spectrum for ocean wavelengths ranging from 18 to 150 m. The curve marked with points gives the Doppler shift as a function of ocean wavelength obtained from the power spectra shown in the upper part of the figure. Note that the theoretical Doppler shift (phase velocity of gravity waves given by $V = \sqrt{\frac{gL}{2\pi}}$ where L is the ocean wavelength (ref 4)) is also shown. The systematic shift of the experimental points towards higher velocity is probably due to tidal currents on which the wave motion is superimposed. We obtain one such set of spectra for each azimuth direction of the antenna system. This set of wave height spectra allows one to present the two-dimensional irregularity spectrum. Figure 3.9 shows the isolines of spectral intensity in the K_x and K_y plane (Fourier space). K_x and K_y refer to the wavenumber of the ocean irregularity structure in the x and y directions respectively ($K = \frac{2\pi}{L}$ and $L = \frac{c}{2\Delta F}$).

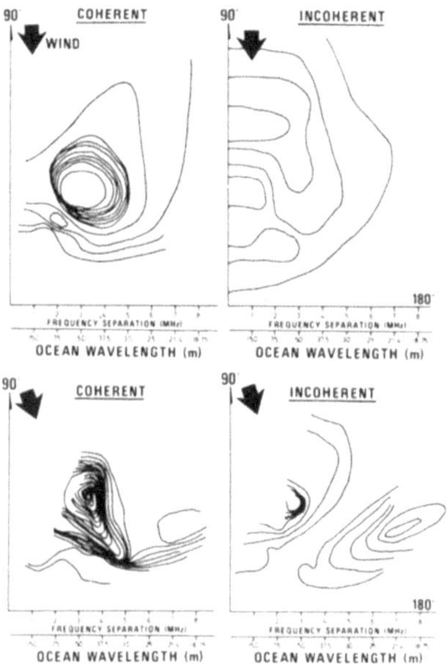

Figure 3.9

Isolines of spectral intensity in the K_x-K_y plane. Note the effect of changes in wind direction on the coherent and on the incoherent irregularity spectrum (for details see ref (4)).

As will be seen, we distinguish between coherent and incoherent components. The incoherent is the "residue intensity" obtained from the set of Doppler spectra shown in Figure 3.8 when the contribution of the "coherent" spectral line is subtracted. It can be visualized as all the random turbulent velocity contributions outside the short velocity interval dominated by the gravity wave.

We see that as far as the coherent waves are concerned, long ocean waves are associated with a small angular spread (they are plane) whereas the converse is the case for short ocean waves.

Radar signature of a ship against a sea clutter background. Having established the characteristic properties of the sea surface with respect to clutter signature, the basis is formed for studying ship targets. As an introduc-

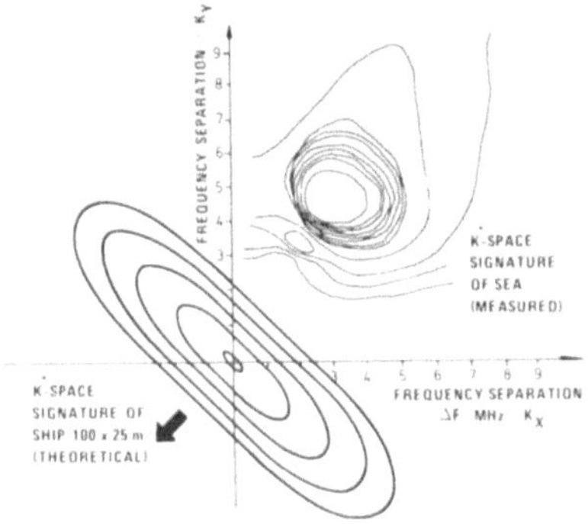

Figure 3.10

K-space representation of the sea surface (isolines of spectral intensity $\Phi(\vec{K})$ as obtained from experiments is compared with the K-space signature of a ship with Gaussian distribution of the scattering elements.

tion to this chapter Figure 3.10 is presented. Here the sea surface signature is plotted in Fourier space in the same way as in Figure 3.9 above based on experimental results. In Figure 3.10, the radar signature of a 100 x 25 m ship is shown. We have here assumed that the scatterers contributing to the backscattered signal are distributed in a Gaussian manner in all directions. Note, as illustrated in Figure 3.2, that for a target with a non-periodic distribution of the scatterers, the correlation function in the frequency domain has its maximum for zero frequency separation. We see from Figure 3.10 that there is no signature overlapping for the particular sea surface structure, and the particular 100 x 25 m ship.

Then let us focus the attention on ship targets obtained experimentally. Figure 3.11 shows the time-record of the frequency covariance function $V(F,t)\ V^*(F+\Delta F,t)$. To illustrate the essential features three values of ΔF (out of the total ensemble of 15 frequencies) have been selected. We see that when the radar beam illuminates the sea surface only, the time-record shows an irregular structure. When the ship enters the radar beam, a Doppler signature which is proportional to the frequency separation ΔF is clearly shown.

Before we change the subject from sea surface targets to aircraft, let us present another two sets of experimental results on ship signatures against a sea clutter background. With reference to Figure 3.11, Figure 3.12 shows the Doppler shift for the various frequency separations caused by the ship and by the sea surface, respectively. Note that the ship gives a linear (non-dispersive) relationship between frequency separation and Doppler shift, whereas the sea surface gives results which are in reasonably good agreement with the theoretical relationship for deep water gravity waves.

Figure 3.13 shows the normalized correlation as a function of frequency separation for a particular cargo-ship and also for the

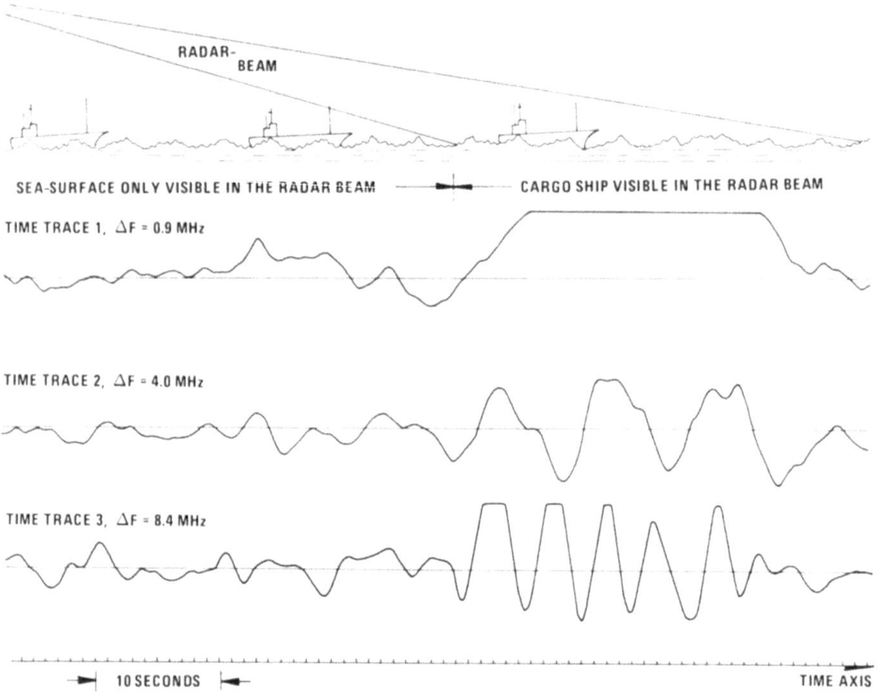

Figure 3.11 Time-record of the covariance function in the frequency domain $V(F,t\ V^{*}F+\Delta F,t)$ for three different values of the frequency separation ΔF.

Figure 3.12

Doppler shift as a function of frequency separation for a rigid target and sea clutter, respectively. Note that deep sea gravity waves are dispersive: there is a square-root relationship between phase velocity and wavelength. A rigid body moving at constant velocity gives a Doppler shift that is proportional to the frequency separation and the velocity of the object.

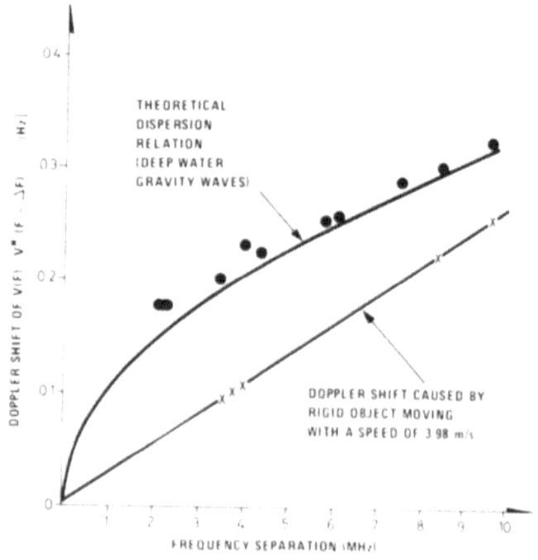

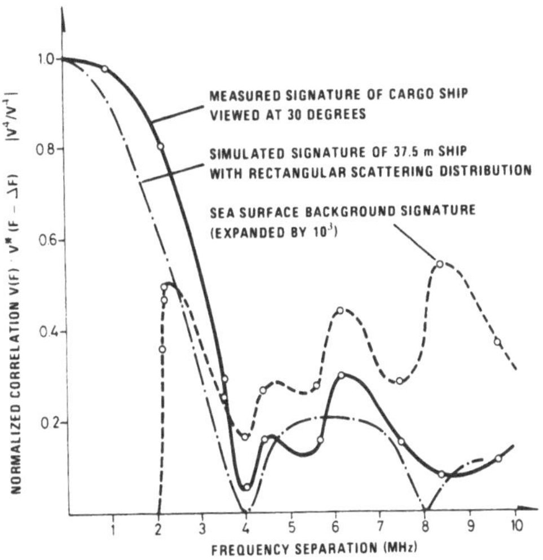

Figure 3.13 Normalized correlation in the frequency domain plotted as a function of frequency separation for a 37.5 m ship and for the sea surface background. Note the good agreement between measured and simulated signatures.

sea surface background. Note that the sea surface signature is expanded by a factor 10^3 relative to the signature of the ship. Note also that there is a remarkably good agreement between experimental results and the theoretical ones if the assumption is made that the scatterers are distributed evenly over the scattering body 37.5 m long.

Radar signature of a rigid airplane. Finally, we shall present a very brief high-lighting of the radar signature investigations which were carried out for various types of air targets. In this brief presentation we shall focus the attention on a particular aircraft, the Cessna 172 airplane. Tests were also performed on airplanes of comparable size, but different structure. These produced drastically different signatures both as regards the Doppler spectrum (flutter and vibrations superimposed on a translatory motion) and as regards the frequency covariance function.

Figure 3.14 shows the signature for the Cessna airplane. Note that by-and-large the agreement with theory when the assumption is made that the scatterers are distributed in a Gaussian manner is reasonably good. There is clear evidence, however, of a small number of scattering centers which dominate over the Gaussian distribution. For details the reader is referred to ref (3).

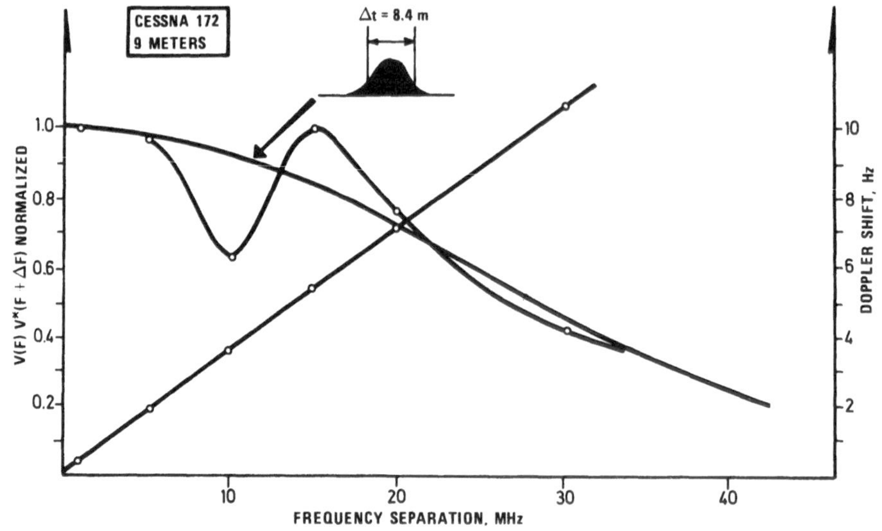

Figure 3.14 Doppler shift and scattering center distribution for a 9 m long Cessna 172.

Figure 3.15 shows time-record of the frequency covariance function for 15 different values of ΔF for this aircraft. Note that the sequence of Doppler spectra suggests that we are dealing with an aircraft with a very flexible structure, as can be seen from the rapid fluctuations in the relative amplitudes of the various ΔF channels.

When these records are subjected to a more detailed mutual coherence analysis (see page 48 of ref (2)) many interesting features suggesting that the aircraft has a well-defined flutter component are brought out.

Polarimetric multifrequency investigations of aircraft target models. The multifrequency radarsystem reported here has full polarimetric capability. That implies that full polarization information is available at each microwave frequency. Two orthogonal polarizations are transmitted simultaneously and on the receiving side the contributions from the co- and crosspolarized returns for each of the two receiving channels are decoded by means of a special coding technique. The scattering matrises of the radar target is thus measured and presented as a two by two complex scattering matrix.

The multifrequency signature is therefore available as five sets of signatures, each one representing one of the five independent elements of the scattering matrix. The scattering matrix can be

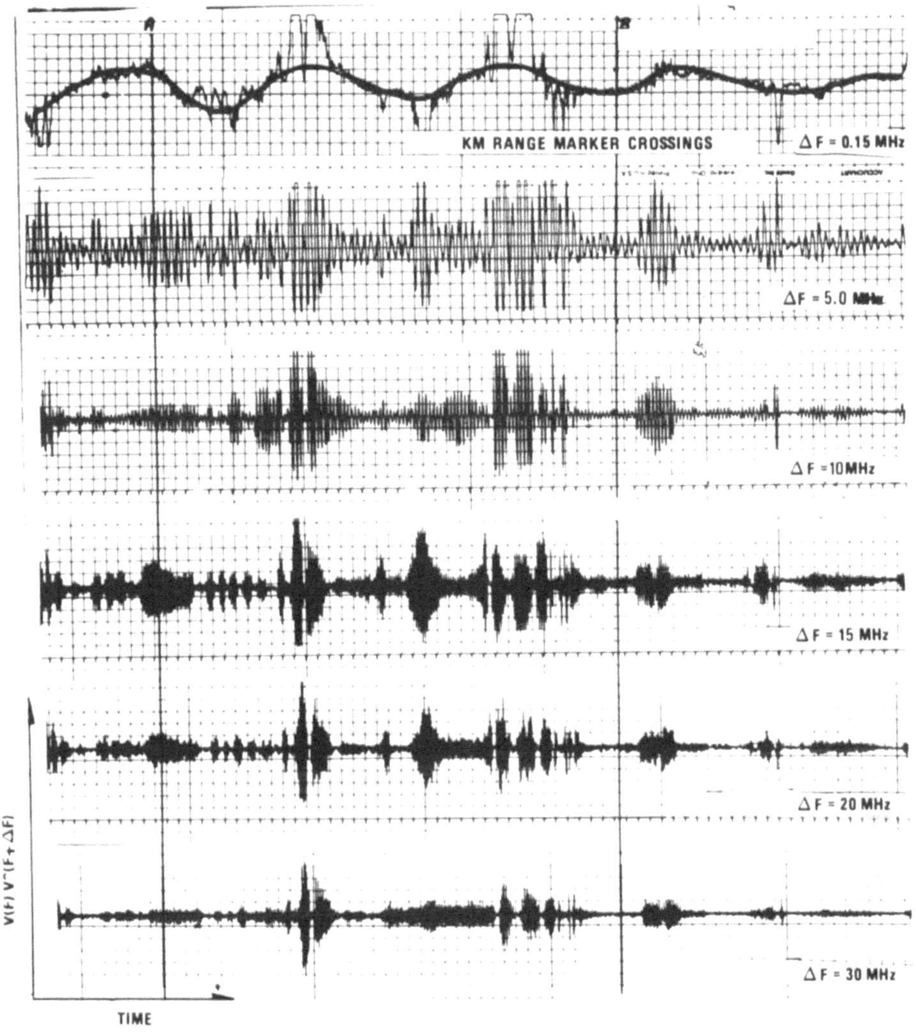

Figure 3.15 Time record of the frequency covariance function for the Cessna aircraft viewed head on as it passes the radar beam. The top record provides information about range marker crossings whereas the five records below give information about motion pattern for five different irregularity scales.

measured by any set of orthogonal polarizations.

It is known from the theory of electromagnetic waves that polarization reveals the directional distribution of scatterers in the plane normal to the look direction of the radar. Features such as the degree of symmetry of the target, the symmetry angle (roll angle of an aircraft), the curvature and other characteristics of the individual scattering centers are embedded in the polarization signatures.

When considering the relative merit of polarization information versus multifrequency (down range-) spatial information with respect to target characterization, the result will dramatically depend on the type of object being illuminated.

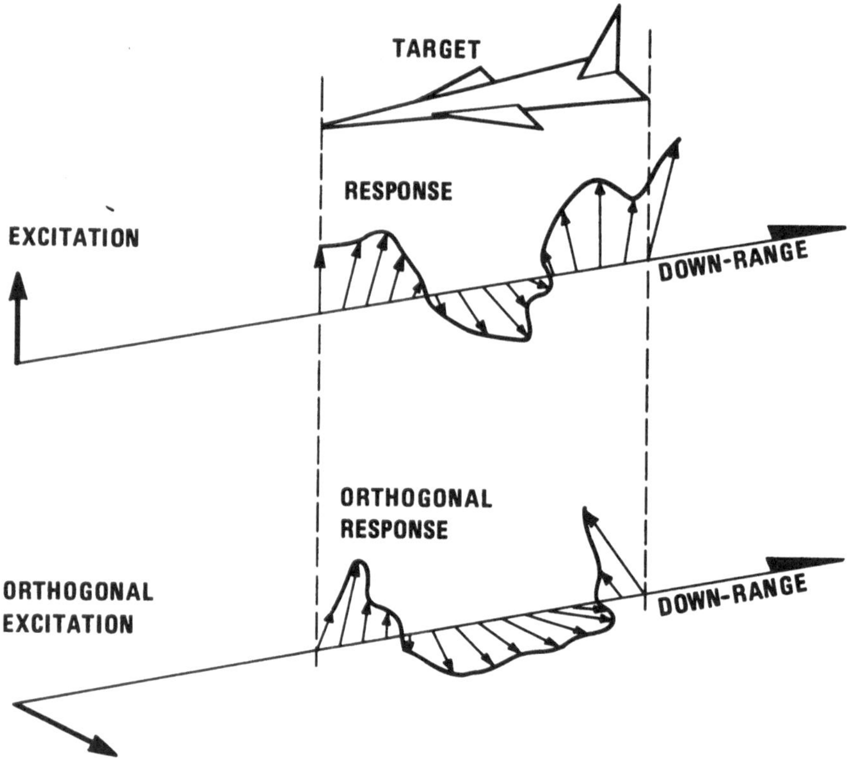

Figure 3.16 The ultimate signature of a target consists of two trajectories in space of electromagnetic scattered field, one for each of two orthonormal excitations.

To illustrate this, consider a typical target as sketched in figure 3.16. The ultimate monostatic radar signature of such a target is the trajectory in three-dimensional space as probed by a set of two orthonormal Dirac delta-functions. At each point on the target the reflected wave will be characterized by an amplitude and direction in space for each of the two polarizations. In the case of resonance phenomena within a resolution cell on the target, also a time delay will have to be taken into account. However, for an extended target, these delayed echoes will contribute to the echoes from the following down-range cells, and will not add complexity to the target response. Note that the cross-polarized returns from each of the two orthogonal excitation are identical, due to the symmetry of the scattering matrix for monostatic imaging through resiprocal transmission media.

For object characterization, a sufficiant set of features will have to be extracted from the target response. Polarimetric in-

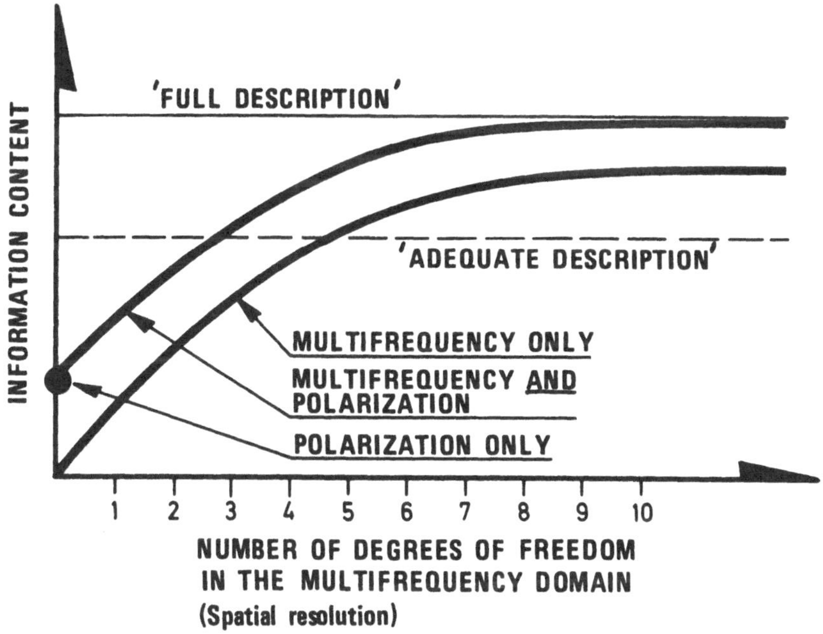

Figure 3.17 An illustration of the relative merit of polarization and multifrequency information.

formation only can in some cases be used. This corresponds to no spatial information, and the downrange signature is projected into the polarization transversal plane giving the degrees of freedom of the scattering matrix.

The same number of degrees of freedom can also be obtained from multifrequency information by selecting a number of spatial resolution elements or multifrequency responses. For very simple targets, like a spheroid, the information content in the polarimetric and multifrequency domain will be the same. For more complicated real targets which require a large number of spatial resolution cells for characterization, it is to be expected that multifrequency information most readily provides adequate description of the target. This is illustrated in figure 3.17, which is an intuitive description of the information content in the two domains for extended targets. The information content acquired by a polarimetric radar, a multifrequency polarimetric radar and

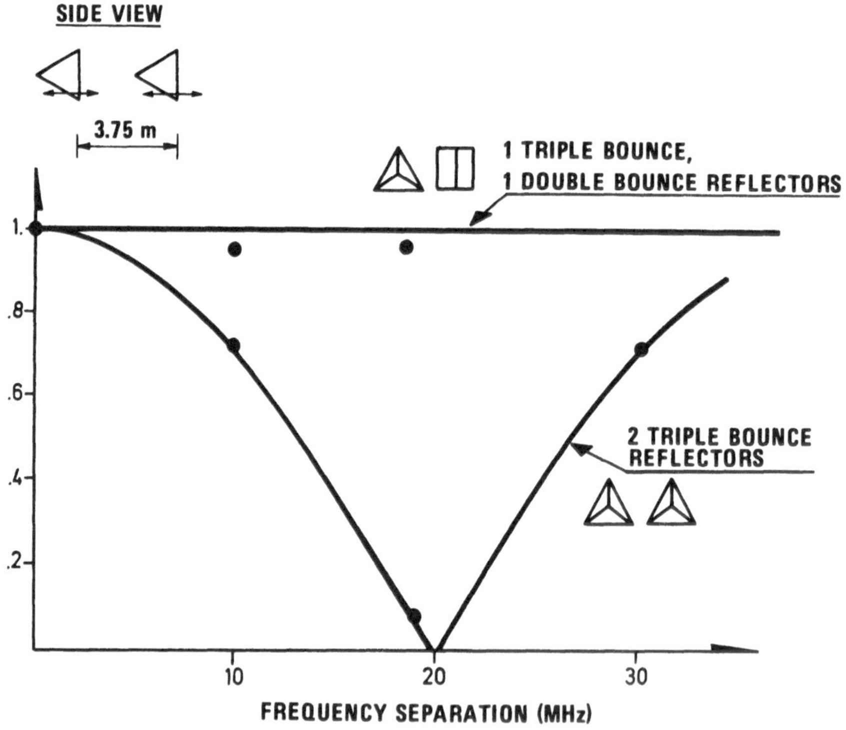

Figure 3.18 Polarimetric multifrequency signatures of a target consisting of two scattering elements.

a non-polarimetric radar is plotted against spatial resolution. For the non-polarimetric multifrequency radar the span of the scattering matrix is used, thereby making the radar insensitive to the roll angle of the target.

The polarimetric multifrequency radar has been used to investigate the signature of targets consisting of discrete polarization sensitive scattering elements. Figure 3.18 shows the signature of a simple target with two reflectors spaced 3.73 meters. The lower trace is the multifrequency signature when the two reflectors are identical triple bounce reflectors, while in the upper trace, one of the reflectors is substituted by a dihedral. This is the multifrequency signature with identical transmit and receive linear polarization (the diagonal elements of the scattering matrix), and the basis is chosen so as to give zero return from the 45 degrees oriented dihedral. Therefore, the multifrequency signature is identical to the signature of one tri-

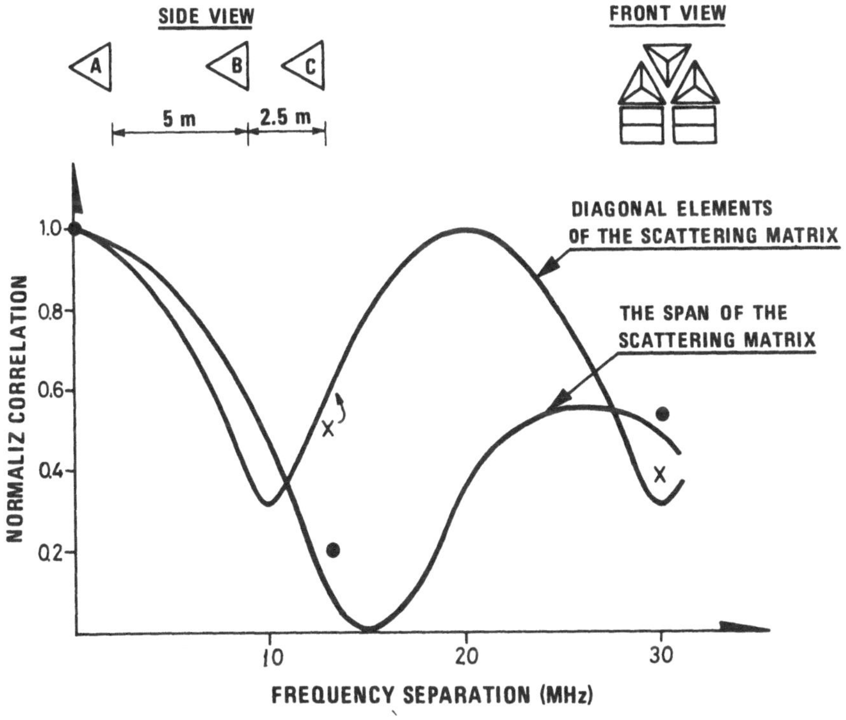

Figure 3.19 Polarimetric multifrequency signatures of a target consisting of five scattering elements.

hedral only. With the two trihedrals, the signature is periodic, with period equal to the velocity of light divided by twice the separation of the reflectors. The fact that the correlation reaches zero, indicates that the reflectors give the same echo strength.

A slightly more complicated target is shown in figure 3.19. Here, the two rear scattering elements (A) are 45 degree oriented dihedrals. Separated by five meters are two colocated trihedrals followed by a single trihedral 2.5 meters ahead. The upper trace in figure 3.19 represents the diagonal elements in the scattering matrix, while the lower one represents the span of the scattering matrix. (The span of the scattering matrix is the sum of all the four elements, and is equal to the total polarimetric power).

As can be seen from these measurements, the dramatic dependence on polarimetric information as seen in the simple target shown in figure 3.18 is somewhat reduced in figure 3.19.

ACKNOWLEDGEMENTS

The authors would like to acknowledge the very valuable services rendered by Dagfin Brodtkorb and Knut Bakke of A/S Informasjons-kontroll who have developed computer software and participated in system design and measurement campaigns.

The authors would also like to thank Erik Olsson and Øystein Sandholt of Eidsvoll Electronics AS who have had the responsibility for the development and production of the radar prototype.

The multifrequency radar is based on a 960 channel radio relay link generously provided by Mr Standahl of A/S Elektrisk Bureau, Division NERA.

Chief scientist Karl Holberg of the Norwegian Defence Research Establishment, the previous affiliation of the authors, has been a continuous source of stimulation and encouragement, and we are also adept to Mr Per Ersdal of the Norwegian Defence Communication Administration who provided antennas and test facilities.

This work has been partly supported by the Royal Norwegian Air Force.

Part of this material first appeared in the Proceedings of the RADAR 82 Conference, and is included in this text with the kind permission of Institute of Electronic Engineers.

REFERENCES

1. Gjessing, D.T. 1978, "Remote Surveillance by Electromagnetic Waves for Air-Water-Land", Ann Arbor Publishers Inc, Ann Arbor, USA.
2. Gjessing, D.T. 1981, "Adaptive Radar in Remote Sensing", Ann Arbor Publishers Inc, Ann Arbor, USA.
3. Gjessing, D.T., Hjelmstad, J., Lund, T. 1982, "A Multifrequency Adaptive Radar for Detection and Identification of Objects. Results on Preliminary Experiments on Aircraft Against a Sea Clutter Background", IEEE Transactions on Antennas and Propagation, AP-30, No. 3, pp. 351-365.
4. Gjessing, D.T., Hjelmstad, J., Lund, T., "Directional Ocean Spectra as Observed with a Multifrequency CW Doppler Radar System", Submitted for publication.
5. Gjessing, D.T. 1981, "Adaptive Techniques for Radar Detection and Identification of Objects in an Ocean Environment", IEEE Journal of Ocean Engineering, OE-6.1, pp. 5-17.
6. Boerner, W.-M., Chan, C.-Y., Mastoris, P., El-Arini, M.B. March 1981, "Polarization Dependence in Electromagnetic Inverse Problems", IEEE-ATP.
7. Huynen, J.R. 1970, "Phenomenological Theory of Radar Targets", Ph D Dissertation, Drukkerij Bronder-Offset, N.V., Rotterdam.
8. Poelmann, A.J. 1979, "Reconsideration of the target detection criterion based on adaptive antenna polarization", Tijdschrift van het Nederlands Electronica en Radiogenootschap, 44.
9. Deschamps, G.A. 1951, "Geometrical representation of the polarization of a plane electromagnetic wave", Proceedings IRE, 39.
10. Bass, F.B., Fuks, I.M. 1979, "Wave Scattering from Statistically Rough Surfaces, Pergamon, New York.
11. Plant, W.J. Jan 1977, "Studies of backscattered sea return with a CW dual-frequency X-band Radar", IEEE Transactions on Antennas and Propagation, Ap-25.
12. Weissmann, D.E., Johnson, J.W. Jan 1977, "Dual frequency correlation radar measurements of the height statistics of ocean waves", IEEE Journal of Oceanic Engineering.
13. Kennaugh, E.M., Moffat, D.L. 1965, "Transient and Impulse Response in Approximations", IEEE Proc. 53 (8), pp. 893-901.
14. Kjelaas, A.G., Nordal, P.E., Bjerkestrand, A. 1977, "Multiwavelength scintillation effects in a long-path CO_2 laser absorption spectrometer", Proc. URSI Comm F Symposium. La Baule, France, April 28 - May 6, 1977.

IV.6 (RS.3)

INVERSE METHODS FOR OCEAN WAVE IMAGING BY SAR

S. Rotheram and J.T. Macklin

Marconi Research Centre,
Chelmsford, Essex, U.K.

ABSTRACT

A Synthetic Aperture Radar (SAR) can form high resolution images of the sea surface. The direct problem for such an imaging system is now reasonably well understood and is summarised here for both the SAR image and its power spectrum. The inverse problem has only recently been explored and some simple approaches based on linear demodulation and speckle removal are described. These are applied to a number of SEASAT images.

1. INTRODUCTION

A Synthetic Aperture Radar (SAR) is able to form high resolution two dimensional images of the sea surface. Fine range resolution is achieved with short pulses using pulse compression. Fine azimuth resolution is achieved using the motion of the platform to coherently synthesise a long antenna. Typical resolutions of a few metres from aircraft and a few tens of metres from spacecraft are achieved. The SAR geometry is shown in Figure 1.

Following SEASAT in 1978 and SIR-A in 1982, a number of spaceborne SARs are being planned including SIR-B in 1984 and ERS-1 in 1987. Like SEASAT, ERS-1 is primarily an oceanographic satellite. The main instrument on ERS-1 will be a SAR which can operate in an imaging mode for obtaining large images of the sea surface, and a wave scatterometer mode for obtaining image spectra of 5 x 5 km patches of the sea surface.

W.-M. Boerner et al. (eds.), Inverse Methods in Electromagnetic Imaging, Part 2, 907–930.

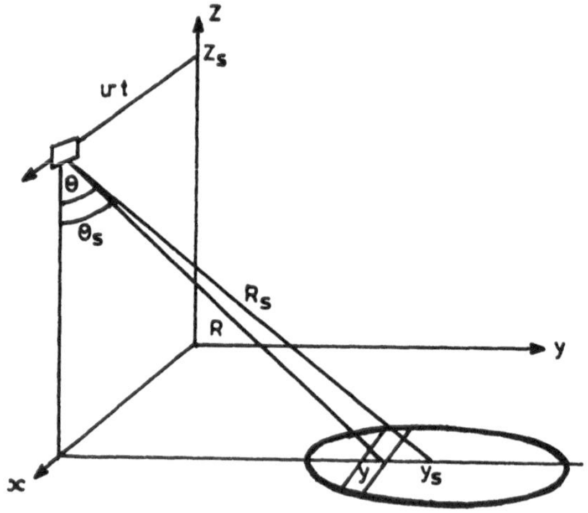

Figure 1. SAR geometry.

The main object of forming images and image spectra of the sea surface is to obtain quantitative information about the sea surface such as wave height and direction, wavenumber spectrum, storm evolution, surface winds etc. To obtain this quantitative information, the relationship between the sea surface and SAR images and spectra needs to be defined as fully as possible. There are two aspects to this relationship. These are the direct and inverse sea imaging problems.

When an imaging radar operating at a wavelength of a few centimetres illuminates the sea surface, most of the backscattered energy is Bragg scattered from water waves of similar wavelength. The resolution of the radar is typically a few metres to tens of metres and so the Bragg waves cannot be resolved. Waves of wavelength longer than the resolution are imaged through a variety of secondary interactions between long waves and Bragg waves. As there is a large separation in wavenumber space between the imaged long waves and the short Bragg waves, this leads to the idea of a two-scale theory of sea surface scattering which was originally proposed by Wright (1). These ideas are illustrated in figure 2 in which the wavenumber plane of ocean waves is divided arbitrarily into long and short waves. The Bragg waves lie in the short wave region whilst the imaged waves lie in the long wave region.

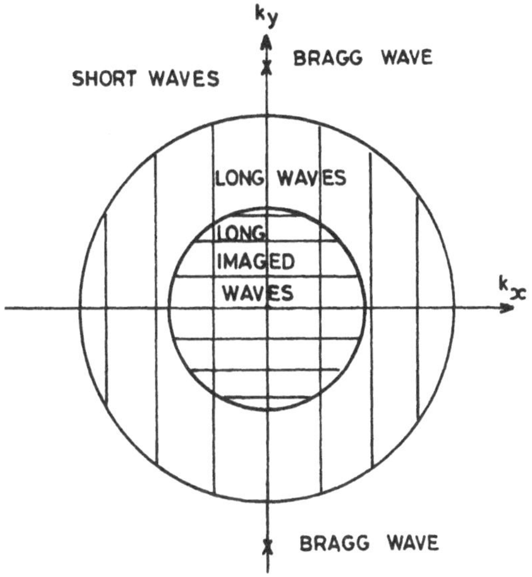

Figure 2. Wavenumber diagram.

The main secondary interactions producing imaging mechanisms have come to be known as tilting, straining, velocity bunching and Doppler splitting (2,3,4). Tilting occurs because the long waves tilt the short waves whilst straining occurs because the short waves propagate in the slowly varying long wave field. These can be described by a perturbation theory (4,5,6). Velocity bunching, which has been studied by a number of approaches (7-13), occurs because the orbital motions of the long waves Doppler shift the Bragg scattered field. A SAR determines azimuthal position by the Doppler shift, broadside being at zero Doppler, and so these periodic Doppler shifts lead to periodic scatterer displacements and thus periodic image modulations. The motion of the short waves also produces an effect known as Doppler splitting (2,4). This is because there are two Bragg waves travelling in opposite directions.

The direct problem for the SAR image is to describe the sea S that corresponds to a given image I. This can be expressed by a functional relation I = N[S]. The functional N is nonlinear due to the effects of velocity bunching although the tilting and straining contributions are essentially linear. It is also stochastic because the Bragg scattering short wave field is random and this is the origin of speckle in SAR imagery. The direct problem for the image is described in section 2. The linear part of the relation is described in detail using a linear modulation transfer function whilst the nonlinear aspects are discussed using a computed example.

The direct problem for the SAR image power spectrum is to describe the sea surface power spectrum E that corresponds to a given image power spectrum P. This can be expressed by a functional relation P = M[E]. As well as many aspects in common with the image I, this functional contains additional aspects attributable to the stochastic nature of both the short and long wave fields. P is a quadratic functional of the image and a fourth order functional of the complex amplitude image before detection. Ensemble averaging over the short waves introduces an additional speckle contribution to P in the form of a background proportional to the static system transfer function of the SAR. Ensemble averaging over the long waves leads to an azimuthal banding due to nonlinear velocity bunching. These aspects are summarised in section 3.

Some tentative inverse methods are given in section 4 and applied to SEASAT imagery. For the image power spectrum the static system transfer function is determined from a featureless speckled image (14). This is then divided out of spectra to leave the wave information on a flat, exponentially distributed, background component. This is then subtracted using a thresholding technique at various confidence levels. The resultant is demodulated using a regularised modulation transfer function to give an estimate of the sea surface power spectrum. Analogous, though less soundly based, techniques are applied to the image. These techniques are exploratory in nature and are the subject of current research. It is hoped that optimum methods will be developed from these based on Bacchus-Gilbert theory.

2. THE SAR IMAGE

To first order in the rms waveheight σ_ℓ, the long waves can be represented by

$$z = \sigma_\ell \int \frac{d^2 \underline{k} d\omega}{(2\pi)^3} \tilde{\zeta}_\ell(\underline{k},\omega)\, e^{-i\underline{k}.\underline{\rho}+i\omega t} \tag{1}$$

$$\tilde{\zeta}_\ell(\underline{k},\omega) = 2\pi\delta(\omega-g^{\frac{1}{2}}k^{\frac{1}{2}})\, \tilde{\zeta}_\ell(\underline{k})+2\pi\delta(\omega+g^{\frac{1}{2}}k^{\frac{1}{2}})\, \tilde{\zeta}_\ell{}^*(-\underline{k}) \tag{2}$$

in which z is height, $\underline{\rho}$ = (x,y) is horizontal position, x is azimuth, y is range, t is time, $\underline{k} = (k_x,k_y)$ is the wavevector of the long waves, ω is the angular frequency, $\tilde{\zeta}(\underline{k},\omega)$ is the wavenumber-frequency amplitude spectrum, $\zeta(\underline{k})$ is the wavenumber amplitude spectrum, g is the acceleration of gravity and eqn. (2) is the deep water gravity wave relation. If the long waves form a stationary and homogeneous random field

$$< \zeta_\ell(\underline{\rho},t) > \; = \; 0 \quad , \quad < \zeta_\ell(\underline{\rho},t)^2 > \; = \; 1 \tag{3}$$

$$< \tilde{\zeta}_\ell(\underline{k},\omega)\, \tilde{\zeta}_\ell{}^*(\underline{k}',\omega') > \; = \; (2\pi)^3\, \delta(\underline{k}-\underline{k}',\omega-\omega')\tilde{E}_\ell(\underline{k},\omega) \tag{4}$$

$$\tilde{E}_\ell(\underline{k},\omega) \; = \; 2\pi\delta(\omega-g^{\frac{1}{2}}k^{\frac{1}{2}})\tilde{E}_\ell(\underline{k}) + 2\pi\delta(\omega+g^{\frac{1}{2}}k^{\frac{1}{2}})\tilde{E}_\ell(-\underline{k}) \tag{5}$$

in which $\tilde{E}_\ell(\underline{k},\omega)$ is the wavenumber-frequency energy spectrum and $\tilde{E}_\ell(k)$ is the wavenumber energy spectrum. The short waves obey a set of similar relations with ℓ changed to s. These are just the linear parts of the wave field.

The long waves amplitude modulate and Doppler shift the short waves. The amplitude modulations are known as straining. In backscattering from this surface, the radio wave is Bragg scattered from the short waves with amplitude and phase modulations from the long waves. In addition to straining, the amplitude modulations are caused by the tilting of the short waves by the long waves. The phase modulations amount to a Doppler shift by the orbital velocity component of the long waves in the look direction of the radar. Details of the short wave representations and the backscattered field are in reference 4.

After SAR processing the complex amplitude image intensity is $W(\underline{\rho})$. The SAR image intensity is $I(\underline{\rho}) = |W(\underline{\rho})|^2$ which is the quantity displayed in a SAR image. It can be expanded in a long wave perturbation expansion

$$I(\underline{\rho}) \; = \; \sum_{n=0}^{\infty} \sigma_\ell{}^n\, I_n(\underline{\rho}) \tag{6}$$

In the following < . > denotes an assembly average over the short waves. The zeroth order term in eqn. (6) is the pure Bragg scattered component given by

$$< I_0 > \; = \; C\,[\tilde{E}_s(\underline{k}_b) + \tilde{E}_s(-\underline{k}_b)] \tag{7}$$

$$\underline{k}_b \; = \; (0,\, -2\, k_s \sin\theta) \quad , \quad \omega_b \; = \; g^{\frac{1}{2}} k_b{}^{\frac{1}{2}} \tag{8}$$

in which C is a constant, $\underline{k}_b$ is the Bragg wavevector, ω_b is the Bragg angular frequency, θ is the angle of incidence to the vertical (see figure 1) and $k_s = 2\pi/\lambda_s$ is the radar wavenumber.

The first order term $I_1(\underline{\rho})$ represents linear modulations and can be written

$$< I_1(\underline{\rho}) > \; = \; < I_0 > \int \frac{d^2\underline{k}d\omega}{(2\pi)^3}\, \tilde{\zeta}_\ell\, (\underline{k},\omega)\; \tilde{T}(k,\omega)\; e^{-i\underline{k}.\underline{\rho}+i\omega t} \tag{9}$$

in terms of a linear modulation transfer function $\tilde{T}(\underline{k},\omega)$. For simplicity the dispersion relation in eqn. (2) has been left understood in eqn. (9), as in eqn. (1) and elsewhere. It is useful to separate $\tilde{T}(\underline{k},\omega)$ into two components

$$\tilde{T}(\underline{k},\omega) = \tilde{M}(\underline{k},\omega)\ \tilde{Q}(\underline{k},\omega) \tag{10}$$

Here $\tilde{M}(\underline{k},\omega)$ is the scattering transfer function and $\tilde{Q}(\underline{k},\omega)$ is the dynamic system transfer function. The transformation from the sea surface ζ_ℓ to the SAR image spectral density $\tilde{\zeta}_\ell\tilde{T}$ can be thought of as the two stage process $\tilde{\zeta}_\ell \rightarrow \tilde{\zeta}_\ell\tilde{M} \rightarrow \tilde{\zeta}_\ell\tilde{M}\tilde{Q}$ meaning sea surface → scatterer density → SAR image density (3).

The scattering transfer function $\tilde{M}(\underline{k},\omega)$ is given by

$$\tilde{M}(\underline{k},\omega) = \sum_{p=\pm 1} \frac{2\tilde{E}_s(p\underline{k}_b)\,e^{-ip\omega_b t_x}}{\{\tilde{E}_s(\underline{k}_b)+\tilde{E}_s(-\underline{k}_b)\}} \left[\tilde{L}_p(\underline{k})\cos(\omega t_x/2) + i\tilde{A}(\underline{k})\sin(\omega t_x/2)\right] \tag{11}$$

$$t_x = \frac{TXk_x}{4\pi}\ , \quad T = \frac{R\lambda_s}{Xv} \tag{12}$$

$$\tilde{A}(\underline{k}) = -2k_s \sin\theta .\ \frac{k_y}{k} + 2i\ k_s \cos\theta \tag{13}$$

$$\tilde{L}_p(\underline{k}) = \frac{k}{2} - \frac{k_x^2}{4k} - \frac{ik_y}{2}\cos\theta - 2i\ k_y \tan\theta\left[1-\mathrm{Re}\left\{\frac{\cos\theta}{(n^2-\sin^2\theta)^{1/2}}\right\}\right] - \frac{\tilde{A}(\underline{k})}{2}\ \underline{k}\cdot\frac{\partial}{\partial \underline{k}_b}\ \ell n\ \tilde{E}_s(p\underline{k}_b) \tag{14}$$

in which X is the length of the antenna in azimuth, v is the velocity of the SAR, T is the one look integration time and n is the complex refractive index of the sea. In brief, tilting and straining are given by the term in $\tilde{L}_p(\underline{k})$, velocity bunching by the term in $\tilde{A}(\underline{k})$, and Doppler splitting by the phases $e^{\mp i\omega_b t_x}$. These are discussed in detail elsewhere (2,3,4).

Figure 3 shows $|\tilde{M}(\underline{k},\omega)|$ for $\omega = g^{\frac{1}{2}}k^{\frac{1}{2}}$ and SEASAT parameters (R/v = 128s, θ = 20°) in an approximation in which $\tilde{E}_s(\underline{k})$ is given by an isotropic Phillips spectrum, Doppler splitting is ignored and the moduli of $\tilde{A}$ and $\tilde{L}_p$ are formed separately. It is plotted against k for various values of the azimuth angle φ where $\underline{k} = k(\cos\varphi, \sin\varphi)$ so φ is the angle that the wavevector makes with the azimuth or x direction. Figure 4 shows the same date but with $|\tilde{M}|^{-1}$ plotted as a contour map in the wavenumber plane.

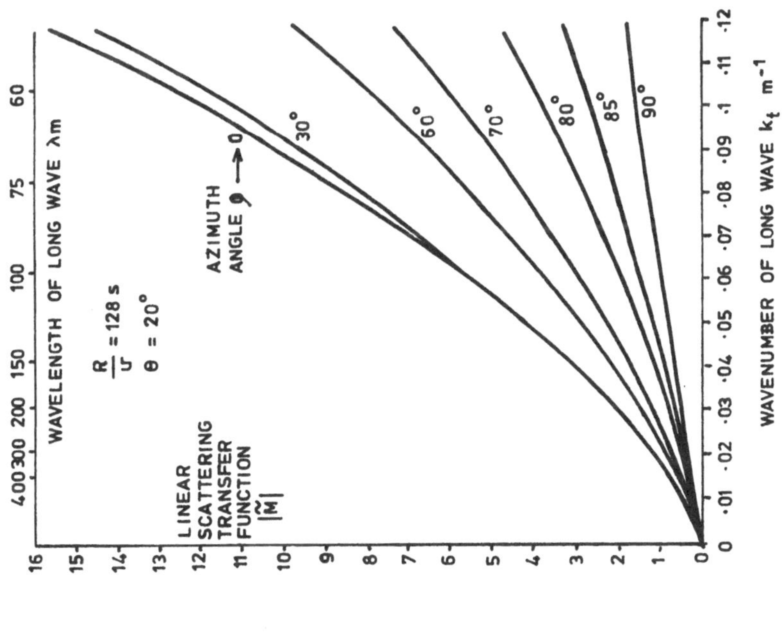

Figure 3 Modulus of scattering transfer function $|\tilde{M}|$. SEASAT parameters

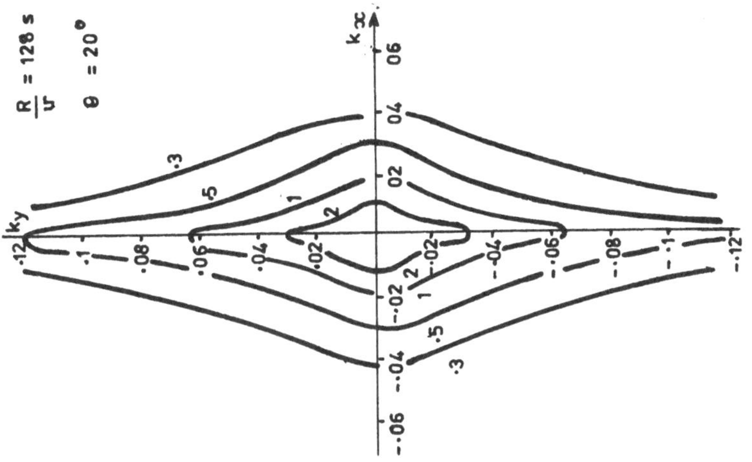

Figure 4 Contour map of $|M|^{-1}$ SEASAT parameters

The dynamic system transfer function $\tilde{Q}(\underline{k},\omega)$ is given by

$$\tilde{Q}(\underline{k},\omega) = \exp\left[-\frac{1}{32\pi}\left(k_x^{\,2}N^2\ X^2 + \frac{k_y c^2\tau^2}{\sin^2\theta} + \frac{4\omega^2\ T^2}{N^2}\right)\right] \qquad (15)$$

in which N is the number of azimuth looks, c is the velocity of light and τ is the compressed pulse length. It expresses the resolution of the system in azimuth, range and time by cutting off waves of large wavenumber and frequency.

A serious complication in SAR imaging theory is that velocity bunching becomes strongly nonlinear as the wave amplitude increases. The effects are described by the higher order terms in eqn. (6) and details are given in reference 4. For a single long wave of amplitude a and wavelength λ, figure 5 shows the linear and nonlinear imaging regimes in the a - λ plane for various azimuth angles φ and SEASAT parameters.

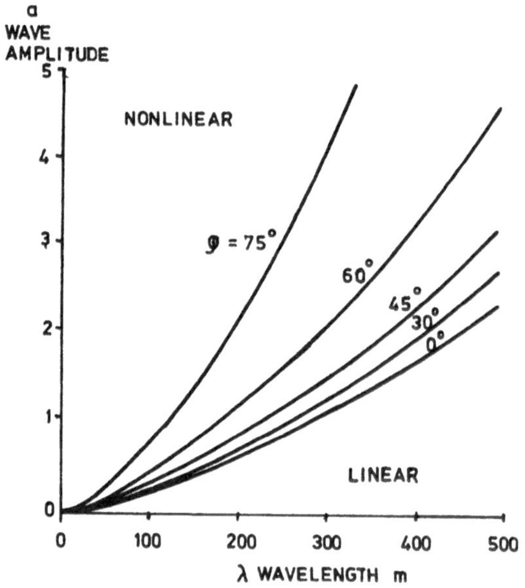

Figure 5. Linear-nonlinear transition in velocity bunching.

Figure 6 shows a computed example for λ = 200m, φ = 0 and increasing values of a. As the wave amplitude a increases, the image modulations increase at first but then decrease as nonlinearity becomes important.

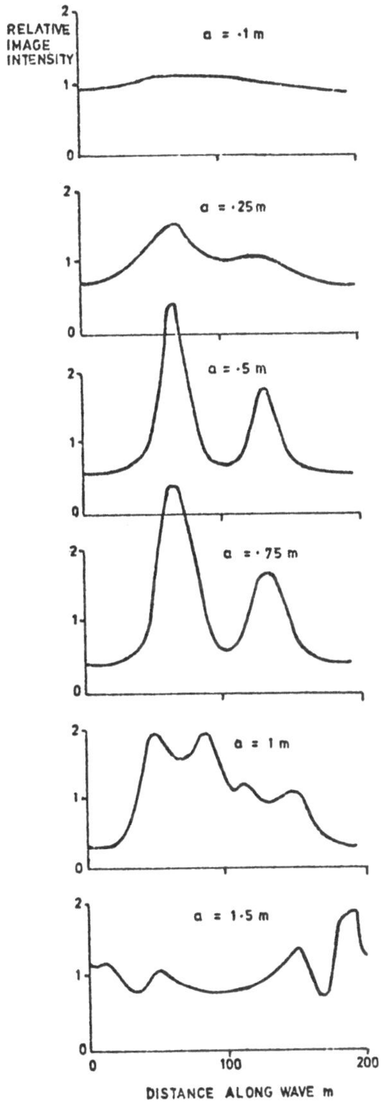

Figure 6. Image intensity for different wave amplitudes a. 200m azimuth waves, 4-look SEASAT.

3. THE SAR IMAGE POWER SPECTRUM

For ocean wave fields, the image power spectrum is in some respects more interesting than the image. This is reflected in the wave scatterometer mode of ERS-1. Some useful studies of the image power spectrum have been published recently (14,15). The theory requires a number of significant and difficult steps beyond that for the SAR image and these are summarised here with fuller details given elsewhere (4).

Consider the effects of sampling and finite image size. Assume that $\underline{\rho}_0 = (x_0, y_0)$ are the coordinates of the bottom left hand corner of the image which is rectangular with L x J pixels each of size u x w. The sampled image is

$$I_{\ell j} = I(x_0+\ell u, y_0+jw) = \int d^2\underline{\rho}\, I(\underline{\rho})\, \delta(x-x_0-\ell u, y-y_0-jw) \qquad (16)$$

with $\ell = 0,1,\ldots,L-1$ and $j=0,1,\ldots,J-1$. The true image spectrum $\tilde{I}(\underline{k})$ is the Fourier transform of $\tilde{I}(\underline{\rho})$ whilst the sampled image spectrum $I_s(\underline{k})$ is the DFT of $I_{\ell j}$ and these are related by

$$\tilde{I}_s(\underline{k}) = \int \frac{d^2\underline{k}'}{(2\pi)^2}\, \tilde{I}(\underline{k}')\, \tilde{K}(\underline{k},\underline{k}') \qquad (17)$$

$$\tilde{K}(\underline{k},\underline{k}') = \left[\frac{1-e^{i(k_x-k_x')Lu}}{1-e^{i(k_x-k_x')u}}\right]\left[\frac{1-e^{i(k_y-k_y')Jw}}{1-e^{i(k_y-k_y')w}}\right] \qquad (18)$$

The sampled image power spectrum $\tilde{P}_s(\underline{k}) = |\tilde{I}_s(\underline{k})|^2$ and so

$$\tilde{P}_s(\underline{k}) = \int \frac{d^2\underline{k}'d^2\underline{k}''}{(2\pi)^4}\, \tilde{I}(\underline{k}')\, \tilde{I}^*(\underline{k}'')\tilde{K}(\underline{k},\underline{k}')\tilde{K}^*(\underline{k},\underline{k}'') \qquad (19)$$

This is the quantity displayed in a SAR image power spectrum.

One is thus led to study the quantity $\tilde{I}(\underline{k}')\tilde{I}^*(\underline{k}'')$. This is related to the complex amplitude $W(\underline{\rho})$ by

$$\tilde{I}(\underline{k})\tilde{I}^*(\underline{k}') = \int d^2\underline{\rho}\, d^2\underline{\rho}'\, |W(\underline{\rho})\, W(\rho')|^2\, e^{i\underline{k}\cdot\underline{\rho}-i\underline{k}'\cdot\underline{\rho}'} \qquad (20)$$

The short wave assembly average of this requires evaluation of the fourth moment $\langle |W(\underline{\rho})W(\underline{\rho}')|^2 \rangle$. For Gaussian $W(\underline{\rho})$ this gives three terms, one of which is negligible. The other two terms are

$$\langle \tilde{I}(\underline{k})\, \tilde{I}^*(\underline{k}') \rangle_1 = \int d^2\underline{\rho}\, d^2\underline{\rho}' \langle |W(\underline{\rho})|^2\rangle\langle |W(\underline{\rho}')|^2\rangle\, e^{i\underline{k}\cdot\underline{\rho}-i\underline{k}'\cdot\underline{\rho}'} \qquad (21)$$

$$\tilde{I}(\underline{k})\ \tilde{I}^*(\underline{k}')\ >_2 = \int d^2\underline{\rho}\, d^2\underline{\rho}'\ |< W(\underline{\rho})W^*(\underline{\rho}')>|^2\, e^{i\underline{k}\cdot\underline{\rho}-i\underline{k}'\cdot\underline{\rho}'} \tag{22}$$

The first term is $< \tilde{I}(\underline{k}) >< \tilde{I}^*(\underline{k}') >$ and gives a term $|< I_s(\underline{k}) >|^2$ in the sampled image power spectrum. Its properties are simply the Fourier transform of those in section 2. The second term given by eqn. (22) is a new feature arising from the crossed terms in the fourth moment. It is purely a consequence of the coherent speckle in SAR imagery.

For the SAR image one is interested in a particular realisation of the sea surface. For the SAR image power spectrum one can average over an ensemble of realisations of the long wave field. A long wave assembly average will be represented by an overbar. For the term in eqn. (22) this leads simply to

$$\overline{< \tilde{I}(\underline{k})\tilde{I}^*(\underline{k}') >_2} = < I_o >^2\ (2\pi)^2\,\delta(\underline{k}-\underline{k}')\ \frac{Xc\tau}{4\sin\theta}\ \tilde{Q}_o(\underline{k})^2 \tag{23}$$

$$\tilde{Q}_o(\underline{k}) = \tilde{Q}(\underline{k},0) = \exp\left[-\frac{1}{32\pi}\left(k_x^2 N^2 X^2 + \frac{k_y^2 c^2 \tau^2}{\sin^2\theta}\right)\right] \tag{24}$$

Here $\tilde{Q}_o(\underline{k})$ will be called the static system transfer function. It is the response to a random structureless scene. Substituting eqn. (24) in eqn. (19) gives the contribution to the sampled image power spectrum

$$< I_o >^2\ \frac{Xc\tau}{4\sin\theta}\ \frac{L\,J}{4u^2 w^2}\ \tilde{Q}_o(\underline{k})^2 \tag{25}$$

so this represents a background contribution that falls off like $\tilde{Q}_o(\underline{k})^2$. It contains no information about the long waves.

The contribution of the term in eqn. (21), after long wave averaging, is much more complicated. It can be written as a long wave perturbation expansion in which only even powers of σ_ℓ appear. An exponential series can be separated from this series with the result

$$\overline{< \tilde{I}(\underline{k})\tilde{I}^*(\underline{k}') >_1} = < I_o >^2\ (2\pi)\ \delta(\underline{k}-\underline{k}')\ \exp\left[-\frac{X^2 k_x^2 \nu^2}{16}\right] \sum_{n=0}^{\infty} \sigma_\ell^{2n} \tilde{F}_n(\underline{k}) \tag{26}$$

$$\nu^2 = \frac{\sigma_\ell T^2}{\pi} \int \frac{d^2\underline{k}\, d\omega}{(2\pi)^3}\ \tilde{E}_\ell(\underline{k},\omega)\ |\tilde{A}(\underline{k})|^2\,\omega^2 \tag{27}$$

The exponential gives an azimuthal cut-off as described below. The zeroth order term in eqn. (26) is

$$\tilde{F}_o(\underline{k}) = (2\pi)^2 \delta(\underline{k}) \tag{28}$$

and arises from the Fourier transform of the mean intensity in the SAR image. Its contribution to the sampled image power spectrum is found from eqns. (19), (26) and (28) to be

$$< I_o >^2 \ |\tilde{K}(\underline{k},0)|^2 \tag{29}$$

This is usually sampled at $\underline{k} = 2\pi\ (\ell/Lu,\ j/J\omega)$ giving zero except at $\underline{k} = 0$ for which eqn. (29) gives

$$< I_o >^2 \ \frac{L\ J}{4u^2 w^2} \tag{30}$$

The first order term $\tilde{F}_1(\underline{k})$ has only been determined in an approximation in which tilting and straining have been ignored. The result is

$$\tilde{F}_1(\underline{k}) = \left| 2\tilde{A}(\underline{k})\tilde{P}(\underline{k}) \sin\left(\frac{\omega t_x}{2}\right) \right|^2 \left[\tilde{E}\ (\underline{k})\tilde{Q}(\underline{k},\omega)^2 + \tilde{E}_\ell(-\underline{k})\tilde{Q}(-\underline{k},\omega)^2\right] \tag{31}$$

$$\tilde{P}(\underline{k}) = \sum_{p=\pm 1} \frac{\tilde{E}_s(p\underline{k}_b)\, e^{-ip\omega_b t_x}}{\{\tilde{E}_s(\underline{k}_b) + \tilde{E}_s(-\underline{k}_b)\}} \tag{32}$$

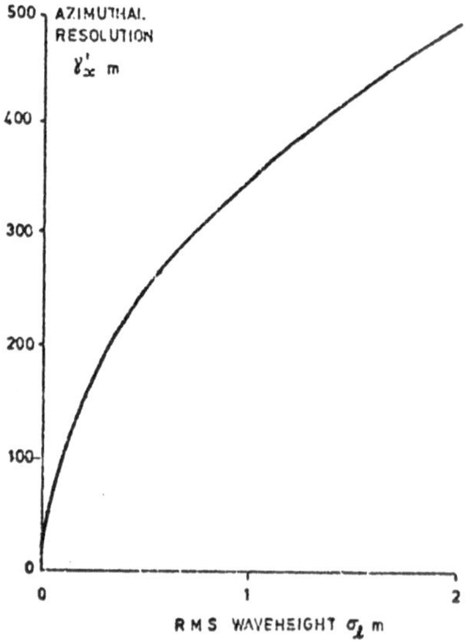

Figure 7. Azimuthal resolution against r.m.s. wave-height. 4-look SEASAT.

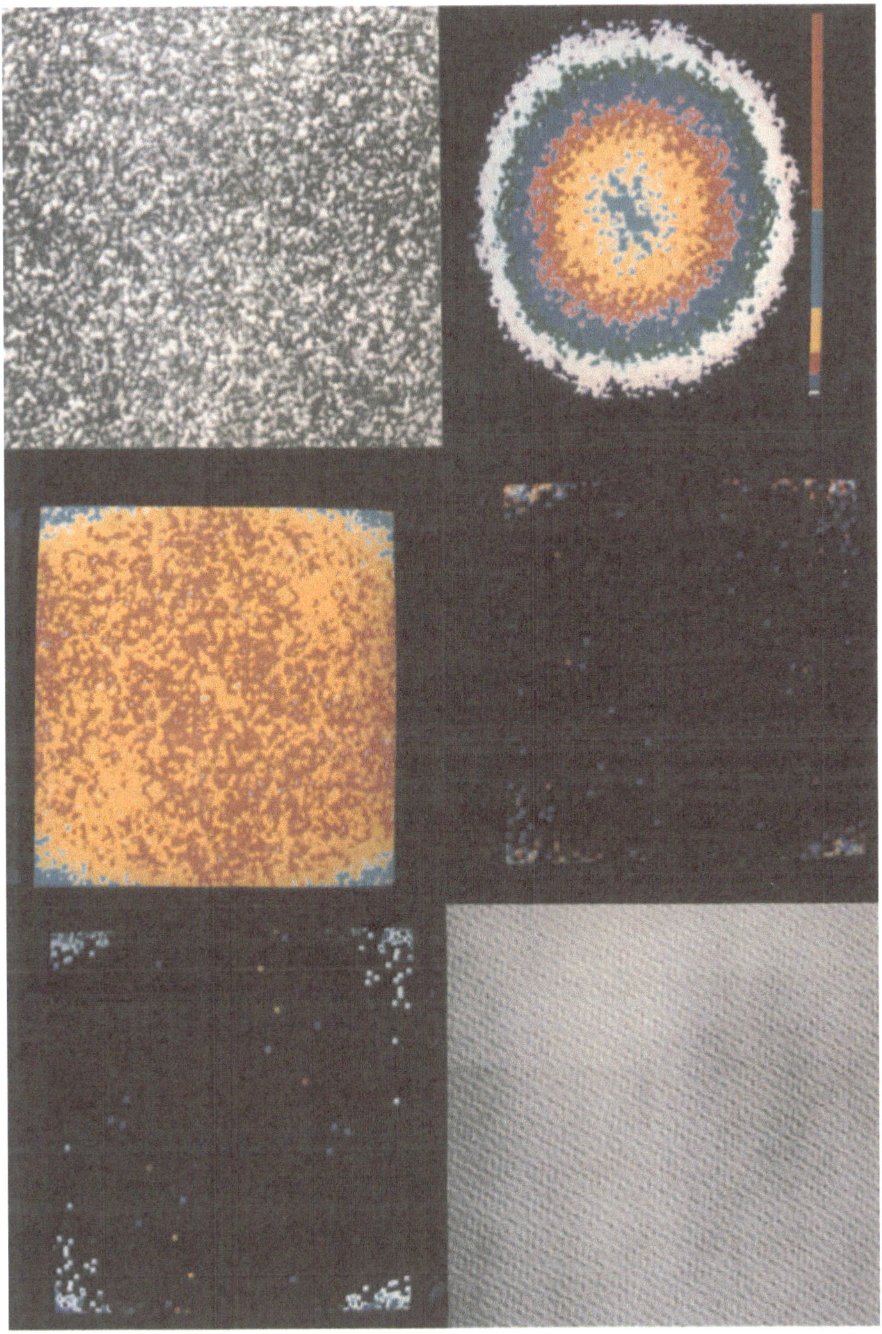

Figure 8. SEASAT image and spectra. Orbit 762.
Latitude $58^{\circ}26'$N. Longitude $5^{\circ}60'$W.

The function $\tilde{P}(\underline{k})$ describes Doppler splitting. The contribution of the first order term to the sampled image power spectrum is found from eqns. (19), (26) and (31) to be

$$< I_o >^2 \frac{L\ J}{4u^2 W^2} \sigma_\ell^2 \exp\left[- \frac{X^2 k_x^2 \nu^2}{16} \right] \tilde{F}_1(\underline{k}) \quad (33)$$

As $\tilde{F}_1(\underline{k})$ is a linear functional of the wavenumber energy spectrum $\tilde{E}_\ell(\underline{k})$, it is the term containing the information in a SAR image power spectrum.

The expression in eqn. (33) is incomplete because of the neglect of tilting and straining. One consequence is that $\tilde{F}_1(\underline{k})$ in eqn. (31) is incomplete. The expression between the modulus signs should be similar to $\tilde{M}(\underline{k},\omega)$ in eqn. (11) but the precise details have not been worked out. In addition to the exponentially decreasing term with k_x in eqn. (33) we also believe that there is an exponentially increasing term of the form $e^{\beta k_y^2}$. However its form and magnitude have not been determined.

Higher order terms in eqn. (26) are a consequence of nonlinearity in velocity bunching. The nth term is an nth order functional of $\tilde{E}_\ell(\underline{k})$. For a narrow band swell wave system $\tilde{E}_\ell(\underline{k})$ is sharply peaked and the higher order terms give contributions that are harmonics of the main peak.

However the principal contribution of nonlinearity is the azimuthal cut-off in eqns. (26) and (33) with ν given by eqn.(27). For a static scene the azimuthal resolution γ_x of a SAR is

$$\gamma_x = \frac{XN}{2} \quad (34)$$

The azimuthal cut-off modifies this to

$$\gamma_x' = \frac{X}{2} (N^2 + \nu^2)^{\frac{1}{2}} \quad (35)$$

For a wind-wave system one finds approximately (4)

$$\frac{X}{2} \nu \sim 2.7 \frac{R}{V} \sigma_\ell^{\frac{1}{2}} \quad (36)$$

in MKS units. This agrees well with the empirical value $2(R/v)\sigma_\ell^{\frac{1}{2}}$ given by Beal (14). For SEASAT eqns. (35) and (36) give

$$\gamma_x' = 6(N^2 + 3400\ \sigma_\ell)^{\frac{1}{2}} \quad (37)$$

Figure 7 shows γ_x' for N = 4 plotted against σ_ℓ. The azimuthal resolution is drastically reduced which is the reason that azimuthal waves appear less often on SAR imagery than one would expect.

4. INVERSE METHODS

4.1. SEASAT images

Before describing some simple inverse methods, it is helpful to introduce the three images and spectra to which these methods are to be applied. These are shown in Figures 8, 9 and 10. They are SEASAT images with the orbits and geographical locations given in table 1. The images are in the top left of the figures.

Table 1

Figure	Orbit No.	Date	Latitude	Longitude	Description
8	762	19/8/78	$58^{\circ}26'$N	$5^{\circ}50'$W	No waves
9	1087	11/9/78	$60^{\circ}32'$N	$12^{\circ}58'$W	Azimuthal waves
10	762	19/8/78	$60^{\circ}11'$N	$6^{\circ}41'$W	Range waves

and consist of 256 x 256 pixels with 12.5 m pixel spacing and 25 m resolution. Although this is 4-look azimuth resolution, only one look is displayed. SAR processing was carried out at the Royal Aircraft Establishment, U.K. The image power spectra are in the top right of the figures. They have been smoothed with a 4 x 4 running average for display purposes and colour coded as shown. In all the images and spectra, the x(azimuth) and k_x coordinates are horizontal and the y(range) and k_y coordinates are vertical. The images are 3 km square whilst the boundaries of the spectra are at k_x, $k_y = \pm 2\pi/25$ m^{-1}.

The image in figure 8 contains virtually no waves. This is indicated by the lack of structure in the power spectrum which consists almost entirely of the background speckle contribution with the expected value in eqn. (25). This is the origin of the radial decrease in the image power spectrum. There are two small peaks indicating some very weak waves.

The image in figure 9 contains near azimuth waves. This is shown by the two large peaks in the image power spectrum which show that the wavelength is about 300 m with azimuth $\varphi = 28^{\circ}$. The two small peaks near to the origin correspond to larger scale structure with a scale size of about 1 km. Note that the power spectra are symmetric about the origin and so there is a 180° ambiguity in the direction of the waves.

The image in figure 10 contains near range waves. The image power spectrum shows that there are two wave systems as the spectra are double peaked. The wavelength is about 200 m with azimuths $\varphi = 80^{\circ}$ and 100°. Another feature of this

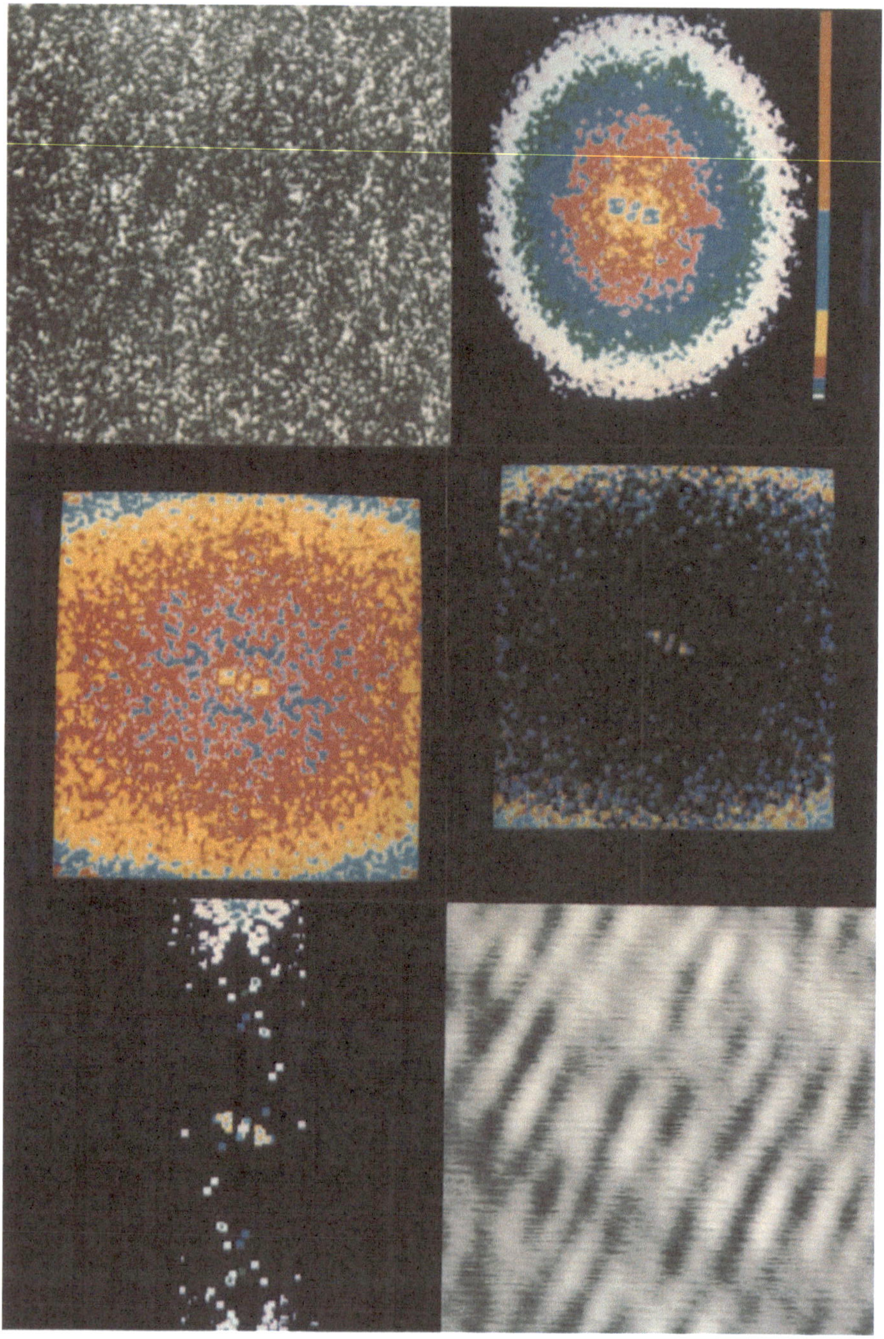

Figure 9. SEASAT image and spectra. Orbit 1087. Latitude 60°32'N. Longitude 12°58'W.

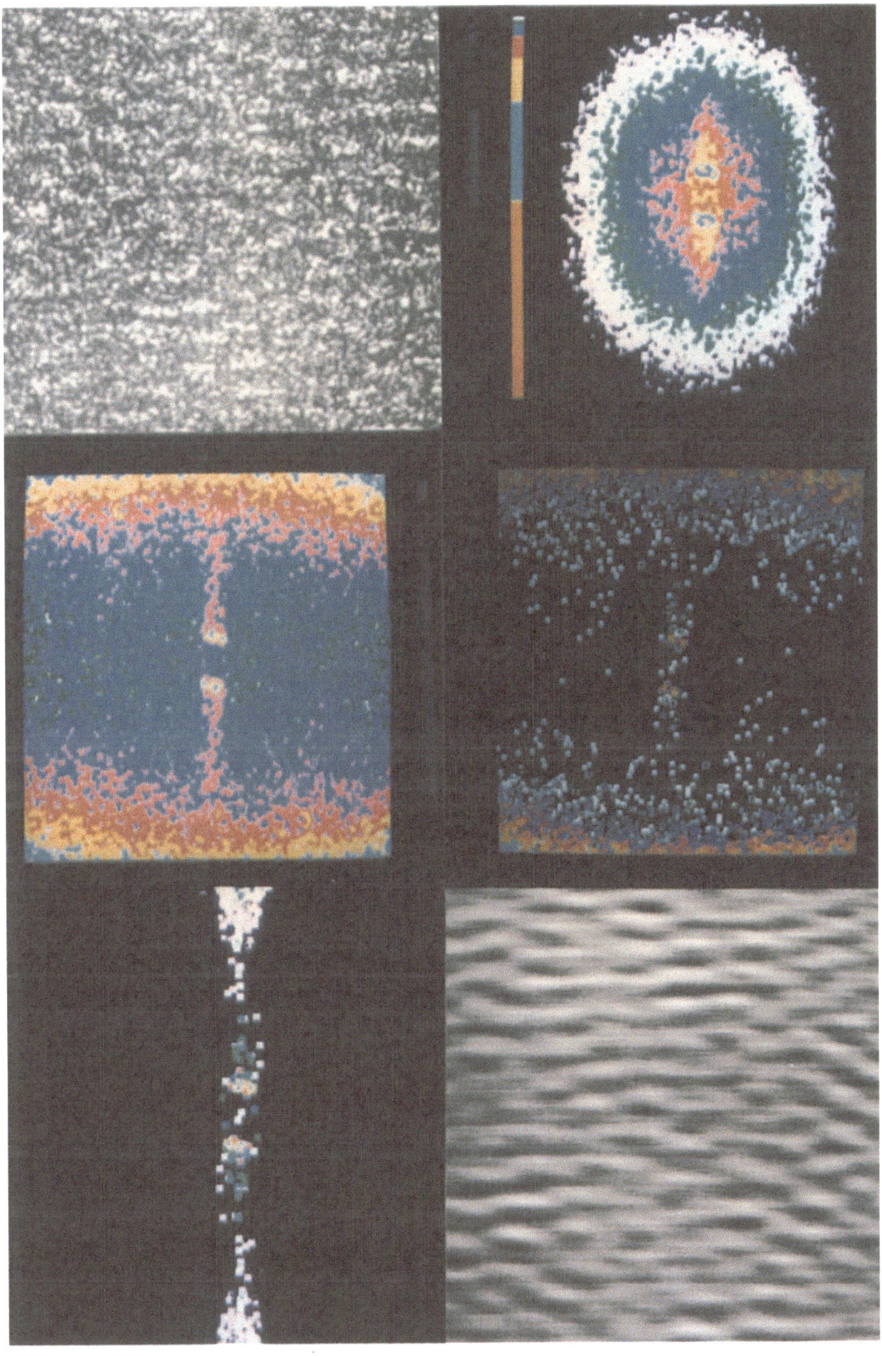

Figure 10. SEASAT image and spectra. Orbit 762.
Latitude 60°11'N. Longitude 6°41'W.

spectrum is the azimuthal banding due to nonlinear velocity bunching and given by the exponential in eqn. (33). There are also lesser peaks at wavenumbers which are approximately twice the main peaks. It is very likely that these are the harmonics of the main peaks corresponding to the n = 2 term in eqn. (26).

4.2. Inverting the image power spectrum

Consider the image power spectrum. The three main contributions are

a) a spike at the origin, $\underline{k} = 0$, given by eqn. (30).

b) a background speckle contribution proportional to $\tilde{Q}_0(k)^2$ given by eqn. (25).

c) a term containing information about the wavenumber energy spectrum $\tilde{E}_\ell(\underline{k})$ given by eqn. (33).

An elementary inverse method involves the removal of the terms in (a) and (b) and the demodulation of the term in (c). The term in (a) is simply removed by setting the central pixel to zero.

The removal of the background speckle contribution requires a determination of the static system transfer function $\tilde{Q}_0(\underline{k})$. The formula in eqn. (24) is based on a rather simple model of a SAR system and in practice one should measure $\tilde{Q}_0(\underline{k})$ for a given SAR system and SAR processor. To determine $\tilde{Q}_0(\underline{k})$ we use a method described by Beal (14). The image power spectrum of a scene devoid of long waves or other large scale structure should have the expected value in eqn. (25) which is proportional to $\tilde{Q}_0(\underline{k})$. It may be shown that the individual pixels are exponentially distributed with this expected value and so considerable averaging is required. If one assumes that the k_x and k_y variations have the independent forms

$$\tilde{Q}_0(\underline{k}) = \tilde{Q}_x(k_x)\tilde{Q}_y(k_y) \ , \quad \tilde{Q}_x(0) = \tilde{Q}_y(0) = 1 \tag{38}$$

an estimate of $\tilde{Q}_x(k_x)$ can be found by integrating $\tilde{P}_s(\underline{k})$ over k_y to give

$$\tilde{Q}_x(k_x) = \frac{\tilde{J}_x(k_x)}{\tilde{J}_x(0)} \ , \quad \tilde{J}_x(k_x) = \int \frac{dk_y}{2\pi} \ \tilde{P}_s(\underline{k}) \tag{39}$$

and similarly for $\tilde{Q}_y(k_y)$ and $\tilde{J}_y(k_y)$.

The image in figure 8 is suitable for determining $\tilde{Q}_0(\underline{k})$. The image power spectrum has been integrated over k_x and k_y to give $\tilde{J}_y(k_y)$ and $\tilde{J}_x(k_x)$ as in eqn. (39). Figure 11 shows $\ell n\ \tilde{J}_x(k_x)$ plotted against k_x^2 and figure 12 shows $\ell n\ J_y(k_y)$

plotted against k_y^2. The results are evidently close to the straight lines drawn showing that $Q_o(\underline{k})$ is well approximated by the Gaussian expression

$$\tilde{Q}_o(\underline{k}) = \exp\left[-42.6\, k_x^2 - 37.0\, k_y^2\right] \qquad (40)$$

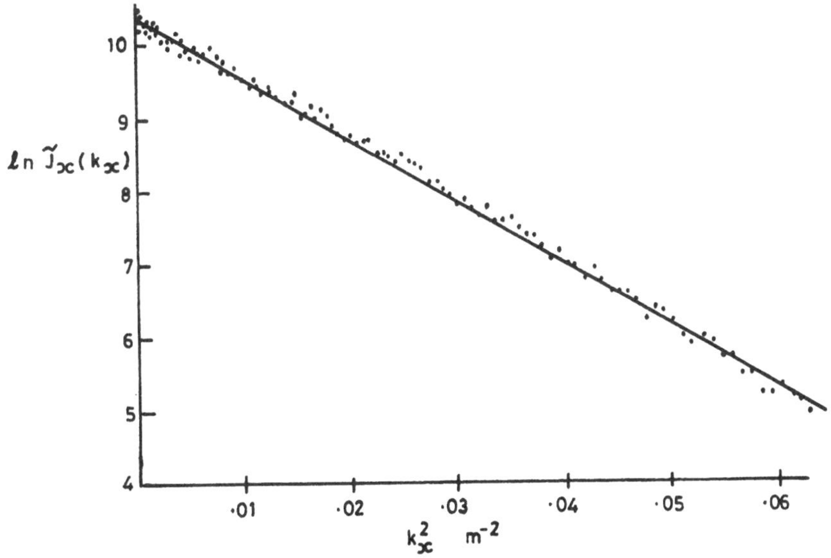

Figure 11. Azimuthal variation of the static system transfer function.

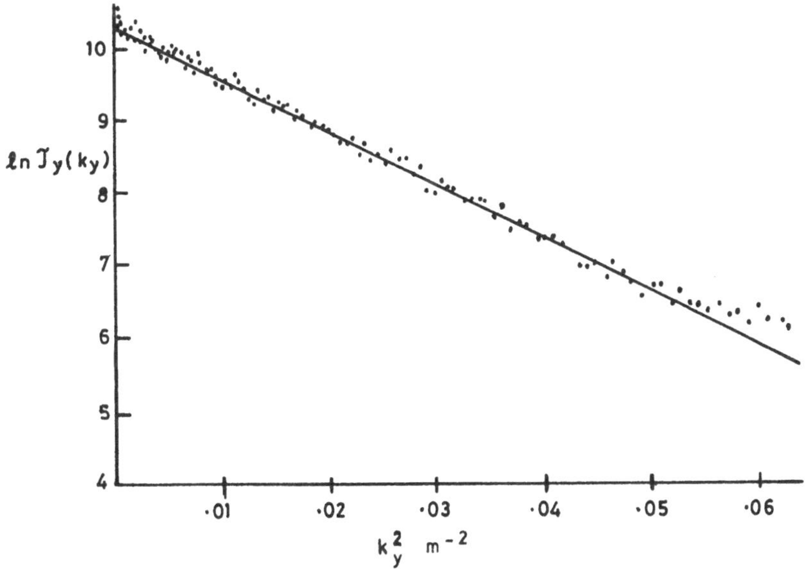

Figure 12. Range variation of the static system transfer function.

The next stage in the processing is to divide the image power spectrum by $\tilde{Q}_o(\underline{k})^2$. This should convert the speckle component into a flat exponentially distributed background. This is shown in the middle left of figures 8, 9 and 10. In figure 8 the spectrum is indeed flat apart from the corners. This residial variation comes from the tail of $\ell n\, \tilde{J}_y(k_y)$ in figure 12 which deviates slightly from a straight line. This can be avoided by fitting a more complicated expression than the Gaussian in eqn. (40). Figure 9 is similar to 8 but with the long wave structure superimposed. Figure 10 shows the azimuthal banding due to the exponential in eqn. (33). There is also a much stronger variation in the k_y direction due to the term $e^{\beta k_y^2}$ mentioned in section 3.

The next stage in the processing of an image power spectrum is the removal of the flat exponentially distributed background. To determine the mean value of this speckle background it is not safe simply to average over the spectrum because of the presence of the long wave structure. A safe procedure is to plot out the intensity distribution of the spectrum. This is shown in figure 13 in which the data marked A comes from figure 8 and that marked B comes from figure 10. For the exponential distribution, the probability density f(x) and mean value $< x >$ are

$$f(x) = \alpha\, e^{-\alpha x} \quad , \quad < x > \;=\; 1/\alpha \tag{41}$$

and so the slope of the distribution is simply related to the mean value. As the long wave structure is concentrated in the high intensity part of the distribution it suffices to measure the slope in the low intensity part of the distribution. The data marked A in figure 13 do not show this feature because of the absence of long wave structure but it is very evident in the data marked B.

Having determined the mean value in the background, it can be removed using a thresholding technique. For the sampled image power spectrum $\tilde{P}_s(\underline{k})$ (top right in figures 8, 9 and 10), the spectrum after removal of the static system transfer function (middle left of figures 8, 9 and 10) will be denoted $\tilde{P}'_s(\underline{k}) = \tilde{P}_s(k)\tilde{Q}_o(k)^{-2}$. For a threshold T the thresholded spectrum will be denoted $\tilde{P}''_s(\underline{k})$ given by

$$\begin{aligned} \tilde{P}''_s(\underline{k}) &= \tilde{P}'_s(k) - < x > \quad \text{for } \tilde{P}'_s(k) > T \\ &= 0 \qquad\qquad\qquad\quad\;\; \text{for } \tilde{P}'_s(\underline{k}) < T \end{aligned} \tag{42}$$

In other words the background is simply subtracted from the spectrum when the spectrum is above the threshold T but the spectrum is set to zero when it is below the threshold T. Data above the threshold is deemed to come from the information

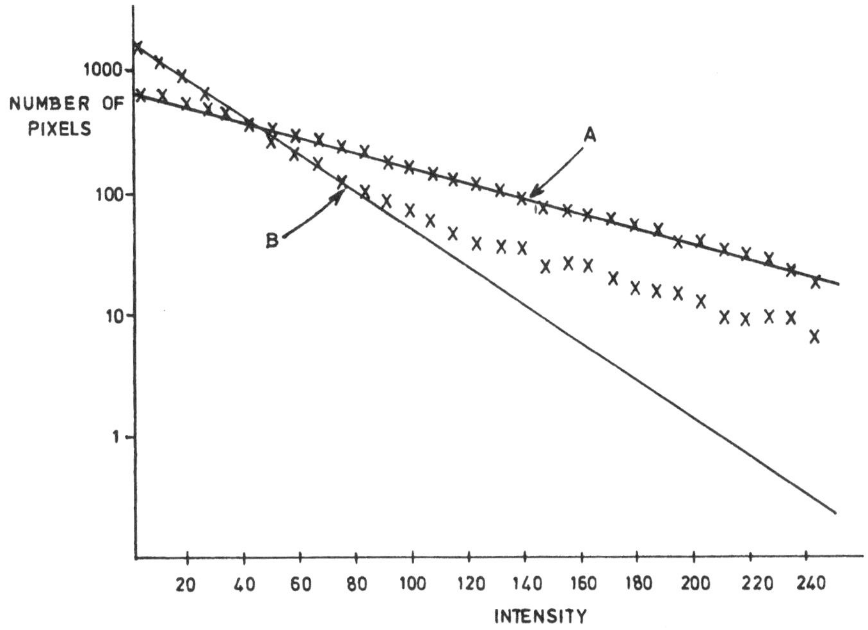

Figure 13. Intensity distributions of image power spectra.

carrying term in eqn. (33) whilst data below the threshold is deemed to come from the exponentially distributed background. It remains to fix T. For the exponential distribution the cumulative distribution p(x) satisfies

$$p(x) = 1 - e^{-x/< x >}, \quad x = -< x > \ln [1 - p(x)] \qquad (43)$$

We have used p = .977 and .999 which are the one sided 2σ and 3σ confidence levels of a Gaussian distribution. This gives $T = 3.8 < x >$ and $T = 6.9 < x >$. The thresholded spectra are shown in the centre right of figures 8, 9 and 10. Figure 8 uses $T = 6.9 < x >$. The result is a blank spectrum containing a few false alarms where the speckle distribution exceeds the threshold. Figures 9 and 10 use $T = 3.8 < x >$. The result is the information carrying term on a blank background plus randomly distributed false alarms. For reasons already given the edges of the spectra are poorly handled by these procedures.

Having isolated the information carrying term, the final stage is the demodulation of this term to extract $E_\ell(k)$. However the expressions in eqns. (31) and (33) are incomplete and this stage has not yet been carried out. The analogous stage for the image is given below. It amounts to linear demodulation.

4.2. Inverting the image

An inverse theory can be constructed for the SAR image based on the linear part of the imaging theory given in section 2. The first stage is to form the DFT of the sampled image $I_{\ell j}$, as in eqn (16). This gives the sampled image spectrum $\tilde{I}_s(\underline{k})$. This is complex but Hermitian. The zeroth order term $<^s\tilde{I}_o>$ is removed by setting the central pixel to zero. The next stage is linear demodulation. This is achieved by dividing by the linear modulation transfer function $\tilde{T}(\underline{k},\omega)$ as in eqn. (9). This function is given by the two terms in eqn. (10). The factor $\tilde{Q}(k,\omega)$ has been approximated by $\tilde{Q}_o(\underline{k})$ in eqn. (40) as the final term of the exponential in eqn. (15) is negligible for N = 4. The factor $\tilde{M}(\underline{k},\omega)$ is given by the approximation in figures 3 and 4.

The function M^{-1} is singular at the origin. A straightforward inversion tends to blow up the central part of the spectrum which corresponds to very large scale structure in the image. To avoid this a regularised version is used in which $\tilde{M}$ is replaced by $\tilde{M} + 1/(4\tilde{M})$. No justification for this ad hoc form will be given here but this form seems to be reasonably successful.

Having demodulated the sampled image spectrum in this way, an inverse DFT produces a reconstructed image. However the result is highly speckled. To reduce the speckle a further stage has been carried out on the demodulated spectrum, denoted $\tilde{I}'_s(\underline{k})$. This is set to zero at the same points as the power spectrum as in eqn. (42) to give $\tilde{I}''_s(\underline{k})$ given by

$$\begin{aligned} \tilde{I}''_s(\underline{k}) &= \tilde{I}'_s(\underline{k}) \quad \text{for} \quad \tilde{P}'_s(\underline{k}) > T \\ &= 0 \quad \text{for} \quad \tilde{P}_s(\underline{k}) < T \end{aligned} \tag{44}$$

It is difficult to display this function as it is complex. The bottom left of figures 8, 9 and 10 shows $|\tilde{I}''_s(\underline{k})|^2$. Note however that the expectation $< |\tilde{I}''_s(\underline{k})|^2 >$ of the displayed images is not the same as $|< I''_s(\underline{k})^s>|^2$, there being an additional term in the former analogous to the background speckle term in eqn.(22).

The final stage in the processing is to form the inverse DFT of $\tilde{I}''_s(\underline{k})$ to give the inverse image $\tilde{I}''_s(\underline{\rho})$. This is shown in the bottom right of figures 8, 9 and 10. The fine structure in figure 8 is meaningless; it simply arises from the corners of $\tilde{I}''_s(\underline{k})$ in the bottom left of figure 8. As described in section 4.1, these are poorly treated and avoid being removed by the thresholding. The inverse images in figures 9 and 10 are much clearer than the originals. This can be attributed mostly to the speckle removal procedure.

5. DISCUSSION

No claim is made that the inverse methods described here are in any sense complete or optimum. They form the first stages in the investigation of such methods. By means of the step by step processes described here, not only are inverse methods being developed but the imaging theories in sections 2 and 3 are being further explored and, in some respects, confirmed. Some parts of the imaging theory, particularly for the power spectrum, remain to be completed. Some further refinement in the step by step inverse methods can confidently be expected in the short term. If this gives confidence in the overall theory, the next stage should be the development of optimum methods. SAR images are stochastic through both the short and long wave fields. Resolution and speckle are traded off by varying the number of looks. A framework for the development of an optimum method is probably Bacchus - Gilbert theory which provides for such a trade-off, and this is the subject of our current research.

6. ACKNOWLEDGEMENT

Part of this work was supported by ESA.

7.REFERENCES

1. Wright, J.W., 1968, A new model for sea clutter. IEE Trans. Ant. Prop. AP-16, pp. 217-223.
2. Alpers, W.R., Ross, D.B. and Rufenach, C.L. 1981, On the detectability of ocean surface waves by real and synthetic aperture radar. J. Geoph. Res. 86, pp.6481-6498.
3. Vesecky, J.F. and Stewart, R.H. 1982, The observation of ocean surface phenomena using imagery from the SEASAT synthetic aperture radar - an assessment. J. Geoph. Res.87, pp. 3397-3430.
4 Rotheram, S. Ocean wave imaging by SAR. AGARD Conference on "Propagation factors affecting remote sensing by radio waves". Oberammergau, 24-28 May, 1983.
5. Keller, W.C. and Wright, J.R. 1975, Microwave scattering and straining of wind generated waves. Radio Science, 10, pp. 139-147.
6. Alpers, W.R. and Hasselmann, K. 1978, The two frequency microwave technique for measuring ocean wave spectra from an airplane or satellite. Boundary Layer Meteorology, 13, pp. 215-230.
7. Larson, R., Moskowitz, L.I., and Wright, J.W. 1976, A Note on SAR imagery of the ocean, IEEE Trans. Ant. Prop. AP-24. pp. 393-394.

8. Alpers, W.R. and Rufenach, C.L. 1979, The effect of orbital motions on synthetic aperture radar imagery of ocean waves. IEEE Trans. Ant. Prop. AP-27, pp. 685-690.
9. Swift, C.T. and Wilson, L.R. 1979, Synthetic aperture radar imagery of moving ocean waves. IEEE Trans. Ant. Prop. AP-27, pp. 727-729.
10. Valenzuela, G.R. 1980, An asymptotic formulation for SAR images of the dynamical ocean surface. Radio Science, 15, pp. 105-114.
11. Rufenach, C.L. and Alpers, W.R. 1981, Imaging ocean waves by synthetic aperture radars with long integration times. IEEE Trans. Ant. Prop. AP-29, pp. 422-428.
12. Raney, R.K. 1981, Wave orbital velocity, fade and SAR response to azimuth waves. IEEE J. Oceanic Eng. OE-6, pp. 140-146.
13. Ivanov, A.V. 1982, On the synthetic aperture radar imaging of ocean surface waves. IEEE J. Oceanic Eng. OE-7. pp. 96-103.
14. Beal, R.C. 1982, The value of spaceborne SAR in the study of wind wave spectra. Submitted to Reviews of Geophysics and Space Physics.
15. Alpers, W.R. and Hasselman, K. 1982, Spectral signal to clutter and thermal noise properties of ocean wave imaging synthetic aperture radars. Int. J. Rem. Sens. 3. pp.423-446.

IV.7 (RS.2)

INVERSE METHODS IN ROUGH-SURFACE SCATTERING

Adrian K. Fung

Remote Sensing Laboratory
University of Kansas Center for Research, Inc.
Lawrence, Kansas USA 66045

ABSTRACT

The general approach to inversion for a specific rough-surface parameter through average power measurements is to maximize the sensitivity of the return power to the parameter of interest and minimize its sensitivity to other surface parameters. This is usually achieved by an appropriate choice of coherent or incoherent measurements, polarization, look angle and frequency.

To invert for rms surface height, sensing wavelength should be large compared with the surface height and coherent measurements at normal incidence are usually made. To recover the surface-roughness spectrum or surface correlation, incoherent measurements should be made over a wide range of incidence angles at a sufficiently low frequency or frequencies and at vertical polarization. On the other hand, to determine the rms slope of a randomly rough surface, it is better to choose an exploring wavelength small compared with both the horizontal and vertical roughness scales. Incoherent measurements should be carried out over a range of angles near vertical incidence. The sensing of the average surface permittivity for a natural terrain is currently done at C-band with an incidence angle around 10° for slightly rough to very rough terrains.

For unknown surfaces which can only be studied with remote sensors, it is possible to determine their roughness scales relative to the incident wavelength by examining their angular scattering characteristics at both vertically and horizontally polarized states. In general, an angular backscattering curve with no polarization dependence indicates scattering by roughness scales larger than the incident wavelength. Over angular regions where vertical and horizontal returns are well separated

W.-M. Boerner et al. (eds.), Inverse Methods in Electromagnetic Imaging, Part 2, 931–942.

the scattering must be dominated by roughness scales small compared with the wavelength.

1.0 INTRODUCTION

The characterizing quantities associated with a randomly rough surface are its rms surface height, σ, rms surface slope, σ_s, roughness spectrum, W, or surface height correlation, $\rho(\xi)$, and the surface permittivity, ε. The practical approach to invert for any one of the above parameters from remote sensing measurements is to choose the polarization, frequency and look angle in such a way that the average measured power has the largest possible sensitivity to the parameter of interest and the smallest sensitivity to other parameters. Mathematically, this is an optimization problem of some kind. However, since mathematical models for randomly rough surfaces are available only for special types of surfaces and appropriate constraints for the surface parameters are not well known, the optimization process has not been carried out mathematically. Instead, the type of rough surface is first examined through data analysis to decide whether there is an applicable scattering model. If there is, the model may be used to determine the optimum polarization, frequency and look angle. If not, some regression analysis is usually applied to measured data to establish sensitivity trends for various parameters. Based upon the obtained sensitivity curves, the choice of optimum polarization, frequency and look angle for the particular parameter of interest is then made.

In this paper, the forward or backscattering characteristics of rough surfaces with rms surface height small, comparable and large compared with the incident wavelength are illustrated. For a slightly rough surface, the approaches to invert the measurements for rms surface height and roughness spectrum are discussed. For a rough surface with large rms heights an approach to estimate their rms slope is shown. Also discussed is an approach to estimate the surface permittivity for any rough soil surface.

2.0 SCATTERING CHARACTERISTICS OF RANDOMLY ROUGH SURFACES

For a randomly rough surface the major surface parameters are the rms surface height, σ, surface correlation length, ℓ, and surface permittivity, ε. When these parameters or other surface attributes are to be recovered from data taken by remote sensors operating at a given frequency, the estimates for these parameters become frequency dependent to various degrees. In fact, the degree of roughness of a random surface is meaningful only relative to the incident wavelength. Thus, we shall speak of a slightly rough surface when $k\sigma<0.3$, where k is the wavelength in air. A medium-rough surface is characterized by $0.3<k\sigma<3$, and a very rough surface by $k\sigma>3$. For theoretical analysis the surface must be sufficiently smooth. Therefore, the radius of curvature at any point on the surface must be larger than a wavelength. In what follows surface characteristics relevant to inversion for surface properties are discussed.

(a) A Slightly Rough Surface ($k\sigma<0.3$)

For a slightly rough surface, coherent scattering is significant near the specular direction while incoherent scattering is important in off-specular directions. It has been shown that the normalized coherent scattering coefficient σ_c^o is proportional to the magnitude square of the characteristic function $\chi(q_z)$ of the rough surface $z(x,y)$, where $q_z=2k\cos\theta$; θ is the incidence angle (1,2). An illustration of σ_c^o versus $2k\sigma\cos\theta$ for four different surfaces are shown in Fig. 1 (3). The theoretical curves in Fig. 1 were generated from either the Gaussian surface height model,

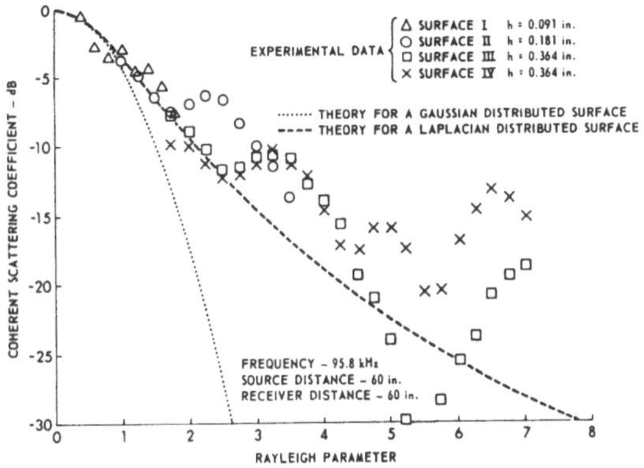

Figure 1. Coherent scattering coefficients versus the parameter $2k\sigma\cos\theta$, for four different random surfaces. Theoretical curves using eqs. (1) and (2), with $\nu=1$, are dots and dashes, respectively (3).

$$\sigma_c^o = \exp\left(-4k^2\sigma^2\cos^2\theta\right) \tag{1}$$

or a class of exponential surface height model,

$$\sigma_c^o = \left(1+2k^2\sigma^2\cos^2\theta/\nu\right)^{-2\nu} \tag{2}$$

where $0.5<\nu\leq 1$. The measurements were taken in the specular direction. Clearly, coherent scattering is governed by the surface height distribution and the $k\sigma$ parameter.

For incoherent scattering in the backscatter direction, the first-order scattering coefficient in $k\sigma$ has been shown to be (4,5)

$$\sigma_{pp}^o(\theta) = 8k^4\sigma^2\cos^4\theta|\alpha_{pp}|^2W(2k\sin\theta) \tag{3}$$

where p=v (vertical) or h (horizontal) polarization; W( ) is the Fourier transform of surface height correlation coefficient, and

$\alpha_{hh} = R_\perp$, Fresnel reflection coefficient for horizontal polarization,

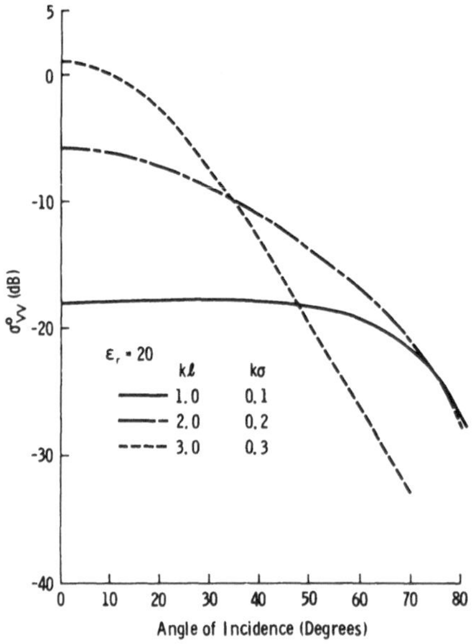

(a) vertical polarization

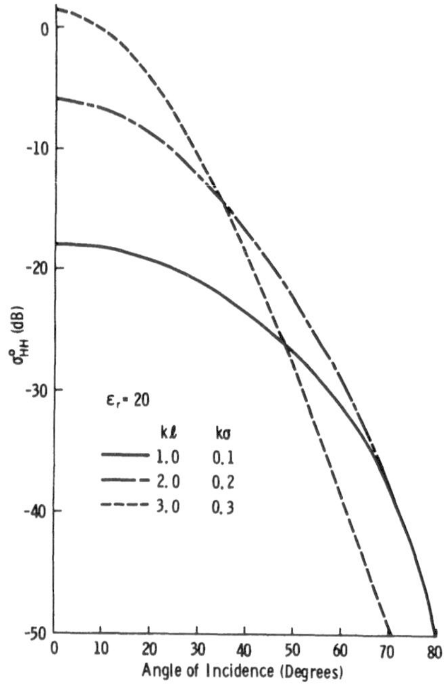

(b) horizontal polarization

Figure 2. Polarized scattering coefficient for a slightly rough surface versus the incidence angle with roughness scales as parameters.

$$\alpha_{vv} = \frac{\left(\varepsilon_r-1\right)\left[\sin^2\theta-\varepsilon_r\left(1+\sin^2\theta\right)\right]}{\left[\varepsilon_r\cos\theta+\left(\varepsilon_r-\sin^2\theta\right)^{\frac{1}{2}}\right]^2}$$

where ε_r is the relative surface permittivity. Fig. 2 shows the behavior of $\sigma^o_{pp}(\theta)$ as a function of $k\sigma$ and $k\ell$ using the same σ/ℓ ratio when $W=(\ell^2/s)\exp[-(k\ell\sin\theta)^2]$. In all cases, $\sigma^o_{vv}(\theta)$ curves drop off much more slowly than the corresponding $\sigma^o_{hh}(\theta)$. Since the roughness spectrum, $W(2k\sin\theta)$, depends only on one spatial wave number at a given incidence angle, θ, and exploring wave number, k, it follows that in backscattering from slightly rough surfaces, only one spatial frequency component is important.

(b) Medium-Rough Surfaces ($0.3<k\sigma<3$)

For medium-rough surfaces, the coherent scattering component can be computed by the same models as for the slightly rough surfaces. The incoherent scattering component is usually computed by the Kirchhoff method (6-8). It shows that the backscattering coefficient is proportional to the Bessel transform of the difference between the joint surface characteristic function $\chi(q_z,-q_z)$ and the magnitude square of the characteristic function $\chi(q_z)$ of the surface, z(x,y), i.e.,

$$\sigma^o_{pp}(\theta) = A\int_0^\infty J_0(2k\xi\sin\theta)\left[\chi(q_z,-q_z)-|\chi(q_z)|^2\right]\xi\, d\xi \tag{4}$$

where $A=2k^2|R_p|^2/\cos\theta$, $q_z=2k\cos\theta$, p=v or h, R_p is the Fresnel reflection coefficient for p polarization, and

$$\chi(q_z,-q_z) = \left\langle\exp\left[jq_z\left(z(x,y) - z(x',y')\right)\right]\right\rangle .$$

For Gaussian height distribution,

$$\chi(q_z,-q_z) = \exp\left\{-g[1-\rho(\xi)]\right\} \tag{5}$$

$$\chi(q_z) = \exp(-g) \tag{6}$$

where $g=4k^2\sigma^2\cos^2\theta$ and $\rho(\xi)$ is the surface correlation coefficient. For a class of exponential distributions (9),

$$\chi(q_z,-q_z) = \left\{1+g[1-\rho(\xi)]/2\nu\right\}^{-2\nu} \tag{7}$$

$$\chi(q_z) = (1+g/2\nu)^{-\nu} \tag{8}$$

In view of the above equations, we expect $\sigma^o_{pp}(\theta)$ to depend strongly on both the surface height distributions and surface correlation coefficient, $\rho(\xi)$. However, as illustrated in Figs. 3a and 3b, the shape of the angular curves for $\sigma^o_{pp}(\theta)$ is controlled more strongly by $\rho(\xi)$ than the surface height distribution. Fig. 4 (10) shows measured backscattering coefficients from three bare soil surfaces. The shapes of the curves suggest that the surface correlation coefficients are much closer to an exponential function than a Gaussian function.

(c) Very Rough Surfaces ($k\sigma>3$)

The scattering model for this type of surface can be obtained by taking the limiting case of the model given by eq. (4). This follows from the fact that for large $k\sigma$ values, $\rho(\xi)$ makes important contributions to the integral in eq. (4) only over a small interval of ξ near $\xi=0$. Hence, one can expand $\rho(\xi)$ about $\xi=0$ and keep only the first two non-zero terms in evaluating the integral in eq. (4) yielding

$$\sigma^o_{vv} = \sigma^o_{hh} = \pi|R_p(0)|^2\sec^4\theta\, p(\tan\theta) \tag{9}$$

where $p(\tan\theta)$ is the probability density function for surface slope. As an example, a Gaussian slope distribution will give

$$\sigma^o_{pp}(\theta) = \left(|R_p|^2/2\sigma_s^2\cos^4\theta\right)\exp\left(-\tan^2\theta/2\sigma_s^2\right) \tag{10}$$

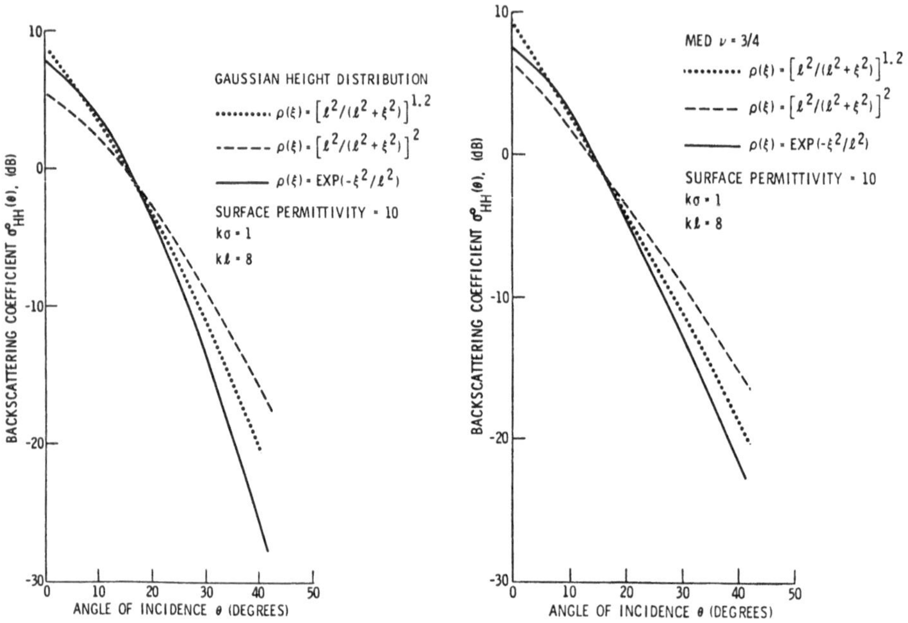

(a) Gaussian height distribution (b) a class of exponential distribution

Figure 3. Comparison of angular characteristics of $\sigma^0_{pp}(\theta)$ for three different autocorrelation functions using eq. (4).

where $\sigma_s^2=\sigma^2|\rho''(0)|$, the variance of surface slope. If we let $\nu=0.75$ in eq. (7) and expand $\rho(\xi)$ and then substitute eq. (7) into eq. (4), we get

$$\sigma^0_{pp}(\theta) = \left(3|R_p|^2/2\sigma_s^2\cos^4\theta\right) \cdot \exp\left(-\sqrt{3}\tan\theta/\sigma_s\right) \tag{11}$$

Obviously, the important point to remember in this case is that $\sigma^0_{pp}(\theta)$ is now proportional to the surface slope distribution. An illustration of eq. (10) is shown in Fig. 5.

(d) Two-Scale Rough Surfaces

Intuitively large returns around the specular direction imply the presence of large-scale roughnesses while a difference between vertically and horizontally polarized returns along nonspecular directions indicates the effects of small-scale roughness. When both types of roughnesses are present on a surface, the backscattering characteristics will appear as shown in Figs. 6 and 7. Fig. 6 shows σ^0_{pp} for a perfectly conducting surface, while Fig. 7 is for a dielectric surface of the same surface statistics. Its relative permittivity is 3.7. The significant difference in the dielectric property is clearly reflected by the difference in spacings between σ^0_{vv} and σ^0_{hh} at large angles of incidence. This change in polarization characteristics is inherent to slightly

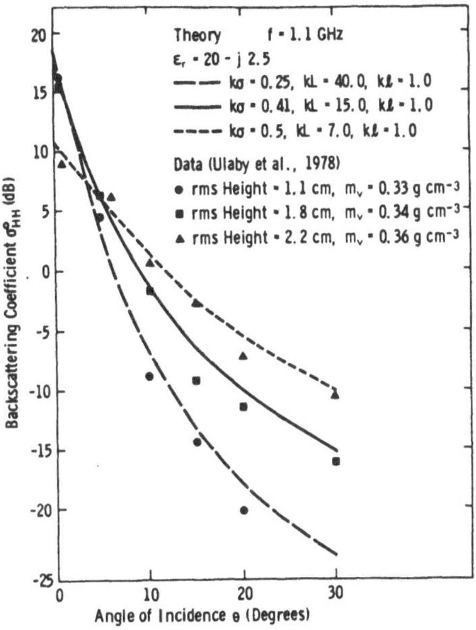

Figure 4. Backscattering coefficients for different bare soil surfaces measured at 1.1 GHz and HH polarization.

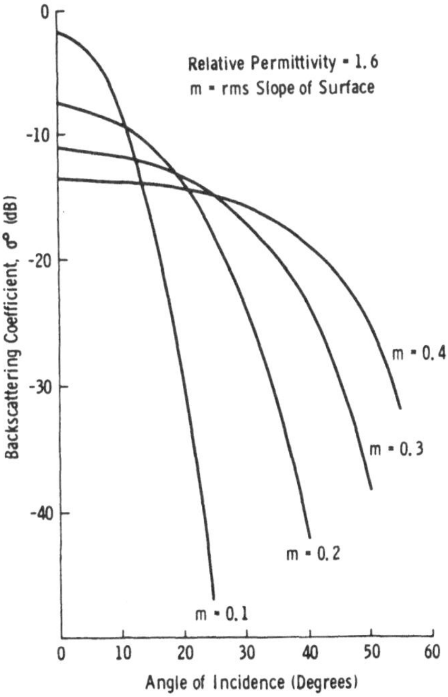

Figure 5. Backscattering coefficients for surface with different rms slopes when the slope distribution is Gaussian.

rough surfaces. It is also noted that except for the level difference, near vertical scattering is the same in both Figs. 6 and 7. This is characteristic of scattering by roughness scales larger than the incident wavelength. The theoretical predictions in these figures follow the two-scale theory by Wu and Fung (11). Measurements were taken at 25 GHz.

3.0 INVERSION FOR SURFACE ATTRIBUTES

Since the degree of surface roughness is frequency dependent, within practical constraints it may be possible to select the incident wavelength such that the surface in question will appear to be slightly, medium or very rough. Thus, the rough-surface scattering properties presented in the previous section can be used to determine the rms surface height, the surface-roughness spectrum, rms surface slope and surface permittivity.

(a) Surface RMS Height

The method of determining the surface height distribution or rms height has been considered by Porteus (1), Bennett (12) and Clay et al. (2). Their idea is to use coherently scattered measurements at many incident and scattered angles or many frequencies. Then eq. (1) or (2) is used to fit the data. Since frequency and angles are known,

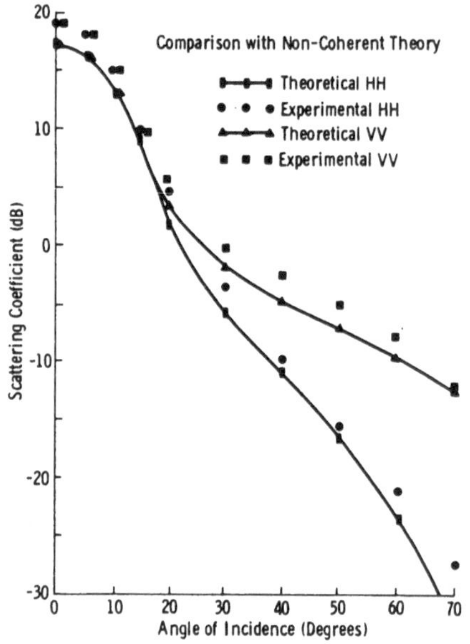

Figure 6. Backscattering from a perfectly conducting man-made two-scale random surface. The large-scale roughness has a rms slope of 0.123 radians; the small-scale roughness has correlation length of 6.2 mm and a standard deviation of 0.85 mm and $\rho(\xi)\cong[1+(\xi/\ell)^2]^{-3/2}$.

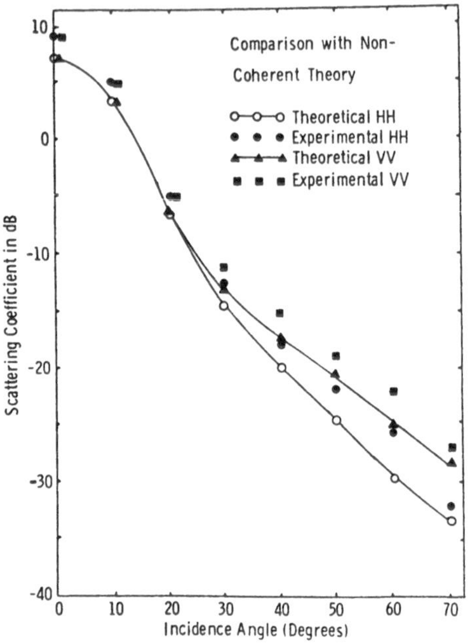

Figure 7. Backscattering from a dielectric surface with ε_r=3.7 and the same statistical roughness properties as the surface in Fig. 6.

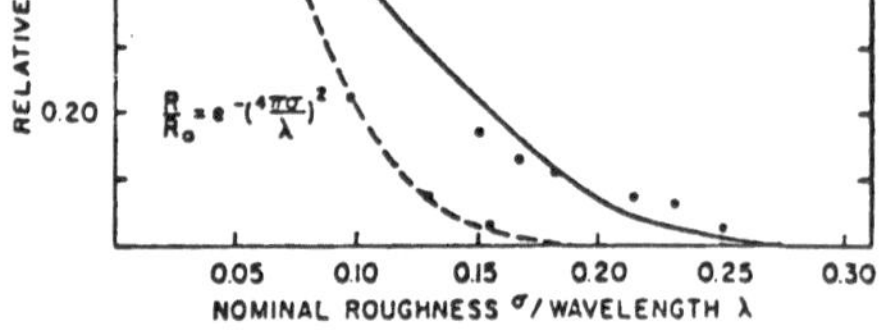

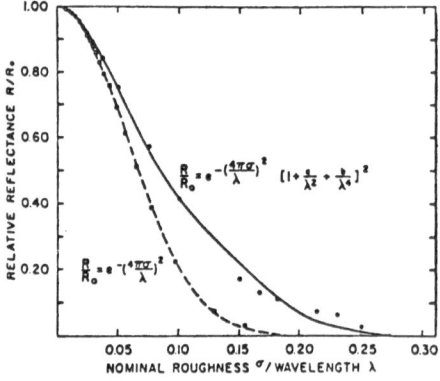

Figure 8. Normalized coherent reflectance measurements versus (σ/λ) for two different random surfaces (12).

the rms height, σ, can be determined. Fig. 8 shows an example of this approach. In the event when the surface distribution is symmetric about the mean surface, the surface height density function can be estimated by taking the inverse Fourier transform of the square root of the coherent scattering coefficient. This has been done by Clay et al. (2) (see Fig. 9). In Fig. 9, W_T is the calculated surface height density from power measurements, B curves are W_T curves after accounting for shadowing, and the dots are from wave staff measurements. In summary, low-frequency ($k\sigma$<0.3), multi-angle or multi-frequency coherent measurements along the specular direction are needed for estimating the surface rms height and possibly surface height density function.

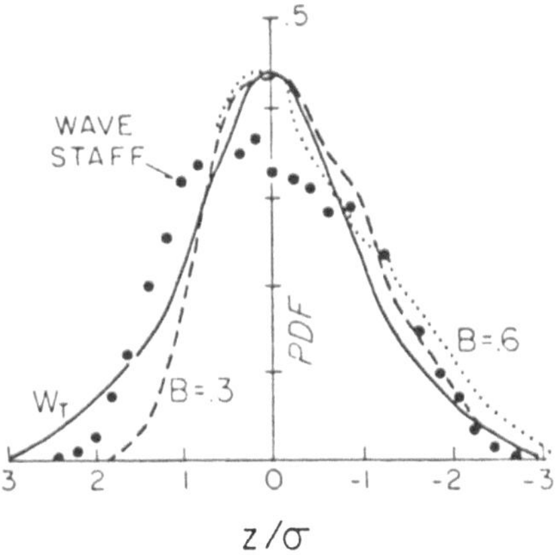

Figure 9. Comparison between surface height density function recovered from inverse Fourier transform of coherent measurements of a water surface and direct wave staff measurements (2).

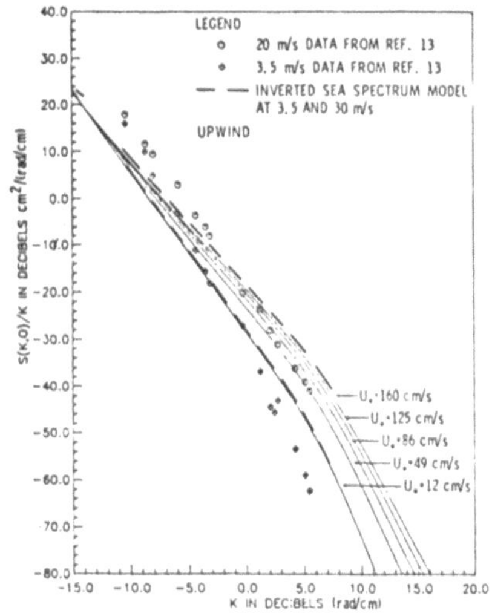

Figure 10. Comparison of the high-frequency portion of the sea spectrum inverted from radar measurements by Valenzuela et al. (13) with a theoretically estimated sea spectrum (14).

(b) Surface-Roughness Spectrum

From eq. (3) it is seen that multi-frequency or multi-angle incoherent measurements along nonspecular directions can be used to determine surface-roughness spectrum provided $k\sigma<0.3$ in all cases.

Natural occurring surfaces rarely consist of only one scale of roughness. When small-scale roughnesses are superimposed upon large-scale roughnesses, the nonspecular measurements are dominated by the small-scale roughness contributions averaged over the slope distribution of the large-scale roughness. This averaging process can produce significant changes on horizontal polarization but relatively little change on vertical polarization. For this reason vertically polarized measurements should be used for recovering the surface spectrum of the small-scale roughness from two-scale surfaces with known relative permittivity. One way to determine whether the condition $k\sigma<0.3$ is satisfied in practice is to take both horizontal and vertical measurements and compare with eq. (3). An application of this idea to validate data and to invert the data for the roughness spectrum has been carried out by Valenzuela et al. (13). An example of the high-frequency portion of the sea spectrum recovered from vertically polarized radar data is shown in Fig. 10, where the dotted lines are theoretical results computed by Fung and Lee (14). The solid lines represent the sea spectrum model used in the radar scatter model (14).

(c) Surface RMS Slope

Surface rms slopes can be found by taking high-frequency ($k\sigma>3$) measurements at many incidence angles from a surface with known permittivity. According to eq. (9), the slope density function can be determined from

$$p(\tan\theta) = \sigma^{o}_{pp}/\left(\pi|R_p(0)|^2\sec^4\theta\right) \quad .$$

Whether the data points satisfy the requirement, $k\sigma>3$, can be determined from polarization measurements. When this condition is satisfied, there should be no distinction between horizontal and vertical measurements. Once the slope density function is known, the rms slope can either be computed or estimated from fitting eq. (10) or eq. (11) to the measurements.

(d) Surface Permittivity

From eq. (10) or eq. (11) it is seen that radar measurements are proportional to surface reflectivity but are confused by surface roughness. Measurements from bare soil surfaces (Fig. 4) indicate that there is a small angular region where the backscattering curves for different surface roughnesses cross one another. This angular region is, therefore, a good choice to establish a calibration curve which relates surface permittivity (or surface moisture) to backscatter measurements. An example of this is shown in Fig. 11. The relation between reflection coefficient at normal incidence and volumetric moisture normalized to its 1/3 Bar value is shown in Fig. 12 (15).

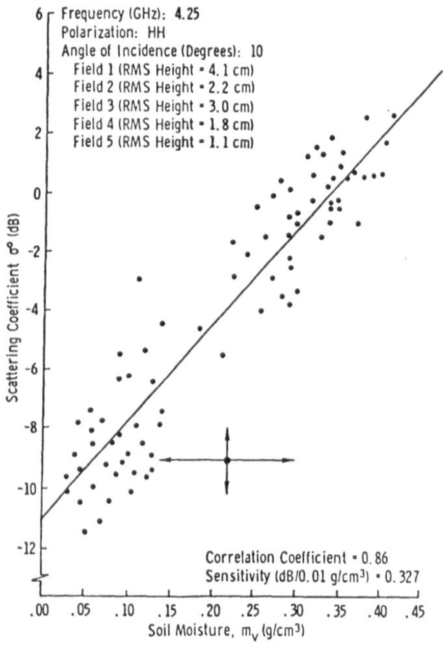

Figure 11. A calibration curve for soil moisture versus backscatter measurements at 4.25 GHz (10).

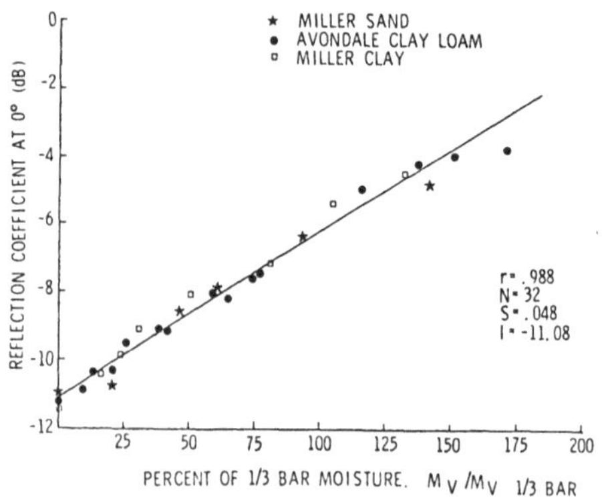

Figure 12. Relation between reflectivity at normal incidence and normalized soil moisture (15).

Studies by Ulaby et al. (10) show that the best choice of angle and frequency for permittivity or soil moisture estimation is 10°-15° and C-band (4.25 GHz).

4.0 CONCLUSIONS

To invert for surface properties, it is necessary first to determine whether the surface is acting as a slightly, medium or very rough surface. This can be ascertained from vertical and horizontal polarization measurements. The same surface may act as a slightly, medium or very rough surface depending upon the incident wavelength and the incidence angle. An illustration of frequency dependence is shown in Fig. 13. The extinction method curves shown in the figure were computed by Fung and Chen (16). Surface rms height has been found from coherent measurements and surface-roughness spectrum from incoherent measurements for slightly rough surfaces. Surface slope distribution can be found from very rough surface and surface permittivity can be estimated from calibrated backscattering versus moisture and moisture versus reflectivity relations established at C-band and 10° incidence angle.

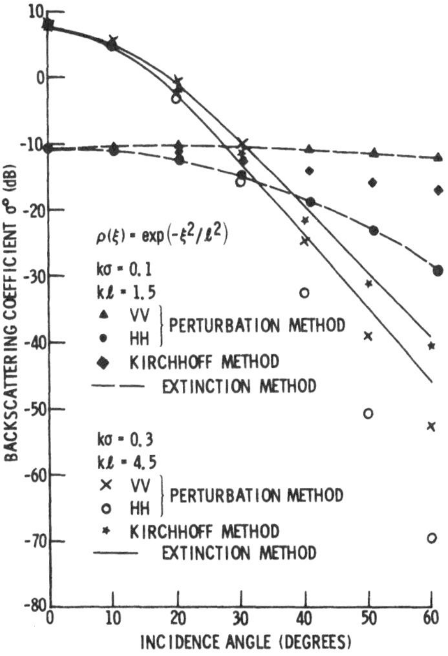

Figure 13. Frequency and angular dependence of a Gaussian distributed random surface.

ACKNOWLEDGMENTS

This work was supported in part by NASA Goddard Space Flight Center under Grant NAG5-268, and in part by the U.S. Army Research Office under Grant DAAG29-80-K-0018.

REFERENCES

(1) Porteus, J.O.: 1963, Relation between the height distribution of a rough surface and the reflectance at normal incidence, J. Opt. Soc. Am. 58, pp. 1394-1402.
(2) Clay, C.S., Medwin, H., and Wright, W.M.: 1973, Specularly scattered sound and the probability density function of a rough surface, J. Acoust. Soc. Am. 53(6), pp. 1677-1682.

(3) Welton, P.J.: 1975, A theoretical and experimental investigation of the scattering of acoustic waves by randomly rough surfaces, Tech. Rep. ARL-TR-75-30, Applied Research Laboratory, University of Texas, Austin, Texas.
(4) Rice, S.O.: 1951, Reflection of electromagnetic waves from slightly rough surfaces, Comm. Pure Appl. Math. 4(2/3), pp. 351-378.
(5) Ulaby, F.T., Moore, R.K., and Fung, A.K.: 1982, Microwave Remote Sensing, Vol. 2, Chapter 12, Addison-Wesley, Reading Massachusetts.
(6) Beckmann, P., and Spizzichino, A.: 1963, The Scattering of Electromagnetic Waves from Rough Surfaces, Pergamon, New York.
(7) Fung, A.K.: 1967, Theory of cross-polarized power returned from a random surface, Appl. Sci. Res. 18, pp. 50-60.
(8) Sancer, M.I.: 1969, Shadow-corrected electromagnetic scattering from a randomly rough surface, IEEE Trans. Ant. and Prop. AP-17, pp. 577-597.
(9) Eom, H.J., and Fung, A.K.: 1982, Scattering coefficients of Kirchhoff surfaces with Gaussian and non-Gaussian surface statistics, Tech. Rep. RSL TR 4601-2, Remote Sensing Laboratory, University of Kansas, Lawrence, Kansas.
(10) Ulaby, F.T., Batlivala, P.P., and Dobson, M.C.: 1978, Microwave backscatter dependence on surface roughness, soil moisture, and soil texture: part 1 - bare soil, IEEE Trans. Geosci. Electr. GE-16(10), pp. 286-295.
(11) Wu, S.T., and Fung, A.K.: 1972, A noncoherent model for microwave emissions and backscattering from the sea surface, J. Geophys. Res. 77(30), pp. 5917-5929.
(12) Bennett, H.F.: 1963, Specular reflectance of aluminized ground glass and the height distribution of surface irregularities, J. Opt. Soc. Am. 53, pp. 1389-1394.
(13) Valenzuela, G.R., Laing, M.B., and Daley, J.C.: 1971, Ocean spectra for the high-frequency waves as determined from airborne radar measurements, J. Marine Research 29, pp. 69-84.
(14) Fung, A.K., and Lee, K.K.: 1983, Variation of sea wave spectrum with wind speed, Digest of IGARSS'83, San Francisco, California, August 31 - September 2, 1983.
(15) Dobson, M.C., and Ulaby, F.T.: 1981, Microwave backscatter dependence on surface roughness, soil moisture, and soil texture: part III - soil terrain, IEEE Trans. Geosci. and Remote Sensing GE-19(1), pp. 51-61.
(16) Fung, A.K., and Chen, M.F.: 1983, Scattering from a perfectly conducting random surface - extinction method, Tech. Rep. RSL TR 592-3, Remote Sensing Laboratory, University of Kansas, Lawrence, Kansas.

IV.8 (SI.2)

ON THE OPTIMUM DETECTION OF SURFACE CHEMICAL COMPOUNDS

Terje Lund

Royal Norwegian Council for Scientific and Industrial Research, Environmental Surveillance Technology Programme

ABSTRACT

Quantitative remote sensing of the material composition of natural surfaces has to utilize models of the spectral reflection and/or emission properties of the different materials as they appear in nature. Surface roughness, local incidence angle variations, material mixtures, layers etc have to be considered to perform accurate measurements. Coherent optical scatterometry offers possibilities to improve the usefulness of existing reflection and scattering models by providing added information about surface structural features by utilizing the decorrelation properties of the interference pattern at the receiver (specle).

1. INTRODUCTION

This contribution will discuss, in broad terms, the problem of remote detection and quantitative characterization of natural surface material composition. In particular, it will focus on the possibilities offered by optical scatterometry i e use of lasers as illumination sources in systems capable of recording polarized intensity fluctuations due to backscattering from surface structural features.

Most of the discussion can be regarded as an extrapolation of general scatterometry to the optical region i e scaling the wavelength.

Optical scatterometry or optical radar is a field of rapidly growing interest, in particular in the military community, of

W.-M. Boerner et al. (eds.), Inverse Methods in Electromagnetic Imaging, Part 2, 943–953.

several reasons: The first and obvious one is angular resolution. In many useful cases the beam spread can easily be made smaller than the target to produce a $1/R^2$ received signal dependance. Another is increased contrast between natural background and man-made objects in the visible and infrared region of the electromagnetic spectrum.

Optical scatterometry can, at least potentially, provide unique means of characterization of surfaces.

We will discuss a number of these separately both in idealized terms and related to possible technical realizations and reported experimental observations.

2. OPTICAL SCATTEROMETRY

Scatterometry is an active electromagnetic remote sensing technique exploiting the maximum information content of a target.

Ideally scatterometry compares the backscattered field from an illuminated surface with the *a priori* known illuminating field.

A scatterometer system can record relative field amplitudes and phases as functions of frequency, time and location (of antennas) for any combination of two orthogonal transmitter and receiver polarizations.

Amplitude changes with time can be related to shape and aspect variations, - with frequency to radial structure and material dielectric properties and with location of sensor to the transverse structure of the target. Relative phase changes with time gives information about radial velocity of the target, - with frequency they record delay or range to the target and again - probing relative phase with more than one or one moving sensor can reveal wavefront curvatures which is associated to the location of strong target backscatter features. In addition target depolarization effects is due to orientation of the surface elements relative to the incident field.

Microwave scatterometry is in an advanced development stage. The mere possibility of performing the inverse transformation from the measured properties of the scattered field to recreate target properties confirms this.

In optical scatterometry the wavelengths are reduced by a factor of $10^4 - 10^6$. Most target surfaces therefore constitute very deep phase screens, that is: the surfaces are rough in terms of wavelengths.

Normal air turbulence causes refractive index variations that appear as local phase and amplitude fluctuations in the illuminating beam giving rise in particular to angle of incidence fluctuations. In optical scatterometry we therefore in general only know some statistical properties of the illumination.

The sensitivity to motions of scatterers within the target and relative to the sensor is correspondingly very high.

In the discussion we will focus on target chemistry - how do we utilize the features of optical scatterometry to obtain as much information as possible about the material composition of the target? We are not interested in target topography *per see* but added information about target roughness, average incidence angle and lateral structure parameters may increase the accuracy of a quantitative material composition measurement.

In the optical and infrared region surfaces exhibit pronounced but broad spectral reflectivities. Furthermore, how the Fresnel reflection varies with incidence angle can be described properly and used to develop models for rough surface reflections.

The ability to observe remotely depends critically on the transmission properties of the atmosphere. The 10 μ-atmospheric window for example provides for a good view of thermal emission distributions with IR-radiometers. Most natural surfaces, however, have an emissivity close to unity partly due to water on or imbedded in the surface. Employing an active system increases the contrast between different surfaces dramatically.

The most common argument for use of the optical wavelength range is nevertheless angular resolution, given by the well known antenna relation: $\theta \approx \lambda/D$ where θ is the angle resolved, D is the antenna aperture and λ is the wavelength.

A 0.5 m diameter "antenna" - diffraction limited, gives for instance the beam from a green dye laser a minimal collimated divergence of 2 μradians or it can, from an altitude of 1000 m be focussed to a 2 mmϕ spot, neglecting atmospheric refractive index inhomogeneities.

What this implies is that we in optical scatterometry have the technical flexibility to probe a surface with an area sample (pixel) comparable to or, if we choose so, much larger than the width of the lateral surface structure autocorrelation function. Surface glints (area of strong specular reflection) may be resolved and more important - single material sub-areas of a composite natural surface can in many cases be probed exclusively.

3. NATURAL SURFACE REFLECTANCE MODELS

Natural surfaces are almost always composites of a number of constituents of more common materials in a variety of spatial arrangements.

We have to calculate the reflective properties of a composite from the properties and geometric arrangement of the constituent materials.

We must conceive the composite as an ensemble or model of a (known but idealized) arrangement of constituents.

Some commonly used and simple models are:

The Plane Mixture Model

The spectral reflectance of the composite C consisting of area-patches of two different materials A and B is

$$\rho(C) = f\ \rho(A) + (1-f)\ \rho(B)$$

where f is the size of the area covered by material A relative to the total area.

Probing with a beam much larger than the individual single material sub-area gives information about relative amounts of materials.

The resulting reflection spectrum is a linear sum of the individual single material spectra and relative amounts (area coverage) can be determined by probing with a number of wavelengths at least equal to the number of different materials and solving the corresponding set of linear equations.

If unknown materials are present they constitute added noise. Moreover this noise will most certainly be coloured in an unpredictable way and attempts to reduce the errors by application of common signal processing techniques may prove difficult.

One way to go is to reduce the size of the target area by collimation/focussing. This increases the possibility of single material illumination on the expense of a greatly increased number of required measurements. In addition to the large number of pixels per area the fluctuations due to interference between spatially separated target scatterers (specle) will often require averaging. This will be discussed later.

The Plane Stacking Reflectance and Transmittance Mode

This model takes into account multiple reflections with two stacked perfectly diffuse plane surfaces of different materials.

The Wet-Dry Plane Stacking Reflectance Model

This is a useful model for a common condition in nature: A thin plane of liquid (usually water) covering and wetting an opaque diffusely reflecting surface.

Bulk and Surface Reflectance

If the material or aggregate of materials are partly transparent we must take into account both surface reflectance (at the interface of the material with air) and internal backscatter of radiation transmitted into the material (bulk reflectance). This component is altered by the spectral absorption properties of the material. The transition from a completely-dry to a water-saturated surface can be calculated by application of the model.

Surface reflectance tends to be single surface reflectance, so that the degree of polarization of incident radiation is largely preserved.

Surface reflectance also tends to produce most of the non-Lambertian character of the reflectance. Bulk reflectance tends to be randomly polarized (multippel scatter) and more nearly Lambertian in directional distribution.

Close to a resonant absorption frequency in a material the impedance changes rapidly and increases the mismatch between air and surface. Hence the surface reflectance increases sharply near the reststrahlen frequency. Increased absorption reduces the bulk reflectivity and the spatial or directional and polarization properties resembles that of a pure surface reflector.

Aggregates

Most natural materials are mixtures, aggregates and solutions. Minerals are generally solid mictures with impurities. Rocks are aggregates of minerals. Sea water is a solution containing organic and inorganic particles.

The surface reflectance of these materials is largely determined by the solvent while the bulk term comes from particulate scattering and solute absorption.

Both size parameters, refractive index and absorption influence the reflective character of an aggregate.

Vegetative Canopy Reflectance Model

Characterization of vegetation is a major task for remote sensing and a large number of both deterministical and statistical models have been developed, among these the Kubelka-Munk model and refinements of this.

To optical scatterometry the most relevant parameters are depth or delay distribution of elemental reflectors and element motions.

4. SPECLE - NOISE AND INFORMATION

Interference between field contributions from a multitude of spatially separated scatterers on the target surface produce a noise-like spatial intensity distribution in the region of the antenna (1,2,3).

If we first assume that all the elemental fields have random phases, i e that the scatterers are randomly distributed in the illuminated area, by the central limit theorem, the inphase and quadrature components of the field both have independent Gaussian distributions with a common variance.

An envelope or intensity detector will therefore exhibit a Rayleigh-distributed output provided it samples a small enough area, a single "specle".

This granularity in the reflected field from a diffuse surface which is observed with coherent illumination is a striking manifestation of the properties of optical scatterometry.

Depending upon the application the appearance of specle may be regarded as useful information or as noise. A number of instruments utilize the properties of specles to characterize the target and its motion.

It may easily be shown that the transverse autocorrelation function of the intensity distribution at the receiver is proportional to the Fourier transform of the aperture limiting the surface. If we illuminate an extended surface this aperture equals the beam width.

The average size of a specle is thus given by the radius of the Fraunhofer diffraction pattern of the limiting aperture.

In microwave scatterometry transmitter and receiver most often use identical or the same antenna. Since the target therefore always is smaller than or fills the beam the average specle size is larger or equal to the antenna aperture.

In optical scatterometry, however, many applications allow us to illuminate a much larger area than the diffraction limited spot of the receiver and to receive with an aperture much larger than the specle.

The receiving optics may resolve the target such that on the detector contributions from subapertures (identical in size to the diffraction limited resolution cell) no longer interfere.

This is aperture averaging. It reduces the multiplicative noise due to specle by an amount equal to the square root of the number of resolution cells. In most cases this is equal to the ratio between the receiver optics diameter and the diameter of the transmitted laser beam.

Another way to reduce specle noise in optical scatterometry is to use frequency decorrelation, i e shift the frequency of the laser and average the received signal.

The amount of shift necessary to decorrelate the return $\Delta\nu$ is inversely proportional to the depth of the target Δz.

$$\Delta\nu = \frac{c}{2\Delta z}$$

Statistically Δz may be regarded as the r.m.s. roughness (deviation from surface mean). This discussion is limited to frequency domain techniques. Time domain scatterometry is of increasing importance in remote sensing because of the recent availability of picosecond-laser pulses.

One easy way to reduce or eliminate specle noise by frequency decorrelation is of course to use incoherent illumination. This, however, removes all information about the target depth distribution.

Target depth distribution is an extremely important property, in particular of course in vegetation classification but also to support surface models (local incidence angle distribution).

Note that if we use a look angle off NADIR on a horizontal target the depth increases and reduces the amount of frequency shift necessary to decorrelate the return.

The third type of decorrelation is available because of time variations. Very important, in particular with static sensors, is relative motion of target scatterers.

Translating the receiver and/or transmitter aperture relative to the target also causes decorrelation. With extended target the amount of motion necessary equals the diameter of the laser beam at the transmitter.

Another, often unwanted decorrelation prosess is due to the propagation medium. Atmospheric turbulence causes perturbations of the illuminating wavefront. One of those is wavefront tilt, the action of which may be regarded as random angular motion of the sensor as seen from the target. Because of propagation reciprocity a monostatic system with common transmitter and receiver aperture is more tolerant to turbulent induced fluctuations than a bistatic one (4).

Beam tilt may also be caused by comparatively slow vertical thermal gradient changes in the atmosphere. The spectrum of turbulence induced perturbations is in most cases limited to a few hundred Hertz allowing for time-correlated measurements within a few milliseconds.

So far we have discussed specle from diffuse targets composed of randomly distributed scatterers with independent phases distributed over several optical cycles. In the optical region most but not all the targets behave that way.

Plane facets, large compared to the optical wavelength, produce marked deviations from the Lambertian cosine-law and may alter the statistics of the received field (5).

Decorrelation effects have recently been extensively studied experimentally to establish accuracy limits for lidar systems (6,7).

To sum up: Coherent light scattered from a rough target interfere to produce a spatial intensity distribution known as "specle". Accurate reflectance information requires a large number of uncorrelated measurements or aperture averaging.

In addition; how the signal decorrelates with frequency and time gives statistical information about target properties that can be used to further improve the reflectance accuracy.

Optical scatterometry offers practical means for this as well as possibility to control essential parameters like illuminated area, detector field of view and effective collecting aperture.

5. TECHNIQUES OF OPTICAL SCATTEROMETRY

a) Sources

Lasers are sources of coherent light. Adequate spatial coherence and frequency stability for scatterometry requires transverse and axial mode selection and stabilised operation. Many lasers exhibit frequency shifts, in particular in pulsed mode because of induced thermal refractive index variations. Running the laser in several axial modes simultaneously can be an interesting alternative because of the fixed frequency and phase relationship between the modes. The propagating modes interfere and provide a spatial intensity modulation that enhances the total reflectivity from structures with a corresponding spatial (depth) periodisity.

Lasers easily produce extremely clean linear polarization and polarization contrast of targets can therefore be measured with high accuracy.

Lasers are available, at least in the laboratories, with outputs in a large spectral region, from the ultraviolet to the far infrared. Practical sources for optical scatterometry though are at the moment only a few.

In our laboratory we are developing a system based on two independently controlled CO_2 lasers.

A low pressure CO_2 laser can be tuned to a number (typically $\sim$ 80) of lines in the 9 - 11 μ region.

These lines correspond to isolated vibration-rotational transitions. At low pressure they are Doppler broadened with a halfwidth of approximately 50 MHz. By tuning the cavity length with a piezoelectric crystal mounted on one of the mirrors, the exact frequency can be shifted across the line.

Amplitude and frequency stabilization loops can be applied reducing the short term frequency deviation to a few kilohertz.

In the fully developed system the two lasers will be locked together with a controllable difference frequency. Together with means for short pulse generation, polarization control and a system for heterodyne detection this will provide a flexible tool for optical scatterometry.

b) Receivers

The receiver consists of an optical antenna (parabolic mirror), filters to reduce added background optical noise, a field of view limiting aperture, detector optics (to match to detector size and field of view) and a square law (intensity) detector plus electronics.

Two different detection techniques may be employed:

- The optical heterodyne technique in which a local oscillator is introduced. This may also be derived from the transmitting laser itself (homodyne).

  Efficient mixing of signal and local oscillator requires complete overlap of the two wavefronts on the detector and the heterodyne receiver thus acts as spatial (angular) filter.

Systems with wide illumination (small transmitting aperture), large aperture receiver and wide detector field of view can be given angular scan capabilities by controlling the local oscillator wavefront incidence angle. Multippel angle local oscillators with one large or several detectors (detector matrix) give the possibility of both resolved and aperture averaged signals.

The local oscillator or transmitter frequency is shifted to give unambigous Doppler returns. One interesting way to do this in aircraft mounted systems is to point the transmitter a few degrees off NADIR in the flight direction.

One main virtue of the optical heterodyne system in addition to its Doppler capability is the increased sensitivity due to suppression of noise other than that associated with the local oscillator-induced fluctuations in the detector current.

- The direct detection technique where the total intensity incident on the receiver aperture is detected. Field of view is only restricted by detector size or other limiting apertures within the optical system.

  Aperture averaging of backscatter is achieved automatically when the transmitter beam diameter is smaller than the receiver aperture and the field of view is sufficient. Doppler information about relative axial motion between target and probe (transmitter/receiver) is not available.

  Direct detection is easier to realize and if system requirements aloow for a high degree of aperture averaging the resulting accuracy of a system may often be better than in the heterodyne case (7).

6. CONCLUSIONS

The possibility of achieving information about target characteristics other than coarse spectral ones with optical scatterometry has the potential of improving surface reflectance models necessary for accurate measurements of the material composition of natural targets.

In practice though most of this is still not exploited to any reasonable extent. Experimental work is needed to support and improve the models, in particular utilization of coherent target depolarization characteristics.

REFERENCES

1. Bramley, E.N. et al 1967, "Diffraction by deeply modulated random phase screen", Proc. IEE, 14, pp. 553-556.
2. Welford, W.T. 1977, "Optical estimation of statistics of surface roughness from light scattering measurements", Optical and Quantum Electronics, 9, pp. 269-287.
3. May, M. 1977, "Information inferred from observation of specles", Journal of Physics E: Scientific Instruments, 10, pp. 849-864.
4. Clifford, S.F.Lading, L. 1983, "Monostatic diffraction limited lidars: the impact of optical refractive turbulence", Applied Optics, 22, pp. 1696-1701.
5. Jakeman, E., Pusey, P.N. 1973, "The statistics of light scattered by random phase screens", J. Phys. A: Math., Nucl. Gen., Vol 6, L88-L92.
6. Menyuk, N., Killinger, D.K. 1983, "Assessment of relative error sources in IR dial measurement accuracy", Applied Optics, 22, pp. 2690-2698.
7. Bufton, J.L. et al 1983, "Frequency-doubled CO_2 lidar measurement and diode laser spectroscopy of atmospheric CO_2, Applied Optics, 22, pp. 2592-2602.

IV.9 (SI.3)

INVERSION PROBLEMS IN SAR IMAGING

Richard W. Larson
David R. Lyzenga
Robert A. Shuchman

Radar Science Laboratory
Radar Division
Environmental Research Institute of Michigan
P.O. Box 8618
Ann Arbor, Michigan 48107

1. INTRODUCTION

Many remote sensing problems may be described as an inversion process, in which information about some process or structure is inferred from a set of indirect measurements at a remote distance. Mathematically, the problem may be stated simply in one dimension as an attempt to solve an equation of the form

$$g(x) = \int_a^b F(x') \, K(x, x') dx' \tag{1}$$

for the function F(x'), given a set of measurements of g(x) and some knowledge of the kernel function K(x, x'). A classical example is the determination of atmospheric temperature and density profiles from satellite improved radiation measurements (Westwater and Strand, 1972).

The difficulty of the inversion is obviously related to the form of the kernel K(x, x') as well as the function F(x') to be estimated. If the kernel is a delta function, i.e.,

$$K(x, x') = \delta(x - x') \tag{2}$$

W.-M. Boerner et al. (eds.), Inverse Methods in Electromagnetic Imaging, Part 2, 955–967.

the inversion is trivial, because F(x) and g(x) become identical. This case is approached when the kernel is narrow compared with the scale of variations in F(x').

The SAR imaging process can be described in the same form as Eq. (1) where F(x') is the reflectivity distribution, g(x) is the image, and K(x, x') is the impulse response function. In this case x refers to the image coordinate and x' refers to the object or surface coordinate. Normally, attempts are made to reduce the imaging problem to the trivial inversion case referred to above by making the impulse response function narrow with respect to the scale of the features to be imaged. Nontrivial inversion problems are encountered, in this sense, only when pushing a given imaging system to its limits in terms of spatial resolution, or when the impulse response function is limited by uncontrollable factors such as the motion of the surface. The latter case is central to the problem of imaging the ocean surface with SAR.

For a SAR system in which the resolution is smaller than the scale of the features to be imaged, the output image is a one-to-one mapping of the reflectivity distribution of the scene. The inverse of Eq. (1) can then be expressed in simplified form for a SAR system as:

$$\sigma(x, y) = H_S^{-1} P_I(x', y') \tag{3}$$

where $P_I(x', y')$ is the SAR output power as measured on the image, H_S is the total SAR system transfer function and $\sigma(x, y)$ is the estimated reflectivity value of the imaged scene. The total system transfer function H_S can, in principle, be obtained if the system is calibrated.

In this paper an example of a SAR calibration algorithm is presented in terms of the inversion of the SAR equation for particular situations. Next, the inversion of SAR data from moving scatterers, in particular the ocean surface, is discussed. Finally, conclusions and recommendations are given.

2. SAR IMAGE INVERSION

The total SAR system transfer function includes the antenna, transmitter, receiver, recorder, signal processor, and output measurement. Calibration must encompass all these effects. A review of SAR calibration techniques is given in Larson et al., (1983). An example of a calibration procedure that is utilized in the inversion process is given in this section.

The data flow diagram for the digital SAR calibration algorithm is given in Figure 1. In this example, X-band SAR signal data recorded on HDDT during a set of data gathering flights over Alberta, Canada in 1980 were used*. The signal data are first converted into a CCT format, compatible with the ERIM digital SAR processor and used to obtain a digital image.

Absolute calibration references were generated from SAR images obtained during flights at the ERIM reference reflector calibration site near Ann Arbor, Michigan. A calibration file was assembled consisting of SAR image measurements of a series of reference reflector signals. An on-board calibration signal generator subsequently provides reference signals which are used to relate image measurements to the calibration file. During subsequent data gathering, calibration signals are recorded on each SAR signal tape. The SAR data to be calibrated are normalized to the reference file from image measurements of calibration signal generator data input.

The SAR images are also normalized using the stored antenna pattern look-up table as illustrated in Figure 1. Data for the antenna pattern were obtained from a series of data gathering measures over a reflector site (in the Ann Arbor area) designed for the purpose of measuring the in-flight antenna pattern.

Normalization to the calibration data file is completed using recorded values of SAR system and geometry parameters for interpolation at different ranges. Image measurements are averaged over a pixel area to obtain an estimate of the calibrated scattering coefficient, σ.

An example of digitally processed SAR data and the calibrated SAR image, normalized to the system calibration, is shown in Figure 2. This example of a site in Red Deer, Alberta, Canada, was obtained during flights on 19 September 1980. An example of imagery from the Michigan calibration site is given in Figure 3.

This digital SAR calibration procedure can easily be adapted to any SAR system. This is accomplished by recording the appropriate calibration signals and reference file of values from the calibration signal generator.

Some problems remain in this calibration process. Special procedures are required for the in-flight measurement of the SAR antenna pattern. At present, uncertainties in the antenna pattern and in the antenna pointing angle are the largest single error source in the overall system transfer function. Reference reflectivity values for absolute calibration are obtained using calibrated retroreflectors and calibrated distributed scenes. Additional research in both antenna pattern measurement techniques and in reflectivity references is being carried out.

---

*The data gathering as supported by funding from DARPA in a contract with Lincoln Laboratory during 1979-1980 time period.

Repetitive measurements are required to improve overall calibration precision.

3. OCEAN WAVE SPECTRUM INVERSION

One of the primary oceanographic applications of synthetic aperture radar is the measurement of ocean wave parameters, including the dominant wavelength and direction, and possibly the wave height. SAR is useful for this application because the spatial resolution is independent of range; consequently resolutions which are fine enough to resolve typical ocean wavelengths are achievable from spaceborne platforms using reasonable antenna sizes. As an example, to design a real-aperture radar with the same performance as Seasat and operating at the same altitude would require an antenna approximately 8 km long. However, since SAR is a range-Doppler imaging device, the image is distorted by surface motions, as described in the following section. Such distortion implies that the estimation of wave parameters from the imagery is a non-trivial inversion problem. This problem is treated in Section 3.2.

3.1 EFFECTS OF SCATTER MOTION ON SAR IMAGING

Since the position of a given scatterer in the along-track direction is determined by its Doppler history, a moving scatterer may be imaged by a synthetic aperture radar at an incorrect position and possibly with a decreased resolution in the along-track direction. This mis-mapping of scatterers is particularly important in the case of the ocean surface, where it leads to two kinds of effects depending on the scale of the motions.

Large-scale ordered motions due to the orbital velocity of long waves on the surface cause a systematic displacement of scatterers which may result in a wave-like image even if there is no variation in the radar cross section across the wave. This effect, which is referred to as "velocity bunching" (Alpers and Rufenach, 1979) is most important for waves traveling in or near the along-track direction. For relatively low-amplitude waves the wave contrast, or modulation index, is proportional to the wave amplitude. Thus, a measurement of the image modulation index could in principle be used to infer the wave height. However, as the wave amplitude increases this process becomes highly nonlinear and the inversion of the wave height becomes difficult or impossible.

The second type of motion effect is due to random surface motions associated with shorter wavelengths on the surface. These motions cause random scatterer displacements, which have the effect of degrading the effective spatial resolution in the along-track direction. The resulting resolution may be comparable to the wavelength of the ocean waves and, therefore, may have an important influence on the imaging process. The impact of those motion effects on the wave spectrum inversion problem are outlined in the following section.

3.2 WAVE SPECTRAL ESTIMATION

To describe the effect of scatterer motions on the wave spectral estimate, we can break down the SAR imaging process into three parts, as indicated in Figure 4. We can also analyze the process in either the spatial domain, as shown on the left side of Figure 4, or the (spatial) frequency domain, as shown on the right.

In the spatial domain, the surface may be described in terms of an elevation $\zeta(x, y)$ which is the sum of two components: one which varies smoothly on the scale of the radar wavelength and describes the large-scale structure of the surface, and a second which describes the small-scale perturbations of the surface on the scale of the radar wavelength. The counterpart of the description in the frequency domain is the wave height spectrum $W(k_x, k_y)$ which may be similarly broken down into a large-scale or low-frequency component $W_\ell(k_x, k_y)$ and a small-scale or high-frequency component $W_s(k_x, k_y)$, where the cutoff wavenumber is typically several times the radar wavenumber (Brown, 1978).

Knowing the small-scale and large-scale structure, it is possible to compute the radar cross section $\sigma_0(x, y)$ at each point on the surface using, for example, a two-scale or composite surface model (Wright, 1968; Valenzuela, 1978). This would then be mapped linearly into the image intensity, if the surface were imaged by an ideal (i.e., infinite-resolution) real-aperture radar. With SAR, however, there are two additional effects as mentioned at the beginning of Section 1.

The effect of large-scale motions (where the scale is now determined by the SAR resolution, rather than the radar wavelength) may be incorporated using the azimuth displacement formula

$$\Delta x = \frac{R}{V}\, \overline{V}_r(x, y) \tag{3}$$

where R is the range, V is the platform velocity, $\overline{V}_r(x, y)$ is the average radial velocity of the scatterers within the resolution cell at (x, y), and Δx is the displacement of the scatterers in the SAR image.

Equation (3) may be used to calculate the re-distribution of the scattered microwave energy within the imaged region, thereby resulting in an effective radar cross section $\sigma_0'(x,y)$ as observed by an infinite-resolution SAR. As noted earlier in Section 3.1, $\sigma_0'(x, y)$ may contain sinusoidal variations having the period of the dominant wave (i.e., a wave image) even if $\sigma_0(x, y)$ is constant across the surface.

The effect of small-scale motions may be described as a resolution effect. This can be seen intuitively by considering that each resolution cell contains a large number of scatterers moving at different velocities. Each of these scatterers is then imaged at a different position, according to Eq. (3), which means that the scattered energy from that resolution cell is spread out over a distance

$$\delta x = \left[\rho_x^2 + \frac{R}{V}\left(\delta V_r\right)^2\right]^{1/2} \tag{4}$$

where δV_r is the spread in the radial velocities and ρ_x is the nominal resolution in the along-track direction. δx is, therefore, the effective along-track resolution of the SAR image. The magnitude of δx is in general a function of position in the image, but for most purposes it may be considered a constant for a given scene.

Defining an effective impulse response function m(x, y), which has a width δx in the along-track direction and a width equal to the nominal resolution in the across-track or range direction, the image intensity may be written as

$$s(x, y) = m(x, y) * \sigma_0'(x, y) \tag{5}$$

where * denotes the convolution operation. In the frequency domain, the equivalent expression is

$$S(k_x, k_y) = M(k_x, k_y) \Sigma_0' (k_x, k_y) \tag{6}$$

where $S(k_x, k_y)$ is the SAR image spectrum, $M(k_x, k_y)$ is the squared Fourier transform of the effective impulse response function m(x, y), and $\Sigma_0'(k_x, k_y)$ is the squared Fourier transform of $\sigma_0'(x, y)$. Under a limited range of conditions $\Sigma_0'(k_x, k_y)$ is proportional to the wave height spectrum $W(k_x, k_y)$, although the constant of proportionality is a function of the wave direction relative to the radar look direction. This constant of proportionality R(x, y) is the sum of three terms, including two terms which account for the so-called "real modulation" due to tilt effects and hydrodynamic effects, and one term which accounts for the velocity bunching mechanism (Alpers et al., 1981).

The modulation transfer function $M(k_x, k_y)$ is dependent on the spectrum of waves smaller than the SAR resolution, which is in turn dependent on the wind speed and other environmental parameters.

Combining all these effects together, we may write

$$S(k_x, k_y) = M(k_x, k_y)\ R(k_x, k_y)\ W(k_x, k_y) \tag{7}$$

under a limited set of conditions, including a sufficiently small wave height. With the transfer functions M and R, an estimate of wave height spectrum can be made from a measurement of the SAR image spectrum. Unfortunately, these transfer functions are dependent upon environmental conditions which may not be known. Therefore, current research is seeking to determine ways of relating these conditions to observable properties of the SAR image or other remote sensor data. Possible leads include the relationship between the long wave spectrum and the short wave spectrum. If the latter relationship were established and found to be significant, an iterative approach to the spectrum inversion process might be indicated, in which a first estimate of the long wave spectrum is made using an assumed modulation transfer function.

A more modest application of SAR is to estimate just the dominant wavelength and direction, rather than the full wave height spectrum. Indeed, several experimental verifications of this capabiity have already been made (e.g., Gonzalez et al., 1979; McLeish et al., 1980; Pawka et al., 1980; Shuchman et al., 1983). Theoretically, however, there is reason to believe that biases may exist in these estimates under certain conditions.

Two cases may be considered. In the first case the long-wave spectrum is very narrow, approaching a delta-function in the frequency domain in the extreme case of a well-defined swell. In this case it can be seen that as long as the wave imaging is a linear process, as described by Eq. (7), the image spectrum will also contain a strong peak at the same location. Thus, the dominant wavelength and direction can be accurately estimated even if the modulation transfer function is not well known. Nonlinearities introduced primarily by the velocity bunching mechanism are still a concern, but have not yet been demonstrated experimentally.

The second case is that of a locally wind-generated sea, which has a much broader wave height spectrum. In this case it can be shown that even a linear process as described by Eq. (7) can result in a shift in the peak of the image spectrum, as compared to the peak of the wave height spectrum. Thus, the apparent dominant wavelength and direction indicated by the SAR image may differ from the actual. This effect is illustrated in

Figure 5, which shows only the resolution effect described by $M(k_x, k_y)$. If the width of this transfer function (which is inversely proportional to the effective resolution δx) is comparable to the spectrum width, an appreciable shift in the spectrum can occur, as indicted by the dashed curve in the lower part and the dashed contour on the upper part of Figure 5. Assuming a Gaussian shape for both the spectrum and the transfer function, the shift in the peak wavenumber is given by

$$\Delta k = \frac{B_S^2}{B_S^2 + B_M^2} k_0 \tag{8}$$

where B_S is the spectrum width, B_M is the width of the transfer function, and k_0 is the peak wavenumber. Since B_M may depend on the wave spectrum, an iterative procedure similar to that described above might be needed to account for this effect. Experimental verification of this effect has been suggested by a slight lengthening and rotation of the waves toward the range direction, but a full evaluation has not yet been performed.

CONCLUSIONS

Two classes of inversion problems in SAR imaging have been discussed in this paper: fine resolution imagery of stationary scenes and imaging of a moving surface. For the situation when the scattering scene is stationary the kernel function or total SAR system transfer function can be measured and used for calibration. To obtain absolute values of scattering coefficients from the output image requires the use of reference reflectors. The reference reflectivity must be calibrated by independent measurements to determine the system transfer function. It is important that consistent bandwidth (i.e., resolution) be utilized in the calibration and inversion of SAR data. This avoids the necessity of applying corrections to the power measurements that would require detailed knowledge of the SAR system frequency response. SAR calibration has been successfully demonstrated. Additional work is required, however, to improve measurement techniques to obtain total system transfer function. Particular problems are the in-flight antenna pattern and reference reflectivity.

Inversion of SAR data obtained from a moving surface requires the consideration of motion effects in addition to those described above for the fixed scatterer situation. Systematic displacements on a scale greater than the resolution result in azimuth shifts in the mapping of the scene onto the SAR image. Fine scale random motion effects on a scale equal to the radar wavelength will degrade along-track resolution. The inversion

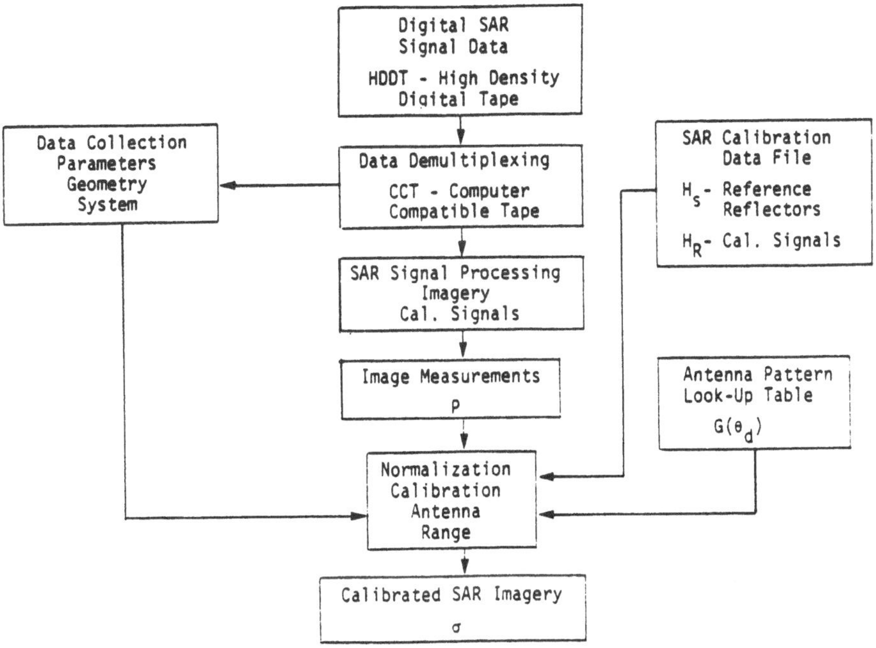

Figure 1. Data Flow Diagram for ERIM Digital SAR Calibration Algorithm.

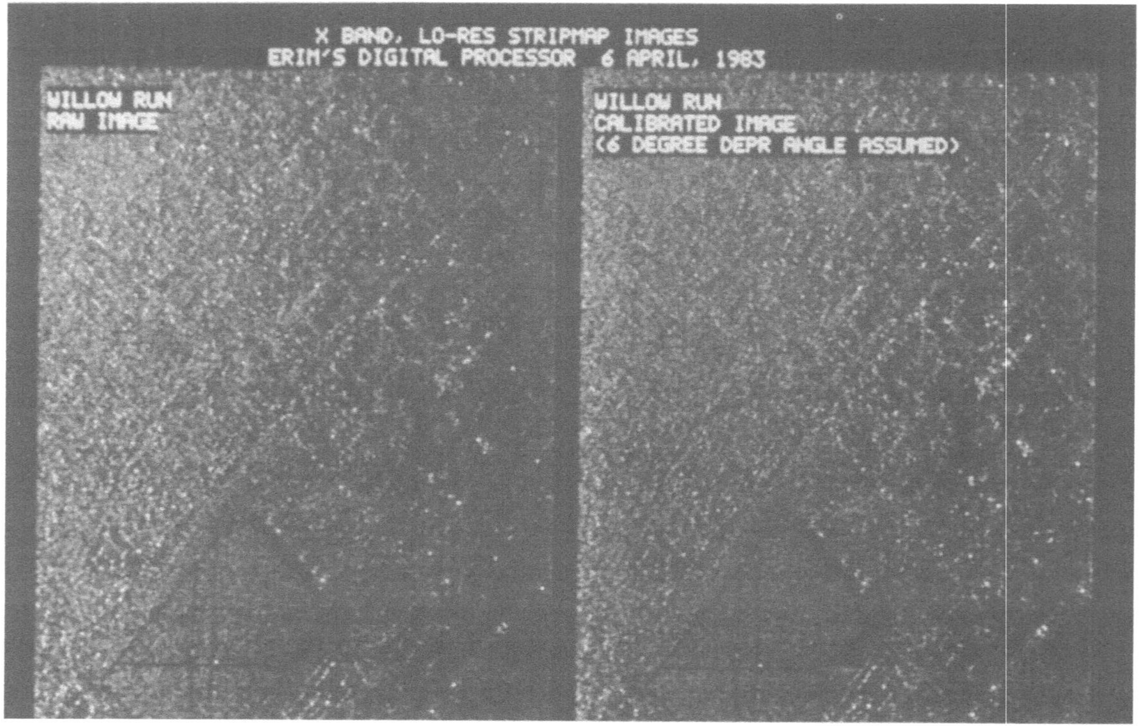

a. Uncalibrated

b. Calibrated By Normalization to the Willow Run Calibration Reflector Array

Figure 2. SAR Imagery of the Red Deer Area, Horizontal Polarization 3.2 cm Wavelength

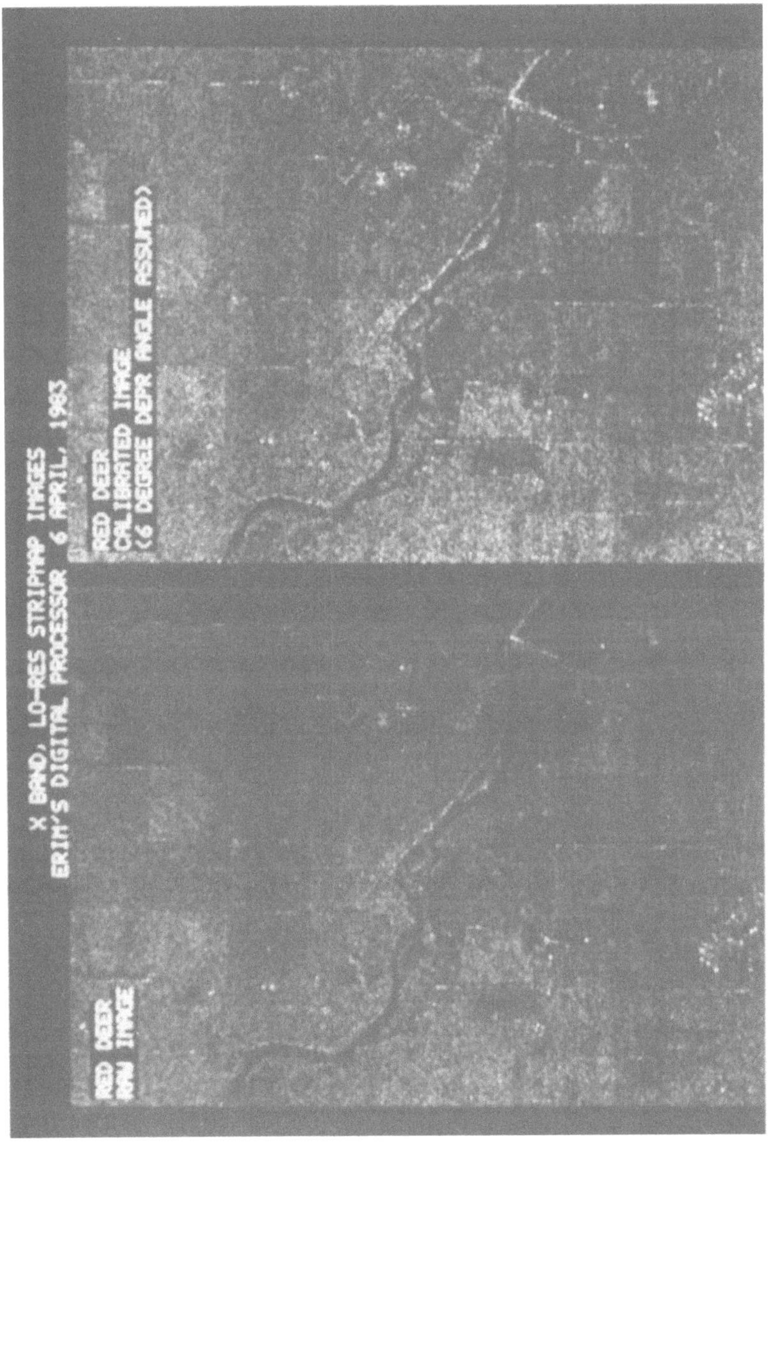

a. Uncalibrated b. Calibrated SAR Image

Figure 3. SAR Image of Willow Run Area Including Calibrated Reflector Array, Horizontal Polarization, 3.2 Wavelength

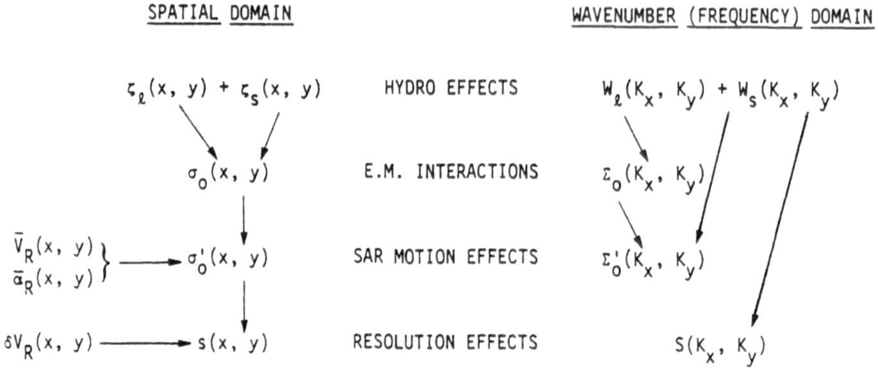

Figure 4. Diagram of SAR Imaging Process for Ocean Waves

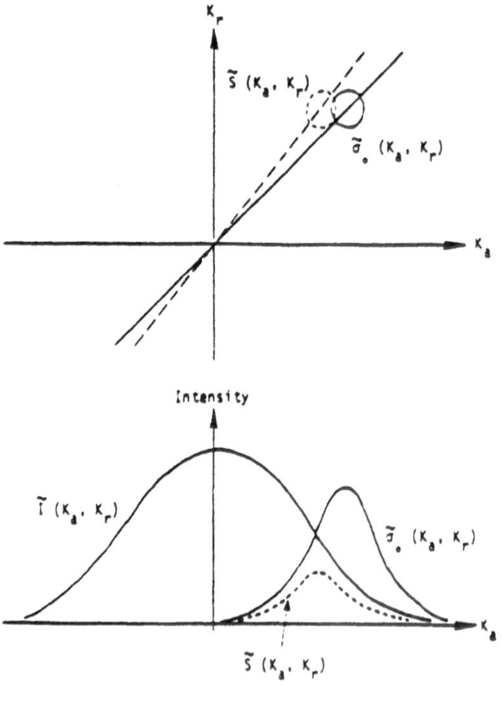

Figure 5. Schematic Diagram Showing the Degraded Azimuth Resolution on SAR-Derived Spectra

is complicated by motion effects in the imaged scene that are not as yet understood. Current research is being directed toward improving the understanding of modulation transfer function of the sea surface.

ACKNOWLEDGMENTS

This paper is based on work carried out at ERIM for various sponsors, including the Office of Naval Research (Contract N0014-81-C-0692), NASA Headquarters (Ocean Processes Branch) and the Air Force WPAFB.

REFERENCES

Alpers, W.R. and C.L. Rufenach, The effect of orbital motions on synthetic aperture radar imagery of ocean waves, IEEE Trans. Antennas Propagat., AP-27, 685-690, 1979.

Gonzalez, F.I., R.C. Beal, W.E. Brown, Jr., D.S. DeLeonibus, J.F.R. Gower, D. Lichy, D.B. Ross, C.L. Rufenach, J.W. Sherman, III, and R.A. Shuchman, Seasat synthetic aperture radar: ocean wave detection capabilities, Science, 204, 1418-1421, 1979.

McLeish, W., D. Ross, R.A. Shuchman, P.G. Teleki, S.V. Hsiao, O.H. Shemdin, and W.E. Brown, Jr., Synthetic aperture radar imaging of ocean waves - comparison with wave measurements, J. Geophys. Res., 85, C9, 5003-5011, 1980.

Pawka, S.S., S.V. Hsiao, O.H. Shemdin, and O.L. Inman, Comparison between wave directional spectra from SAR and pressure sensor arrays, J. Geophys. Res., 85, C9, 4987-4995, 1980.

Brown, G.S., Backscattering from a Gaussian-distributed perfectly conducting rough surface, IEEE, AP-26, No. 3, p. 472, May 1978.

Westwater, E.R., and O.N. Strand, Inversion Techniques, Chap. 16 in Remote Sensing of the Troposphere, V.E. Derr, ed., NOAA, Wave Propagat. Lab., Boulder, Colorado, August 1972.

Wright, J.W., A New Model for Sea Clutter, IEEE Trans. Antenna and Propagat., AP-16, 554-568, 1968.

Valenzuela, G.R., Theories for the Interaction of Electromagnetic and Oceanic Waves - A Review, Boundary Layer Meteorology, 13, 61-85, 1978.

Shuchman, R.A., W. Rosenthal, J.D. Lyden, D.R. Lyzenga, E.S. Kasischke, H. Gunther, and H. Lime, Analysis of MARSEN X-Band SAR Ocean Wave Data, J. Geophys. Res., 88, 9757-9768, 1983.

IV.10 (SI.5)

INVERSE METHODS IN RADIO GLACIOLOGY

Dr M E R Walford

H H Wills Physics Laboratory, University of Bristol

ABSTRACT

Since the early sixties, Radio Glaciology has been carried out over polar ice, over sea ice and over water-laden temperate glacier ice. Initially, the principal observable in pulsed echo sounding was the echo-delay time and the inversion problem centred around the need to know both the velocity of propagation of electromagnetic waves in ice and snow and the position of the observation platform, accurately enough for depth sounding. The Antarctic Atlas, in publication this year, demonstrates that a wealth of geographical knowledge of the Antarctic has been obtained in the past 20 years principally by such means.

However, glaciologists have also sought to exploit radio echo sounding in other ways. Polarization has been utilized, for example, in the study of flow-induced orientation effects in Greenland ice. Spatial fading patterns have provided limited information about the roughness of the glacier beds and have also been used to establish a frame of reference fixed with respect to a glacier bed against which small horizontal or vertical displacements of the observer's platform may be monitored, independent of satellite or surveyor's benchmarks.

Perhaps the most ambitious attempt to solve an inversion problem in Radio Glaciology consists in a "radio echo microscope" experiment: a phase-sensitive pulsed echo sounder is used to sample echoes over a finite aperture in the snow surface and subsequently, in a large computer, an image of the echoing glacier bed is computed, using a scalar, pulsed, diffraction theory. This method has been used with some success over ice approximately 43 wavelengths thick.

W.-M. Boerner et al. (eds.), Inverse Methods in Electromagnetic Imaging, Part 2, 969–983.

Scattered echoes from within the ice have been observed from stratifications in Polar Ice which can be followed sometimes for hundreds of kilometres. In temperate ice, on the other hand, heavy scattering from irregular water bodies can totally obscure the glacier bed unless monocycle radar with decameter wavelengths is employed. Both of these problems present inversion problems of great interest which have not yet been satisfactorily solved: we have yet to relate stratifications observed in boreholes to the stratification echoes in other than a few cases; we would like to learn something useful about the distribution of water bodies in temperate ice from their radio echo scattering cross-sections.

INTRODUCTION

Radio glaciology began in the early 1960s. There were two reasons. Airborne radio altimeters operating over the Antarctic Ice Sheet were found to register echoes from the underlying bedrock as well as from the snow surface (Waite 1962). Secondly, unusual interference bands in ionspheric spread echoes were obtained at a station on Ellesworth Ice Shelf, Antarctica. They were successfully interpreted on the assumption that radiation was being reflected from the sea underlying the ice shelf (Evans 1961). Evans subsequently designed a radio-echo sounder optimized for Polar Ice sounding based on a systematic analysis of the physical properties of polar ice masses (Robin et al 1969). From these beginnings Radio Glaciology has blossomed abundantly. Pulsed-echo sounding instruments have been designed for probing sea ice, for sounding water impregnated "temperate" glaciers and for exploring permanently frozen ground. The band of frequencies in common use is from 5MHz to 1GHz, being limited by antenna size at low frequencies and by heavy scattering in the ice at high frequencies. Some instruments are airborne, some mounted on surface vehicles and one at least is back-packed. However, space satellites do not provide suitable platforms.

Impressive results of the first twenty-one years of radio-glaciology in Antarctica are presented in "Antarctica: Glaciological and Geophysical Folio", a map folio edited byD.J. Drewry at the Scott Polar Research Institute in Cambridge (Drewry 1983). Selected folio maps of the Antarctic Ice Sheet show upper and lower surface topography, internal layering, subglacial lakes, calculated ice velocities, mass balances, driving stresses, isostatic depression and major tectonic units. Ross Ice Shelf Maps show the absorption, bottom roughness and ice streamlines. All these maps could not have been produced without the information provided by Radio Glaciology: it plays an important role in unlocking the secrets of Antarctica.

VARIOUS RADIO ECHO TECHNIQUES

Much of the foregoing work was based on airborne measurements of echo amplitudes and on pulsed echo arrival times averaged over many transmissions. However, in attempting to understand glacier behaviour we often require to know more than such simple measurements afford. Hence it is that a wide variety of radar techniques have been applied to problems in Radio Glaciology. There is enormous potential here because polar ice is a remarkably stable, clear and relatively isotropic, homogeneous medium of propagation with which to work. To illustrate this we refer to the polarization studies by Hargreaves (Hargreaves, 1977). He measured the state of polarization of the weak echoes returned from stratifications at different depths within the Greenland ice sheet. A step-wise data inversion procedure was applied and it was shown that there is a small systematic radio anisotropy in the ice sheet of order $|\varepsilon/\sigma - 1| \approx 3.\ 10^{-4}$. This is probably the result of orientation introduced by the slow flow of the ice. Laboratory measurements show that single crystal radio anisotropy is less than 3 10^{-3}.

Reflection at the glacier bed itself often produces strongly depolarized echoes. However, this is in may cases to be ascribed to bedrock geometry rather than dielectric anisotropy. The bed may be boulder-strewn, cracked, striated and rough on many scales including the radio wavelength scale. Under these circumstances we should expect the echoing wavefield to include a randomly polarized component arising from diffuse scattering, edge diffraction and/or multiple scattering events. An interesting example however, of a systematic polarization effect occurs in the reflection of radio waves from the bottom of the King George VI Ice Shelf in Antarctica (Walton, 1981). One speculates that this is ultimately caused by water flow beneath the ice shelf systematically controlling the structure of the interface as has been observed in sea ice and in the laboratory (Langhorne, 1983).

FADING PATTERNS IN RADIO ECHOES

Limited statistical information about the roughness of the reflecting glacier bed can be obtained from a study of the statistical properties of radio-echo spatial fading patterns. It should be emphasized that usually rather low gain antennas are used and so rapid, deep spatial fading is the norm. In three papers Berry develops a theory of the diffraction of quasimonochromatic pulses from rough surfaces, based on Kirchhoff theory. In his 1972 paper he deals with the forward problem and shows that in the case of paraxial geometry and with the assumption that no shadowing or multiple scattering occurs at the bed, the returned echo $\psi(t)$ can be written essentially as a convolution of two functions, one representing the transmitted pulse and one involving only properties of the reflecting surface. In his 1973 paper he proceeds to develop a statistical theory centred around an expression for the two point distribution function for $\psi(t, \bar{r})$. The 1975 paper discusses practical applications to glaciology and relates his work to somewhat parallel treatments by Harrison (1970-71) and by Oswald (1975). In respect of the fading patterns of echoes Berry finds that very small departures from bedrock smoothness of order $\lambda/10$ where λ is the central radar wavelength in ice, serve to produce significant spatial fading. No information about bedrock roughness can be inferred from the spatial fading rate except in a limited regime of very gentle bed topography. (Such a regime was investigated theoretically by Oswald (1975).) However, one can infer bedrock statistical information from the statistics of the echo envelope. In particular one can identify a "coherent" echo component which predominates in echoes returned from the rather smooth bed of most ice shelves and a long "incoherent" echo tail which is important over rough glacier beds. Suppose that the glacier bed can be approximately described as an isotropically rough surface with a gaussian height distribution describable by a single horizontal scale L and a vertical scale S. Then according to Berry the parameters L and S can be derived from mean echo shape parameters S and T which measure the "degree echo coherence" and the "half power length" respectively (Berry, 1975). There has been a somewhat limited number of field studies of bed roughness.

In one study Oswald identified the occurrence of lakes under the Antarctic Ice Sheet on the basis of fading characteristics. He also claims to be able to identify geologically different regimes beneath Devon Island Ice Cap, Canada (Oswald, 1975). Neal studied the character of echoes returned from the bottom of the Ross Ice Shelf and tentatively relates them to ice shelf topography and the distribution of currents in the underlying sea (Neal, 1979).

Another view of the spatial fading pattern of radio echoes is that it provides a kind of surveyor's bench mark: a frame of reference fixed with respect to the glacier bed against which one can measure the horizontal movement of a flowing glacier (Nye et al, 1972a, Walford 1972). Furthermore, if one builds an echo sounder sensitive to the phase of echoes, one can measure vertical displacements of the echo sounder with respect to this radio echo frame of reference (Nye et al, 1972b, Walford et al, 1977). Displacements can be measured with a precision of a small fraction of a radar wavelength - a few centimeters say - being limited by the signals-to-noise ratio or in the long term by slow changes in spatial fading patterns produced by shear distortion of the reflecting glacier bed (Doake, 1975). We have no measurement of any residual radio-echo shimmering due to the passage of the slowly flowing ice stream, although I believe this small effect should be observable under suitable conditions.

We have an experiment under way to measure, using the phase-sensitive radio echo sounder, the vertical strain rate over a ten-year period of the Devon Island Ice Cap. Also we hope at a future opportunity to use the phase-sensitive radio echo sounder as a measuring rod to study periodic geometrical distortions produced by the action of ocean tides, in a major glacier where it goes afloat to become an ice stream within an ice shelf.

INVERSE METHODS IN POLAR ICE

The phase-sensitive echo sounder which we discussed briefly above, also opens the door to a number of interesting inverse methods in Radio Glaciology. Let us first consider instruments (figure 1). Our machine was built at Bristol University. It transmits 300 ns pulses of 60 MHz waves locked in phase with respect to a crystal-controlled oscillator. Radio echoes are usually received on the same simple dipole antenna used for transmission and we record not only the echo amplitude a(t) as a function of delay time t, but also the phase $\phi(t)$ with respect to the continuously running oscillator. Let the voltage signal at the aerial terminals be written

$$\psi(t) = a(t) \cos\{\omega_0 t + \phi(t)\}$$

where $\cos \omega_0 t$ represents the reference oscillator signal. a(t) and $\phi(t)$ vary more slowly with time than does $\cos \omega_0 t$. We may record a(t) and $\phi(t)$ or display them immediately on an oscilloscope. A particularly interesting display mode is obtained if we apply $a(t) \cos \phi(t)$ to the X plates and $a(t) \sin \phi(t)$ to the Y plates of the oscilloscope, for then the instantaneous radial displacement of the spot from the centre of the display is proportional to a(t) and the instantaneous azimuth angle is $\phi(t)$. A returning echo thus produces a characteristic closed curve on the display which we call an echo signature. It is a simple loop if the echo is returned from a plane mirror; it is a convoluted figure with bend points where echoes are reflected from a complicated glacier bed. This display mode is in itself of value to the field glaciologist who can observe the signature responding to antenna rotations and displacements. If the signature is not too complicated, useful inferences about the target nature and distribution can be made directly (Walford et al, 1981).

We shall next consider two inverse problems in polar Radio Glaciology. Firstly consider the echo $\psi(t)$ observed at a fixed point P on the glacier surface. Following Berry (1972) it may be regarded as a convolution of two functions F'(t) and G(t) thus:

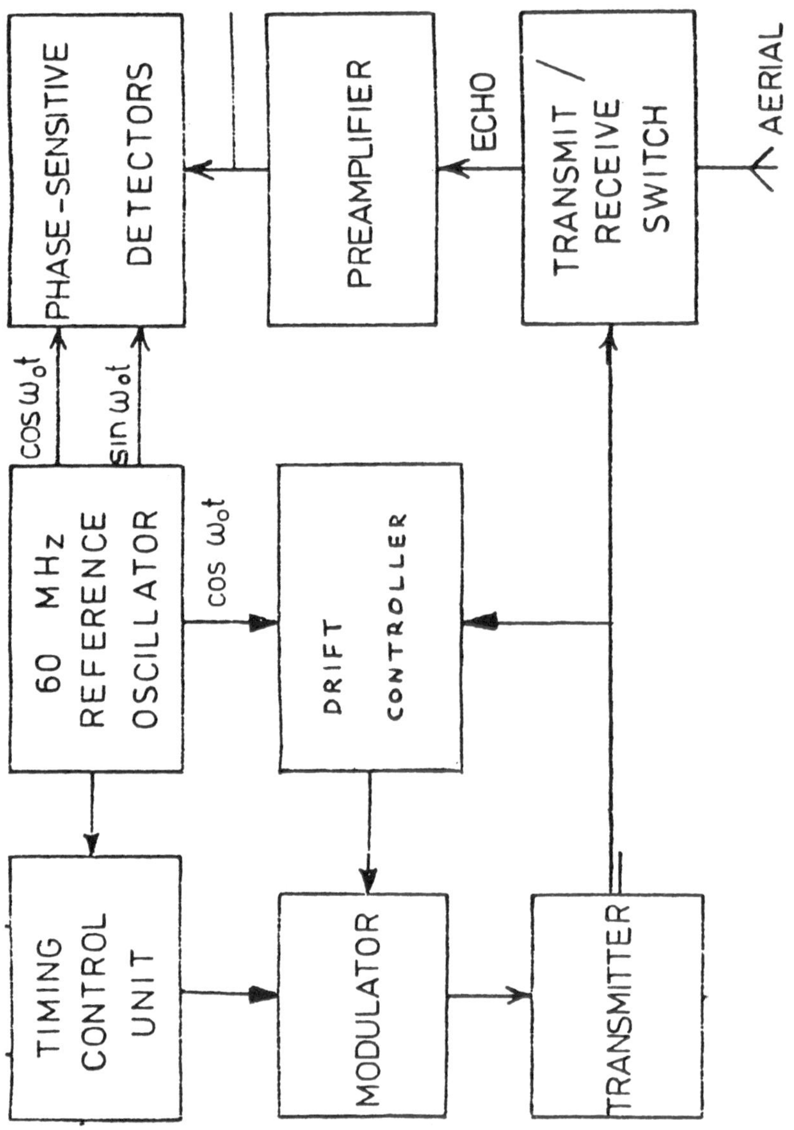

Figure 1. Block diagram of phase-sensitive radio echo sounder.

$\psi(t) = F'(t) * G(t)$. Here F(t) represents the time-dependence of the transmitted signal. G(t) depends on the velocity of radio wave in ice and on a function g(r) which logs the reflection coefficient of targets as a function of their range r from the point of observation P. Berry gives a paraxial expression for g(r), in the form of an r-contour integral, which properly takes into account the obliquity factor (Berry, 1972).

In general, a glacier bed has structure on all scales: it is a fractal surface (Mandelbrot, 1977) and so g(r) is a fractal function of time. It is real and positive if the reasonable assumption is made that the reflection coefficient of the glacier bed is real and positive. g(r) is the function we want to know: it is the target of inversion processing.

We have dabbled with this problem! We took measurements of $\psi(t)$ and F(t) from an ultrasonic model glacier-sounder in the laboratory. We found frequency-domain deconvolution techniques easier to handle conceptually and in practice than time-domain methods and so obtained estimates of g(r) which were severely limited by bandwidth. Our estimates still have the oscilliatory character of radio wavelets in fact. We prefer to think of them, therefore, as "ideal echoes" in the sense of being the shortest simply-structured echoes obtainable within a given bandwidth. We achieved a pulse compression ratio of about 4:1 in the model experiments. We shall have to tackle this problem more seriously in the future if we are to sound where echoes are known to be very weak.

Radio echoes observed at a fixed point tell us something about the function g(r) but nothing about the directions in which reflecting targets lie: for this we must move the echo sounder about on the snow surface. We now face the second inversion problem: that of reconstructing as far as possible the three-dimensional geometry of reflecting targets from echoes sampled within a limited aperture at the snow surface. A fundamental limitation is set by refraction: no rays travel in the ice at more than 35° to the vertical.

Our approach to this inversion problem centres around the concept of the simple microscope (figure 2). There is an equivalence between the radio echo sounder situation and the object space of the simple microscope. Suppose the radio echo sounder is placed at point P, a wave with time variation F(t) is launched and an echo $\psi(t)$ is returned. Now consider the simple microscope: suppose a coherent radio wave is launched upwards from the glacier bed O and sampled by a receiver placed at P, giving $\psi_m(t)$. It can be shown that $\psi(t) = \Psi_m(t)$ to a good approximation if the following conditions are met:-

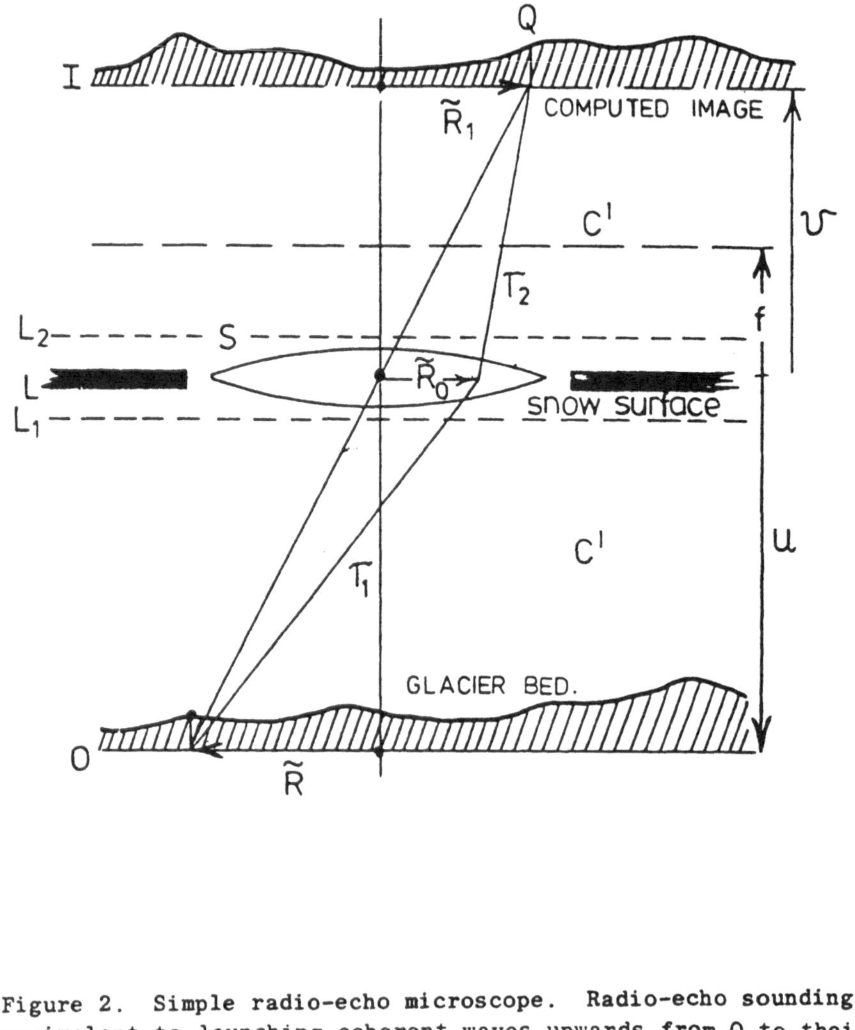

Figure 2. Simple radio-echo microscope. Radio-echo sounding is equivalent to launching coherent waves upwards from O to their observation in plane L_1. Propagation from L_1 to the image plane 1 is carried out in the computer.

1. The geometry of the glacier bed is identical in each case.

2. The velocity of propagation in the microscope space is half of that in the radio echo space.

3. The coherent wave launched from O has time variation F(t) and its local amplitude is proportional to the local value of the reflection coefficient of the glacier bed.

4. The geometry is paraxial.

Given this principle of equivalence (Walford et al, 1981) we may realize the object space of a simple microscope by making field measurements of radio echoes over an aperture in the snow surface. The lens and image space of the simple microscope is then realized in the computer; we should obtain a diffraction-limited image at I of the glacier bed O.

One interesting question is how does one focus this radio echo microscope. If we were constrained to use a single continuous-wave frequency, we would have no a priori knowledge of the ice depth and so focussing would be a trial and error procedure, as it is indeed in the optical microscope. Fortunately in the case of polar ice sounding we do know the approximate ice depth from pulsed echo delay times and so can calculate where the focussed image should lie. Another advantage of using wideband pulses is that it affords some measure of verification of the imaging process. As long as the system is non-dispersive, the image produced in plane I should consist of a wave whose signature at every point is a simple loop of the same shape as that produced by the transmitted wave F(t). The amplitude of the loop is proportional to the local glacier bed reflection coefficient and its orientation to the local elevation of the glacier bed with respect to the object plane, O.

We have carried out such an experiment using field data collected at the Mer de Glace Agassiz in Ellesmere Island, Canada. The ice was 130m (43 wavelengths) thick and we sampled over an aperture 60m square. Figure 3 and 4 show the results of imaging. It appears to have been fairly successful and reveals a very gently undulating glacier bed with a reflection coefficient varying spatially by about 10 dB.

This imaging technique rapidly becomes prohibitively laborious over thicker ice, but we hope to test it further at a site where the bedrock geometry is somewhat more complex.

Ultimately one will wish to consider imaging using a vector-wave, pulsed, radio-echo system. In this context the scalar function

Figure 3. Part of the array of imaged signatures produced in plane of the radio-echo microscope (figure 2). Samples are plotted on a 5 metre grid.

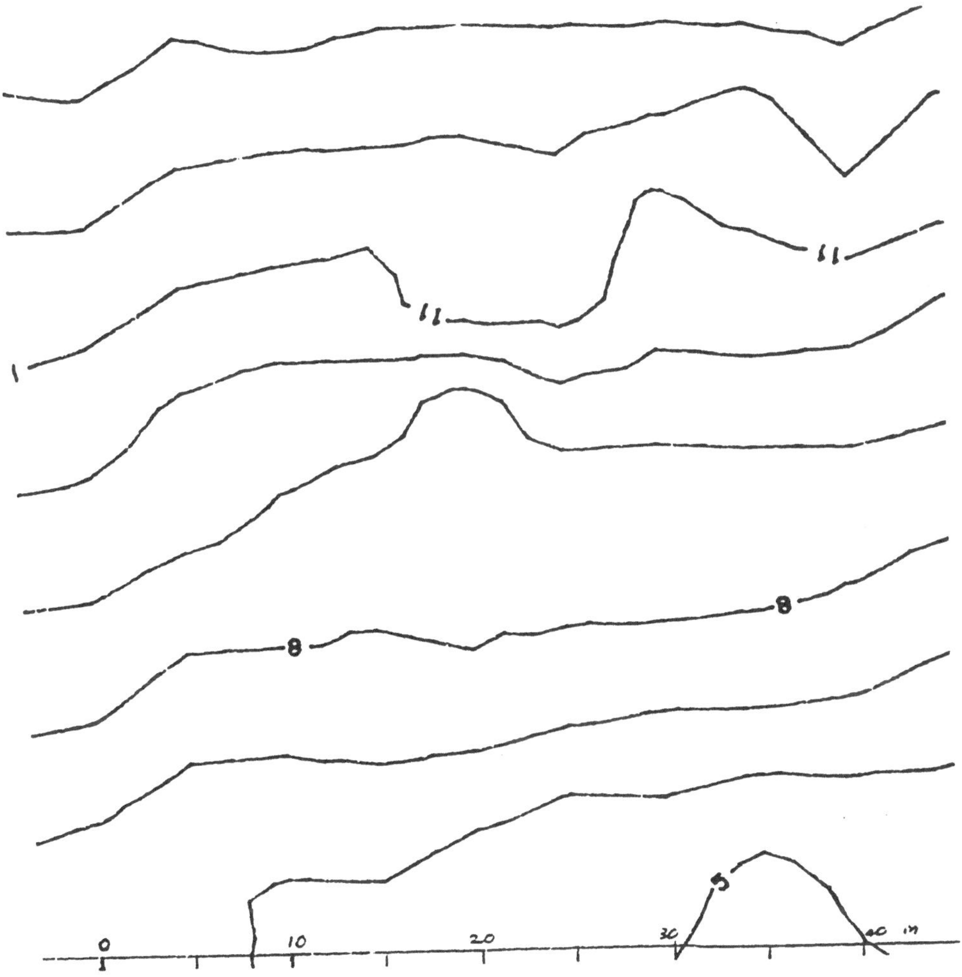

Figure 4. Contour map of part of the glacier bed as calculated from the orientations of image signatures in figure 3. Contour lines are 0.26m apart.

g(r) would be replaced by a back-scattering matrix S(r) which, whilst invarient physically, can be expressed with respect to arbitrary orthogonal polarization axes (Boerner, 1980) as:

$$\begin{bmatrix} S_{11}(r) & S_{12}(r) \\ S_{21}(r) & S_{22}(r) \end{bmatrix}$$

The S(r) matrix would then be the target for frequency domain deconvolution at a fixed point of observation P. Similarly $S(\vec{r})$ would replace the scalar reflection coefficient as the parameter to be imaged in the pulsed vector radio-echo microscope.

TEMPERATE GLACIER SOUNDING

We briefly consider the problems peculiar to radio echo sounding in temperate glaciers. These arise because of scattering by the free water distribution in them (Ewan Smith et al, 1972). The glaciologist would like firstly to observe the glacier bed in detail despite clutter arising from the water distribution, and secondly to learn something of this water distribution itself. For example, we know very little about the relative amounts of water present in the different scale sizes of channel or cavity, ranging from large moulins and intraglacial rivers down to intergranular veins. We know that the water regime varies temporally in response to input fluctuations and also that it varies spatially within a glacier even to the extent that some parts may be temperate and others polar.

The usual method adopted to sound the temperate glacier bed is to drastically reduce the centre frequency so that a monocycle of 30m wavelength radiation is transmitted. This reduces the effect of scattering and enables the depth of the glacier to be estimated albeit with low resolution (Watts et al, 1976). An alternative technique which can sometimes be helpful is to use a higher frequency than normal and make use of high gain aerial structures to reduce the clutter contributions.

The radio echo microscope offers a third possible technique which would permit one to sound a temperate glacier with say 3 metre wavelengths. It should be successful as long as internal scattering is not so heavy that the transverse coherence of the transmitted wave is severely reduced. The proper image distance would be found from independent low-frequency observations or by trial and error: one anticipates that "focussing" the microscope down through the glacier one might discover the image of the glacier bed. It would be characterized by large simple image signatures which would behave systematically when explored laterally. This would contrast with the weak random images to be expected when the radio echo microscope is focussed at a point within or beneath the glacier. If the pulsed echo radio-microscope appears to work, producing a convincing image of the glacier bed, we could then focus it upon the interior of the glacier with some confidence of producing meaningful results. There would result an image of the water distribution which we could characterize by a suitable scalar scattering density function $\rho(\bar{r})$ or a scattering matrix density function $S(\bar{r})$. It would be interesting to examine these for spatial and temporal variation, and for evidence of structure, orientation and dispersive behaviour. We are exploring some of these possibilities in a series of experiments at Storglaciaren, Sweden.

SUMMARY

We have briefly reviewed the development of radio glaciological techniques, discussing aspects of the inverse problem in polar and temperate glaciers. It is clear that there is much scope remaining and that the future development of the subject should be extremely interesting.

REFERENCES

Berry, M.V. 1972. On deducing the form of surfaces from their diffracted echoes. Journal of Phys. A, Vol.5, No.2, pp.272-291.

Berry, M.V. 1973. The statistical properties of echoes diffracted from rough surfaces. Phil.Trans., Ser.A, Vol.273, No.1237, pp.611-54.

Berry, M.V. 1975. Theory of radio echoes from glacier beds. J. Glaciology, Vol.15, No.73, pp.65-74.

Boerner, W.-M. 1980. Polarization utilization in electromagnetic inverse scattering. Topics in Current Physics, No.20, pp.237-303.

Doake, C.S.M. 1975. Glacier sliding measured with a radio echo technique. J. Glaciology, Vol.15, No.73, pp.89-93.

Drewry, D.J. 1983. Antarctica : Glaciological and Geophysical Folio. Scott Polar Research Institute, University of Cambridge, U.K.

Ewan Smith, B.M. and Evans, S. 1972. Radio echo sounding : absorption and scattering by water inclusions and ice lenses. J. Glaciology, Vol.11, No.61, pp.133-147.

Evans, S. 1961. Polar ionospheric spread echoes and radio frequency properties of ice shelves. J.Geophys.Res., Vol.66, No.12, pp.4137-41.

Hargreaves, N.D. 1977. The polarization of radio signals in the radio echo sounding of ice sheets. J.Phys. D: Applied Phys., Vol.10.

Harrison, C.H. 1970. Reconstruction of subglacial relief from radio echo records. Geophysics, Vol.35, No.6, pp.1099-1115.

Harrison, C.H. 1971. Radio echo sounding : focussing effects in wavy strata. Geophysical Journal of the Royal Astro. Soc., Vol.24, No.4, pp.383-400.

Langhorne, P.J. 1983. Laboratory experiments on crystal orientation in sea ice. Annals of Glaciology No.4, pp.163-169.

Mandelbrot, B.B. 1977. Fractals, form, chance and dimensions. Freeman. London.

Neal, C.S. 1979. The dynamics of the Ross Ice Shelf revealed by radio echo sounding. J.Glaciol., Vol.24, No.90, pp.295-307.

Nye et al, 1972a. Nye, J.F., Berry, M.V. and Walford, M.E.R. Measuring the change in thickness of the Antarctic Ice Sheet. Nature, Physical Science, Vol.240, pp.7-9.

Nye et al, 1972b. Nye, J.F., Kyte, R.G. and Threlfall, D.C. Proposal for measuring the movement of a large ice sheet by observing radio echoes. J.Glaciology, Vol.11, No.63, pp.319-325.

Oswald, G.K.A. 1975. Investigation of sub-ice bedrock characteristics by radio echo sounding. J.Glaciology, Vol.15, No.73, pp.75-87.

Robin et al, 1969. Robin, G. de Q., Evans, S. and Bailey, J.T. Interpretation of radio echo sounding in polar ice sheets. Vol.265, No.1166, pp.437-505.

Waite, A.H. and Schmidt, S.J. 1962. Gross errors in height indication from pulsed radar altimeters operating over thick ice or snow. Proc.IRE., Vol.50, No.6, pp.1515-1520.

Walford, M.E.R. 1972. Glacier movement measured with a radio-echo technique. Nature, Vol.239, No.5367, pp.93-95.

Walford et al, 1977. Walford, M.E.R., Holdorf, P.C. and Oakberg, R.G. Phase sensitive radio echo sounding at the Devon Island Ice Cap. J.Glaciology, Vol.18, No.79.

Walford, M.E.R. and Harper, M.F.L. 1981. The detailed study of glacier beds using radio echo techniques. Geophys. J.R. Astro.Soc., No.67, pp.487-514.

Walton, J.L.W. 1981. Unpublished contribution to the British Branch of the International Glaciological Society, Norwich.

Watts, R.D. and England, A.W. 1976. Radio echo sounding of temperate glaciers : ice properties and sounder design criteria. J. Glaciology, Vol.17, No.75, pp.39-48.

IV.11 (MI.1)

FAST MILLIMETER WAVE IMAGING

Hans H. Brand

Lehrstuhl für Hochfrequenztechnik
Universität Erlangen-Nürnberg, Cauerstr. 9
D 8520 Erlangen, Federal Republic of Germany

ABSTRACT

The contribution deals with discussing the conditions that microwave imaging systems for traffic applications have to fulfil. Regarding moving objects the most important requirement of movement recognition and multiple target capability leads to very short times tolerable for collecting and processing the backscattered signals. High resolution with moderate aperture size requires short wavelength utilization. Hence, fast millimeter wave system should be a solution. Experimental results from a 70 GHz-imaging system are presented.

1. INTRODUCTION

Non-visual long-wavelength imaging techniques have gained increasing attention in the past two decades for different fields of applications. In most cases the wavelength λ is much smaller than the size L of the object intended to be recognized (approximately $L \gtrsim 100\ \lambda$), however, large compared to the roughness ΔL of the object surface. This wavelength region is generally called microwaves, and the only justification for their use in imaging (compared to the nearly perfect visual-wavelength region) is the fact that sometimes the medium embedding the objects is impenetrable for visual optical waves but transparent for microwaves. Most applications can be subdivided in two groups. In the air-embedding case fog, smog or other adverse weather is the reason for optical invisibility and in this case electromagnetic microwaves (300 MHz up to 300 GHz) are the best countermeasure of choice. When the objects are embedded in blurred water or water-like biological media, acoustic microwaves (10 kHz up to 10 MHz) have the

W.-M. Boerner et al. (eds.), Inverse Methods in Electromagnetic Imaging, Part 2, 985–995.

better chance to penetrate the embedding.

Apart from laboratory experiments where objects can be made stationary, real outdoor application has to regard moving object situations. It is the movement of the objects that constrains observing some conditions in the implementation of an imaging system, otherwise the image reconstruction process is disturbed and movement - in particular in a multiple target situation - cannot be recognized.

Typical examples for potentially useful applications of microwave imaging systems with realtime and multiple target capability are assisting helicopter landing manoevres in life-saving-mission or navigating land traffic on the apron of an airport for the purpose of avoiding collisions in fog or dust (/1/, /2/).

2. FUNDAMENTAL IMAGING RELATIONS AND DEFINITIONS

The purpose of this contribution is discussing the conditions that have to be fulfilled by a microwave imaging system in order to obtain information from a moving object scenery. First of all the notions used in this context should be clarified.

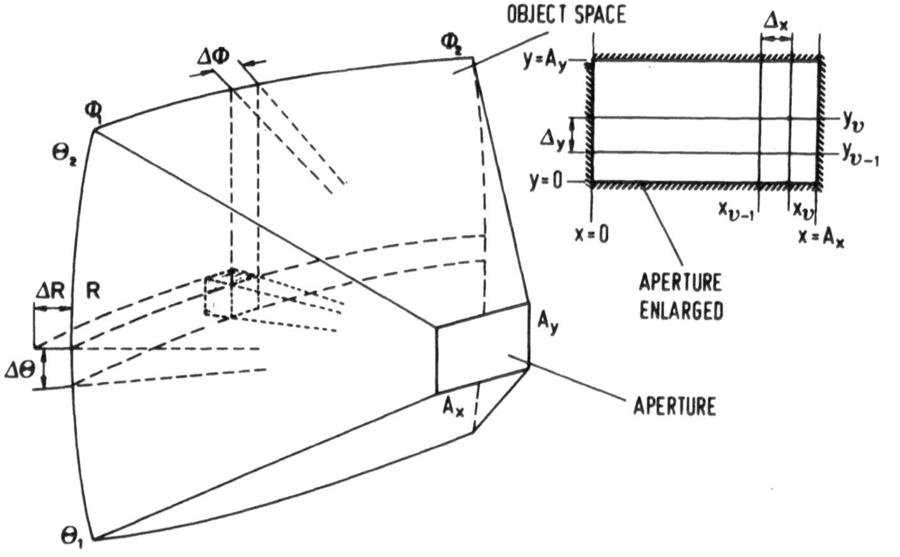

FIG.1 OBJECT SPACE, APERTURE AND RESOLUTION CELL

The key element connecting the object space with the image-formation system is the so-called a p e r t u r e which is used as a measure for the amount of radiation coming from the object space into the imaging system. According to Fig. 1 the aperture is assumed to form a plane window having the width A_x and the height A_y.The object scenery may be illuminated with microwave radiation by a point source from inside the aperture or by a source placed outside but very close to the aperture boundary. This will be called the monostatic case. In the bistatic case, the illuminating source is plaçed far away from the receiving aperture(compared with the aperture length A_x or A_y), but this case is not further considered here. A portion of the illuminating radiation is backscattered by the targets and only the small part of this backscattered radiation that is catched by the aperture can be used for the image generation. It is this limitation, or the cut-off-function of the window called "aperture" that prohibits improving resolution in the imaging process more and more. From diffraction theory we obtain for the lowest angle difference $\Delta\emptyset$ or $\Delta\theta$ that can be resolved by the given aperture

$$\Delta\emptyset = \frac{\lambda}{A_x} \quad (1a) \quad ; \quad \Delta\theta = \frac{\lambda}{A_y} \quad (1b) \quad ,$$

where λ is the free space wavelength of the radiation.

A time-modulation of the illuminating waves or utilizing a spread spectrum signal is a well known tool for resolving range (/3/, /4/), /5/). When B is the frequency bandwidth of the illumination signal and the same value B can be handled in the band-limited receiver, then from radar theory we obtain for the lowest range difference ΔR that can be resolved by the given spectral "aperture"

$$\Delta R = \frac{c}{2B} \quad (2)$$

where c is the free space velocity of the radiation.
In eqns. (1) and (2) correction coefficients of the order of magnitude of unity have been neglected for simplifying the estimations. From the geometry of Fig. 1 it can be seen that at the distance R from the aperture a resolution cell in the object space can be defined which is indicated by the lateral dimensions

$$\Delta X = R \cdot \Delta\emptyset \quad (3a) \quad ; \quad \Delta Y = R \cdot \Delta\theta \quad (3b)$$

and described by the volume $V_R = \Delta R \cdot \Delta X \cdot \Delta Y$, yielding

$$V_R = \frac{c}{2} \frac{R^2 \lambda^2}{A_x A_y B} \quad . \quad (4)$$

The difference between microwave imaging and conventional radar can now be described by means of this resolution cell. The term i m a g i n g seems suitable in the optical sense where the objects are recognized in size and shape (and possibly other parameters) provided that the object dimensions are very large compared to the appropriate dimensions of the resolution cell according to eqns. (1) - (4). Otherwise, if the object is very small compared to the resolution cell, it looks like a p o i n t s c a t t e r e r and the term r a d a r is used.

In most cases radar point targets do not really scatter isotropically. This characterizing property, i.e. the dependence of the radar cross section versus aspect angle, is summarized by the term target s i g n a t u r e.

In microwave imaging the term signature makes no sense for the objects as a whole, but when we restrict it to the resolution cell, or precisely to its lateral cross section, this partial signature (including polarization /6/) may help to better recognize the shape, even in the case of glittering and deflecting surfaces.

At last the term m u l t i p l e t a r g e t c a p a b i l i t y has to be regarded. In most cases the radiation entering the aperture can only be received at discrete points x_ν, y_ν. The equidistant spacing of the probes between adjacent points may be Δx and Δy according to Fig. 1. Then, from sampling theory, we can learn, that there is a periodic lateral ambiguity in the object space, i.e. only a lateral region of the width

$$X_p = \Delta X(N_x - 1) \quad (5a) \quad ; \quad Y_p = \Delta Y(N_y - 1) \quad (5b)$$

can be recognized unambiguously /4/. Here N_x and N_y are the numbers of the probes in the aperture, related to the spacings of the probes by

$$N_x = \frac{A_x}{\Delta x} \quad (6a) \quad ; \quad N_y = \frac{A_y}{\Delta y} \quad (6b).$$

Targets in a distance of multiples of X_p or Y_p cannot be distinguished from each other. In a similar way we have to accept a periodic ambiguity in range of the range period

$$R_p = \Delta R(N_f - 1) \quad (7)$$

where N_f is the number of discrete spectral lines (monochromatic frequencies) in the frequency bandwidth B of the illumination signal /4/.

3. IMAGING SYSTEM REQUIREMENTS

Regarding E/M-imaging for traffic applications or similar tasks we have to consider a specific arrangement where the object space is illuminated from nearly the same point as the backscattered radiation is received in the aperture later on. In this case the image of the object space is the spatial distribution of reflectivity corresponding to the objects viewed from the imaging system and at the frequency used. There are different ways for imaging system implementation. Its m o v e m e n t r e c o g n i t i o n c a p a b i l i t y depends on how this implementation is achieved.

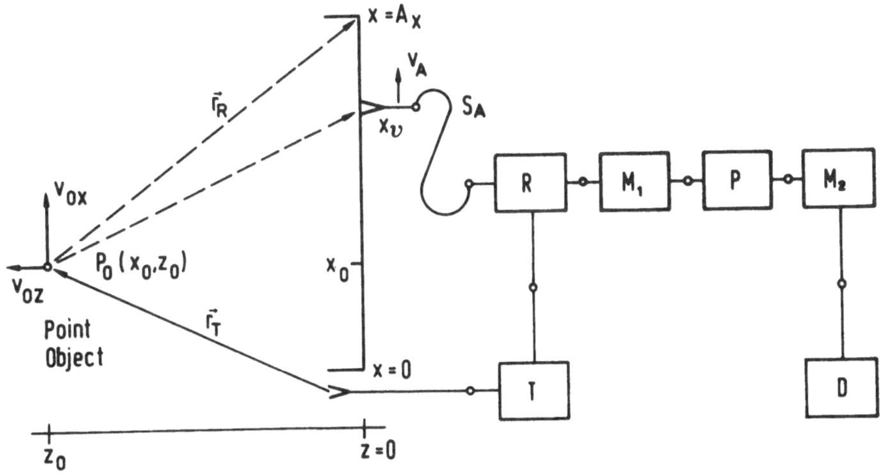

FIG. 2 IMAGE FORMATION BY SEQUENTIAL SIGNAL PROCESSING

Fig. 2 shows the block schematic of a system that generates the image sequentially at least in two stages. This is the typical system model for true microwave holography /7/. There is only one receiver R capable of detecting complex rf signals by virtue of a reference signal from the transmitter T. The incident radiation in the aperture is measured at one instant only in one point x_ν by a probe that scans the aperture with the aperture scan velocity v_A. There may be a number of N_x complex values in a 1-dimensional aperture, that have to be stored in the memory M_1 fed by the receiver R. (In the 2-dimensional case of an aperture area $A_x \cdot A_y$, other N_y-1 scans may be performed, hence $N_x \cdot N_y$ values have to be stored in a matrix format in M_1). In optical holography M_1 and R coincide as the light sensitive film that covers the complete aperture. The term h o l o g r a m is used for this memory, and in contrast to the procedure used in microwaves, in the optical case the data are written into the memory simultaneously. In microwave holography the data storage in M_1 is today usually implemented in a digital electronic memory, combined with a digital processor P (/5/, /7/, /8/). This processor has to calculate the image data from the hologram data, the result "image" may be subsequently stored in a second memory M_2. This picture memory M_2 then feeds a display which in the usual case of a CRT can only build up the picture again sequentially.

It takes a certain time to write and read the memories and to transform the hologram data to the picture data in the processor mostly using Fourier algorithms. The sum of these partial times (including A/D- and D/A-conversion) may be called the time for sequential signal processing τ_{ssp}. It is obvious that the lowest achievable value of τ_{ssp} in a system must be

$$\tau_{ssp} \leq \tau_r \qquad (8)$$

compared to the picture repetition time τ_r that is required by an application. Otherwise the image formation and its presentation run out of synchronism. This is a strong requirement for digital signal processing equipment regarding high resolution pictures (10^6 elements) and moving object applications with $\tau_r \leq 40$ ms. However, the real bottleneck for movement recognition capability seems to be a limitation in the aperture scan velocity.

As worst case estimation, the object is assumed to be a point scatterer moving only in z-direction with velocity v_{oz}. In a distance of $z_o \geq 10 \cdot A_x$, according to practical situations, the ray path lengths r_T and r_R between transmitter-antenna and object and receiver-antenna, respectively, can be approximated by the range R. While the receiver probe scans the aperture with the velocity v_A, the time required to scan from $x = 0$ to $x = A_x$ is $\tau_A = A_x/v_A$. In this time intervall the ray path sum should not have changed more than a quarter wavelength (corresponding to a phase difference of $\pi/2$) between

$$(x = 0,\ z = 0) \text{ and } (x_0,\ z_0) \text{ and } (x = A_x,\ z = 0)$$

due to object movement in order to avoid phase distortion of the hologram. The time for this range displacement is $\tau_{oz} = \lambda(8 \cdot v_{oz})^{-1}$. Equalizing τ_A and τ_{oz}, we obtain the highest admissible object velocity v_{oz} for a given aperture scan velocity v_A

$$v_{oz} \leq v_A \frac{\lambda}{8A_x} \qquad (9)$$

This equation let us learn: the better the lateral resolution the lower is the object velocity admissible. Assuming a mechanical scanning of a 1-dimensional aperture, a mean value of $v_A = 1$ cm/s seems adequate. Then with a moderate resolution of $\lambda/A_x = 0{,}01$ the highest tolerable object velocity is $v_{oz} \leq 45$ mm/h which is far from realistic traffic.

Regarding an aperture scan in two dimensions the requirements are still stronger. When there are N_y rows scanned all with the same velocity v_A as in the 1-dimensional case (e.g. in a meander-like trace) and supposing the requirement that the phase distortion along the ray path between (x = 0, y = 0, z = 0) and (x_0, y_0, z_0) and (x = A_x, y = A_y, z = 0) is less than a quarter wavelength due to movement, then we obtain from an extension of eqn. (9) for the highest admissible object velocity

$$v_{oz} \leq v_A \frac{\lambda}{8 \cdot A_x \cdot N_y} \quad (10a) \quad \text{or} \quad v_{oz} \leq v_A \frac{\lambda}{8 \cdot A_y \cdot N_x} \quad (10b) .$$

There is only small phase distortion from lateral movement because the ray path length changes only little, in particular when the object moves on the ellipse that spans over the aperture corners (e.g. x = 0 and x = A_x). Hence, the radial velocity component remains the critical value to judge the real-time capability of a sequential signal processing system.

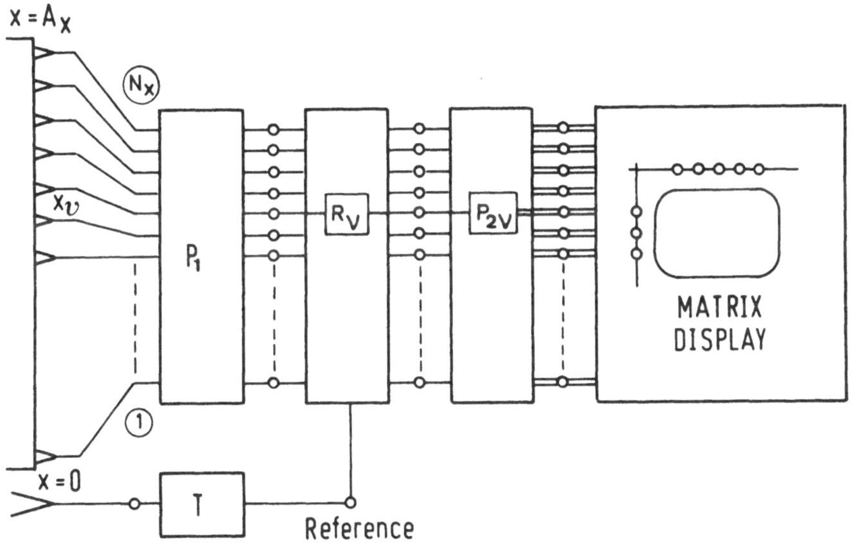

FIG.3 IMAGE FORMATION BY PARALLEL SIGNAL PROCESSING

Fig. 3 indicates the concept of a system with a consistent parallel processing down to the display. In this case the processor P_1 indeed performs its transformation from the aperture-plane (hologram) to the output "lateral image" in real time. P_1 can be implemented by a lens /9/ or a mirror system or by a microwave network /10/. In the next stage the "lateral image"-signal is detected and amplified by a finite number of parallel receivers R_ν. Hence, spatial quantization must be introduced because of the finite number of receivers even in the case of a lens P_1 which usually works continuously in space. The outputs of the parallel operating receivers R_ν feed another set of parallel working processors $P_{2\nu}$ that filters the frequency spectrum, if the image of range is desirable. Finally the 2-dimensional output of the processor $P_{2\nu}$ feeds a matrix display where all picture elements are generated independently and simultaneously. This concept yields the shortest time for image formation, the parallel signal processing time τ_{psp} which is equivalent to the natural signal propagation time through the complete system, ranging in the order of some nanoseconds. The prize one pays for this high speed is the huge expense of paralleling the various components.

In this case no signal processing delay affects the movement recognition. The only limitation for the highest tolerable object velocity v_{oz}(or v_{ox} or v_{oy}) is the fluorescence decay time of the display, that determines the picture repetition time τ_r as well as the highest rates of image sequences that human observers can recognize.

If we assume that the object displacement by moving during the picture repetition time τ_r is just one resolution cell, then the maximum object velocity is given by

$$v_{ox} \leq \frac{\lambda}{A_x} \cdot \frac{R}{\tau_r} \quad (11a) \quad , \quad v_{oy} \leq \frac{\lambda}{A_y} \cdot \frac{R}{\tau_r} \quad (11b)$$

for the lateral movement and

$$v_{oz} \leq \frac{c}{2B \cdot \tau_r} \quad (12)$$

for the radial component.

It should be noted that the admissible v_{ox} and v_{oy} increase with R according to the natural human viewing, and again the better the lateral resolution the lower is the tolerable velocity. The highest admissible radial velocity, however, is independent of R but inversely proportional to the bandwidth B. Assuming the same resolution of $\lambda/A_x = \lambda/A_y = 0{,}01$ as in the former example, and letting R = 100 m, $\Delta R = \Delta X$ at R = 100 m and τ_r = 20 ms, then we obtain for the maximum velocity components $v_{ox} = v_{oy} \leq 180$ km/h, that allows recognition of movement in most cases of application.

A compromise between the high-cost parallel processing and the low-speed sequential processing seems to be a hybrid concept /11/, that is shown in the block schematic of Fig. 4. In a first stage

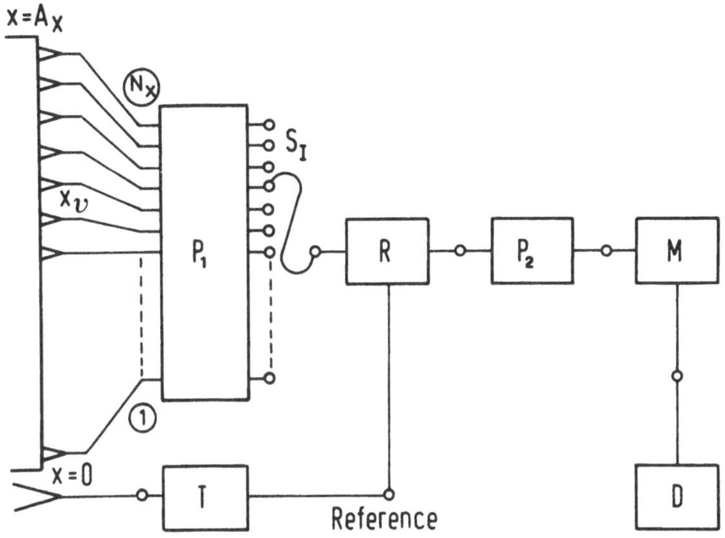

FIG.4 IMAGE FORMATION IN A HYBRID PROCESSING SYSTEM

the aperture signals are transformed into the signals of the lateral image by the parallel processor P_1. Because this is a linear process that neither needs any amplification nor dissipation, it can be implemented passively on a low-cost basis. The second stage consisting of cost-effective active components (i.e. a highly sensitive receiver R, a second processor P_2 for ranging, if necessary, and a memory M to feed a CRT-display), operates in a sequential manner. These two stages are connected by the lateral image switch S_I, that has to handle microwave signals indeed, but these image-signals at the carrier frequency are not as phase sensitive as the aperture switch S_A has to scan in the case of Fig. 2. Because of the parallel processing in P_1, the movement recognition capability is only determined by the picture repetition time τ_r according to equations (11) and (12). In this hybrid version τ_r depends on the image scan velocity of S_I and/or on the spectral processing time of P_2. In any case of realtime applications, τ_r should not exceed the flicker-time limit of 40 ms of the human eye.

4. EXPERIMENTAL IMPLEMENTATION

Following the hybrid concept according to Fig. 4, an experimental fast millimeter-wave imaging system for testing the applicability in land traffic was developed at the university in Erlangen /12/.

Operating at a center frequency of 70 GHz (corresponding to λ = 4,3 mm) the most suitable solution for the processor P_1 was a lens designed and fabricated in a new version of the Luneburg typ (/13/, /14/). This circular symmetric lens recommends a rotating switch to realize S_I. A highly sensitive receiver /15/ then feeds the range processor P_2 consisting of 50 parallel coupled analog filters in the video region. No resolution in elevation was provided. The resolution in azimuth is 1,2 degree or ΔX = 2 m in a distance of R = 100 m. The resolution in range is $\Delta R'$ = 1 m based on spectral bandwidth, but only ΔR = 2 m with regard to filter channel number.

As there is no discretization in the aperture ($N_x = \infty$), the lateral period of ambiguity is $X_p = \infty$. From spectral line density of the FM-modulated signal with a sweep range of 150 MHz and a FM-repetition frequency of 1,6 kHz, and using eqn. (7) one obtains for the range ambiguity period R_p = 180 km, a value that far exceeds the maximum range of 100 m from range processing capability (number and tuning of the filters) and receiver sensitivity. The picture repetition time is τ_r = 40 ms resulting in a maximum object velocity v_{ox} = 180 km/h from eqn. (11) or v_{oz} = 180 km/h from eqn. (12).

5. ACKNOWLEDGEMENT

The research project on the 70 GHz-imaging system was supported by the Deutsche Forschungsgemeinschaft under contract Br 522/7-8, which is gratefully acknowledged. I thank Dr. W. Platte for many discussions and advices.

6. REFERENCES

/ 1/ K.L. FULLER, AVOID, ein Nahbereichsradar mit hohem Auflössungsvermögen. Philips Techn. Rdsch. Vol 32 (1971/72) S. 46 - 50

/ 2/ B. REMBOLD, H.G. WIPPICH, M. BISCHOFF, W.F.X. FRANK, A 60 GHz Collision Warning Sensor for Helicopters. Proc. of the Military Microwave Conference, 1982, 7 B, pp. 344 - 351

/ 3/ D.L. MENSA, High Resolution Radar Imaging. Artech House, Dedham 1981

/ 4/ R. KARG, Multifrequente rechnergestützte holographische Abbildung im Mikrowellenbereich. Dr.-thesis, Univ. Erlangen-Nürnberg 1976

/ 5/ R. KARG, Multifrequency Microwave Holography. Archiv für Elektrotechnik und Übertragungstechnik, Vol 31 (1977), S. 150 - 156

/ 6/ H. GNISS, K. MAGURA, R. KARG, H. ERMERT, H. BRAND, W.-M. BOERNER, Polarisation dependence of image fidelity in microwave mapping systems. Inter. IEEE Antennas and Propagation Symposium Digest May 1978, pp. 38 - 41

/ 7/ G. TRICOLES and N.H. FARHAT, Microwave Holography: Applications and Techniques. Proc. IEEE Vol 65 (1977) No. 1, pp. 108 - 121

/ 8/ J. DETLEFSEN, Abbildung mit Mikrowellen. Fortschr.-Ber. VDI-Z. Reihe 10, No 5, 1979

/ 9/ S. CORNBLEET, Microwave Optics, Academic Press, London 1976

/10/ G. KÖNIG, Entwurf und Untersuchung eines Echtzeit-Mikrowellenabbildungssystems für den Nahbereich. Dr.-thesis, Univ. Erlangen-Nürnberg 1983

/11/ H. BRAND, A 2-Dimensional Realtime Millimeter-Wave Imaging System. 12th Assembly of the Intern. Commission for Optics (ICC12), Graz IOCC 1981, Session A 5, p. 151

/12/ H. BRAND, E. KREUTZER and M. VOGEL, Realtime 70 GHz-Imaging System for Close Range Applications. Proc. of the 13th European Microwave Conference, 1983, A 6.1

/13/ M. VOGEL, Theoretische und experimentelle Untersuchungen zur quasioptischen Abbildung mit Millimeterwellen, insbesondere mit Luneburg-Linsen. Fortschr.-Ber. VDI-Z. Reihe 10, No 14, 1982

/14/ M. VOGEL, A New Kind of Planar Waveguide Luneburg Antenna for the MM-Wave Region. Proc. of the 13th European Microwave Conference, 1983, P. 7

/15/ E. KREUTZER, Entwurf und Realisierung eines mehrzielfähigen 70 GHz-FM/CW-Radarempfangssystems. Dr.-thesis, Univ. Erlangen-Nürnberg 1982

IV. 12 (IM.2)

ELECTROMAGNETIC LOW FREQUENCY IMAGING

S.K. Chaudhuri

Department of Electrical Engineering
University of Waterloo
Waterloo, Ont., Canada, N2L 3G1

Abstract

In this paper characteristic parameters of low frequency scattered signal are analyzed for their dependence on the object's geometry, and specificially the Rayleigh coefficient is identified as a suitable quantity for the imaging and identification applications involving volume and size estimations.

In order to recover the Rayleigh coefficient from the backscattered signals, an inverse scattering model is presented here, which is based on the time-domain concepts of electromagnetic theory. Using the first five moment-condition integrals, the returned pulse response is processed to recover the Rayleigh coefficient, and the next higher order nonzero coefficient of the power series expansion in k (wave number) of the object backscattering response. The Rayleigh coefficients thus recovered are related to the volume, elongation factor (eccentricity), and orientation of an equivalent spheroid by use of a non-linear optimization algorithm.

Numerical simulation of the scheme of this paper has been tested on different scatterers. The result of these simulations are presented.

Introduction

It has been of interest to investigate and develop analytical methods for using received electromagnetic scattering data to establish knowledge on the geometrical and constitutive parameters of the scattering sources. Such methods have dealt with a variety of data such as amplitude, phase, polarization, and frequency. Also, a wide range of scattering models [1, 2] incl-

W.-M. Boerner et al. (eds.), Inverse Methods in Electromagnetic Imaging, Part 2, 997–1007.

uding geometrical optics, physical optics and geometrical theory of diffraction have been utilized. In contrast, relatively few imaging techniques which are based on low frequency (Rayleigh and low resonance region) scattering models have been put forward. Some of these are as follows: (i) A prediction-correlation technique, which uses low frequency natural resonance for target discrimination [3]. (ii) A limited attempt to obtain the geometrical data of a body with variable permittivity by using a eigenmode representation of the Rayleigh scattering has been reported [4]. (iii) A time-domain technique for profile recovery of a conducting body of revolution uses the backscatter pulse response [5] which implicitly involves the low frequency scattering informations (along with the high frequency informations).

In electromagnetic low frequency imaging one is mainly concerned with the recovery of the size (volume) of a given scattering body. This stems from the fact that when the wavelength is much longer than the dimensions of a body, the observed effect depends more on the size of the body than on its shape. In Section II of this paper the characteristic parameters of a low frequency scattered signal are analyzed for their dependence on the object's geometry, and specifically the Rayleigh coefficient is identified as a suitable quantity for the imaging and identification applications involving volume and size estimations.

In order to recover the Rayleigh coefficient from the backscattered signals, an inverse scattering model is presented in Section III. This model is based on the time-domain concepts of electromagnetic theory. Using the first five moment-condition integrals, the returned pulse response is processed to recover the Rayleigh coefficient and the next higher order nonzero coefficient of the power series expansion in k (wave no.) of the object's backscattering phasor response [6]. Numerical simulation of this scheme has been tested and the results are presented.

Finally, in Section IV the Rayleigh coefficients, recovered by the above processing scheme are related to the body's rough size informations [volume, elongation factor (eccentricity), orientation] - in the sense of an equivalent spheroid - by use of a nonlinear optimization algorithm.

II. Characteristic Scattering Paramets of Targets at Low Frequency

According to the low frequency scattering models [7-9], when the physical dimensions of a scatterer is small compared to the wavelength, λ, of the interrogating signal, the electric (or magnetic) field components of the scattered signal can be respresented as

$$G(k) = \sum_{n=0}^{\infty} a_n K^n \; , \; k = 2\pi/\lambda \qquad (1)$$

Here a_n's are the characteristic coefficients of the scattering system. In addition to being functions of the geometrical and constitutive parameters of the scatterer, these coefficients are also dependent on the aspect and the polarization of the incident signal. It has been observed that the knowledge of the size of the body, modified by a rough indication of shape, should suffice for a description of the body in finding the Rayleigh cross-section [9]. Thus, in principle the Rayleigh coefficient should be useful in obtaining an estimate of the scatterer size and in getting a rough indication of its shape. It is the purpose of the present paper to explore this possibility.

As background, it is pointed out that the solution to an e.m. scattering problem can be expressed as a multipole expansion. For a scatterer much smaller than the wavelength, retaining only the dipole terms gives a good approximation to the far field. Specifying dipole moments (related to the Rayleigh co-efficient) of the body does not determine the body uniquely (i.e. different bodies may have the same dipole moments). However, for a given aspect the variation of these dipole moments with the incident polarization (i.e. their polarization-locus) will be different for different scatterers and therefore can be helpful in target discrimination models. The finer details of the structure of the body is exhibited by the higher moments (and seriously affect the cross-section at small wavelengths); therefore, they do not affect the Rayleigh coefficient.

In order to observe the quantitative variation of the Rayleigh coefficient with the polarization and the direction of the incident wave, numerical computations are carried out for a conducting 2:1 prolate spheroid and a 3:2:1 ellipsoid by adopting the solution scheme proposed by Stevenson [8]. Both the objects were centered at the origin of the co-ordinate system, and their axii were along the co-ordinate axii x, y and z. Major axis (which is also the axis of symmetry for the prolate spheroid) in both cases were along the x-axis. For the ellipsoid the smallest axis (=1) was along the z-direction. The direction of incidence was confined to the x-y plane, and two polarizations, TE (incident electric field normal to the x-y planes), and TM (incident magnetic field normal to the x-y plane) were analyzed. The results are tabulated in Table I.

From Table I it is seen that, as the direction of incidence varies from the end-on ($\phi=0^{\circ}$) to the broad-side ($\phi=90^{\circ}$), there is very small change in the Rayleigh coefficient corresponding to the TE polarization, whereas for the TM polarization this coefficient goes through a large variation. This indicates that for a smooth conducting spheroidal object the Rayleigh coefficient is dependent on the physical dimension of the object along the incident electric field direction. This is why in the above examples, for the

TE case, where the physical dimension of the object along the incident E-field direction did not change for the variation in ϕ , the Rayleigh coefficient was relatively insensitive to the direction of incidence. Similarly, for the TM polarization the large variation in the Rayleigh coefficient matches with the large change in the physical dimension along the incident E-field direction as ϕ changes from 0° to 90°.

From the above observations it may be conjectured, that for a given aspect angle, the ratio of the Rayleigh coefficient corresponding to two orthogonal (linear) polarizations will provide a rough measure of the eccentricity of the scatterer geometry in a plane transverse to the direction of incidence. With more polarization data between these two reference orthogonal polarization directions, the accuracy of this eccentricity estimation may be improved. By using this transverse-eccentricity data for different angles of incidence a rough indication of the shape of the scatterer could be obtained.

Another interesting observation made during the calculations is that when all the dimensions of a spheroid were changed by a scale factor (e.g. by 2 or 0.5), the associated Rayleigh coefficient changes by the same scale factor. This contradicts the earlier suggestion that the Rayleigh coefficient increases according to the volume [9] (i.e. when all the dimensions are doubled the Rayleigh coefficient does not increase by a factor of 8 [cubic relation]). These results again suggest that at least for the conducting spheroidal structures the Rayleigh coefficient is linearly proportional to the physical size of the scatterer in the direction of the incident E-field (e.g. in Table I note that for the 3:2:1 ellipsoid for $\phi = 0$ the ratio of the Rayleigh coefficients for TM and TE polarization is 2:1, which is exactly equal to the transverse eccentricity of the target along that direction).

From the above analysis it appears that the extraction of the co-efficients a_n in (1) [in particular a_2 - the Rayleigh coefficient]from the backscattered signal could provide a basis for an electromagnetic low frequency imaging scheme. A method for the extraction of these low frequency target parameters from a time-domain pulse (transient) response is developed in the next section. Such pulse scattered data will have high-frequency scattering informations as well as creeping wave informations. This data can be useful in other high frequency imaging techniques [10,11] also. Thus, the pulse response can be used in an unified imaging technique which uses the low frequency and the high frequency scattering characteristics of a target simultaneously.

III. Coefficient Recovery Technique (CRT).

Let the received backscatter response of the target to an appropriate pulse $E_i(t)$ be given by $R(t) = E_i(t)*F_I(t)$; where $F_I(t)$ is the impulse response of the target for a given orientation and polarization. Using CRT it has been shown that [6],

$$a_2 = - \frac{\int_{-\infty}^{+\infty} t^3 R(t)dt}{\int_{-\infty}^{+\infty} t\, E_i(t)dt}, \text{ and} \qquad (2)$$

$$a_4 = \frac{\int_{-\infty}^{+\infty} t^5 R(t)dt + 20\, a_2 \int_{\infty}^{\infty} t^3 E_i(t)dt}{120 \int_{-\infty}^{+\infty} t\, E_i(t)dt}. \qquad (3)$$

[Note $a_0 = a_1 = 0$ and $\int_{-\infty}^{+\infty} E_i(t)dt = 0$].

From (2) and (3), it is apparent that the determination of a_2 and a_4 are very sensitive to the 'late' values of the received signal, $R(t)$ [due to the t^3, t^5 weighting factors]. This means that numbers that should analytically be zero and are not due to noise, can become significant and cause the CRT scheme of (2) and (3) to fail. This difficulty is alleviated by modifying the CRT with a proper time window as follows:

Time-windowed CRT

Let the signal processed by the CRT algorithm be $R'(t) = R(t)W(t)$, instead of $R(t)$. Here $W(t)$ is a suitable time-window which suppresses the noise (and the signal) for the 'late' values of time, t. Let $F\{W(t)\}=V(k)$, and the moment integrals $\int_{-\infty}^{+\infty} u^n\, V(u)du$; $n = 0, 1, 2, \ldots$ exist. It can be shown that the earlier CRT formulas now become

$$a_2 = - \frac{\int_{-\infty}^{+\infty} t^3 W(t)R(t)dt}{6\int_{-\infty}^{+\infty} t\, W(t)E_i(t)dt}, \text{ and} \qquad (4)$$

$$a_4 = \frac{\int_{-\infty}^{+\infty} t^5 W(t)R(t)dt + 20a_2 \int_{-\infty}^{+\infty} t^3 W(t)E_i(t)dt}{120 \int_{-\infty}^{+\infty} t\, W(t)E_i(t)dt}, \qquad (5)$$

provided

$$\int_{-\infty}^{+\infty} u^n V(u)du = 0 \text{ for } n = 1, 2, \ldots \qquad (6)$$

$$\neq 0 \text{ for } n = 0.$$

The condition (6) interpreted in time domain will yield

$$\left.\frac{d^n W(t)}{dt^n}\right|_{\substack{t=0\\ n\geq 1}} = 0 \quad ; \quad W(t)\Big|_{t=0} \neq 0 \qquad (7)$$

A rectangular function, although satisfies condition in (7), can not be used as a time-window because for such a function moment integrals in (6) do not exist. To the best of authors' knowledge the condition (6) [and therefore (7)] cannot be satisfied exactly by any window function, however, a window function which satisfies these conditions approximately is

$$W(t) = \text{rect}(t/T) * e^{-(at)^2}. \qquad (8)$$

The accuracy of the approximation (and therefore the error in computing a_2 and a_4) is controlled by the proper choice of the constants T , and a .

In order to study the performance of the modified CRT scheme of (4) and (5) with the window function of (8) under noisy conditions, normally distributed noise of different levels (mean = 0, standard deviation = σ) were generated in the computer and added to the return signal R(t). A quantitative measure of the noise level with respect to the signal is obtained by signal-to-noise ratio which is calculated as

$$S/N \text{ (dB)} = 10 \log \frac{\sum_{p=1}^{M} |R_n|^2/M}{\sigma^2}$$

Where $R_n = R(\Delta t)$; Δt is the sampling interval.

The use of the modified CRT scheme under noisy conditions was found to be very effective. A typical recovery result is presented in Fig. 1. The return signal R(t) from a 3:2:1 ellopsoid for the TE case and the direction of incidence $\theta_i = 90^0$, $\phi_i = 30^0$ was corrupted by normally distributed noise at different levels. At each S/N level several trials of the CRT and the modified CRT were attempted to arrive at the upper and the lower bounds, and the mean value (see Fig.1.) Similar results were obtained for various other configurations.

IV Scatterer Dimension and Orientation Recovery

If the scatterer is assumed to be ellipsoidal (including any special case of an ellipsoid) in shape, it is possible to recover its dimensions and the orientation from the recovered values of a a_2 using a simple non-linear optimization scheme. Note that the direct expressions [8] for a_2 in terms of given geometry and orientation is very involved and the direct inversion is usually not possible.

For the iterative optimization solution scheme, a_2 is recovered by CRT for 10 different polarization (covering a 90 degree polarization rotation) over two different look angles (see Fig.2.) A corresponding set of values of a_2 are calculated by using analytical expressions [8] for arbitrary values of the dimensions and the orientation. These values of the dimension and the orientation are adjusted iteratively until the functional

$$F = \sum_{n=1}^{10} [K_n(\Omega_i)_{CRT} - K_n(\Omega_i)_{Cal}]^2 + [K_n(\Omega_i+\Omega_d)_{CRT} - K_n(\Omega_i+\Omega_d)_{Cal}]^2 \tag{9}$$

is minimized to zero. In (9), K_n denotes the value of a a_2 for nth polarization case; 'CRT' and 'Cal' denotes values recovered from (4) and values being continuously calculated by adjusting the unknown geometrical parameters and the orientation of the object, respectively ; $\Omega_i = (\theta_i, \phi_i)$ is the unknown orientation and $\Omega_d = (\theta_d, \phi_d)$ is the known difference between the two look angles. The iterative minimization of F in (9), by adjusting the ellipsoid dimensions and Ω_i, is achieved on computer by a non-linear optimization routine. Note, when the value of F in (9) is minimized to zero, 20 simultaneous non-linear equations involving the unknown dimensions and orientation are satisfied. Thus, although in principal the optimized solution of (9) is not unique, it is not expected to be of major concern for most cases.

The numerical simulation of this scheme of this paper has been applied extensively on various different scatterers. The return signal, R(t), for various different situations were generated [12] on the computer by one operator. These were passed on to a second operator who had no information available about the objects which generated the given data set. Some of these results of the geometry recovery test are presented in Table II. The object dimensions and orientation recovered in Table II were done with the use of only a single look angle (m = 1 in Fig. 2.) The results show that although the recovery of the dimensions was good, the recovery of the orientation was unacceptable. When the experiments were tried with two look angles, the complete recovery (the dimensions and the orientation) was achieved.

In the experiment of Table II, the 'Cal' values of K_n for the case No. 10 were computed using the prolate spheroidal variation only; i,e., the unknown object shape was constrained to a body of revolution (2 - dim. variation). It is interesting to note that the volume of the recovered prolate spheroid and that of the actual ellipsoidal object are approximately equal (within 10%). This demonstrates the potential of the CRT scheme in recovering at least the volume of an unknown, smooth, but arbitrarily shaped conducting scatterer. The case No. 4. in the Table II points out one of the limitations of the rough geometry estimation scheme of (9).

V. Discussion

A basic model for an electromagnetic low frequency imaging scheme has been presented. It is shown that the Rayleigh coefficient can be useful in obtaining an estimate of the scatterer size, and in getting a rough indication of its shape. The numerical tests indicate that the proposed CRT scheme has the potential to recover accurate information about a conducting ellipsoidal scatterer from the limited and experimentally probable input data.

With the accurate and optimum construction of a time-window (which may not always be possible in a practical situation) the mapping scheme's breakdown S/N level was improved by more than 10 dB from 30 dB to below 20dB. In an actual situation the exact duration of the object's time-domain return is difficult to determine from extremely noisy data, and thus the effectiveness of the time-windowed CRT may be reduced.

Although an expression for the a_4 term was derived in CRT, its use in the actual determination of the geometry and orientation is not recommended. This is because its value is dependent on the location of the origin of the time scale (which can be related to the space co-ordinate origin with respect to which the body geometry is described), which cannot be normalized in an inverse scattering scheme.

In principle, the imaging method suggested in this paper can be applied to dielectric scatterers also. This is because the derivation of the CRT expression (4) holds for conducting as well as dielectric scatterers, and analytical expression for the 'Cal' values of K_n in (9) is available for the dielectric ellipsoids from [8]. However, before the scheme can be generalized to dielectric scatterers an additional unknown parameter, dielectric constant ε_r, has to be taken into account. This may be done by considering three or more look angles.

Finally, it is suggested that the method of this paper can be combined with a high frequency imaging technique [10,11] to produce a mapping scheme which uses low-frequency and high - frequency scattering information simultaneously. Since the suggested high frequency technique also uses an equivalent-ellipsoidal concept, the proposed unified-scheme can iteratively switch between the two component schemes, and thus augment the results of each model. Furthermore, if the pulse scattering data is used as the input, as suggested in this paper, then by means of a Fast Fourier Transform (FFT) of this return the input requirements of the high-frequency imaging technique can be met.

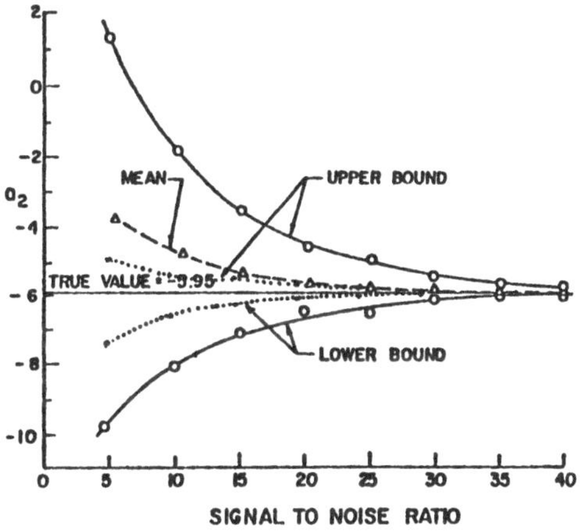

Fig.I. Rayleigh Coefficient Recovery Result of the CRT Scheme, (o——o, Δ---Δ) Without the Time-Window, (*...*) with the Time-Window.

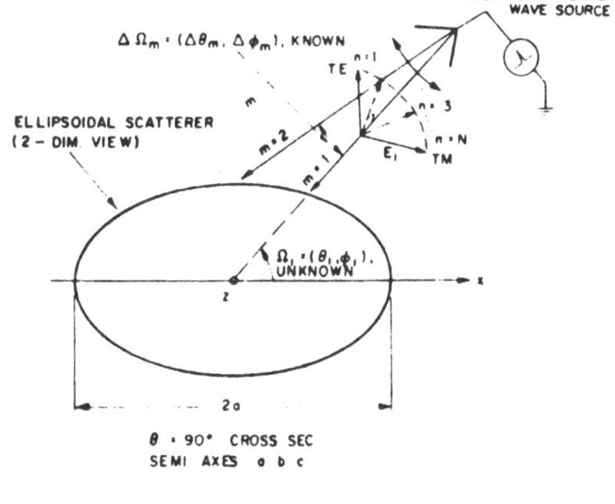

Fig.2. Data Acquisition Scheme for CRT Algorithm.

TABLE I

Backscattering Rayleigh Coefficient; $\theta_i = 90^0$

GEOMETRY	POLZ	$\phi_i = 0^0$	10^0	20^0	30^0	40^0	50^0	60^0	70^0	80^0	90^0
PROLATE SPHEROID (2:1)	TE	-2.749	-2.739	-2.711	-2.667	-2.613	-2.556	-2.502	-2.458	-2.430	-2.420
	TM	-2.749	-2.816	-3.010	-3.306	-3.670	-4.057	-4.420	-4.716	-4.910	-4.977
ELLIPSOID (3:2:1)	TE	-6.180	-6.170	-6.130	-6.090	-6.020	-5.960	-5.900	-5.850	-5.810	-5.800
	TM	-12.39	-12.39	-12.89	-13.65	-14.58	-15.58	-16.51	-17.27	-17.77	-17.94

TABLE II

GEOMETRY RECOVERY TEST

. Single look angle (M=1)

. a_4 information not used [second term on RHS of (10) not used]

No.	ACTUAL VALUES		RECOVERED VALUES	
	Semi-axis ratio	orientation	Semi-axis ratio	orientation
1	2:1:1	10°	2.03:1.026:1.026	6.62°
2	10:1:1	10°	10.04:0.979:0.979	18.32°
3	10:1:1	90°	9.079:1.005:1.005	53.12°
4	100:1:1	60°	unable to recover	
5	3.5:1:1	45°	3.536:0.991:0.991	38.71°
6	5:1:1	30°	5.206:0.959:0.959	31.31°
7	8:1:1	30°	8.351:0.973:0.973	33.53°
8	2:1:1	70°	2.016:1.01:1.01	53.46°
9	1.2:1:1	60°	1.193:0.998:0.998	83.85°
10	3:2:1	30°	3.541:1.364:1.364	53.89°

VI. References

[I] Special Issue on Inverse Methods in Electromagnetics, IEEE Trans. Antennas and Propagation, Vol. AP-29, March, 1981.

[2] W.M. Boerner, "Polarization Utilization in Electromagnetic Inverse Scattering", Chapter 7 in Inverse Scattering Problems in Optics, H.P. Baltes, Ed., Topics in Current Physics, Vol. 20, New York: Springer, 1980, pp. 237-290.

[3] D.L. Moffatt and R.K. Mains, "Detection and Discrimination of radar Targets", IEEE Trans. Antennas and Propagation, Vol. AP-23, pp. 358-367. May, 1975.

[4] H. Chaloupka, "Geometrical Data of Dielectric Objects from Monostatic Backscattering," Radio Science, Vol. 16, No.6, pp. 1053-1058, Nov. - Dec. 1981.

[5] C.L. Bennett, "Time Domain Inverse Scattering", IEEE Trans. Antennas and Propagation, Vol. AP-29, pp. 213-219, Mar.'81.

[6] S.K. Chaudhuri, "Estimation of the Volume and the Size of a Scatterer from Band-Limited Time Domain Signature," IEEE Trans. Antennas and Propagation, Vol. AP-29, pp. 398-400, March, 1981.

[7] A.F. Stevenson, "Solution of Electromagnetic Scattering Problems as Power Series in the Ratio (dimension of scatterer) Wavelength," Jour. Applied Physics, Vol.24, pp.1134-1142, Sept. 1953.

[8] A.F. Stevenson, "Electromagnetic Scattering by an Ellipsoid in the Third Approximation," Journ. Applied Physics, Vol. 24, pp. 1143-1151. Sept. 1953.

[9] K.M. Siegel, "Far Field Scattering from Bodies of Revolution", Appl. Sci. Res. B., Vol. 7, pp. 293-328, 1958.

[10] S.K. Chaudhuri, "Utilization of Polarization-Depolarization Characteristics in Electromagnetic Inverse Scattering", Ph.D. Thesis, Faculty of Graduate Studies, University of Manitoba, Winnipeg, Canada, No. 1976.

[11] S.K. Chaudhuri and W.M. Boerner, "Polarization Utilization in Profile Inversion of a Perfectly Conducting Prolate Spheroid," IEEE Trans. Antennas and Propagation, Vol. AP-25, pp. 505-511, July, 1977.

[12] S.K. Chaudhuri, "A Time Domain Synthesis of Electromagnetic Backscattering by Conducting Ellipsoids," IEEE Trans. Antennas and Propagation, Vol. AP-28, pp.523-531, July,'80

IV.13 (MI.5)

ELECTROMAGNETIC IMAGING OF DIELECTRIC TARGETS

Heinz Chaloupka

Ruhr-Universität Bochum

now: Bergische Universität-Gesamthochschule, FB 13
D-5600 Wuppertal, F.R. Germany

ABSTRACT

The paper deals with the problem of reconstructing dielectric inhomogeneities from scattering data taken for different aspect angles and wavelengths. Starting with the Born approximation, the observable quantities turn out to be simply related to the Fourier transform of the searched object-function and the same spatial frequency component may be obtained via different mapping configurations ("redundant measurements"). The exact scattering can be represented by the sum of the Born term and an additional term stemming from coupling between different parts of the scatterer. Two concepts are proposed, to take the latter term into account. The first uses values obtained from "redundant measurements" to reduce the influence of the coupling term. In the second concept the coupling is considered by means of an approximate integral equation.

1. INTRODUCTION

The present paper is divided into three main parts. In the first part (chapter 2) different representations for the scattering by dielectric inhomogeneities are presented. The second part (chapter 3) summarizes the properties of imaging algorithms based on the Born approximation and introduces the concept of "redundant measurements". In the last part (chapter 4) two concepts for the case where the Born approximation is not sufficient are proposed. In 4.1 redundant measurements are utilized to reduce the effect of multiple scattering, whereas in 4.2 an approximate integral equation for the unknown object-function is derived. The dielec-

W.-M. Boerner et al. (eds.), Inverse Methods in Electromagnetic Imaging, Part 2, 1009–1021.

tric inhomogeneities are assumed to exhibit a permittivity differing in a region of finite extent from unity. In this region $\varepsilon_r(\vec{r})$ is allowed to be an arbitrary function of $\vec{r}$. This representation includes isolated homogeneous and inhomogeneous targets as well as an ensemble of different dielectric targets. In some sections (4.1.3 and 4.2) the discussion is restricted to two-dimensional inhomogeneities. Furthermore, it is assumed throughout this paper that ε_r is real and nondispersive.

2. REPRESENTATION OF THE SCATTERING

2.1 Volume integral equation and the transition operator

The dielectric bodies can, with respect to the scattering, be replaced by the equivalent current densitiy $\vec{J}(\vec{r})$ which is the solution of the volume integral equation [1]

$$\vec{J}(\vec{r}) = j\omega\varepsilon_o\gamma(\vec{r}) \; [\vec{E}^i(\vec{r}) + \int_V \overleftrightarrow{G}(\vec{r},\vec{r}\,') \; \vec{J}(\vec{r}\,') \; d^3\vec{r}\,' \;] \tag{1}$$

where $\vec{r}$ is the point-vector, $\vec{E}^i(\vec{r})$ the incident field and $\overleftrightarrow{G}$ the dyadic Green's function of free space. $\gamma(\vec{r})$ is the "object function" with

$$\gamma(\vec{r}) = \varepsilon_r(\vec{r}) - 1 \tag{2}$$

Separating the singularity of $G(\vec{r},\vec{r}\,')$ at $\vec{r} = \vec{r}\,'$ via the introduction of a "principal value" integral [2] leads to a new integral equation, which can be obtained from eq. (1) by replacing $\gamma(\vec{r})$ by the "modified object function"

$$\gamma_M(\vec{r}) = \frac{3(\varepsilon_r(\vec{r})-1)}{\varepsilon_r(\vec{r})+2} = \frac{3}{\gamma(\vec{r})+3}\;\gamma(\vec{r}) \tag{3}$$

and V by V_- [3]. V_- is the volume, where an infinitesimal sphere with center at $\vec{r}$ has been excluded. The solution of this integral equation for an arbitrary incident field may be written in the form

$$\vec{J}(\vec{r}) = j\omega\varepsilon_o\gamma_M(\vec{r}) \; [\vec{E}^i(\vec{r}) + \int_V \overleftrightarrow{T}(\vec{r},\vec{r}\,',k_o)\vec{E}^i(\vec{r}\,')d^3\vec{r}\,'] = \vec{J}_B(\vec{r}) + \vec{J}_C(\vec{r}) \tag{4}$$

with the dyadic "transition operator" $T(\vec{r},\vec{r}\,',k_o)$, which itself is a function of $\gamma_M(\vec{r})$ but is independent of $\vec{E}^i(\vec{r})$. Eq. (4) may be interpreted by means of a physical picture. If the dielectric inhomogeneities are approximately considered as an ensemble of independent and isotropic point-scatterers only the first term $\vec{J}_B$ results. The second term $\vec{J}_C$ takes the coupling between the different points into account. $\overleftrightarrow{T}(\vec{r},\vec{r}',k_o)\cdot\vec{E}^i(\vec{r}')$ is the contribution to the scattered field at $\vec{r}$, which is originated by coupling to the point-scatterer at $\vec{r}'$ in the presence of all other point-scatterers. If the dependency of $\overleftrightarrow{T}$ on k_o is represented by means of its Fourier transform

$$\overleftrightarrow{T}(\vec{r},\vec{r}',k_o) = \int_{s_{min}}^{\infty} \overleftrightarrow{T}(\vec{r},\vec{r}',s)\; e^{-jk_o s}\, ds \tag{5}$$

the fact that the excitation at $\vec{r}'$ needs (for $\varepsilon_r \geq 1$) at least a time interval $\Delta t = |\vec{r}-\vec{r}'|/c$ to arrive at $\vec{r}$ results in

$$s_{min} \geq |\vec{r}-\vec{r}'|/c \tag{6}$$

For $|\varepsilon_r - 1| < 1$ a Neumann's series solution of the integral equation leads to a first crude approximation

$$J_C(\vec{r}) \approx (j\omega\varepsilon_o)^2\; \gamma_M(\vec{r}) \int_{V_-} \overleftrightarrow{G}(\vec{r},\vec{r}')\; \gamma_M(\vec{r}')\; \vec{E}^i(\vec{r}')\; d^3\vec{r}' \tag{7}$$

This is equivalent to the second order term in a "multiple scattering" model of the inhomogeneities.

For the two-dimensional problem, where the dielectric body is assumed to be cylindrical with axis parallel to z and $\vec{E}^i = E^i(\vec{\rho})\vec{u}_z$, $\vec{\rho} = x\,\vec{u}_x + y\,\vec{u}_y$ the scalar integral equation

$$J(\vec{\rho}) = j\omega\varepsilon_o\; \gamma(\vec{\rho})\; [E^i(\vec{\rho}) + \int G(\vec{\rho},\vec{\rho}')\; J(\vec{\rho}')\; d^2\vec{\rho}'] \tag{8}$$

holds.

Neclecting coupling leads to

$$\vec{J} \approx \vec{J}_B = j\omega\varepsilon_o\; \gamma_M \vec{E}^i \quad \text{and} \quad J \approx J_B = j\omega\varepsilon_o \gamma E^i \tag{9a,b}$$

for the three- and two-dimensional case, respectively. Although eq. (9a) is a modified form both eqs. (9a,b) are in this context refered to as Born approximation.

2.2 Formulation of the integral equation in the spatial frequency domain

As will be shown in the next section the observable quantities, that is the scattered field outside the region where the scatterer is situated, are related to the Fourier transform of the current density distribution. This is one of the reasons why a transformation into the spatial frequency domain offers some advantages.
For the Fourier transform pair the convention

$$\tilde{F}(\vec{K}) = \mathfrak{F}\{F(\vec{r})\} = \int F(\vec{r})\, e^{j\vec{K}\cdot\vec{r}}\, d^n\vec{r} \tag{10a}$$

and

$$F(\vec{r}) = \mathfrak{F}^{-1}\{F(\vec{K})\} = \frac{1}{(2\pi)^n}\int \tilde{F}(\vec{K})\, e^{-j\vec{K}\cdot\vec{r}}\, d^n\vec{K} \tag{10b}$$

with n = 3 for the three- and n = 2 for the two-dimensional case is used. $\vec{K}$ is the spatial frequency with

$$\vec{K} = K_x\vec{u}_x + K_y\vec{u}_y + K_z\vec{u}_z = K\vec{u} \tag{11}$$

If the incident wave is a plane wave in the direction $\vec{u}_i$, which means that the transmitter is in the direction $\vec{u}_t = -\vec{u}_i$ and therefore

$$\vec{E}^i = \vec{e}_t\, e^{-jk_o\vec{u}_i\cdot\vec{r}} = \vec{e}_t\, e^{jk_o\vec{u}_t\cdot\vec{r}} \tag{12}$$

one obtains the spatial domain integral equation

$$\tilde{\vec{J}}(\vec{K}) = j\omega\varepsilon_o\, \Gamma_M(\vec{K}+k_o\vec{u}_t) + \frac{1}{(2\pi)^3}\int \frac{2k_o^2 + \hat{\vec{K}}^2 - 3\hat{\vec{K}}(\hat{\vec{K}}\cdot\tilde{\vec{J}}(\hat{\vec{K}}))}{3(\hat{\vec{K}}^2-k_o^2)}\, \Gamma_M(\vec{K}-\hat{\vec{K}})\, d^3\hat{\vec{K}} \tag{13}$$

$\Gamma_M(\vec{K})$ is the Fourier transform of the object function $\gamma_M(\vec{r})$. In the two-dimensional case with $\vec{K} = K_x\vec{u}_x + K_y\vec{u}_y$

$$\tilde{J}(\vec{K}) = j\omega\varepsilon_o\, \Gamma(\vec{K}+k_o\vec{u}_t) + \frac{k_o^2}{(2\pi)^2}\int \frac{\tilde{J}(\hat{\vec{K}})}{\hat{\vec{K}}^2 - k_o^2}\, \Gamma(\vec{K}-\hat{\vec{K}})\, d^2\hat{\vec{K}} \tag{14}$$

follows from eq. (8).

2.3 Observable quantities

The scattered field at points $\vec{r}$ outside the smallest spherical surface enclosing the object is completely determined by the values of the nonradial components of $\tilde{\vec{J}}(\vec{K})$ at the sphere $|\vec{K}| = k_o$ in the $\vec{K}$-space. From the knowledge of the scattered field only this components can be reconstructed uniquely [4,5]. The far field scattered in the direction $\vec{u}_r = \vec{r}/r$ is directly related to $\tilde{\vec{J}}$. With the unit vector $\vec{e}_r \perp \vec{u}_r$

$$\vec{e}_r \cdot \vec{E}^s(\vec{r}) \rightarrow \frac{-j\omega\mu_o}{4\pi r}\, e^{-jk_o r}\, \tilde{\vec{J}}(k_o\vec{u}_r)\cdot\vec{e}_r \tag{15}$$

holds. Alternatively, the knowledge of the tangential parts of $\vec{E}$ and $\vec{H}$ at a smooth surface enclosing the object in an arbitrary distance yields the components $\tilde{\vec{J}}$ at $|\vec{K}| = k_o$ via

$$-\int_S [((\vec{u}_r \times \vec{e}_r) \times \vec{n}) \frac{\vec{E}}{Z_o} + (\vec{n} \times \vec{e}_r) \cdot \vec{H})] \, e^{jk_o \vec{u}_r \cdot \vec{r}} \, dS = \tilde{\vec{J}}(k_o \vec{u}_r) \cdot \vec{e}_r \tag{16}$$

Again $\vec{u}_r$ and $\vec{e}_r$ are arbitrary unit vectors with $\vec{u}_r \perp \vec{e}_r$ and $\vec{n}$ denoting the normal vector pointing out of the enclosed volume. If the so called Rayleigh-Sommerfeld formula or its vector analog is utilized to reconstruct the current density distribution from the field recorded on S [6,7] the obtained result is the superposition of only those spectral components with $|\vec{K}| = k_o$ (Ewald-sphere). That method is therefore equivalent to a method, where eq. (16) is used to determine those components.

Here we shall introduce an observable quantity C which according to the results mentioned above may be determined either by measurements in the far field or at a surface at nearer distance:

$$C_3(k_o, \vec{u}_r, \vec{e}_r) = \frac{1}{j\omega\varepsilon_o} \vec{e}_r \cdot \tilde{\vec{J}}(k_o \vec{u}_r) \tag{17}$$

For the two-dimensional and scalar case

$$C_2(k_o, \vec{u}_r) = \frac{1}{j\omega\varepsilon_o} \tilde{\vec{J}}(k_o \vec{u}_r) \tag{18}$$

with $\vec{u}_r$ perpendicular to the z-axis is introduced in an analogous manner.

3. IMAGING THEORY BASED ON BORN APPROXIMATION

3.1 Relationships between the scattering and the objectfunction

In this chapter 3 the validity of the Born approximation (eqs. (9a,b)) is assumed and some consequences of this assumption for imaging of dielectric targets are discussed. By inserting eq. (9a) into eq. (17) and introducing a plane incident wave according to eq. (12) one obtains

$$C_3(k_o, \vec{u}_r, \vec{e}_r) = (\vec{e}_r \cdot \vec{e}_t) \int \gamma_M(\vec{r}) \, e^{jk_o(\vec{u}_t + \vec{u}_r) \cdot \vec{r}} \, d^3\vec{r}$$

C_3 is now also a function of $\vec{u}_t$ and $\vec{e}_t$ and the integral expression is recognized to be the Fourier transform at the spatial frequency

$$\vec{K} = K\vec{u} = k_o(\vec{u}_t+\vec{u}_r) = 2k_o \cos\alpha\ \vec{u} \tag{19}$$

where 2α is the bistatic angle between the transmitter direction u_t and the receiver direction $\vec{u}_r$. We now can write with $\Gamma_M(\vec{K})$ as the Fourier transform of $\gamma_M(\vec{r})$

$$C_3(k_o,\vec{u}_r,\vec{e}_r,\vec{u}_t,\vec{e}_t) = (\vec{e}_r\cdot\vec{e}_t)\ \Gamma_M(\vec{K}) \tag{20}$$

For the two-dimensional case

$$C_2(k_o,\vec{u}_r,\vec{u}_t) = \Gamma(\vec{K}) \tag{21}$$

holds.

3.2 Redundant measurements with different mapping configurations

Because reconstruction of $\gamma_M(\vec{r})$ is attempted, Γ_M should be determined in a "large" region B_K (aperture) of the $\vec{K}$-space. From eq. (19) is seen, that the same vector $\vec{K}$ belongs to different combinations of k_o, $\vec{u}_r$ and $\vec{u}_t$ (mapping-parameters) and is independent of the chosen polarization ($\vec{e}_t$, $\vec{e}_r$). Hence there exists an infinite number of different "measurements" all leading with respect to eq. (19) to the same $\vec{K}$. These measurements are refered to as "redundant measurements".

(a) Since $K = 2k_o \cos\alpha$, variations in k_o are to a certain extent exchangeable for variations in α. Fig. 1 shows three different acquisition schemes, all leading to the same aperture B_K in $\vec{K}$-space. In this example $\vec{u}_t$ and $\vec{u}_r$ are chosen to remain in the same plane. In case (a) the direction of waveincidence denoted by ϕ is fixed, so that the vector $\vec{K}$ is varied by changing the receiver-direction and the wavelength. Case (b) is the monostatic configuration with $\vec{u}_t = \vec{u}_r$ and in case (c) transmitter- and receiver-direction displaced by a fixed bistatic angle α are moved simultaneously.

Instead of reversing the direction of $\vec{u}$ only the conjugate of C_3 or C_2 has to be taken. Therefore an 180° aperture is sufficient as long as Born the approximation holds.

(b) Because in the three-dimensional case $\vec{u}_t$ and $\vec{u}_r$ are not uniquely determined if $\vec{u}$ and α are given a further possibility of redundant measurements exists.

(c) In addition to (a) and (b) measurements may be excecuted with different polarization vectors $\vec{e}_t$ and different $\vec{e}_r$.

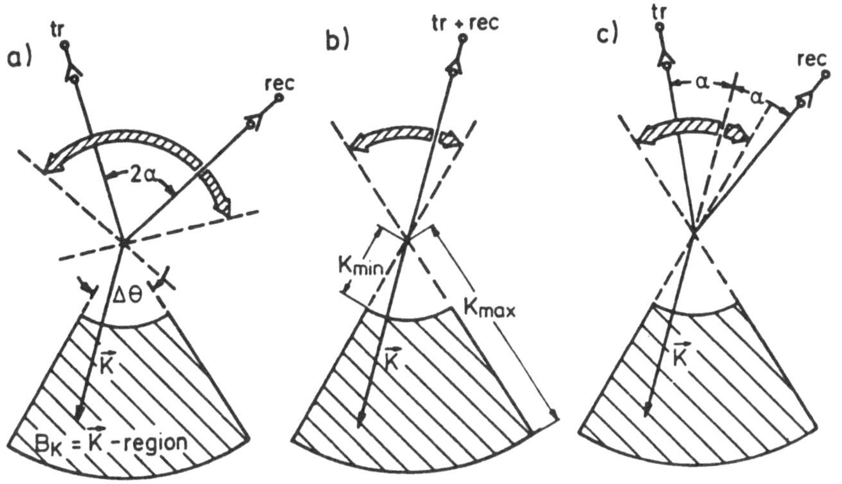

Fig. 1. Different acquisition schemes

3.3 Appropriate algorithms for reconstruction

If $\Gamma_M(\vec{K})$ is known in the bounded region B_K a backtransformation with the "window-function" $W(\vec{K})$, which is zero for $\vec{K}$ outside and sufficiently smooth within B_K, leads to the image-function

$$I(\vec{r}) = \mathfrak{F}^{-1}\{W(\vec{K})\ \Gamma_M(\vec{K})\} = \chi_{BK} * \gamma_M(\vec{r}) \tag{22}$$

According to Born approximation the image function is given by the convolution product of the unknown object-function $\gamma_M(\vec{r})$ and the point-spread-function $\chi_{BK}(\vec{r}) = \mathfrak{F}^{-1}\{W(\vec{K})\}$. The possibility to reconstruct $\gamma_M(\vec{r})$ approximately from $I(\vec{r})$ depends on the properties of χ_{BK} and "a priori" informations about $\gamma_M(\vec{r})$. For dielectric inhomogeneities, where $\varepsilon_r(\vec{r})$ changes abruptly the imaging of grad $\gamma_M(\vec{r})$ instead of γ_M may offer some adventages. One obtains a vector image by taking

$$\vec{I}_g(\vec{r}) = \mathfrak{F}^{-1}\{-j\vec{K}\ W(\vec{K})\Gamma_M(\vec{K})\} = \chi_{BK} * \mathrm{grad}\gamma_M(\vec{r}) \tag{23}$$

Which algorithm is superior for performing the transformation eq. (22) or (23) depends primarly on the form in which the scattering data are available. The backtransformation may also be carried out with the aid of the Radon transform [8], if in a first step from measurements with constant $\vec{u}$ and variable K the functions

$$A(\zeta,\vec{u}) = \int I(\vec{r})\ \delta(\zeta-\vec{u}\cdot\vec{r})\ d^3\vec{r} = \frac{1}{2\pi}\int W(K\vec{u})\Gamma_M(K\vec{u})e^{jK\zeta}\ dK$$

are determined. The second step is the determination of $I(\vec{r})$ from $A(\zeta,\vec{u})$ via Radon transform.
The special case $W(\vec{K}) = \delta(\vec{K}\cdot\vec{u}_z)$, where $\Gamma_M(\vec{K})$ is given in the plane $K_z = 0$ leads with eq. (22) to

$$I(\vec{r}) = I(x,y) = \int_{-\infty}^{\infty} \gamma_M(x,y,z)\, dz$$

This function is proportional to the "high-function h(x,y)" of the scatterer, if ε_r is constant within the scatterer-volume.

4. IMAGING CONCEPTS BASING ON HIGHER SCATTERING MODELS

4.1 Utilization of "redundant measurements"

4.1.1 General consideration: The exact scattering amplitude C can formally be represented by introducing eqs. (4), (5) and (12) into eq. (17)

$$C = C_B + C_{cp} = (\vec{e}_r \cdot \vec{e}_t) \int_V \gamma_M w_1 d^3\vec{r} + \int_V d^3\vec{r}\gamma_M \int_{V'}\int_S w_2 \, ds d^3\vec{r}\,' \tag{24}$$

with

$$w_1 = e^{j\vec{K}\cdot\vec{r}} \tag{25a}$$

and

$$w_2 = e^{j\vec{K}\cdot\vec{r}} (\vec{e}_r \cdot \overleftrightarrow{T}(\vec{r},\vec{r}\,',s)\cdot\vec{e}_t) e^{jk_0[\vec{u}_t\cdot(\vec{r}-\vec{r}\,')-s]} \tag{25b}$$

Whereas w_1 depends on $\vec{K}$ only and therefore is equal for all redundant measurements, w_2 depends on the special form of the used imaging configuration, because k_0 and $\vec{u}_t$ are different functions of $\vec{K}$ for different configurations. For configuration (a) in Fig.1 (bistatic and fixed direction of illumination)

$$u_t = \text{const.} \quad \text{and} \quad k_0 = K/(2\cos\alpha)$$

is valid, but for case (c) (bistatic with fixed angle α and measurement-plane orthogonal to $\vec{v}$)

$$\vec{u}_t = (\vec{K}\cos\alpha + (\vec{v}\times\vec{K})\sin\alpha)/K$$

and $k_0 = K/(2\cos\alpha)$ holds. By using an algorithm for reconstruction, which is based on the Born approximation the point spread function is now combined by two parts: χ^B_{BK} corresponding to w_1 being shift-invariant and depending only on the form of the aperture B_K, and χ^{cp}_{BK} corresponding to w_2, being shift-variant and depending on the special imaging configuration.

$$I(\vec{r}) = \int \gamma(\vec{r})\chi^B_{BK}(\vec{r}-\vec{r})d^3\vec{r} + \int d^3\vec{r}\gamma(\vec{r}) \iint \chi^{cp}_{BK}(\vec{r}-\vec{r},\vec{r},\vec{r}\,',s)d^3\vec{r}\,'ds \tag{26}$$

Properties of χ^{cp}_{BK} for some configurations are found in [3].

4.1.2 Reduction of image distortions by averaging: With "redundant measurements", which belong to the same value of $\vec{K}$ the influence of C_{cp} on C can be reduced. Here two special forms of such measurements will be mentioned:

(a) If $\vec{u}_t$ and $\vec{u}_r$ are changed simultaneously, so that $u = (\vec{u}_t + \vec{u}_r)/|\vec{u}_t + \vec{u}_r|$ and $\vec{v} = (\vec{u}_t \times \vec{u}_r)/|\vec{u}_t \times \vec{u}_r|$ remain constant and furthermore $\vec{e}_t \cdot \vec{e}_r = 1$ holds, C will be a function of the bistatic angle α only:

$$C(\vec{K},\alpha) = C_B(\vec{K}) + C_{cp}(\vec{K},\alpha) \tag{27}$$

Whereas the searched value $C_B(\vec{K}) = \Gamma(\vec{K})$ is independent of α, the unwanted contribution C_{cp} varies with α. This fact can be used to estimate C_B from measurements with different bistatic angles α. For the average

$$C_{av}(\vec{K}) = \frac{1}{\Delta\alpha}\int_{\alpha_o}^{\alpha_o+\Delta\alpha} C(\vec{K},\alpha)\, d\alpha \tag{28}$$

as a result of the triangle inequality

$$|C_{av}(\vec{K}) - C_B(\vec{K})| < |C(\vec{K},\alpha) - C_B(\vec{K})|_{max} \tag{29}$$

always holds. Significant advantages are offered by this method, if C_{cp} is originated by coupling between points, which are situated some wavelength apart. This can be proven by investigating the α-dependence of w_2 in eq. (25b):

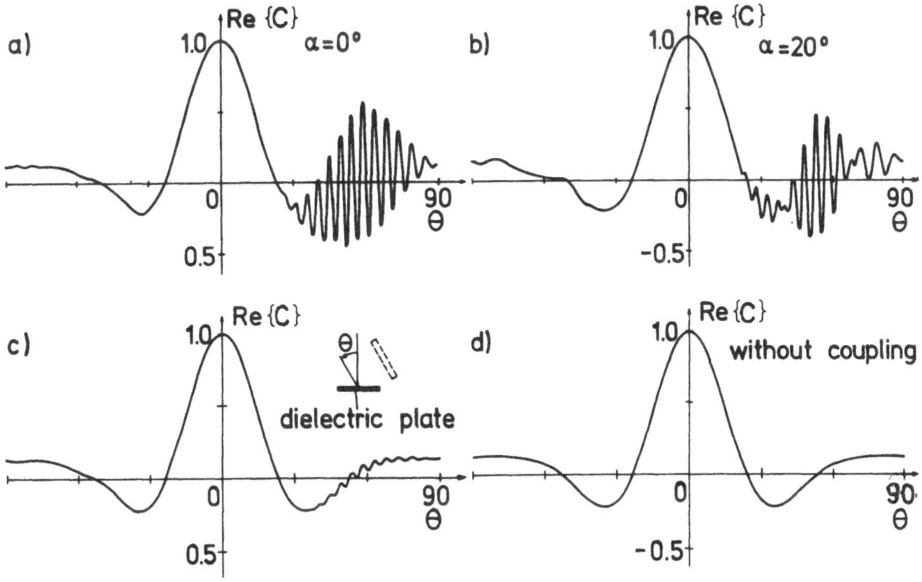

Fig. 2. Scattering coefficient versus Θ for a dielectric plate. (a) and (b): with coupling to a second plate (c): averaged curve (α = 0-20°) (d): without coupling

$$w_2 \sim e^{-jKf(\alpha)} \quad \text{with } f(\alpha) = \frac{s}{\cos\alpha} - \vec{v}\cdot[(\vec{r}-\vec{r}\,')\times\vec{u}]\tan\alpha$$

w_2 is for Ks >> 1 a rapidly oscillating function of the variable $1/\cos\alpha$, so that averaging reduces C_{cp} by a large amount. This is demonstrated in Fig. 2 for the example of two dielectric plates orientated obliquely with respect to each other. Fig. 2 shows the backscattering by the plate with center in the origin of the coordinate-system for K = 50/platewidth as a function of Θ. Without coupling the function Re(C) given in Fig. (2d) results. In (2a) (monostatic case) and (2b) (bistatic with $\alpha = 20^{\circ}$) the coupling contribution is for constant Θ a rapidly oscillating function in $1/\cos\alpha$. Taking advantage of this fact by averaging the result in the region $0 < \alpha < 20^{\circ}$ (1° steps) gives curve (c) which approximates the uncoupled case (Fig. 2d) sufficiently well.

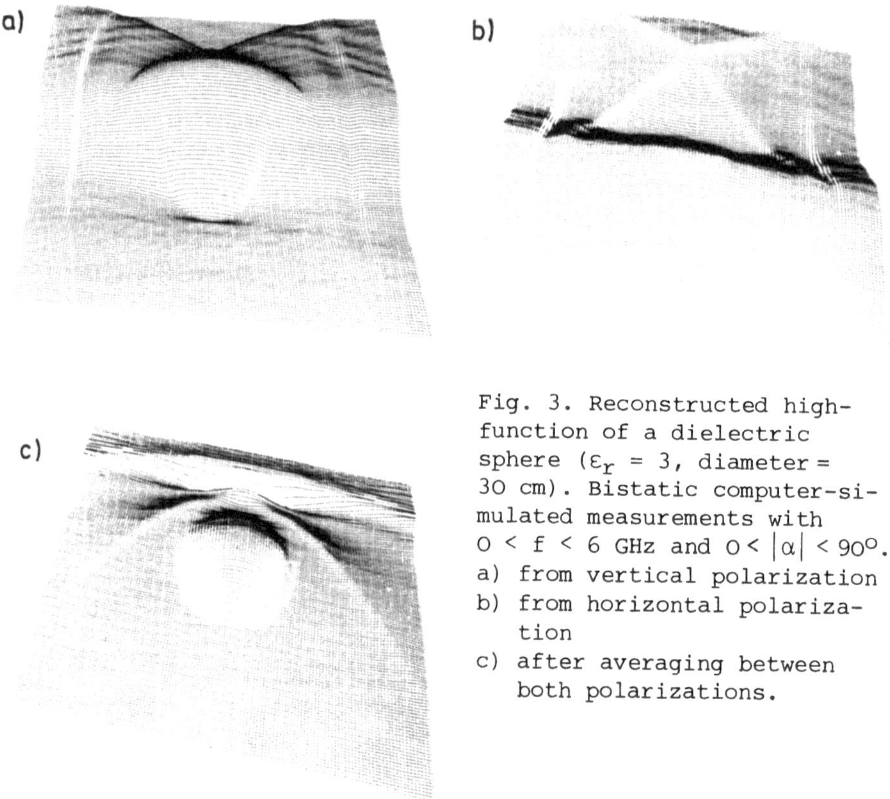

Fig. 3. Reconstructed high-function of a dielectric sphere ($\varepsilon_r = 3$, diameter = 30 cm). Bistatic computer-simulated measurements with $0 < f < 6$ GHz and $0 < |\alpha| < 90^{\circ}$.
a) from vertical polarization
b) from horizontal polarization
c) after averaging between both polarizations.

(b) Another possibility to reduce the influence of C_{cp} is the averaging of measurements, which differ only with respect to the polarization vectors $\vec{e}_t$ and $\vec{e}_r$:

$$C_{av} = C(\vec{e}_{r1},\vec{e}_{t1}) + C(\vec{e}_{r2},\vec{e}_{t2}) \tag{30a}$$

which according to eq. (20) is related to $\Gamma(\vec{K})$ via

$$C_{av} \approx C_B(\vec{e}_{r1},\vec{e}_{t1}) + C_B(\vec{e}_{r2},\vec{e}_{t2}) = (\vec{e}_{r1}\cdot\vec{e}_{t1} + \vec{e}_{r2}\cdot\vec{e}_{t2})\Gamma_M(\vec{K}) \tag{30b}$$

The advantage of this method is demonstrated in Fig. 3, where the high-function (see section 3.3) of a dielectric sphere with $\varepsilon_r = 3$ is reconstructed from computer-simulated multifrequency bistatic measurements (configuration Fig. 1a) in the plane z = O. Fig. 3a and 3b show the results obtained with vertical ($\vec{e}_r = \vec{e}_t = \vec{u}_z$) and horizontal ($\vec{e}_r \perp \vec{u}_z$, $\vec{e}_t \perp \vec{u}_z$) polarization, respectively. With Γ_M determined by averaging between both polarizations (eqs.(30a,b)) considerable improvements are obtained (Fig. 3c).

4.1.3 Estimation of the correct solution via extrapolation: In the two-dimensional case one can with $K = 2k_o \vec{u} \cos\alpha$ write

$$C_2(k_o,\vec{u}_t,\vec{u}_r) = C(K\vec{u},k_o) = C_B(K\vec{u}) + C_{cp}(K\vec{u},k_o)$$

The k_o-dependence of $C(K\vec{u},k_o)$ is for constant $\vec{K}$ because of $\cos\alpha \le 1$ measurable only for values $k \ge K/2$. Furthermore, it is shown by aid of eq. (14) that

$$\lim_{k_o \to 0} C_{cp}(K\vec{u},k_o) = 0 \tag{31}$$

Therefore Born approximation approaches in the two-dimensional case the exact solution in the limit $k_o \to 0$:

$$\lim_{k_o \to 0} C(K\vec{u},k_o) = C_B(K\vec{u}) \tag{32}$$

Because measuring data belonging to this limiting case are not available, attempts can be made to estimate these values from data in the region $k_o \ge K/2$ by extrapolation.

4.2 Formulation of an approximate integral equation for the unknown object-function

The following discussion is again restricted to the twodimensional scalar case. We first have to remember that imaging algorithms based on Born approximation lead to image distortions because the influence of the scattered field is neglected in the scatterer region.
In section 4.1.2 the effects due to the scattered field were considered as "random contributions" which could be reduced by averaging. In this last section the possibility of taking the scatte-

red field in a deterministic sense into account is discussed. Coming back to eq. (14) the influence of the scattered field is seen to be represented by the integral term. Considering eq. (14) as an integral equation for the seached function $\Gamma(\vec{K})$, $\tilde{J}(\vec{K})$ is required to be known in the complete $\vec{K}$-region. Discussing first the case of constant k_o and $\vec{u}_z$ we remember from section 2.3 that from the observable quantity C (see eq. 18) the function $\tilde{J}(\vec{K})$ can be determined for the values $|\vec{K}| = k_o$ (circle in 2-dimensional $\vec{K}$-space). Because of the existence of nonradiation sources of finite extension and with vanishing Fourier transform at $|\vec{K}| = k_o$ [9], the function $\tilde{J}(\vec{K})$ at $|\vec{K}| = k_o$ does in general not allow to extrapolate for the needed values at $|\vec{K}| \neq k_o$.
If we now allow the frequency to be changed, we first have to note that $\tilde{J}(\vec{K})$ is changed, too, and we therefore introduce the notation

$$\tilde{J}(\vec{K}) = \tilde{J}(\vec{K},k_o,\vec{u}_t)$$

The inverse problem for multifrequency observations could be solved rigereously if a relationship between non observable function $\tilde{J}(\vec{K},k_o,\vec{u}_t)$ for $|\vec{K}| \neq k_o$ and the observable function $\tilde{J}(\vec{K},k_o,\vec{u}_t)$ for $|\vec{K}| = k_o$ were known. If such a relationship exists, which does not explicitly contain the unknown function $\gamma(\vec{r})$, is still unknown.
Here we propose an approximation for this relationship, which again utilizes the Born approximation. From eq. (14) follows with the Born approximation

$$\tilde{J}(\vec{K},k_o\vec{u}_t) \approx j\,\frac{k_o}{Z_o}\,\Gamma(\vec{K}+k_o\vec{u}_t) \qquad (33)$$

Whereas in chapter 3 this relation was used to determine Γ directly from $\tilde{J}(\vec{K},k_o\vec{u}_t)$ at $|\vec{K}| = k_o$, it is now used to give an approximate relationship between the current-distribution for different frequencies.
From eq. (33)

$$\frac{\tilde{J}(\vec{K}_1,k_{o1}\vec{u}_t)}{k_{o1}} \approx \frac{\tilde{J}(\vec{K}_2,k_{o2}\vec{u}_t)}{k_{o2}} \quad \text{if} \quad \vec{K}_1+k_{o1}\vec{u}_t = \vec{K}_2+k_{o2}\vec{u}_t \qquad (34)$$

is deduced. In the special case of monostatic measurements only $\tilde{J}(k_o\vec{u}_t,k_o\vec{u}_t) = jk_oC_2(k_o,\vec{u}_t,\vec{u}_t)/Z_o$ is observable and one obtains:

$$\tilde{J}(\vec{K},k_o\vec{u}_t) \approx \frac{2k_o}{|\vec{K}+k_o\vec{u}_t|}\,\tilde{J}\left(\frac{\vec{K}+k_o\vec{u}_t}{2},\frac{\vec{K}+k_o\vec{u}_t}{2}\right) = \frac{2jk_o^2}{Z_o|\vec{K}+k_o\vec{u}_t|}\,C\left(\frac{\vec{K}+k_o\vec{u}_t}{2}\right) \qquad (35)$$

Here $C_2(k_o,\vec{u}_t,\vec{u}_t) = C(k_o\vec{u}_t)$ is used. Inserting eq. (35) into eq. (14) yield the approximative integral equation for $\Gamma(\vec{K})$, where the function $C(k_o\vec{u}_t)$ is measurable.

$$C(k_o\vec{u}_t) \approx \Gamma(2k_o\vec{u}_t) - \frac{k_o^2}{(2\pi)^2}\int \frac{C(\frac{\vec{K}+k_o\vec{u}_t}{2})}{\vec{K}^2 - k_o^2}\Gamma(k_o\vec{u}_t-\vec{K})d^2\vec{K} \qquad (36)$$

Similar equations can be derived for the bistatic case. For numerical purposes the singular kernel, which results from the fact that Green's function is transformed over the infinite space, may be replaced by a nonsingular kernel obtained with a finite space.

REFERENCES

[1] Harrington, R.F., Time-harmonic electromagnetic fields. New York: McGraw-Hill, 1961

[2] Chen, K.M., A simple physical picture of tensor Green's function in source region. Proc. IEEE, Vol. 65, pp. 1202-1204, 1977

[3] Chaloupka, H., Effects of mutual coupling between different points of a scatterer on imaging, in: New Procedures in Nondestructive Testing (Proceedings), Berlin: Springer-Verlag, pp. 305-316, 1983

[4] Hoenders, B.J., The uniqueness of inverse problems, in: Inverse source problems in optics. New York: Springer-Verlag, pp. 41-82, 1978

[5] Bleistein, N. and Cohen, J.K., Nonuniqueness in the inverse source problem in acoustics and electromagnetics. J. of Math. Phys., Vol. 18, pp. 194-201, 1977

[6] Stone, W.R., The inverse medium problem and a closed-form inverse scattering solution to the medium synthesis problem. Radio Science, Vol. 16, pp. 1029-1035, 1981

[7] Berger, M., Brück, D., Fischer, M., Langenberg, K.J., Oberst, J. and Schmitz, V., Potential and limits of holographic reconstruction algorithms. J. of Nondestructive Evaluation, Vol. 2, pp. 85-111, 1981

[8] Boerner, W.M. and Ho, S.M., Use of Radon's projection theory in electromagnetic inverse scattering. IEEE Trans. AP, Vol. 29, pp. 336-341, 1981

[9] Devaney, A.J. and Sherman, G.C., Non uniqueness in inverse source and scattering problems. IEEE Trans. AP, Vol. 30, pp. 1034-1037, 1982

IV.14 (MI.3)

A FOUR-CHANNEL MILLIMETER-WAVE ON-LINE IMAGING METHOD

Jürgen Detlefsen

Technische Universität München, Munich

Microwave imaging is achieved by a method, which uses several homodyne receivers, positioned along a linear aperture in order to evaluate the incident angles of incoming wavefronts. The sampled values of the field distribution along the aperture are submitted to a Fourier transfrom, giving the directions of incidence. As this method is used for short range applications, the reduction of the sensitivity imposed by the homodyne principle can be tolerated, avoiding expensive IF circuitry. Considering moving objects, the changes of the incident field distribution is well described by the doppler effect. This means, that only low frequency data have to be processed and the evaluation of the data can be done on-line by a desk-top calculator. As distance information is obtained by additional switching between two adjacent frequencies, two dimensional images can be demonstrated on-line on the screen of a desk-top calculator.

INTRODUCTION

Imaging systems are required to yield high resolution desirably in three dimensions in order to enable the observer to identify the object. Images are essentially reconstructed intensity distributions of the sources of the scattered field, which are generated by illuminating the objects. All the available information for the reconstruction process has to be drawn from the scattered field which can be measured to some limited extent with respect to its spatial and

W.-M. Boerner et al. (eds.), Inverse Methods in Electromagnetic Imaging, Part 2, 1023–1032.

frequency distribution and to its polarization properties. In this case the angular information concerning the directions of incoming wavefronts will be determined from the two-dimensional distribution of the scattered field emerging from the illuminated scene, whereas the radial distance will be obtained by transmitting a signal of appropriate bandwidth. As only the principles of this imaging process are to be experimentally verified within this context, two-dimensional objects are imaged using a linear array of only four receivers to yield the angle of incidence. Further only a radial distance measurement capability and not resolution is provided by switching between two adjacent frequencies transmitted in a CW-mode. The resulting system can be established on a low cost basis, because of a relatively simple microwave head and because all of the dataprocessing can be accomplished by software. This results in a very effective system, which allows to study the principal properties of the imaging system, laying fundamentals for its extension to more resolution cells and larger imaging areas, which are necessary for planned applications in the field of short range surveillance.

THEORETICAL BACKGROUND

The fundamentals of the method are discussed in detail in [1]. Principally the directions of the incoming wavefronts are determined by equidistantly sampling the field distribution by four receivers along an aperture of 12 wavelengths extent. The geometry of the problem is illustrated in Fig. 1. According to Fig. 2 the linear phase distribution obtained along the aperture for a point-scatterer is equivalent to a sinusoidally variing scattered field amplitude. If this distribution of the scattered field is submitted to a Fourier transform, the result will indicate the angular position α of the scatterer. The lateral resolution is determined by the width L of the aperture

$$\frac{\Delta y}{\lambda} = \frac{r}{L}. \qquad (1)$$

The discrete sampling of the scattered field causes angular ambiguities. The maximum unambiguous range y_B is given by the spacing ΔL between successive sampling points on the aperture, yielding

$$\frac{y_B}{\lambda} = \frac{r}{\Delta L}. \qquad (2)$$

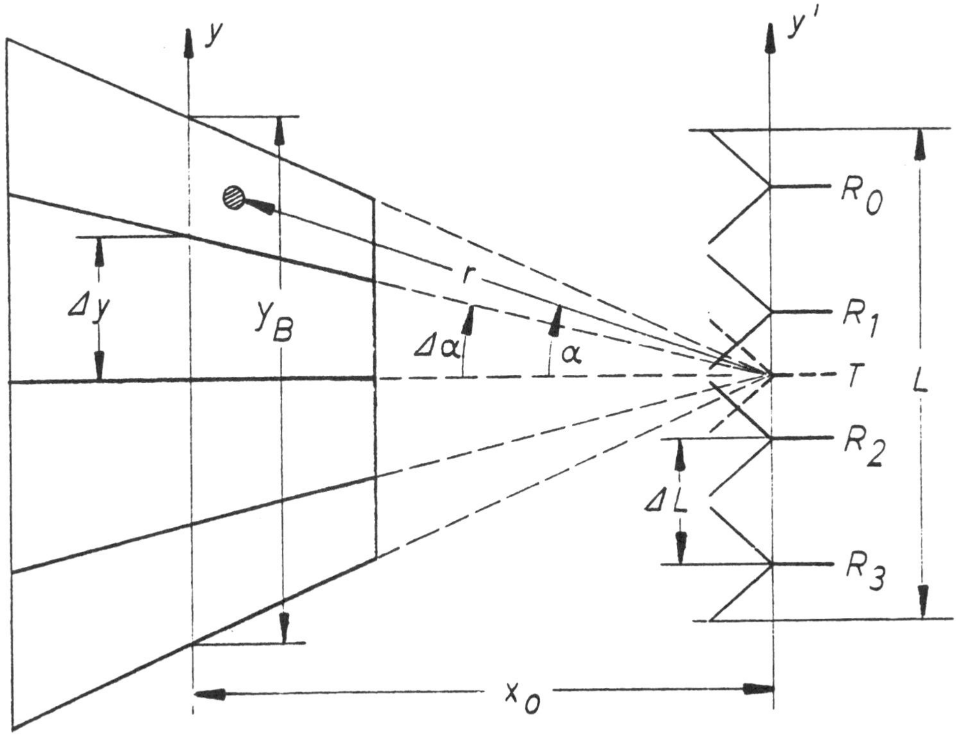

Fig. 1 Geometrical Configuration

$L = 13.45\lambda$, $\Delta L = 3.4\lambda$, $\Delta\alpha = 4.26^{o}$

$x_o = 4.5m$, $y_B = 1,35m$, $\lambda = 8.4mm$

The use of four receiving antennas of width 3.1λ spaced by 3.4λresults in four angular resolution cells of width $\Delta\alpha = 4.26^{o}$ according to a total aperture length of 13.45λ. The ambiguities can be removed by a proper design of the antennas along the aperture, which have to confine the illuminated sector to the extent of the unambiguous range. This means that the total aperture of length L has to be filled in completely with the apertures of the four receiving antennas. Several point scatterers at different lateral positions or the scattering centers of a complex body will be resolved due to the Fourier transform algorithm, if they are separated at least by a resolution cell width.

Regarding the radial distance, the phase information of the received signal is a highly ambiguous measure according to

$$\varphi = 4\pi \frac{r}{\lambda} \qquad (3)$$

This ambiguity of half a wavelength can be reduced to acceptable ranges for laboratory purposes, by evaluating the phase difference $\Delta\varphi$ between the signals, obtained from two different frequencies spaced by Δf

$$\Delta\varphi = 4\pi\left(\frac{r}{\lambda_1} - \frac{r}{\lambda_2}\right) =$$

$$= 2\pi \frac{2 \cdot \Delta f}{c_0} \cdot r \qquad (4)$$

The frequency difference has to be kept small in order to yield unambiguous ranges of sufficient extent. A frequency difference of 15MHz results in an unambiguous range of 10m, which is convenient for laboratory tests.

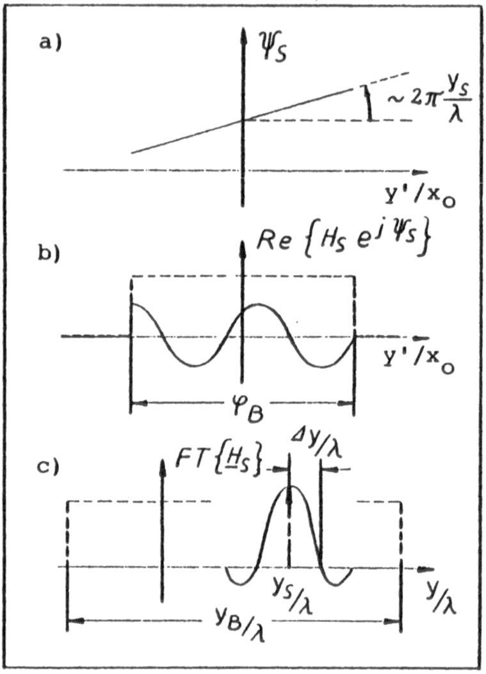

Fig. 2: Reconstruction Process
a) phase distribution
b) scattered field
c) Fourier transform

MICROWAVE PART OF THE IMAGING SYSTEM

The blockdiagram of Fig. 3 principally shows the set-up. An aperture of 12λ with four receivers spaced by 3λ is used to collect samples of the incoming wavefronts, which are processed by four quadrature homodyne receivers, yielding the complex field amplitudes. The reduction of the sensitivity imposed by the homodyne principle can be tolerated, avoiding expensive IF circuitry, as this method is used for short-range applications only ($r<150$m). The oscillator is a continuous wave Gunn oscillator, which is bias modulated to achieve frequency stepping. In order to establish a DC-reference level for the evaluation of the mixer signals, the PIN-modulator is used to switch off the transmitted power for a short period at each frequency. The complete cycle time is 1ms, thus allowing all those objects to be processed correctly which move generating a maximum doppler shift of 500 Hz.

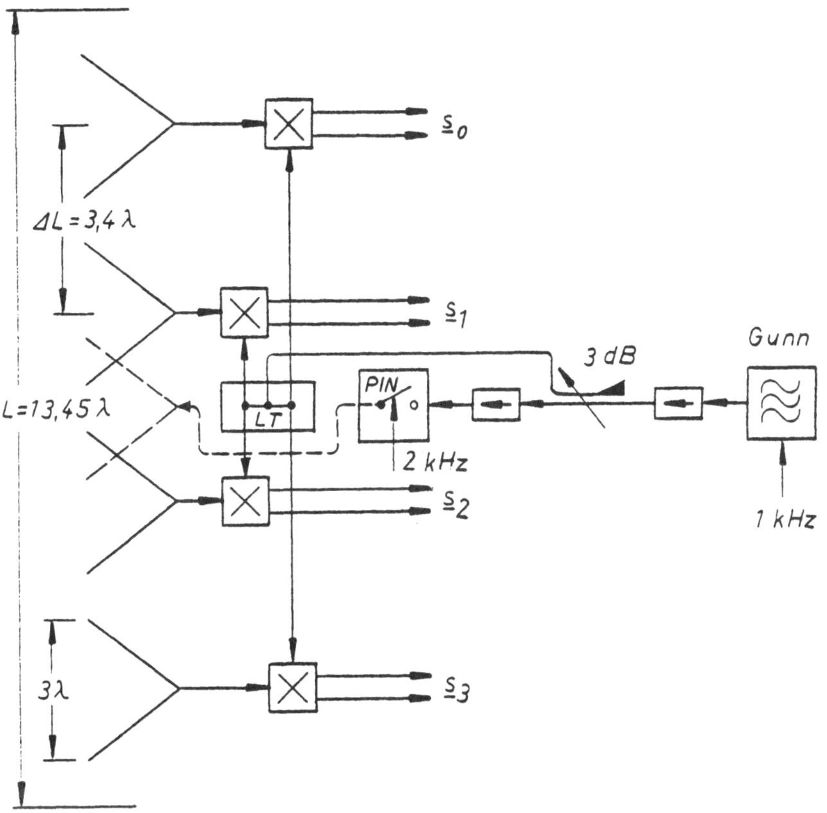

Fig. 3: Microwave Block Diagram
Gunn-osc. : P_s = 19dBm, Δf = 15MHz

SIGNAL PROCESSING

The received signals which are essentially the base band doppler signals of the illuminated scenery are of low frequency type and, according to Fig. 4, have first to be amplified, demultiplexed according to their transmitter frequency and subtracted from their DC-reference levels in order to obtain the continuous response of the original signals at both frequencies for each receiving channel. This procedure afterwards allows the signals to be sampled simultaneously in order to achieve digitized values of the signals pertinent to the same instant of time. The data are processed by a microprocessor, giving flexibility in data handling and display. The flowchart of Fig. 5 indicates, that after some environmental correction and a focussing procedure which compensates for the phase deviations

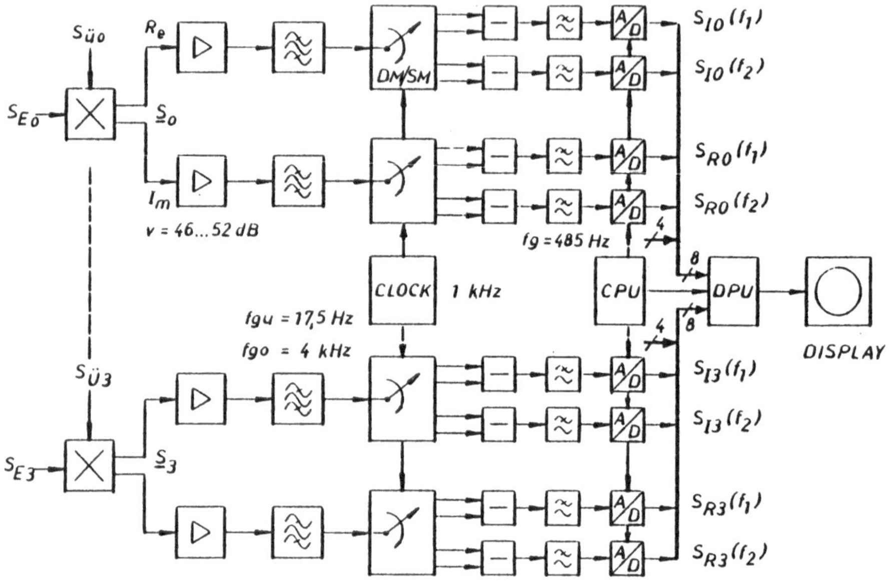

DPU: Data-Processing-Unit
DM/SH: Demultiplex/Sample and Hold-Unit

Fig. 4: Signal-Processing-Blockdiagram

in the microwave head, the FFT is performed on the data sets of both frequencies. The magnitudes of the resulting complex amplitudes represent the angle information, whereas the phase is used to evaluate the pertinent distance. Responses exceeding some given threshold level are registered in a final table of object coordinates and amplitudes which are displayed on a computer screen according to their geometrical positions. This type of data processing, using a machine language code for a 6502 microprocessor allows a renewal of the image every 100ms resulting in a near on-line imaging capability. This time can be substantially reduced by using faster microprocessors and by parallel processing, which will be necessary for an on-line imaging system with sufficient resolution cells.

LABORATORY RESULTS

The main purpose of the constructed imaging system is to verify theoretic properties of the imaging process and to gain experience with limitations of the experimental system. To find out the accuracy limits of the system, measurements were performed on objects moving

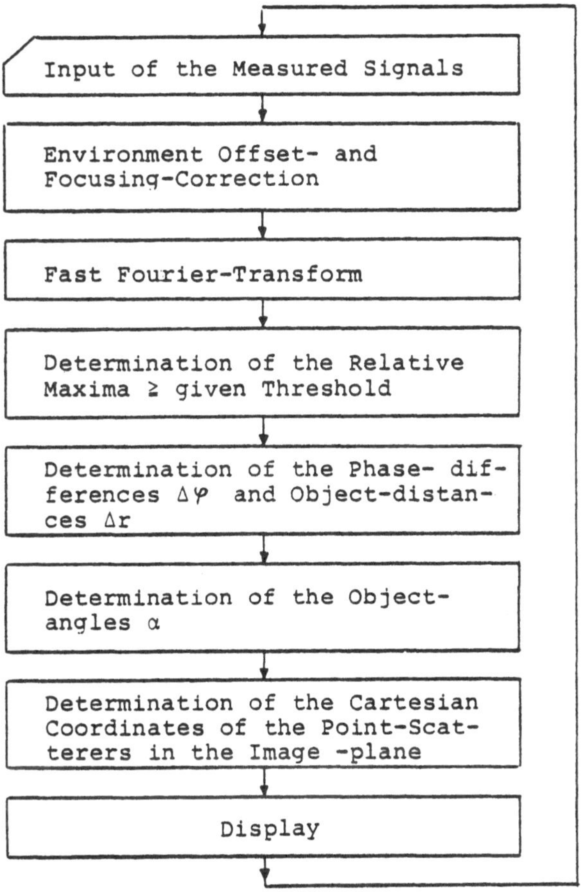

Fig. 5:
Signal Processing
Flow Chart

along well defined tracks. All these results are obtained with large signal to noise ratios (S/N>25dB) for point shape scatterers. In a first step the measured phase distribution along the aperture was examined. The standard deviation of the measured phase of an individual receiver turned out to be 20°, whereas the calculated linear phasefront shows a standard deviation of $\sigma_\alpha = 0.1°$ [2].
In a second step, the resolution properties of the system have been investigated. Looking at only one object, located centrally in a angular resolution cell, results in maximum output from this channel, whereas the levels deviating from zero in the other cells display the measurement errors of the system.
A convenient means to arrive at an exact statement of the angular measurement accuracy and the resolution obtained is to reconstruct the continuous intensity

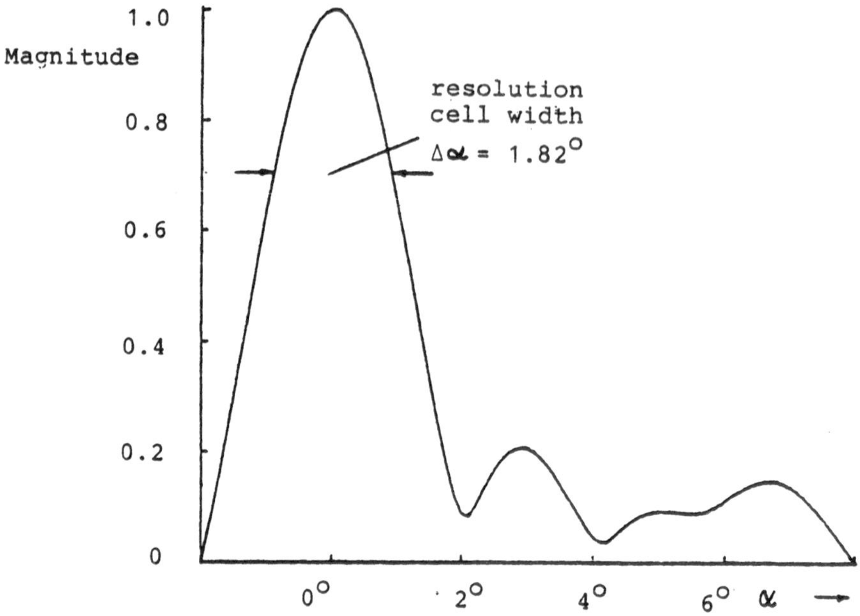

Fig. 6: Reconstructed Continuous Intensity Distribution, L = 27.9λ

distribution from the discrete output samples. Using the continuous response, lateral position and achieved resolution can be evaluated. Fig. 6 shows the reconstructed distribution for a point scatterer. The resulting angular measurement accuracy is $\pm 0.026^{o}$ for a 27.9λ aperture and reaches the order of the errors in the geometric adjustment of the system. The obtained resolution completely meets the theoretically predicted values. The obtained distance measuring accuracy depends on thermal noise, on non ideal adjustment of system components (quadrature mixer) and on the quantisation noise due to a 8 Bit A/D-converter. The total distance uncertainty, which essentially reflects the uncertainty of the phase measurement is $\pm 3.5^{o}$, which means that a distance can be determined to approximatly ± 1% of the unambiguous range, which is 10m in this laboratory set-up [3].
The radar cross section of the smallest scattering center to be processed properly turned out to be $0.045m^2$, using a 80mW Gunn oscillator. This shows that the homodyne principle is a sufficient and efficient means for this type of short-range imaging.

Fig. 7 and 8 show typical results for point shape objects moving in the imaging area, centered 4.5m afar

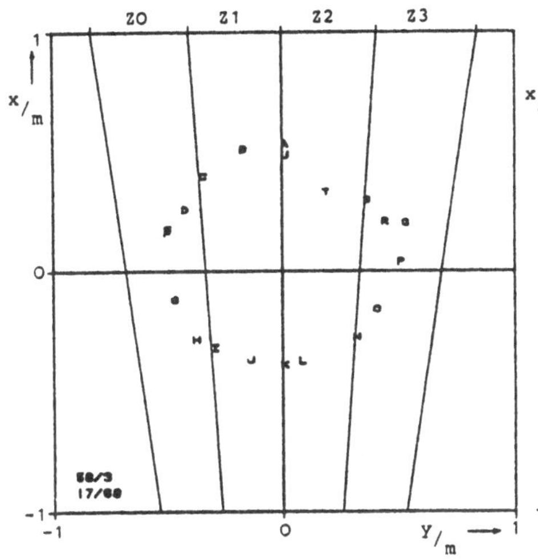

Fig. 7: Microwave-image of a metallic cylinder (diameter 15mm, length 1m) moving around the central point x,y = (0,0) (radius = 512mm)

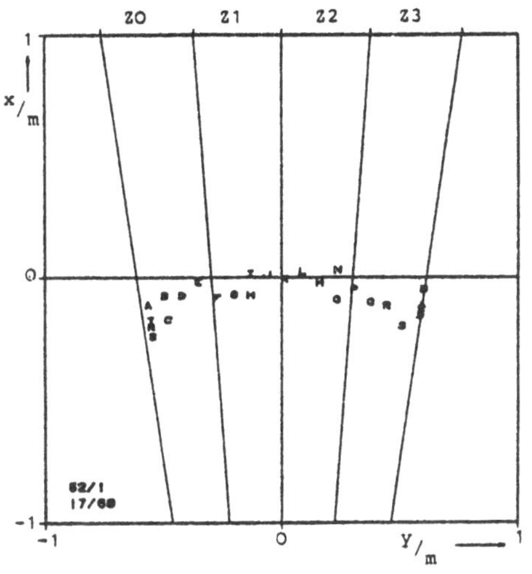

Fig. 8: Microwave-image of a metallic cylinder (diameter 15mm, length 1m) moving from x,y = (0, - 0.68) to x,y = (0,0.6)

from the microwave head. As no distance resolution is provided in this system,problems arise, when more than one scattering center is in the same resolution cell. This is demonstrated in Fig. 9, where a fixed and a rotating metallic cylinder are imaged. Due to the small diameters, only minor shadowing takes place and a single object is indicated at the medium distance of both. When the dimensions of the objects reach the width of the resolution cell, only the object which is closer to the imaging system will be indicated.

CONCLUSION

A relatively simple imaging system has been established using few receivers along a linear aperture. The homodyne principle used for detection gives sufficient sensitivity. The imaging properties can be completely predicted from theory. The results obtained using this laboratory set-up can be applied to a full imaging system with many angular resolution cells, which will be constructed, without increasing the number of receivers by additionally using the synthetic aperture principle for the case of a relative movement between imaging system and the object area.

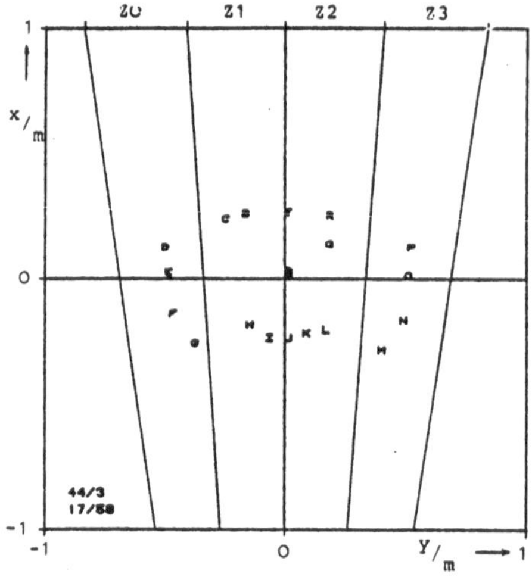

Fig. 9a:

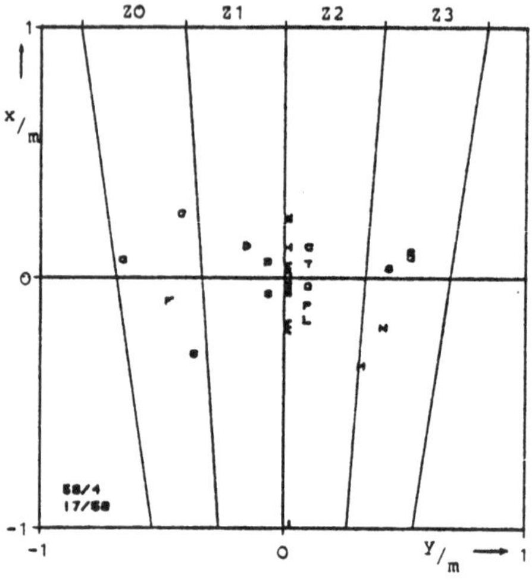

Fig. 9b:

Microwave-images of two identical metallic cylinders (diameter 15mm, length 1m) one fixed at x,y = (0,0), the other moving around the central point (radius r_o = 512mm)
a) reconstruction from simulated data
b) reconstruction from measured data

REFERENCES

[1] Detlefsen, J.: Abbildung mit Mikrowellen, Fortschritt-Berichte der VDI Zeitschriften. Reihe 10, Nr. 5, VDI Verlag Düsseldorf, 1979

[2] Detlefsen, J.: Multistatic CW-Dopplerradar with Lateral-Resolution for Short-Range-Applications, Conference Proceedings, 12th European Microwave Conference, Helsinki 1982, p. 107 - 112

[3] Bockmair, M.; Detlefsen, J.: A Multistatic Two-Dimensional Millimeter Wave Imaging System at 35 GHz, Conference Proceedings, 13th European Microwave Conference, Nürnberg 1983, S. A6.3

Grateful acknowledgement is given to the Deutsche Forschungsgemeinschaft (DFG) for financial support.

V.1 (OI.1)

THE HOLOGRAPHIC PRINCIPLE

A.W. Lohmann

Physikalisches Institut der Universität
Erwin-Rommel-Straße 1, 8520 Erlangen,
Fed. Rep. of Germany

Abstract

At the frequency range of the visible light all detectors are phase-blind. Due to this phase-blindness some information about the electromagnetic signal is lost during recording.

In holography some redundant information is added before recording. This is done so that only uninteresting information is lost due to the phase-blindness.

The clever way of adding ballast information before the detection process will be explained in various ways, as modulation, as multiplexing and as the implementation of the so-called mathematical "polarisation identity".

Not only phase-blindness, but also blindness towards color and (physical) polarisation can be overcome, based on the holographic principle.

§1 The holographic concept as a cure against phase-blindness

Dennis Gabor (1900-1979), the inventor of holography, once said in an interview: "I invented holography when I was 47. But the invention began already 30 years earlier. I was an enthusiastic photographer in my younger years. At age 17, I wondered: what happens to an image on its way from an object to the photographic plate? You cannot see the image until it is trapped in the correct plane. If you try to trap the image in the wrong plane, you recognise very little or nothing at all. But surely, the image must be contained somehow in the light at every plane between object and the correct plane".

W.-M. Boerner et al. (eds.), Inverse Methods in Electromagnetic Imaging, Part 2, 1033–1042.

What Gabor sensed correctly already in 1917 was the continuous flux of information from the object to the correct image plane. What he did not understand at the time, is that all detectors are blind for the phase of the light wave. The phase-blindness makes it impossible to record the image information completely at an arbitrary plane.

Later on Gabor realised, although the direct recording of a phase is impossible, but phase differences can be sensed in interference experiments. It is the crux of the holographic concept, to record some phase differences in such a way that the original phase distribution can be reconstructed, apart from an uninteresting additive constant phase.

Another way of introducing the concept of holography is based on a fairy tale on the art of writing research proposals. Suppose it is known that the research agency will always reduce the proposed budget by a factor 1/2. Obviously, the scientist will propose a budget twice as large as he really wants it. The art of the proposal game is now to specify the various items in the proposal such that the agency will cut out only the ballast, that is not wanted anyway.

The holographic plate is like the research agency. Exposed to information in the form of amplitudes and phases, the photographic plate will recognise only the amplitude information, neglecting the phase information. The phase-blindness of the photographic detector is harmless, if the complex object information is encoded entirely into the amplitude of the optical wave as it arrives in the detector plane.

In §2 we will explain how a complex signal can be converted reversibly into a real-positive amplitude signal, or even into a binary signal. Binary signals are popular with computer-generated holograms.

In §3 we will discuss the holographic concept in the context of four hologram types.

Not only phase-blindness, but also blindness towards the polarisation and/or the color of the light can be overcome by extending the holographic concept properly.

In §4 we will present holography as the implementation of a mathematical concept, called "polarisation identity". This term has nothing to do with what one calls polarisation of transversal waves, or at most very little. It is a concept that shows how a quadratic nonlinearity can be employed for certain signal manipulations. The concept can be extended also to higher order polynomial nonlinearities.

Finally, in §5, we mention briefly where else besides in coherent optics the holographic concept has been used to overcome phase-blindness.

§2 Equivalence of complex, real, positive and binary signals

We consider a complex signal $V_o(x) = A(x)\exp[i\varphi(x)]$ and its Fourier transform $\tilde{V}_o(u)$:

$$\tilde{V}_o(u) = \int V_o(x)\exp(-2\pi iux)dx \quad (1)$$

We will call $\tilde{V}_o(u)$ the frequency spectrum, which we assume to be band-limited, as indicated in the top line of fig. 1.

Adding the complex conjugate, converts the complex signal V_o into a real signal $V_o+V_o^*$. But the process is irreversible, in general, since the positive and negative frequencies are mixed up. That mix-up can be avoided by modulation with the carrier frequency U_R, which has to be at least one half of the bandwidth Δu, if reversibility is desired. (See $\tilde{V}_R(u)$). In other words, for the price of doubling the bandwidth we can replace the complex signal by an equivalent real signal. Positivity can be achieved by adding a bias B. It does not cost any bandwidth but it places a burden on the dynamic range of $\tilde{V}_p(u)$. Binary signals $V_B(x)$ are simple and noise-resistant. We may obtain them by pulse modulation, but at the expense of a significant increase of the bandwidth, possibly by a factor of ten.

Most of the computer generated holograms are in the format of $V_B(x,y)$.

§3 Four types of holograms

3.1. The image plane hologram

The image plane hologram is perhaps not the most fascinating type, but it is quite suitable for starting the discussion. Besides,this type is commonly used in interferometric applications of holography. Fig. 2 shows the recording and the reconstruction processes. The complex wave amplitude $V_o(x)$, coming from the object, is imaged upon the photographic plate, which will be the hologram after photochemical development. A tilted reference wave V_R interferes with the object wave V_o and forms interferometric fringes on the plate. These fringes act like a diffraction grating during the reconstruction process. One of the diffraction orders is a replica of the object wave $V_o(x)$.

Apparently, the angular separation of the diffracted beams depends on the tilt of the reference wave V_R during recording. How much tilt is needed for separating completely the three beams during reconstruction? This question can be translated into

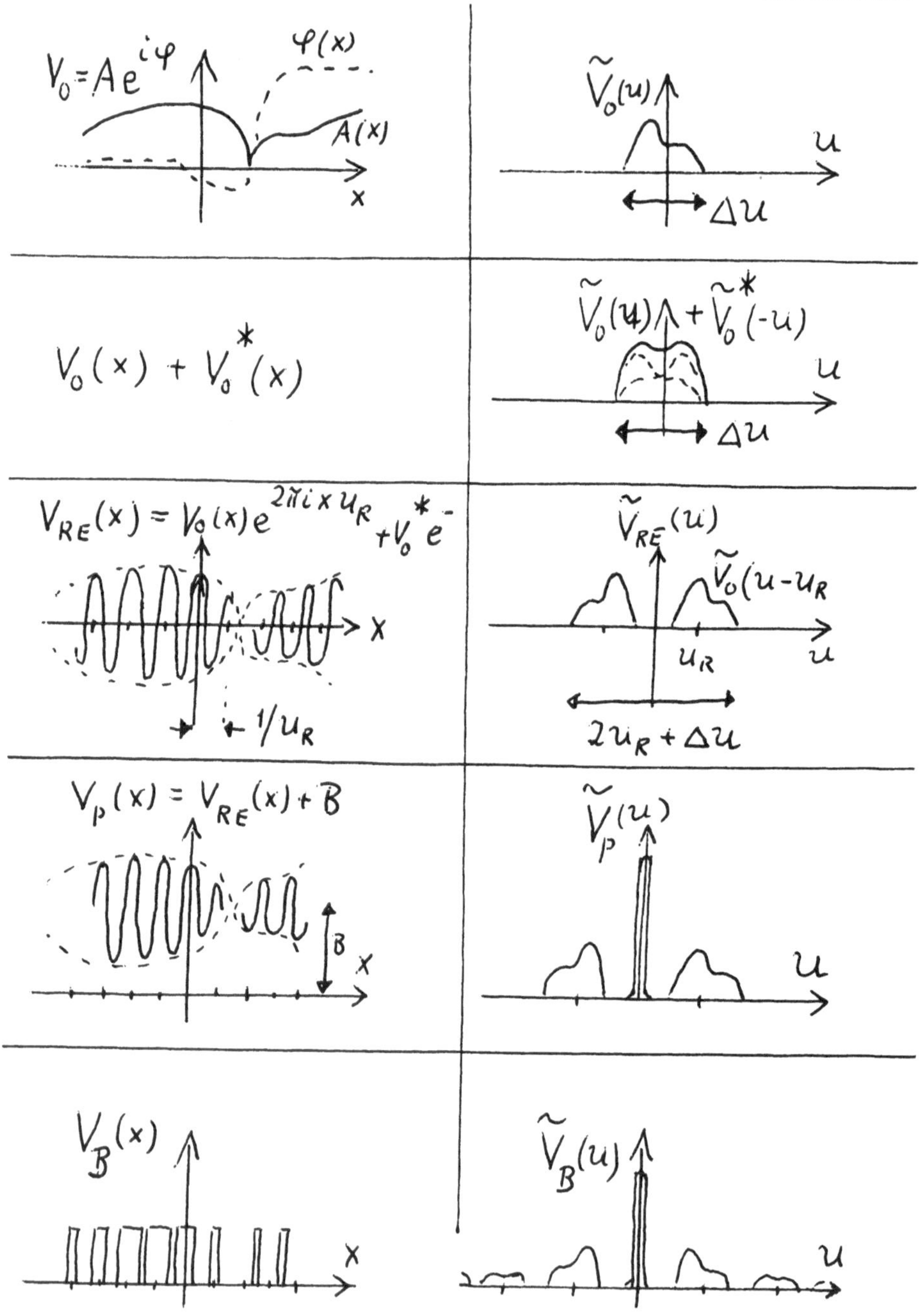

Fig. 1 Representation of some related signals and their frequency spectra. From top: complex; with complex conjugate; modulated, with complex conjugate; as before with bias, now positive; binary pulse modulated.

the language of modulation, as the following equations are supposed to express.

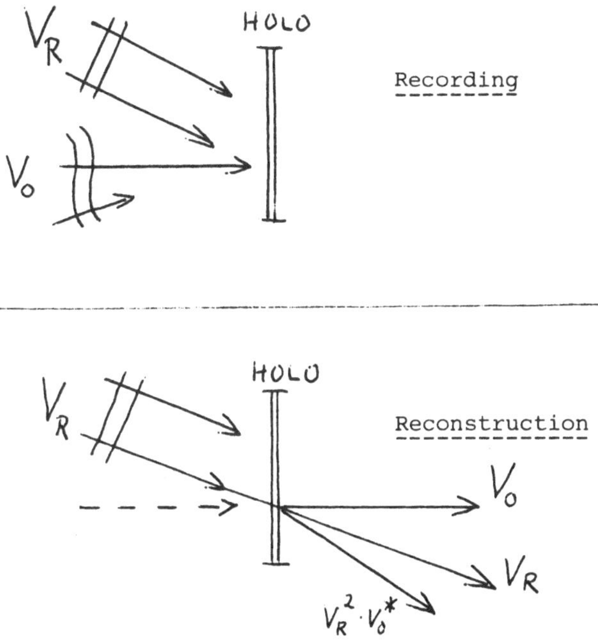

Fig. 2 Recording and reconstructing an image plane hologram

Complex amplitude arriving at hologram:

$$V_o(x) + V_R(x) = V_o(x) + \exp(2\pi i x U_R) \tag{2}$$

Recorded intensity:

$$I(x) = |V_o+V_R|^2 = |V_o|^2+1+V_o\exp(-)+V_o^*\exp(+) \tag{3}$$

The photographic plate is developped such that the amplitude transmittance T(x) is equal to the intensity I(x), apart from an uninteresting constant factor. When illuminated by the reference wave $V_R = \exp(2\pi i x U_R)$, the complex amplitude V(x) immediately behind the hologram is:

$$V(x) = I(x)V_R(x) = V_R\cdot\left[|V_o|^2+1\right] + V_o + V_o^* V_R^2 \tag{4}$$

We now compute the angular spectrum $\tilde{V}(u)$, which will appear as complex amplitude distribution in the Fraunhofer diffraction domain somewhere behind the hologram. For the sake of simplicity we assume the term $|V_o|^2$ in eq. 4 to be negligible:

$$\tilde{V}(u) = \delta(u-u_R) + \tilde{V}_o(u) + \tilde{V}_o^*(-u-2u_R) \quad (5)$$

These three terms will be separable if the tilt parameter u_R is larger than half of the bandwidth Δu of the object spectrum $\tilde{V}_o$. (This requirement would have to be modified, if $|V_o|^2$ were not negligible.) Apart from a lateral shift, the spatial frequency spectrum $\tilde{V}(u)$ is the same as the spectrum $\tilde{V}_p(u)$ of the positive signal $V_p(x)$ (see fig. 1).

We now recognise, we had to pay as price for the positivity of the hologram transmission $I(x)$ a factor 2 in bandwidth requirement. The frequency spectrum $\tilde{I}(u)$ is exactly like $\tilde{V}_p(u)$, at least twice as wide as $\tilde{V}_o(u)$.

3.2. The Fourier hologram

The frame in the object domain can be filled in one half by an arbitrary object per se. The other half has to remain dark, except for a single bright spot at the outer edge (fig. 3)

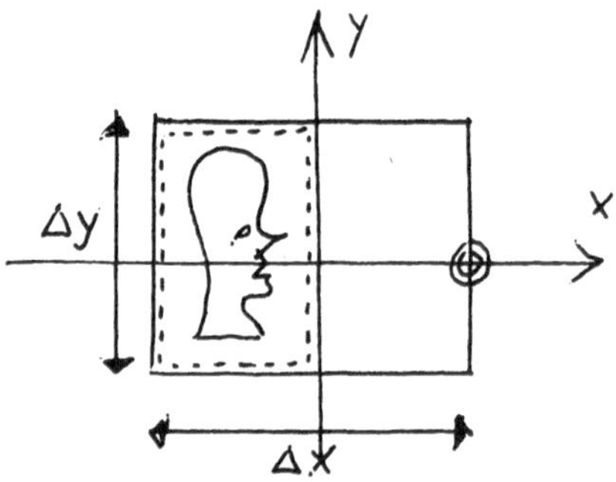

$$V(x) = \delta(x-\frac{\Delta x}{2}) + V_o(x+\frac{\Delta x}{4})$$

Fig. 3 The frame for the input during recording of a Fourier-hologram

The recording process consists of two parts: Fourier transformation by means of Fraunhofer diffraction; photographic recording of the intensity:

$$I(u) = |\tilde{V}(u)|^2 = 1+|\tilde{V}_o|^2 + \tilde{V}_o\exp(-) + \tilde{V}_o^*\exp(+) \quad (6)$$

Looking carefully at the spatial frequency spectrum of the hologram transmittance $I(u)$ reveals, that separability of the terms is feasible, if we are willing to sacrifice one half of the area of the input domain, the one half that has to remain dark except for the single bright spot (fig. 3).

3.3. The polarisation hologram

So far, we paid a factor-2-price when substituting for a complex signal by a real positive signal. Both signals were scalar by nature. Now we want to deal with polarised light, which can be described by a two-component vector $\{E_x, E_y\}$ with complex E_x and E_y. The photographic plate is blind not only for phase data, but also for polarisation data (if we neglect the Weigert effect, that is not very pronounced, usually. See Opt. Comm. 10, 129, 1979).

In terms of signal representation we now wish to convert the two complex signals $E_x(x,y)$ and $E_y(x,y)$ reversibly into one real-positive signal $I(x,y)$. As may be expected, we will have to pay a price with a factor four either in bandwidth or in input space. Specifically, only one quarter of the input domain is free for accepting the complex polarised object. The remaining three quarters have to be dark, except for two bright points at suitable locations. These two bright spots are polarised orthogonally, each one contributing as source of a reference wave to a sub-hologram with the corresponding state of polarisation (fig. 4). Both of these polarised reference waves are needed when reconstructing the polarised complex wavefield.

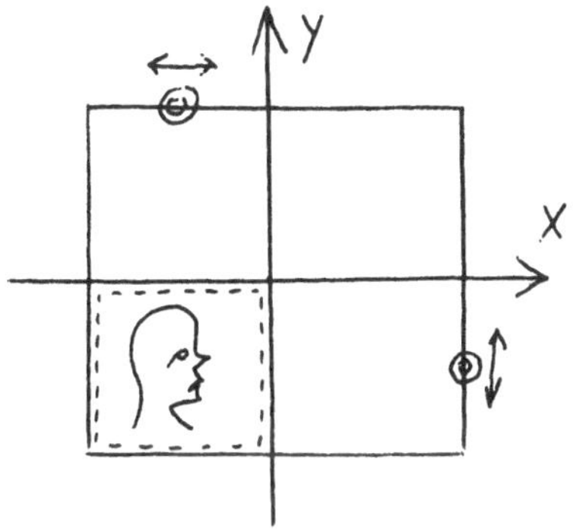

Fig. 4 The assignement of the input plane for recording a polarisation Fourier hologram

3.4. Three-color hologram

Suppose now the object wave consists of three colors, based on three suitable monochromatic wavelengths. High resolution photographic plates as used for holography are color-blind. What do we do? We provide three monochromatic reference waves, coming in from suitable angles. Hence, we record on the photographic plate a single grey positive signal, that supposedly knows everything about three independent complex wavefronts. Such an assignment for reversible recording - or encoding - is feasible, but at a price of a factor six. This factor would be doubled to twelve, if the three complex color signals should be distinguishable also by their states of polarisation.

In summary of §3 we conclude, that phase-blindness, polarisation-blindness and color-blindness can be overcome by a suitable encoding scheme (modulation, multiplexing). But we have to pay a price by a factor that describes the signal manyfold in comparison to the scalar positive signal, which can be "carried" by the photographic plate directly.

§4 Complex operations, based on the "polarisation identity"

Historically, one of the main problems in the early years of holography was the suppression of the unwanted twin image. Equally unwanted is the so-called zero-order term, sometimes referred to as intermodulation or autocorrelation term. In this chapter we want to show, how all unwanted terms can be suppressed.

To introduce this topic, let us remember, how the process of multiplication of a and b can be replaced by the operations of adding, subtracting and squaring:

$$(a+b)^2 - (a-b)^2 = 4ab \tag{7}$$

W.D. Montgomery (Opt. Lett. 2, 120, 1978) has generalised the algorithm of eq. 7, assuming that adding of two complex signals is possible, also phase-shifting by multiples of $\pi/2$, and modulus-squaring. Properly combined, these operations can achieve the multiplication of two complex signals V_O and V_R (eq. 8)

$$4V_O V_R^* = \sum_{m=0}^{3} (i)^m \cdot |V_O + (i)^m V_R|^2 \tag{8}$$

This is the algorithm, called "polarisation identity" by Montgomery. This algorithm can be implemented holographically. Each of the four modulus-square terms can be produced as a hologram. The reconstructions from these four holograms are to be combined by means of beam-splitters, for example. The result contains only the wanted cross term $V_O V_R^*$, not the twin image term $V_O^* V_R$ and the intermodulation terms $|V_O|^2$ and $|V_R|^2$.

This algorithm can be made somewhat more efficient with only three summands. Let W be the third root of unity: $\exp(2\pi i/3)$. Then we may write:

$$3V_O V_R^* = \sum_{m=0}^{2} W^m \cdot |V_O + W^m V_R|^2 \tag{9}$$

We want to go one step further and consider a third-order polynomial nonlinearity as being feasible. On that base we find the following algorithm for a triple product:

$$6V_1 V_2 V_3 = \sum_{m,n=0}^{2} \sum^{2} (W)^{-(n+m)} (V_1 + W^n V_2 + W^m V_3)^3 \tag{10}$$

Nonlinear distortion of the photographic characteristic curve may serve to implement this or any other similar "polarisation identity".

§5 Some other applications of the holographic concept

Let us begin by sketching the mathematical algorithm of the holographic process, including recording and reconstruction processes. The linear operator F may represent a Fourier transformation (in case of Fraunhofer diffraction), or a Fresnel transformation (in case of Fresnel diffraction), or the identity operator (in case of perfect imaging), or possibly something else. The variables of the signals V maybe spatial (x,y,z), temporal (t), polarisation, spectral (λ) or something else. The variables may even change in the course of the processing, for example (R,PHI) before $|\ldots|^2$, and (x,t) behind it.

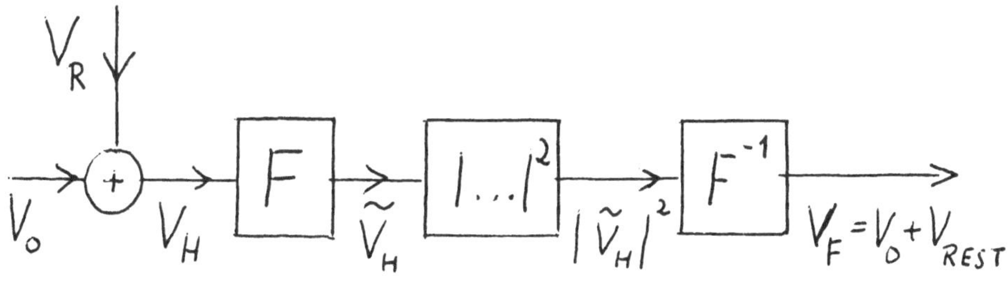

Fig. 5 The algorithm of holography

The quality of the signal may also vary, for example from voltage to photographic transmittance. The main thing is to achieve our goal, which is the recovery of the object signal V_O from the final signal V_F:

$$V_F = V_O + V_{REST} \tag{11}$$

Hopefully, V_O can be isolated from the rest. That, in essence, is the concept of holography.

This concept has been implemented at first with visible light. Electron waves, acoustical waves, x-rays (heavy atom replacement in molecular crystals), IR and microwaves have been used also. When working with visible light, it is not necessary to require coherence. The holographic concept has been utilised also with partially coherent light and even with incoherent light. If the light is very weak our detectors respond to single photons. Even then is it possible to benefit from Gabor's concept, the principle of holography.

Valuable discussions with H.O. Bartelt are gratefully acknowledged.

V.2 (PT.1)

REVIEW OF TOMOGRAPHIC IMAGING METHODS APPLIED TO MEDICAL IMAGING

Schwierz, G.

Siemens AG, Division of Medical Technique,
Erlangen/Germany

ABSTRACT

With the introduction of X-ray CT it became possible to produce contrasty sectional images of the human body free of superimpositions. The indisputable successes of this new radiographic technique prompted the extension of it to other types of radiation used in medical diagnosis as well. But even the elementary requirements for adequate penetration capacity on the one hand and sufficient interaction with the tissue on the other hand exclude the majority of known types of radiation. Although basically applicable to corpuscular radiation of high energy or - after modification - to ultrasound waves, the principle will in the near future first of all conquer the field of extremely long waves in the shape of NMR (nuclear magnetic resonance). Therefore, the reconstruction approaches which have become known for X-ray CT are presented primarily, these being the algebraic and here in particular the iterative as well as the integral transformation procedures, which have achieved greater importance. Two-dimensional display is preferred here. Using NMR as an example, where three-dimensional (3D) image reconstruction appears more likely to be achievable, marked differences between the 3D-reconstruction problem and the 2D-problem can be demonstrated. This is in particular the local character in the 3D case of the convolution operation, which is an important part of image reconstruction, and the derivability of the reconstruction method from principles of potential theory. For wavelength ranges where the nature of the waves may not be neglected, in recent years there have arisen approaches which in a natural way associate themselves with those already known from CT.

W.-M. Boerner et al. (eds.), Inverse Methods in Electromagnetic Imaging, Part 2, 1043–1064.

INTRODUCTION

In 1972, G.N. Hounsfield introduced a new radiographic imaging procedure, X-ray computed tomography (CT). In contrast to conventional X-ray imaging techniques, CT makes possible the display of isolated sections through the body. Interference shadows from structures outside the slice no longer arise. Differences in tissue density are discriminated more than ten times better than with classical procedures, whereby the possibilities for soft-tissue diagnosis are significantly improved.

CT has captured its place amongst imaging procedures in medical diagnostics in a very short time, initially in skull diagnosis and later, after the image recording time of originally some minutes could be reduced to a few seconds, also in the diagnosis of the whole body.

It suggests itself that the range of application of this successful technique should be enlarged by using other types of radiation. In order to be able to estimate which types of radiation are suitable for this, the basic principle is firstly explained with reference to the simplest CT system. (Fig. 1) With this the X-ray tube together with a detector coupled rigidly with it carries out an alternating sequence of uniform translation and rotation movements. All movements take place in a single plane. During one linear scanning process, a finely collimated X-ray beam is led over the slice in the object to be displayed and the radiation intensity let through the object is interrogated at the detector in equidistant steps a few hundred times. After each linear movement, the meas-

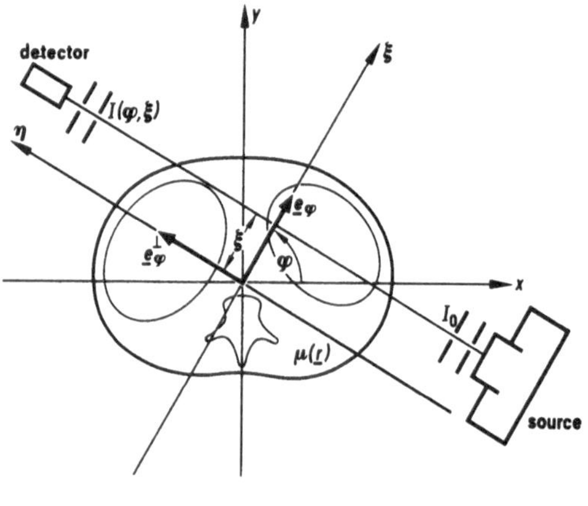

Figure 1. Basic scanning arrangement and system of coordinates $\underline{e}_\varphi$, $\underline{e}_\varphi^\perp$ unit vectors; I_0, $I(\varphi, \xi)$ X-ray intensity in front of and behind the patient; $\mu(\underline{r})$ distribution of attenuation coefficient.

uring system carries out a rotational step (of typically 1/2 degree) and the scanning process is repeated. With the simple scanner type described, it is sufficient in principle to rotate the measuring system by a total of only 180 degrees.

The projections gained during a linear scanning movement are subsequently subjected to certain mathematical operations in the computer of which so-called filtration is the most important, and are reconstructed by superposition into an image matrix in the memory of the computer.

The basic requirements on a beam quality suitable for CT are now easy to recognize.

. The radiation must be able to penetrate biological tissue of approx. 30 cm thickness, on the other hand it must experience a measurable modulation in the tissue. The requirement of penetration capacity already excludes all electromagnetic radiation between the HF-range and that of X-ray radiation. On the other hand, the CT principle can be realized for the two ranges quoted, in the shape of X-ray CT already mentioned as well as of NMR-CT (nuclear magnetic resonance). The application of high-energy electromagnetic radiation, such as of γ-rays, would lead to a reduction in the image contrast achievable and would thus mean relinquishing one of the most important advantages of X-ray CT.

High-energy particles are capable of penetration,such as positrons of 200 - 300 MeV, or heavier ions the application of which would have the advantage of a marked reduction in patient dose. Radioactive pharmaceuticals which are deposited preferably in certain types of tissue, enable organ-specific and functional diagn. statements to be made. With the so-called emission CT, both γ-ray and positron emitters come into consideration. In contrast with single photon ECT, positron ECT permits largely the suppression of the inherent absorption in the tissue and leads to unequivocable determination of the source distribution. However, at present, serious problems with the making available of the pharmaceuticals oppose the application of positrons on a broad basis.

In the non-usable wavelength range between a few meters (NMR) and around 10^{-10} m (X-rays) are still to be found ultrasound waves of roughly 0,5 mm to 2 mm wavelength, which satisfy the above requirements.

. The measured interaction of the radiation with the tissue must be localizable at least in the sense that it can be associated with a straight line or a plane (in the case of three-dimensional CT) through the object. In principle, bent curves of surfaces are also permissible, but then the reconstruction effort naturally increases considerably. The latter case of measurement recording on semi-circles or hemispherical surfaces in Fourier-space occurs in ultrasound CT which is still under development.

. Finally, the contribution of a tissue element to the scan signal should be independent of the location of the element on the curve or surface. This condition is as a rule only approximatively fulfilled, its violation gives rise to image artefacts.

1. X-RAY CT

1.1 Integral equation and approaches to a solution

If a narrowly collimated monochromatic X-ray beam of intensity I_0 penetrates tissue, the intensity

$$I(\varphi, \xi) = I_0 \exp(-\int \mu(\underline{e}_\varphi \xi + \underline{e}_\varphi^\perp \eta) d\eta) \tag{1}$$

is let through (Fig. 1).

The distribution of the attenuation coefficients $\mu(\underline{r})$ is to be determined from the linear integral equation

$$\int \mu(\underline{e}_\varphi \xi + \underline{e}_\varphi^\perp \eta) d\eta = \ln(I_0 / I(\varphi, \xi)) =: p_\varphi(\xi) \tag{2}$$

The known approaches at a solution can be arranged in two main groups, according to if the discretization of the integral equation takes place directly, this therefore being transformed into a normally linear system of equations, or if initially a closed solution formula is derived and the discretely gained sampling values are only inserted in this solution formula. In the first case one has an algebraic reconstruction procedure, in the second case an integral transformation procedure. The most important representatives of each group should be outlined.

1.2 Algebraic reconstruction procedure

The iteration procedure (1)

The object field is divided into small, generally square cells, called pixels (Fig.2), and the attenuation coefficient in each pixel is assumed to be constant. (More sophisticated approaches are not discussed here).

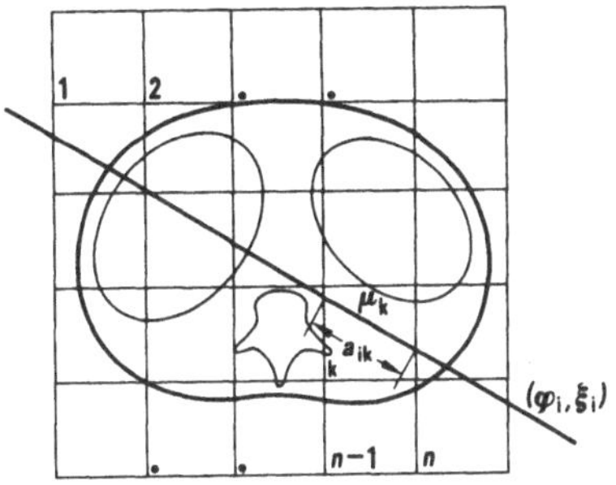

Figure 2. Approximation to the linear integral equation by a linear system of equations. φ_i, ξ_i beam characterization parameters; a_{ik} length of the path of the i-th beam through the k-th pixel with constant attenuation coefficient μ_k.

The integral measured along the path (φ_i, ξ_i) is therefore approximated by a finite sum:

$$p_i \approx \sum_{i,k} a_{ik}\,\mu_k \qquad (1 \leqslant i \leqslant m,\; 1 \leqslant k \leqslant n) \tag{3}$$

n is the number of pixels, m the number of measurements, a_{ik} represents in the simplest case the path-length of the i-th beam in the k-th pixel.

This system of equations can be solved iteratively with the Kacmarz method (2). Let $\tilde{\mu}_1, \ldots \tilde{\mu}_n$ be an approximative solution to the above system of equations and let i be the number of the next measurement to be taken into account. Then the difference

$$p_i - \sum a_{ik}\,\tilde{\mu}_k \tag{4}$$

is distributed according to the weighting a_{ik} on the pixels hit by the i-th beam, i.e. this difference is projected back to the image matrix. One proceeds thus successively with all scan data. The iteration cycle must subsequently be repeated. Geometrically, this is to be interpreted as repeated orthogonal projection of the approximation vectors $\tilde{\underline{\mu}} = (\tilde{\mu}_1, \ldots, \tilde{\mu}_n)^T$ onto the hyperplane which is the next one to be taken into account. Here one recognizes immediately that the iteration with inconsistent scan data shows a cyclical convergence behaviour. On the other hand, with consistent data it converges, namely towards that solution which has the smallest distance to the iteration starting point in the Enclidean metric. Variations of the method, which provide the Gaussian optimal solution (3), are possible.

This procedure was actually realized in the first CT units but it could not prevail against the integral transformation procedures, which is primarily attributable to the inherently long computing time of the iteration procedures.

Estimation theory approach (4, 5, 6)

Let a class K of CT objects $\underline{\mu}$ with an a priori probability density $f_0(\underline{\mu})$ be given. Should a new CT scan be made, then $f_0(\underline{\mu})\, d\mu_1 \ldots d\mu_n$ gives the probability that the image will fall in the region $\underline{\mu}, \ldots, \underline{\mu} + d\underline{\mu}$. However, if a scan data vector $\underline{p}$ is measured, the probabilities change fundamentally since the majority of images $\underline{\mu} \in K$ are incompatible with the scan data. The scan therefore transforms the a priori density f_0 into the a posteriori density $f(\underline{\mu}|\underline{p})$, i.e. into a conditional probability density. This will, as a rule, be much narrower than f_0 the less the reliability of the scan data is impaired by statistical disturbances such as the quantum noise. The optimally estimated image $\underline{\mu}^*$ is to be considered as that one which maximizes $f(\underline{\mu}|\underline{p})$ in the presence of $\underline{p}$. This leads under the limiting assumption of Gaussian distributed attenuation values μ_k to the optimum solution

$$\underline{\mu}^* = \bar{\underline{\mu}} - (A^T E^{-1} A + C^{-1})^{-1} A^T E^{-1} (A\bar{\underline{\mu}} - \underline{p}). \qquad (5)$$

Here $\bar{\underline{\mu}}$ means the expectation values of the images belonging to the class K determined from the a priori probability, C the covariance matrix of the μ_k values and E that of the measurements. A is the matrix assigned to the system of equations (3). The complexity of the solution formula as well as the lack of knowledge concerning the a priori distribution stand in opposition to the practical applicability of this solution. Further simplifying assumptions such as uncorrelatedness of the μ_k values and large signal-to-noise ratio lead to the simplified result

$$\underline{\mu}^* = A^T (AA^T)^+ \underline{p} \qquad (6)$$

Here $(AA^T)^+$ signifies the Moore-Penrose inverse of AA^T; the transformed measurements are to be backprojected to the image matrix by means of A^T. The size of the matrix to be inverted does not depend upon the number of pixels! In principle, this approach therefore permits the transition to infinitely fine pixelling. In this case the discretely sampled image $\underline{\mu}^*$ is transformed to a continuously sampled $\mu^*(x,y)$. This solves the Gaussian optimization problem

$$\int_{K_0} |\mu^*-\mu|^2 dx\ dy = \min_{\tilde{\mu}} \int_{K_0} |\bar{\mu}-\mu|^2 dx\ dy \qquad (7)$$

whereby μ^* possesses minimal norm

$$\|\mu^*\| = \left(\int_{K_0} |\mu^*|^2 dx\ dy\right)^{1/2} \qquad (8)$$

beneath all possible solutions of (7).
(K_0 signifies the object circle).
In contrast to the convolution process customary in CT, the system matrix AA^T couples all scan data with one another and therefore contains 10^9 to 10^{12} elements differing from zero!
Because of the block-cyclical form, the inversion can be carried out in principle but the solution suffers from the typically bad condition of the matrix for Gaussian normal systems of equations. The introduction of suchprocedures in practice therefore appears questionable.

1.3 Integral transformation procedure

On the other hand, integral transformation procedures have gained acceptance, not least because they are based to a very large extent upon system-theoretical approaches and thus make the reconstruction problem at all transparent.

The Radon formula

The Radon formula (7) represents the indeed oldest solution of the image resonstruction problem. It can be derived without the Fourier transformation.

Idea: If $\mu(\underline{r})$ ($\underline{r} \in \mathbb{R}^2$) were rotationally smmetrical about the position $\underline{r} = \underline{r}_0$, then all scan information would be contained in a single projection. The image reconstruction problem would then correspond to the known problem in optics which has been solved there of determining the point spread function of an imaging system from the line spread function. The general reconstruction problem can, however, now be reduced to this special one (Fig. 3) by averaging the object $\mu(\underline{r})$ on all circles about $\underline{r}_o$, i.e. by going over to the substitute object

$$\mu_R(\underline{r}_o, \varrho) := \frac{1}{2\pi} \cdot \int_0^{2\pi} \mu(\underline{r}_o + \varrho\, \underline{e}_\varphi)\, d\varphi, \tag{9}$$

the projection of which can be determined in a simple manner from the measured projections $p_\varphi(\xi)$. For fixed $\varrho \geqslant 0$, from each projection the scan beams are to be selected which are tangential to the circle with $\underline{r}_o$ as center and with radius ϱ. The measurements are then to be averaged over all projections.

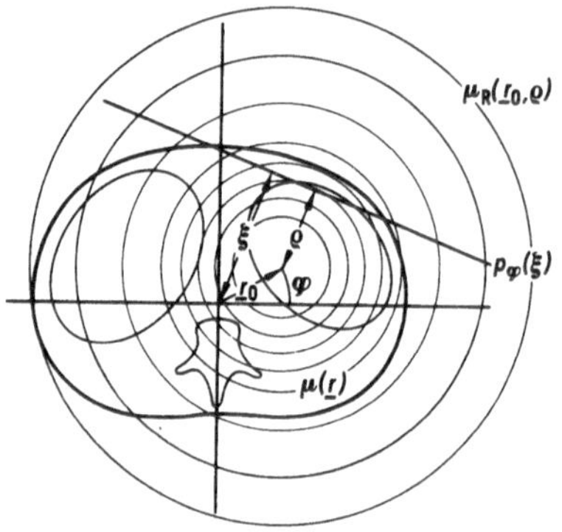

Figure 3. Substitution of the object function $\mu(\underline{r})$ by a rotationally symmetric function $\mu_R(\underline{r}_o, \varrho)$ with symmetry center at $\underline{r}_o$. All measurements $p_\varphi(\xi)$ associated with rays which are tangential to the circle at $\underline{r}_o$ with radius ϱ are averaged.

If μ at $\underline{r} = \underline{r}_o$ is continuous, then because of

$$\lim_{\varrho \to 0} \mu_R(\underline{r}_o, \varrho) = \mu(\underline{r}_o) \tag{10}$$

$\mu_R(\underline{r}_o, \varrho)$ is exactly the value sought for. With the averaged projection $\overline{p}(\underline{r}_o, \varrho)$, it applies according to Radon that

$$\mu(\underline{r}_o) = -\frac{1}{\pi} \int_0^\infty \frac{1}{\varrho} \frac{d}{d\varrho} \bar{p} (\underline{r}_o, \varrho) d\varrho \qquad (11)$$

In practice this procedure is always used with changed over sequence of the two integrations where it assumes the form of a convolution backprojection procedure.

Fourier procedure

Idea: If the object contains only a single δ-shaped element attenuating the radiation, then each projection also is a δ-function. If the scan data p_φ are projected back to the object plane, then with a finite number of projections a star pattern arises (Fig. 4), with continuous sampling, on the other hand, the image plane is covered by the point spread function $1/|\underline{r} - \underline{r}_o|$. This backprojection transfers an object distribution $\mu(\underline{r})$ into the very strongly faded approximation image

$$\bar{\mu}(\underline{r}) = \int \mu(\underline{r}') \frac{1}{|\underline{r} - \underline{r}'|} d^2 r' = (\mu ** \frac{1}{r}) (\underline{r}) \qquad (12)$$

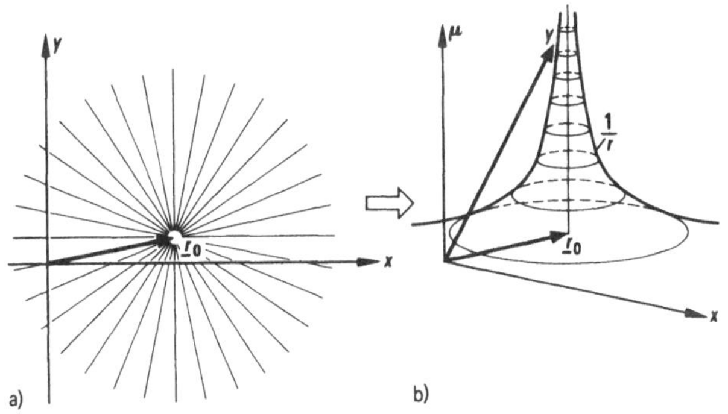

Figure 4. Generation of the point spread function $1/r$, where r is the distance of a given point to the δ-peak at $\underline{r}_o$.
a) Star pattern arising from a finite number of projections.

b) Point spread function arising from continuous sampling.

This convolution product can be resolved according to $\mu(\underline{r})$ by Fourier transformation

$$\mu(\underline{r}) = \mathfrak{F}_2^{-1}\left\{|\underline{\nu}|\mathfrak{F}_2\{\bar{\mu}\}\right\}(\underline{r}). \qquad (13)$$

This is a typical filter equation with which the high spatial frequencies $|\underline{\nu}|$ suppressed by the integration (2) are lifted again.

Convolution and backprojection procedures (8, 9)

Basically the image reconstruction can be carried out even during scan-data acquisition. Obtaining an instant image is possible. The instant image naturally represents a very important advantage for clinical application. It is with dynamic studies in part even indispensable.
Idea: The function p_φ backprojected onto the image plane represents a strip image (Fig. 5). It now suggests itself to carry out the superposition of these strip images only after filtration and thus to make use of the fact that the two-dimensional filtration of the strip image $p_\varphi(\underline{r} \cdot \underline{e}_\varphi)$ is equivalent to a one-dimensional filtration of the function $p_\varphi(\xi)$ with subsequent backprojection. This leads to the algorithm

$$q_\varphi(\xi) := \mathfrak{F}_1^{-1}\{|\nu|\,\mathfrak{F}_1\{p_\varphi\}\}(\xi) \qquad (14)$$

$$\mu(\underline{r}) = \int_0^\pi q_\varphi(\underline{r} \cdot \underline{e}_\varphi)\,d\varphi. \qquad (15)$$

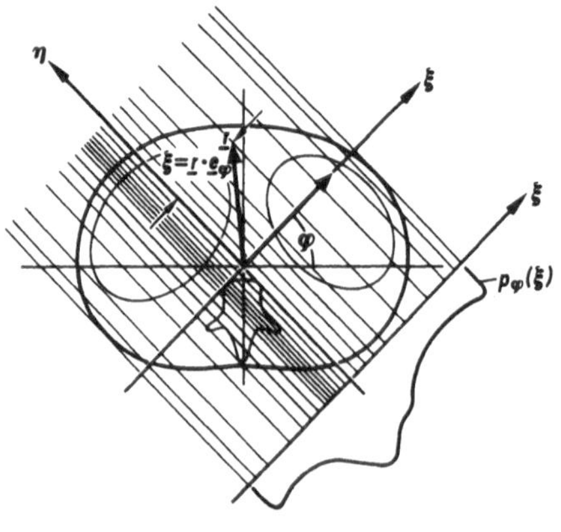

Figure 5. Strip image $p_\varphi(\underline{r} \cdot \underline{e}_\varphi)$

Both sections of this algorithm can be carried out in principle in real time, whereby it should be noted that special computer hardware is required.
Note: With discrete sampling systems, according to the Nyquist theorem, frequencies $\geq 1/(2a)$ cannot be acquired correctly. By prefiltration of the object over the sampling strip width, which must be selected sufficiently large, the Nyquist condition must thus be approximately enforced. Then the filter $|\nu|$ may be continued arbitrarily above $1/(2a)$ and drawn down to 0. However, in practice, $|\nu|$ is modified even below the Nyquist frequency, for instance according to Shepp and Logan (9) to

$$\begin{cases} \left|\sin\left(\pi \frac{\nu}{a}\right)\right| & \text{for } |\nu| \leq 1/(2a) \\ 0 & \text{for } |\nu| > 1/(2a) \end{cases} \tag{16}$$

This having been determined, $q_\varphi(\xi)$ results by interpolation from the data

$$q(ka) = \sum_{-\infty}^{+\infty} h_j \cdot p_\varphi((k-j)a) \tag{17}$$

with

$$h_j := -\frac{2}{\pi^2 a} \cdot \frac{1}{4j^2-1} \quad (-\infty < j < +\infty). \tag{18}$$

h_j $(-\infty < j < +\infty)$ is the Shepp and Logan kernal.

The pipeline principle

The division (14) and (15) of the reconstruction algorithm into two main steps is very well suited for the application of a special data processing technic.

The data obtained during a projection are immediately transferred to the memory of the computer. While projection No. i is subjected to A/D-conversion, the previously measured projection No. i-1 is already undergoing some necessary pre-processing steps. At the same time, the projection No. i-2 is back-projected. In this mode of operation, the slowest functional unit determines the processing speed of the entire system. By its nature, back projection is the operation that is most critical with respect to time. For this, a special hardware solution must be worked out.

It should be mentioned, that the pipeline principle can only be applied to CT-machines, where the projections are processed in the sequence they are measured. This is the case for instance with the simple scanner described, the so called dual motion machine, and with a certain type of fan-beam machine.

The Cormack procedure (10)

Although this procedure is certainly not applicable in its original form, because of lack of stability it should be explained briefly here, since it is distinguished by a very interesting property.

The projection function

$$p_\varphi(r) \quad (o \leq \varphi < 2\pi, \ o \leq r \leq 1) \tag{19}$$

is developed into a Fourier series on circles centered on $r = o$. (The radius of the object circle is set = 1 here).

The Fourier coefficients

$$p_n(r) = \frac{1}{2\pi} \int_o^{2\pi} p_\varphi(r) \ e^{-jn\varphi} d\varphi \tag{20}$$

contain for a fixed r only scan information on the part of the object located externally to the circle with radius r centered on o. This applies then also for the integral

$$\mu_n(r) := -\frac{1}{\pi} \frac{d}{dr} \int_r^1 \frac{r \ p_n(r')T_n(r'/r)}{\sqrt{r'^2 - r^2}} \cdot \frac{dr'}{r'} \tag{21}$$

(T_n = n-th Tschebyscheff polynomial). According to Cormack, these are just the Fourier coefficients of the development of μ.

If only scan data from an object are available for the part external to a circle, then it is accordingly possible to reconstruct the object there. This statement can, however, already be deduced from the algebraic reconstruction procedure.

As a consequence unambiguous reconstruction is possible from hollow projections (11), which sometimes occur in clinical applications, for instance if the projections are disturbed by highly radio-opaque materials such as artificial metal denturs or other types of implantation or surgical clips.

2. NMR-RECONSTRUCTION PROCEDURES (12, 13, 14)

Compared with X-ray CT, NMR offers a considerably larger multiplicity of possibilities for scan-data acquisition. Apart from the measurement of line integrals, here the measurement of plane integrals of the object function is possible with less difficulty which permits the reconstruction of three-dimensional images. Such images must subsequently be made accessible to observation by selection of suitable two-dimensional sections. Since the two-dimensional case has been treated exhaustively in section 1 and in principle each of the procedures discussed can also be used in NMR, the threedimensional image reconstruction should preferably be described.

Generation of signals in NMR (15)

The patient is located in an extremely homogeneous magnetic field B_0 of approx. 0.1 to 2 T (Tesla). The area of homogeneity encloses the volume of a skull, i.e. it is a sphere of 40 to 50 cm diameter.

In interaction with the thermal motion, a measurable magnetization $\underline{M}$ of the tissue in the direction of the magnetic field $\underline{B}_0$ is set up. It arises because of the alignment of the magnetic moments of the hydrogen nuclei abundantly present in the biological tissue, which alignment is preferentially in this direction. The individual nuclei precess obout the direction of the field with the Larmor frequency

$$\omega_0 = \gamma B_0 \tag{22}$$

with the gyromagnetic ratio $\gamma = 2\pi \cdot 42$ MHz/T.

If now a magnetic radiofrequency field $\underline{B}_1(t)$ of frequency ω_0 circularly polarized about the field direction, is beamed in, then the magnetization vector rotates away from the field direction about the momentary axis of the $\underline{B}_1$ field (Fig. 6).
It can be flipped over by 90° by the selection of suitably dimenioned pulses. This process is not selective. In order to obtain measurements depending on location, the basic field is superimposed directly after the 90° pulse in a prescribed spatial direction by a weak spatially constant field gradient $G\underline{n}$ ($\underline{n}$ unit vector). The Larmor frequency increases now linearly in the direction of the gradient, it is constant on planes perpendicular to it.

An induction signal $S(t)$ (t time) can now be received in the RF-coil which is switched to receive. This signal is composed as follows of the contributions of the individual planes:

$$S(t) = \int \varrho(\underline{r}) \exp(j\gamma(B_o + G\underline{n} \cdot \underline{r})t)d^3r \tag{23}$$

$\varrho(\underline{r})$ signifies the density of the hydrogen nuclei. (A multiplicative equipment constant is left out here). The Larmor basic frequency ω_o is already eliminated during measurement with a phase-sensitive detector so that with

$$\underline{\omega} \; := 2\pi\underline{\nu} \; := (\gamma Gt)\underline{n} \tag{24}$$

directly

$$\tilde{\varrho}(\underline{\nu}) = \int \varrho(\underline{r}) e^{j\underline{\omega}\cdot\underline{r}} d^3r = \mathcal{F}_3\{\varrho\}(\underline{\nu}) \tag{25}$$

i.e. the Fourier transform of ϱ is measured.

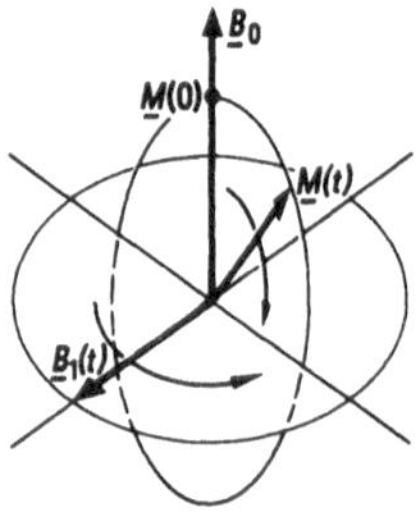

Figure 6. Rotation of the magnetization vector $\underline{M}(t)$ about the momentary rotation axis of the RF-field $\underline{B}_1(t)$. $\underline{B}_1(t)$ is circularly polarized about the homogeneous field $\underline{B}_o$.

With a more accurate description, relaxation processes must be taken into account and these are discussed at the end.

Thus, the image reconstruction in NMR is basically solvable by Fourier backtransformation. The apparently trivial step of the required conversion from the polar grid to a Cartesian grid adapted to the FFT is, however, numerically so critical that, although this method suggests itself, it is not used as a rule.

The central slice theorem

This theorem is fundamental for all reconstructive procedures.

Let

$$p(\xi, \underline{n}) := \int \varrho(\underline{r}) \cdot \delta(\xi - \underline{n} \cdot \underline{r}) d^3 r \qquad (26)$$

$(0 \leq \xi < \infty, \underline{n}$ unit vector$)$

be a projection ($\underline{n}$ is a three dimensional generalization of the unit vector $\underline{e}_\varphi$ of figure 1).

Then:

$$\mathcal{F}_3\{\varrho\}(\nu \underline{n}) = \mathcal{F}_1\{p(\cdot, \underline{n})\}(\nu). \qquad (27)$$

So the one-dimensional Fourier transform of the projection $p(\xi, \underline{n})$ turns out to be a cut through the three-dimensional Fourier transform of $\varrho(\underline{r})$.

The mapping (26) is called Radon transform.
(Analogously, this sentence applies naturally also in the $\mathbb{R}^2$ and can also there be used as starting point for the derivation of reconstruction procedures).

The local character of the convolution in three-dimensional image reconstruction

We are following here the representation of Barret and coworkers (16). According to the central slice theorem the plane integrals, i.e. the projection values $p(\xi, \underline{n})$, can be determined directly from the measurements by Fourier transformation. If these are backprojected into the $\mathbb{R}^3$, then the " strip image " $p(\underline{r} \cdot \underline{n}, \underline{n})$ oriented in direction $\underline{n}$ is produced and by superposition of all strip images, the faded approximation image

$$\bar{\varrho}(\underline{r}) = \frac{1}{2\pi} \int p(\underline{r} \cdot \underline{n}, \underline{n}) d\Omega \ . \qquad (28)$$

$d\Omega$ denotes the solid angle element. The range of integration is a hemisphere.

One can show that the point spread function of this imaging process is again proportional to $1/r$! The integral can therefore be converted to

$$\bar{\varrho}(\underline{r}) = \frac{1}{2}\int \varrho(\underline{r}')\frac{1}{|\underline{r}-\underline{r}'|}\,d^3r' = \frac{1}{2}(\varrho *** \frac{1}{r})(\underline{r}) \qquad (29)$$

and assumes the familiar form (12) of section 1.3.
Despite the great similarity of the integral equations (12) and (29), there is an essential difference. The point-spread function $1/r$ can be interpreted in the $\mathbb{R}^3$ as potential of a point charge, in the $\mathbb{R}^2$ on the other hand not! There it would have to be $\ln(1/r)$. $\bar{\varrho}(r)$ can therefore be considered as potential of a charge density $\varrho(\underline{r})$, which leads with the Laplace operator Δ directly to the solution formula

$$\varrho(\underline{r}) = -\frac{1}{2\pi}\Delta\bar{\varrho}(\underline{r}) = \frac{1}{4\pi^2}\Delta\int p(\underline{r}\cdot\underline{n},\underline{n})\,d\Omega \qquad (30)$$

By bringing the Laplace operator into the integral, one obtains again a convolution backprojection procedure. Since the Laplace operation to a strip image is equivalent to a two-fold differentiation in the direction of $\underline{n}$, the reconstruction formula (30) simplifies to

$$\varrho(\underline{r}) = \frac{-1}{4\pi^2}\int \frac{\partial^2}{\partial\xi^2}\,p(\xi,\underline{n})\Big|_{\xi=\underline{r}\cdot\underline{n}}\,d\Omega \qquad (31)$$

Because of the shift-invariance of the differential operator this this may again be interpreted as convolution. In contrast to the solution in the $\mathbb{R}^2$, the convolution operator here has local character. Its discrete form is reduced with the sampling interval a to the three-point formula

$$h_j \sim \begin{cases} -2/a^2 & j = 0 \\ 1/a^2 & |j| = 1 \\ 0 & |j| > 1 \end{cases} \qquad (32)$$

By Fourier transformation of equation (31), one obtains a filter equation analogous to equation (13)

$$\mathfrak{F}_3\{\varrho\}(\underline{\nu}) = -\frac{1}{2\pi}\mathfrak{F}_3\{\Delta\bar{\varrho}\}(\underline{\nu}) = 2\pi\,|\underline{\nu}|^2\,\mathfrak{F}_3\{\bar{\varrho}\}(\underline{\nu}) \quad (33)$$

i.e.

$$\varrho(\underline{r}) = 2\pi\,\mathfrak{F}_3^{-1}\{|\underline{\nu}|^2\,\mathfrak{F}_3\{\bar{\varrho}\}\}(\underline{r}) \quad (34)$$

This time filtration must be done with $|\underline{\nu}|^2$ corresponding to the Laplace operation in the object space. On the other hand, the filtration with $|\nu|$ in (13) cannot be construed simply as differentiation. Rather corresponding to the decomposition $|\nu| = \nu \cdot \text{sign}(\nu)$ the filtration with $|\nu|$ is to be described as differentiation with prior or subsequent Hilbert transformation.

The image reconstruction in $\mathbb{R}^2$ therefore is proven to be basically more complicated than in $\mathbb{R}^3$. This phenomenon carries on with problems of higher dimensions, whereby a difference must be made between even and uneven dimensions (17).

Variants to the NMR-image reconstruction

Variations in the scanning procedure always bring about corresponding variations in the image reconstruction algorithm. Thus, for example, it is possible to record the Fourier transform of ϱ directly on a cartesian grid, this being the case both three-dimensionally and two-dimensionally on selected section planes, so that the image reconstruction is reduced to a Fourier backtransformation (14). Even the point-by-point sampling of the object is possible with entering of the measurement results in the image matrix as trivial reconstruction process. However, with scan times in the hours region, the latter procedure is extremely slow and moreover supplies no particularly good spatial sharpness. The selection of a suitable reconstruction procedure in NMR is made therefore, in no way primarily under the aspect of simple implementation but rather according to more decisive criteria such as scan time and image noise and the technical and equipment possibilities.

Relaxations

Interactions between neighbouring magnetic moments lead to fading out of the components of magnetization perpendicular to the field direction - i.e. without the energy of the magnetization in the external field changing - and thus to fading away of the induction signal with the relaxation time T_2. This spin-spin relaxation is superimposed by the spin-lattice relaxation which leads to increasing the components of magnetization in the field direction and is to be understood as conversion of the precession energy into heat.

It takes place with the relaxation time T_1. These processes lead to line propagation. However, it is of more relevance that instead of the density $\varrho(\underline{r})$ of the nuclei, approximately the magnitude

$$\varrho(\underline{r})\exp(-\tau/T_2(\underline{r}))(1-\exp(-T_r/T_1(\underline{r}))) \tag{35}$$

is measured. Here τ means the time between two excitations of the nuclear resonance system and T_r the repetition time with which the individual projections follow one another. Therefore the physical character of the image can be specifically influenced with the aid of the parameters τ and T_r. In particular, the display of the T_1 time could become of considerable importance since tumorous tissue possesses a significantly higher T_1 time than healthy tissue.

3. REMARKS ON ULTRASOUND CT

Apart from the classical high-resolution radiographic imaging procedures with a resolution capacity of fractions of a millimeter, imaging procedures with less favourable spatial resolution have also proven to be thoroughly interesting in medical applications. Mention is made here only of nuclear diagnostic procedures with typically a few millimeters, ultrasound procedures with approx. 1 mm and X-ray CT with 0,5 to 1 mm spatial resolution.

As mentioned at the beginning, electromagnetic waves in the wavelength range of millimeters cannot be made use of for medical diagnosis because of their inadequate penetration capacity. On the other hand, possibilities exist for ultrasound transmission CT (18). Both the sound intensity let through by the body and the time for the passage of short sound pulses can be considered as magnitude to be measured. The latter case refers to the reconstruction of the distribution of the refractiv index in the tissue.

The method has good chances of success with freely accessible soft-tissue organs such as the mamma or the testicles and, with certain restrictions, the extremities as well. The extremely poor US-permeability of bones as well as the practically complete reflection at the tissue/air or air/tissue interfaces are bound to represent a decisive impediment to transmission CT of the trunk of the body. (In the case of the extremities, the way out by hollow projections is conceivable).

For this reason, possibilities of US-image reconstruction in the conventional reflection mode have also become subjects of discussion. These approaches make use of the fact that the US image itself is already a tomographic image and aim at the symmetrization of the

" point-spread function " which is narrow in the direction of the wave normals but wide perpendicular to this direction. The US images recorded from different directions are overlapped after possible frequency filtration which is perfectly comparable with that used in CT (19).

A proposal originates from Müller and coworkers (20) for the approximate solution of the wave equation in weak inhomogeneous media. (This approach also excludes air or bone pockets).

This method cannot be presented here but the similarity with the reconstruction methods discussed should be illustrated. The interference function $F(\underline{r})$ should be determined from the time-separated wave equation

$$\Delta\psi(\underline{r}) + k^2\,\psi(\underline{r}) = F(\underline{r}) \cdot \psi(\underline{r}) \tag{36}$$

with $F(\underline{r}) = k^2(1 - C_0/C(\underline{r}))$, whereby $C(\underline{r})$ implies the sound velocity depending on the location, C_0 a reference velocity and k the wave number. Here it is assumed that after beaming in a plane wave from a prescribable direction, the entire wave field disturbed by the body is known on a large but limited detector area. The detector should be rotatable in relaxation to the object as well. Then the measuring procedure can be considered again to be measurement of the Fourier transform of the filtered object function, whereby the measuring process itself acts as a frequency filter. The measured data arise here on spherical surfaces in the Fourier space. These spheres touch - like the Ewald spheres in solid-state- physics - the coordinate origin, their center is given by $-\underline{k}$, i.e. by the wave-number vector. If one tends here to extremely small wavellength, i.e. to $k \longrightarrow \infty$, then these spheres degenerate into planes through the zero point and the same measurement situation is present as in the case of X-ray CT which indeed represents the limiting case of infinitely short waves. Devaney (21) could show that then the image reconstruction algorithm also is transformed into the well-known convolution backprojection algorithm. But even without the boundary transition, there is similarity between the algorithms. Primarily the backprojection is to be converted into an operation termed by Devaney as " backpropagation ".

Thus, an important step in the direction of a general image reconstruction theory, applicable to all wavelength ranges, has been taken.

SUMMARY

Various approaches to the reconstruction of tomographic images where projections are available have been presented. The possibilities of three-dimensional CT have been illustrated using NMR as an example and the characteristic differences compared to the two-dimensional problem are displayed. A view of more recent work on ultrasound CT shows that it is possible to transfer the CT reconstruction principle to types of radiation with which the fact of finite wavelength may not be neglected.

REFERENCES

(1) Gordon, R.: A tutorial on ART (algebraic reconstruction techniques). IEEE Trans. on Nuclear Science, Vol. NS-21, (1974) 78 - 93

(2) Guenther, R.B., et al.: Reconstruction of objects from radiographs and the location of brain tumors. Proc. Nat. Acad. Sci. USA 71 (1974), 4884 - 4886

(3) Huebel, J.G. and Lantz, B.: A converging algebraic image reconstruction technique incorporating a generalized error model. Conference record of the 9-th Asilomar Conference on Circuits, Systems and Computers held in Pacific Grove, California, Nov. 3 - 5, 1975, 571 - 576

(4) Herman, G.T., Hurwitz, H. and Lent, A.: A bayesian analysis of image reconstruction. In " Reconstruction tomography in diagnostic radiology and nuclear medicine " edited by Ter-Pogossian M.M. et al. University Park Press, Baltimore, London, Tokyo (1977)

(5) Buonocore, M.H. et al: A natural pixel decomposition for two-dimensional image reconstruction IEEE Trans. on Biomedical Engineering, Vol. BME-28 No. 2 (1981), 69 - 78

(6) Natterer, F.: Efficient implementation of " optimal " algorithms in computerized tomography Math. Meth. in the Appl. Sci. 2 (1980) 545 - 555

(7) Radon, J.: Über die Bestimmung von Funktionen durch ihre Integralwerte längs gewisser Mannigfaltigkeiten. Ber. Verk. Sächs. Akad. 69 (1917) 262 - 277

(8) Ramachandran, G. and Lakshminarayanan, A.: Three-dimensional reconstruction from radiographs and electron micrographs:

application of convolutions instead of Fourier transform
Proc. Nat. Acad. Sci. USA 68:9 (1971) 2236 - 2240

(9) Shepp, L.A. and Logan, B.F.: The fourier reconstruction of a head section
IEEE Trans. on Nucl. Sci. Vol. NS - 21 (1974) 21 - 43

(10) Cormack, A.: Representation of a function by its line integrals, with some radiological applications
J. Appl. Physics 34:9 (1963) 2722 - 2727

(11) Lewitt, R.M. and Bates, R.H.T.: Image reconstruction from projections
Part I - IV: Optik 50 (1978) 19, 85, 189 and 269

(12) Lauterbur, P.C.: Image formation by inducing local interactions: Examples employing nuclear magnetic resonance
Nature 242 (1973) 190 - 191

(13) Mansfield, P. and Pykett, I.L.: Biological and medical imaging by NMR
J. of Magn. Res. 29 (1978) 355 - 373

(14) Kumar, A., Welti, D. and Ernst, R.R.: NMR Fourier Zeugmatography
J. of Magn. Res. 18 (1975) 69 - 83

(15) Loeffler, W. and Oppelt, A.: Physical principles of NMR tomography
Europ. J. Radiol. 1 (1981) 338 - 344

(16) Ming-Yee Chin, Barrett H.H. and Simpson, R.G.: Three dimensional reconstruction from planar projections
J. Opt. Soc. Am., Vol. 70, No. 7 (1980) 755 - 762

(17) Ludwig, D.: The Radon-transform on Euclidean space
Commun. on pure and appl. math. XIX (1966) 49 - 81

(18) Greenleaf, J.F. et al.: Reconstruction of spatial distributions of refraction indices in tissue from time of flight profiles
From the technical digest of the conference on " Image processing for 2-D and 3-D reconstruction from projections: Theory and practice in medicine and the physical sciences " held at Stanford University, California, Aug. 4 - 7, 1975

(19) Hundt, E.E. and Trautenberg, E.A.: Digital processing of ultrasonic data by deconvolution
IEEE Trans. on Sonics and Ultrasonics, Vol. SU - 27, No. 5 (1980) 249 - 252

(20) Mueller, R.K., Kaveh, M. and Wade, G.: Reconstructive tomography and applications to ultrasonics
Proc. IEEE, Vol. 67, No. 4 (1979) 567 - 587

(21) Devaney, A.J.: A filtered back propagation algorithm for diffraction tomography
Submitted to IEEE Trans. Biomed. Eng., Nov. 1981.

V.3 (MM.1)

INVERSE METHODS IN MICROWAVE METROLOGY

Walter K. Kahn

Department of Electrical Engineering and Computer Science
The George Washington University
Washington, D.C. 20052 U.S.A.

ABSTRACT If the various ways of computing the response (output) of a system (circuit) due to any excitation (input) from completely known constitution of the system (given circuit) are categorized as direct, then, by contrast, the various ways in which the constitution (circuit characterization) of a system may be inferred from sets of corresponding excitations and responses fall into the category of inverse methods. [3] To provide the focus on microwave measurements as conventionally understood, we limit the discussion to single-frequency excitation. We then consider how significant features of the system can be inferred through consideration of

Resonance Condition
Transformation Invariants
Physical Symmetry

from data gained when a system is observed in this limited way.

1. INTRODUCTION

Just what sort of thing might constitute a "significant feature" of a microwave circuit is, in the final analysis, a matter of subjective judgement. In the same spirit in which we may decide that the significant features of a car we are inspecting are precisely those features which do not change as we examine it at closer range or from different angles (a most inappropriate decision, however, if our objective is to safely cross a highway), we decide that significant features of a network will not depend on the precise lengths of the uniform waveguide leads used to make network connections. In fact, much more general peripheral elements, such as arbitrary reciprocal network for example, may be

W.-M. Boerner et al. (eds.), Inverse Methods in Electromagnetic Imaging, Part 2, 1065–1075.

ignorable if our purpose is to probe the essential nonreciprocal aspect of some network. Each of the techniques to be described has some such feature of a microwave network as its focus.

2. RESONANCE

The cavity system and corresponding network representation are illustrated in Fig. 1a and 1b respectively. The M straight lines in Fig. 2.2a are uniform waveguides in each of which any finite number of modes may propagate. The simple closed surface T, shown dashed, encloses the junction S; so the remainder of the cavity system is the junction P. (The same symbol is employed for a junction and its normalized-voltage scattering matrix.) [13]

The interconnection of the junctions S and P is expressed by

$$\underline{b} = S\underline{a} \tag{2.1a}$$

and

$$\underline{a} = P\underline{b} \tag{2.1b}$$

where $\underline{a}$ and $\underline{b}$ are N dimensional column matrices, the elements of which are normalized-voltage, rms, incident and reflected wave phasors, one pair for each of the NxM propagating modes. Reciprocity and the conservation of energy imply $S^{-1}=S^*$ and $P^{-1}=P^*$, where * denotes the complex conjugate.

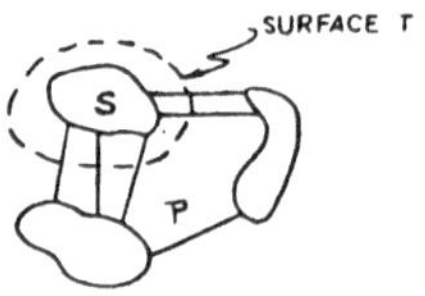

Fig. 2.1a

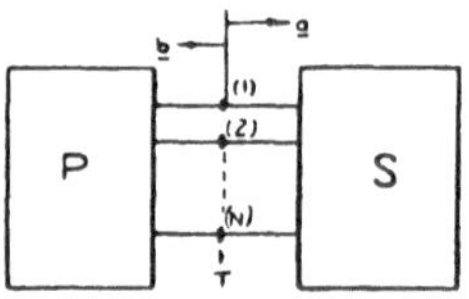

Fig. 2.1b

Eliminating either $\underline{a}$ or $\underline{b}$ from (1) yields

$$(S-P^*)\underline{a} = 0 \tag{2.2a}$$

and

$$(S-P^*)\underline{b}^* = 0 \tag{2.2b}$$

The condition for the existence of non-zero solutions of (2.2) is, of course,

$$\det(S-P^*) = 0 \tag{2.3}$$

The parameters descriptive of a lossless reciprocal N-port at one frequency may be determined by observing those positions of short circuits placed in each of its N (modal) transmission line leads which render the cavity thus formed resonant. [6] The network situation conveniently described as the interconnection of S with a special junction P, P diagonal,

$$P = -[p_n \delta_{mn}] \;, \tag{2.4}$$

where $p_n = \exp(-j2\theta_n), \theta_n = \kappa_n \ell_n, \kappa_n$ is the propagation constant of the nth mode, and ℓ_n is the distance measured from the reference plane in the nth transmission line towards the modal short circuit in P. Initially the distances ℓ_n are to be regarded as independently variable, though various physically significant constraints will be imposed subsequently.

The resonance relation (2.3) is an N-linear expression in the p_n, being a sum of all possible products of these quantities, each with a principal minor of S as coefficient. The general results are exemplified by the simplest case N=2. Through appropriate choice of reference planes the scattering matrix of any lossless reciprocal 2-port may have Im S_{11}=0 and Re S_{11}-0. The resonance relation (3) then becomes

$$p_1 p_2 - S_{11} - 1 + S_{11} p_2 - 1 = 0 \qquad (2.5a)$$

or, in terms of the more familiar [15] variables θ_1 and θ_2,

$$\tan \theta = \gamma \tan \theta_2 \qquad (2.5b)$$

where $-\gamma = (1+S_{11})/(1-S_{11})$.

Now consider that the two ports of S are in fact merely distinct modes of a single waveguide, M-1. Then, if an ordinary plunger is inserted into the waveguide, the positions of the corresponding modal short circuits are constrained by

$$\theta_1 = \kappa_1 \ell + c_1 \quad , \quad \theta_2 = \kappa_2 \ell = c_2 \; . \qquad (2.6)$$

The periodicity of the resonance relation equation (2.5) permits all intersections with the constraint line (2.6) to be conveniently viewed by reducing all parts of (2.6) to the first zone $0 \le \theta_1 < \pi$, $0 \le \theta_2 < \pi$.

Assuming $\kappa_1/\kappa_2 = q_1/q_2$, where q_1 and q_2 are relatively prime integers yields the number of interactions of (2.5) and (2.6) is

$$v = q_1 + q_2 . \qquad (2.7)$$

If κ_1/κ_2 is irrational, the intersections are everywhere dense.

3. THE DIRECTIONAL COUPLER CORE OF AN ARBITRARY, LOSSLESS, RECIPROCAL 4-PORT

The ideal directional coupler which might lurk within a particular 4-port has appeared somewhat reluctant in the sense that no formula was available which would yield the characteristics of this coupler in terms of the conventional parameters of the given 4-port. We present the necessary formulas, [11] and note the generalization of the original result which became apparent in the derivation.
A 4-port, Fig. 3.1, may be described by its normalized voltage-scattering matrix, $S = [S_{ij}]$,

$$b_i = \sum_j S_{ij} a_j \qquad (3.1)$$

or, equivalently, by a scattering-transfer matrix, $T = [T_{ij}]$,

$$\begin{bmatrix} b_3 \\ a_3 \\ \hline b_4 \\ a_4 \end{bmatrix} = T \begin{bmatrix} a_1 \\ b_1 \\ \hline a_2 \\ b_2 \end{bmatrix} \tag{3.2}$$

Both Fig. 1 and (2) indicate a particular grouping of the four ports of (1) into in-ports and out-ports: other groupings are, of course, possible; one formal way of obtaining these is a simple renumbering of terminals.

Now consider the scatter-transfer matrix representation (2) of an ideal directional coupler. There are three possible combinations

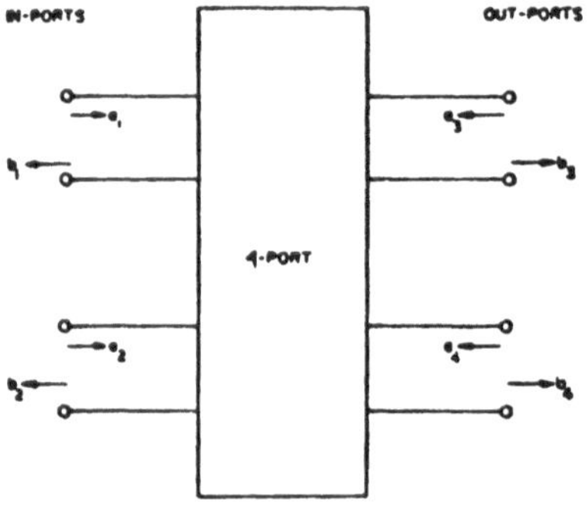

Fig. 3.1 General 4-port

of decoupled ports. Corresponding to the three possible forms of the scattering matrix for these ideal cases (at suitable reference planes),

$$S_1 = \begin{bmatrix} 0 & 0 & \alpha & j\beta \\ 0 & 0 & j\beta & \alpha \\ \alpha & j\beta & 0 & 0 \\ j\beta & \alpha & 0 & 0 \end{bmatrix}, \quad S_2 = \begin{bmatrix} 0 & \alpha & 0 & j\beta \\ \alpha & 0 & j\beta & 0 \\ 0 & j\beta & 0 & \alpha \\ j\beta & 0 & \alpha & 0 \end{bmatrix}, \quad S_3 = \begin{bmatrix} 0 & j\beta & \alpha & 0 \\ j\beta & 0 & 0 & \alpha \\ \alpha & 0 & 0 & j\beta \\ 0 & \alpha & j\beta & 0 \end{bmatrix}; \tag{3.3}$$

one has three forms of T.

Conservation of energy would imply $\alpha^2 + \beta^2 = 1$. Now it is clear that on interconnection with 2-ports at each port a transfer matrix, partitioned as shown,

$$T_i = \begin{bmatrix} t_{11} & t_{12} \\ t_{21} & t_{22} \end{bmatrix} \tag{3.4}$$

goes over into

$$T = \begin{bmatrix} At_{11}C & At_{12}D \\ Bt_{21}C & Bt_{22}D \end{bmatrix} \tag{3.5}$$

wherein A, B, C and D are transfer representations of the 2-ports. Thus. the subdeterminants are invariants provided det(AC) = det(AD) = det(BC) = det(BD) = 1. The latter is true for any reciprocal 2-ports, lossless or dissipative. [7] The invariants are, by inspection,

$$\alpha^2 \ , \quad \beta^2 = 1-\alpha^2 \ ; \tag{3.6a}$$

$$\left(\frac{\alpha}{\beta}\right)^2 = \frac{\alpha^2}{1-\alpha^2} \ , \qquad 1 + \left(\frac{\alpha}{\beta}\right)^2 = \frac{1}{1-\alpha^2} \ ; \tag{3.6b}$$

$$\frac{\alpha^2+\beta^2}{\alpha^2} = \frac{1}{\alpha^2} \ , \qquad \left(\frac{\beta}{\alpha}\right)^2 = \frac{1-\alpha^2}{\alpha^2} \ ; \tag{3.6c}$$

Evidently, the transfer matrices corresponding to a general scattering matrix have subdeterminants which correspond to the invariants (3.6). For the grouping of the ports indicated in (3.2), the determinants of the sub-matrices are

$$\frac{S_{24}S_{31} - S_{21}S_{34}}{S_{24}S_{13} - S_{23}S_{14}} \qquad \frac{S_{14}S_{32} - S_{12}S_{34}}{S_{14}S_{23} - S_{13}S_{24}}$$

$$\frac{S_{23}S_{41} - S_{21}S_{43}}{S_{23}S_{14} - S_{24}S_{13}} \qquad \frac{S_{13}S_{42} - S_{12}S_{43}}{S_{13}S_{24} - S_{14}S_{23}} .$$

The above relations are valid for arbitrary 4-ports, but the interpretation in terms of an imbedded ideal directional coupler is limited to the lossless reciprocal case. As previously mentioned, the analogous forms for the different groupings of ports may be obtained formally by permuting the port designations. The new expressions thus obtained are also invariant. The magnitudes of the several invariants in a specific case, in particular whether sums of differences equal unity, determine the proper choice of directional coupler terminals [see (3.6)]. Of the 2-ports A,B,C,D which convert the lossless, reciprocal 4-port to an ideal directional coupler, one may be chosen arbitrarily. The other three may then be obtained from (3.5). In particular one port may remain untuned.

4. THE ESSENTIAL NON-RECIPROCITY AND ESSENTIAL LOSS

In the evaluation of particular two-ports as components for use in a microwave system, it is pointed to distinguish certain "essential properties" of the component from those which may be adjusted by the addition of lossless reciprocal networks (tuners) at the ports. [9,10,16] These "essential properties" of a two-port, dissipation and nonreciprocal behavior, are explicitly exhibited by distinct bilaterally-matched elements in the modified Wheeler representation. [1] The general linear, nonreciprocal (not necessarily passive) two-port can always be decomposed into a reciprocal two-port in tandem with a nonreciprocal two-port, as shown in Fig. 4.1 and (4.2).

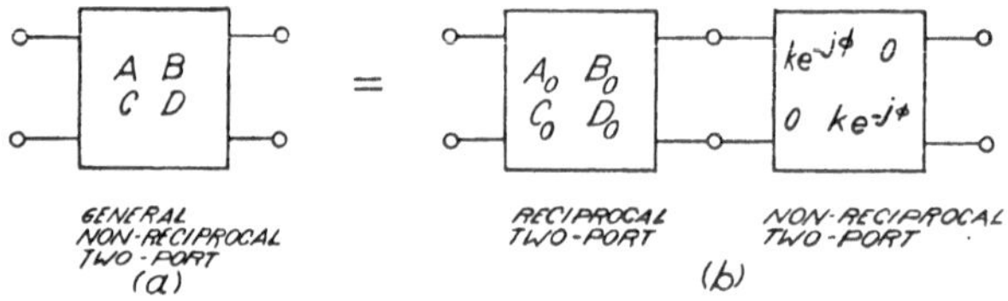

Fig. 4.1 Decomposition of general two-port

While this particular break-up is not the only one possible, it is unique when, in addition, it is required that the matrix representing the nonreciprocal two-port commute with all other matrices, i.e., that it be a scalar matrix. It can be seen that this requirement can, in general, be met when one recalls that reciprocal two-ports must obey the condition

$$A_0D_0-B_0C_0 = 1, \tag{4.1}$$

and when the matrix representing the general non-reciprocal two-port is decomposed as follows:

$$\begin{bmatrix} A & B \\ C & D \end{bmatrix} = \begin{bmatrix} A_0 & B_0 \\ C_0 & D_0 \end{bmatrix} \begin{bmatrix} ke^{-j\Phi} & 0 \\ 0 & ke^{-j\Phi} \end{bmatrix} \tag{4.2}$$

This decomposition assures that $A_0D_0-B_0C_0 \equiv 1$ and yields the definition $k \exp\{-j\Phi\} \equiv \sqrt{AD-BC}$. The ambiguity in phase angle introduced by the square root is trivial.

The nonreciprocal two-port defined by the scalar matrix in (4.2) has been termed "ideal amplifier phase shifter," and in a somewhat different context, "ratio repeater." Its effect on the properties of a cascade of networks is completely independent of its position within the cascade.

The modified Wheeler network is a suitable representation of dissipative, reciprocal, passive, linear two-ports and consists of three lossless transmission lines ℓ_1, ℓ_2, and ℓ, two ideal transformers n_1:1 and 1:n_2, and an attenuator, scattering matrix

$$\begin{bmatrix} 0 & K \\ K & 0 \end{bmatrix}, \quad 0 \leq K \leq 1. \tag{4.4}$$

Passive non-reciprocal two-ports can now be conveniently represented as shown in Fig. 4.2 a or b. It is recognized that the particular break-up employed is only one of a variety of possible ones.

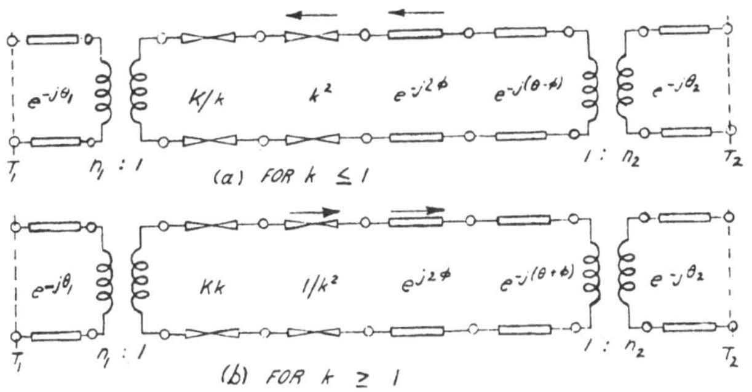

Fig. 4.2 Extension of the modified Wheeler network to the nonreciprocal case.

These concepts have been generalized to the 2N-port case [7,8].

5. SYMMETRIES

A systematic evaluation of waveguide junctions from the standpoint of their conformance to certain symmetries may be developed. While ideal symmetrical junctions have received extensive treatment in recent literature, little account is taken, in these papers, of the fact that all actual junctions are, to some degree, asymmetrical. On specialized consideration of a particular junction, engineers have commonly improvised parameters descriptive of junction asymmetries. It generally remained uncertain whether or not a set of asymmetry parameters, introduced ad hoc, was either complete (in the sense that any arbitrary asymmetry could be described) or minimal (in the sense that no linear relations subsisted among elements of the set). These questions are resolved simply and, we believe, naturally in terms of the same theoretical framework which has been successfully employed in the analysis of ideal symmetrical junctions; i.e., the theory of linear transformations and representations of finite point groups. [2,14]

Alternative (scattering) descriptions will be developed, entirely equivalent in point of generality to the conventional scattering matrix, but especially appropriate to junctions conforming to particular symmetries. The N^2 complex parameters entering into such a description fall into one of two categories: 1) those parameters which necessarily vanish when the junction represented actually conforms to the particular symmetries which determined the description; and 2) the parameters which do not necessarily vanish in that case. Those in the first category are denoted asymmetry parameters, and those in the second, symmetry parameters. The elements of the conventional scattering matrix will be expressed (linearly) in terms of the asymmetry and symmetry parameters, and conversely. [4]

Consider a hybrid-T junction such as, for example, shown in Table

TABLE I

HYBRID T JUNCTION F SYMMETRY PLANE

$$F = \begin{bmatrix} 0 & 1 & 0 & 0 \\ 1 & 0 & 0 & 0 \\ 0 & 0 & 1 & 0 \\ 0 & 0 & 0 & -1 \end{bmatrix} \qquad T = \frac{1}{\sqrt{2}} \begin{bmatrix} 1 & 0 & 1 & 0 \\ 1 & 0 & -1 & 0 \\ 0 & \sqrt{2} & 0 & 0 \\ 0 & 0 & 0 & \sqrt{2} \end{bmatrix}$$

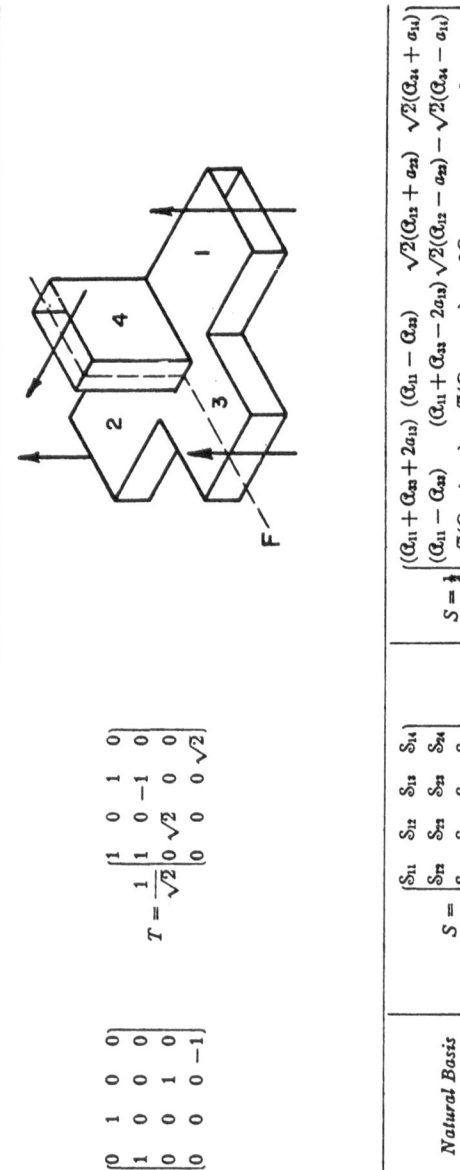

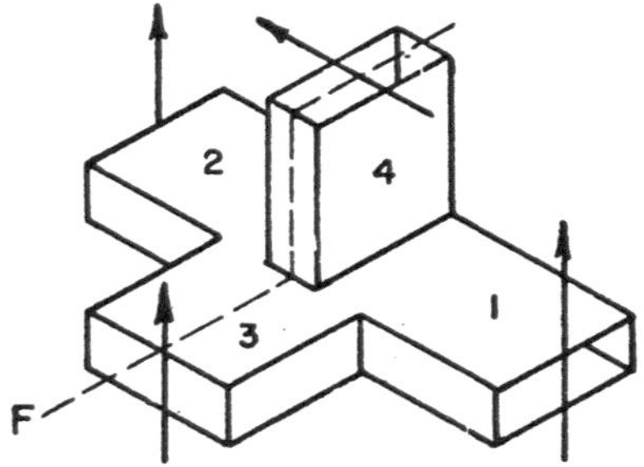

Natural Basis	$S = \begin{bmatrix} \mathcal{S}_{11} & \mathcal{S}_{12} & \mathcal{S}_{13} & \mathcal{S}_{14} \\ \mathcal{S}_{12} & \mathcal{S}_{22} & \mathcal{S}_{23} & \mathcal{S}_{24} \\ \mathcal{S}_{13} & \mathcal{S}_{23} & \mathcal{S}_{33} & \mathcal{S}_{34} \\ \mathcal{S}_{14} & \mathcal{S}_{24} & \mathcal{S}_{34} & \mathcal{S}_{44} \end{bmatrix}$	$S = \frac{1}{2} \begin{bmatrix} (\mathcal{A}_{11} + \mathcal{A}_{33} + 2a_{13}) & (\mathcal{A}_{11} - \mathcal{A}_{33}) & \sqrt{2}(\mathcal{A}_{12} + a_{23}) & \sqrt{2}(\mathcal{A}_{34} + a_{14}) \\ (\mathcal{A}_{11} - \mathcal{A}_{33}) & (\mathcal{A}_{11} + \mathcal{A}_{33} - 2a_{13}) & \sqrt{2}(\mathcal{A}_{12} - a_{23}) & -\sqrt{2}(\mathcal{A}_{34} - a_{14}) \\ \sqrt{2}(\mathcal{A}_{12} + a_{23}) & \sqrt{2}(\mathcal{A}_{12} - a_{23}) & 2\mathcal{A}_{22} & 2a_{24} \\ \sqrt{2}(\mathcal{A}_{34} + a_{14}) & -\sqrt{2}(\mathcal{A}_{34} - a_{14}) & 2a_{24} & 2\mathcal{A}_{44} \end{bmatrix}$
Transformed Basis	$S_F = \begin{bmatrix} \mathcal{A}_{11} & \mathcal{A}_{12} & a_{13} & a_{14} \\ \mathcal{A}_{12} & \mathcal{A}_{22} & a_{23} & a_{24} \\ a_{13} & a_{23} & \mathcal{A}_{33} & \mathcal{A}_{34} \\ a_{14} & a_{24} & \mathcal{A}_{34} & \mathcal{A}_{44} \end{bmatrix}$	$S_F = \frac{1}{2} \begin{bmatrix} (\mathcal{S}_{11} + \mathcal{S}_{22} + 2\mathcal{S}_{12}) & \sqrt{2}(\mathcal{S}_{13} + \mathcal{S}_{23}) & (\mathcal{S}_{11} - \mathcal{S}_{22}) & \sqrt{2}(\mathcal{S}_{14} + \mathcal{S}_{24}) \\ \sqrt{2}(\mathcal{S}_{13} + \mathcal{S}_{23}) & 2\mathcal{S}_{33} & \sqrt{2}(\mathcal{S}_{13} - \mathcal{S}_{23}) & 2\mathcal{S}_{34} \\ (\mathcal{S}_{11} - \mathcal{S}_{22}) & \sqrt{2}(\mathcal{S}_{13} - \mathcal{S}_{23}) & (\mathcal{S}_{11} + \mathcal{S}_{22} - 2\mathcal{S}_{12}) & \sqrt{2}(\mathcal{S}_{14} - \mathcal{S}_{24}) \\ \sqrt{2}(\mathcal{S}_{14} + \mathcal{S}_{24}) & 2\mathcal{S}_{34} & \sqrt{2}(\mathcal{S}_{14} - \mathcal{S}_{24}) & 2\mathcal{S}_{44} \end{bmatrix}$

I. The asymmetry parameters for this junction may be determined by measuring the elements of the scattering matrix $S = (S_{ij})$ and then substituting in the second matrix listed in the second column,

$$a_{13} = \frac{1}{2}(S_{11} - S_{22}) = \text{reflection difference},$$

$$a_{14} = \frac{1}{2}(S_{14} + S_{24}) = \text{E-arm balance depth},$$

$$a_{23} = \frac{1}{2}(S_{13} - S_{32}) = \text{H-arm balance depth},$$

$$a_{24} = S_{34} = \text{E-H arm isolation}$$

Fig. 5.1 indicates how these asymmetry parameters might be measured directly provided the network $S\{T_F\}$, were available.

$$S\{T_F\} = \frac{1}{\sqrt{2}} \left[\begin{array}{cccc|cccc} 0 & 0 & 0 & 0 & 1 & 1 & 0 & 0 \\ 0 & 0 & 0 & 0 & 0 & 0 & \sqrt{2} & 0 \\ 0 & 0 & 0 & 0 & 1 & -1 & 0 & 0 \\ 0 & 0 & 0 & 0 & 0 & 0 & 0 & \sqrt{2} \\ \hline 1 & 0 & 1 & 0 & 0 & 0 & 0 & 0 \\ 1 & 0 & -1 & 0 & 0 & 0 & 0 & 0 \\ 0 & \sqrt{2} & 0 & 0 & 0 & 0 & 0 & 0 \\ 0 & 0 & 0 & \sqrt{2} & 0 & 0 & 0 & 0 \end{array}\right], \qquad (5.1)$$

The equivalent circuit of $S\{T_F\}$, shown in Fig. 5.1, consists of an ideal hybrid-T and two direct connections, as may be verified by inspection. Thus, if a suitable high-quality hybrid-T junction is available, the asymmetry parameters of a second hybrid-T junction may be measured directly by connecting these two as required by the terminal markings for Fig. 5.1.

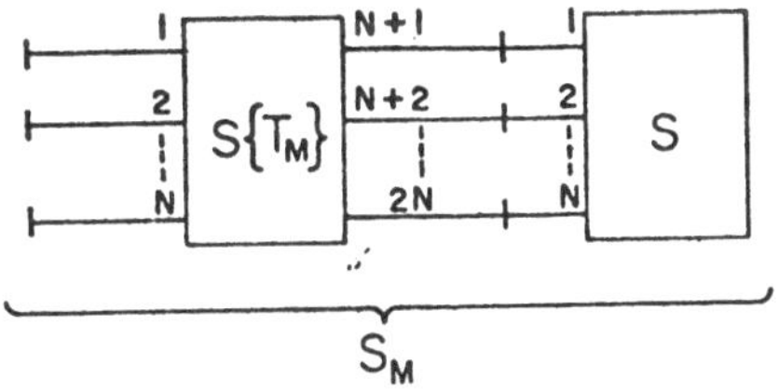

Fig. 5.1 Network representation of S_M

6. CONCLUDING REMARKS

The several techniques described in this paper do not represent an exhaustive survey. From the outset, pulse techniques and frequency domain techniques were ruled beyond the scope of this paper. Interesting methods for providing phase and amplitude information from power measurements alone, 6-port methods [5], are now being developed.

7. REFERENCES

[1] H.M. Altschuler and W.K. Kahn, "Nonreciprocal Two-Ports Represented by Modified Wheeler Networks," Trans. I.R.E. PGMTT, Vol. MTT-4, pp. 228-233, Oct. 1956

[2] B.A. Auld, "Applications of Group Theory in the Study of Symmetrical Waveguide Junctions," Stanford University, Stanford, Cal., MLR-157, March 1952

[3] W.-M. Boerner, A.K. Jordan, I.W. Kay, "Special Issue on Inverse Methods in Electromagnetics," Vol. AP-29, No. 2, March 1981

[4] M. Cohen and W.K. Kahn, "Analytical Asymmetry Parameters for Symmetrical Waveguide Junctions," IRE Trans. on Microwave Theory and Techniques, Vol. MTT-7, No. 4, Oct. 1959

[5] G.F. Engen, "The 6-port reflectometer: An alternative network analyser," IEEE Trans. on Microwave Theory and Tech., Vol. MTT-25, pp. 1075-1080, Dec. 1977

[6] L.B. Felsen, W.K. Kahn and L. Levey, "Measurement of Two-Mode Discontinuities in a Multimode Waveguide by a Resonance Technique," IRE Trans. on Microwave Theory and Techniques, Vol MTT-7, No. 1, pp. 102-110, Jan. 1959

[7] L.B. Felsen and W.K. Kahn, "Transfer Characteristics of 2N-Port Networks," Presented at the "Symposium on Millimeter Waves," March 31, April 1 & 2, 1959. Vol. IX, MRI Symposia Series, New York: Polytechnic Press, Polytechnic Institute of Brooklyn, pp. 477-512, 1960

[8] H. Kurss, "Some General Multiport Representations," Proceedings of the Symposium on Millimeter Waves, MRI Symposia Series Vol. IX, New York: Polytechnic Press, Polytechnic Institute of Brooklyn, pp. 577-593, 1960

[9] H.A. Haus and R.B. Adler, "Circuit Theory of Linear Noisy Networks," Technology Press and John Wiley and Sons, Inc., New York, 1959

[10] H.A. Haus, "Equivalent circuit for a passive nonreciprocal network," J. Applied Phys., Vol. 25, pp. 1500-1502, Dec. 1954

[11] W.K. Kahn and R.L. Kyhl, "The Directional Coupler Core of an Arbitrary, Lossless, Reciprocal 4-Port," Proc. IRE, Vol. 49, No. 11, pp. 1699-1700, Nov. 1961

[12] D.M. Kerns, "Analysis of symmetrical waveguide junctions," J. Res., NBS, Vol. 46, pp. 267-282, April 1951

[13] C.G. Montgomery, R.H. Dicke and A.M. Purcell, "Principles of Microwave Networks," Rad. Lab. Series, Vol. 8, New York: McGraw-Hill, 1948

[14] A.E. Pannenborg, "On the scattering matrix of symmetrical waveguide junctions," Phillips Res. Repts., Vol. 7, pp. 133, April 1952

[15] A. Weissfloch, "Schaltungstheorie und Messtechnik des Dezimeter und Centimeter Wellengebietes," Basel and Stuttgart: Birkhauser Verlag, 1954

[16] H.A. Wheeler, Wheeler Monographs, Vol. I, Nos. 9, 10. Great Neck, N.Y., Wheeler Labs, Inc., 1953

V.4 (MI.2)

HOLOGRAPHIC AND TOMOGRAPHIC IMAGING WITH MICROWAVES AND ULTRASOUND

A.P. Anderson* and M.F. Adams+

* Department of Electronic & Electrical Engineering, University of Sheffield.
\+ Formerly with the Department of Electronic & Electrical Engineering, University of Sheffield, now with Eikontech Ltd.

ABSTRACT

Microwave and acoustic holographic imaging experienced considerable interest during the period 1970-80 because of its diagnostic potential. The high lateral resolution of the CW/synthetic aperture approach vis-a-vis echo timing carries the penalty of poor depth resolution so that long wavelength holography was essentially two dimensional at that time. However, it does provide a 'phase image' which is particularly useful in reflector antenna diagnostics. The limitations of quadratic correction focusing and single view imaging (inherited from optical holography) have been removed by the adoption of inverse diffraction processes, for example, plane-to-plane transformation, and multiview data recording. The tomographic technique described is analogous to the back-projection and filtering process used in projection tomography and yields high resolution in three dimensions. Results from microwave and ultrasonic image scanners are presented. The 'phase tomogram' is seen to be, once more, a useful product of this approach and its significance has been investigated by computer simulation.

1. INTRODUCTION AND BACKGROUND

1.1 Long Wavelength Imaging with Coherent Waves

Long wavelength visualisation falls broadly into two classes; techniques which employ echo-timing in some form and therefore require substantial bandwidth capability of the system, and holographic-based techniques which are essentially monochromatic and require phase information. This paper is concerned with the second of these classes to which the restriction 'small scale'

W.-M. Boerner et al. (eds.), Inverse Methods in Electromagnetic Imaging, Part 2, 1077–1105.

Figure 1(a) : 1st microwave holographic image (1965) reproduced from ref. [1:2].

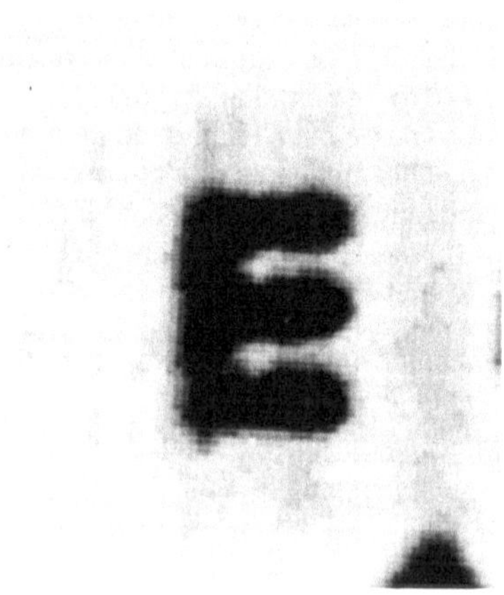

Figure 1(b) : Digitally reconstituted microwave image (1978)

usually applies, i.e. the effective data acquisition aperture will be of the order of tens or hundreds of wavelengths. Despite the implications of this restriction, microwave and acoustic holographic imaging experienced considerable interest during the period 1970-80 because of its diagnostic potential. The high lateral resolution capability vis-a-vis echo-timing systems was achieved with the penalty of poor depth resolution such that long wavelength holography was essentially two-dimensional. Nevertheless the 'natural' image format of the data presented to the human observer enhanced its diagnostic potency. Moreover, it offered another, often overlooked, diagnostic capability - the 'phase image'. Let us rapidly summarise the progress in that period.

1.2 Milestones in Microwave Holography Pre-1979

Historically, the first microwave and acoustic holograms were recorded in 1951 [1:1] before the advent of lasers or digital processing techniques. The first publicised demonstration of small scale microwave holographic imaging occurred in 1965 [1:2] and is reproduced in Figure 1.1(a). The object was a letter 'A' approximately 2 m high, 70 wavelengths under illumination by X-band radiation. An on-axis or 'in-line' Fresnel region hologram with an aperture 3 m x 3 m was mapped by recording the field intensity and converting it to a reduced size optical transparency for image reconstruction by laser light. The optical image processing stage provides first order quadratic phase correction of the Fresnel field. The subsequent development of microwave holography is described in references [1:3] [1:4], the most significant factor being the replacement of optical diffraction by digital image processing. The microwave image of a letter 'E' only eleven wavelengths high seen in Figure 1.1(b) demonstrates the improvement obtained. The inversion algorithms used to reconstitute image fields reproduce the quadratic correction previously performed by the optical system.

Perhaps the most useful diagnostic application of microwave holographic imaging arising from that period is the metrology of large reflector antennas [1:5]. Figure 1.2 shows examples of the results obtained in 1978 [1:6]. The holographic aperture is synthesised by scanning the antenna through a few degrees around boresight and recording the complex received field from a fixed source in the Fresnel region. The amplitude image corresponding to the current distribution and the phase image corresponding to errors in the reflector profile provide valuable diagnostic information. The imaging technique can also be carried out using a synchronous satellite source [1:7]. A recent measurement of the Max Planck Institute 100 m telescope using the OTS satellite shown in Figure 1.3(a)-(d) indicates the sensitivity of the microwave technique and the high quality of this antenna surface.

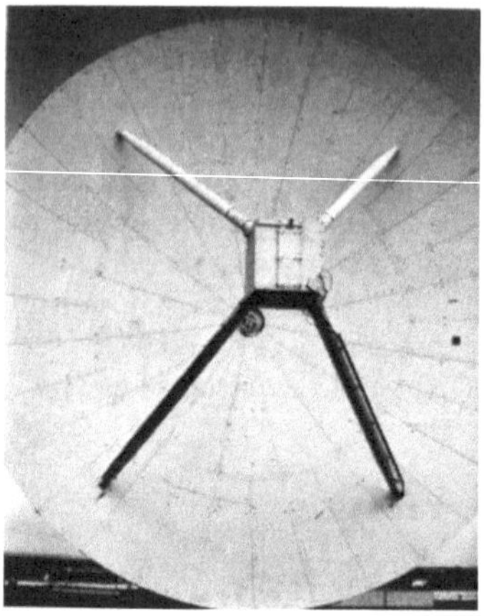

Figure 1.2(a) : The 25 m reflector antenna at Chilbolton, Hampshire, England.

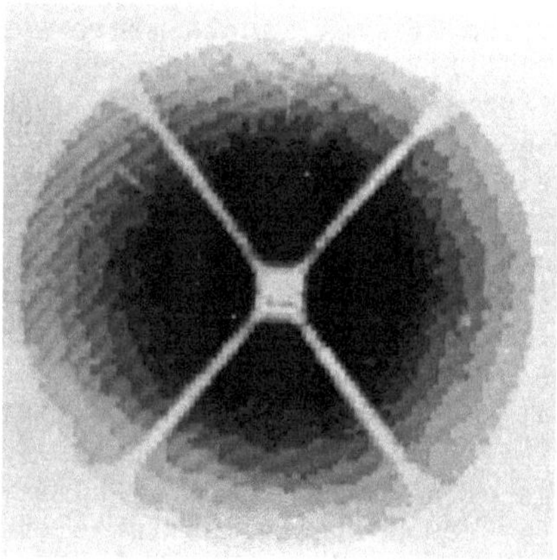

Figure 1.2(b) : The aperture amplitude distribution (16 levels 2 dB $level^{-1}$ measured from a ground based source 1 km distant.

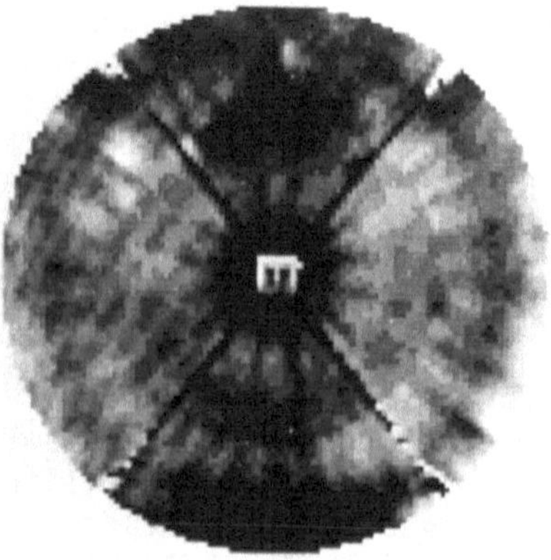

Figure 1.2(c) : The profile error (16 levels over range ± 4.5 mm) showing astigmatic distortion due to gravity.

Figure 1.3(a) : The Max Planck Institute 100 m reflector antenna at Effelsburg, W. Germany.
See over for Figures (b)-(d)

1.3 Limitations of Quadratic Correction Focusing and Single View Imaging

The inversion of a diverging wavefront into a focused image using the Fourier transform and quadratic correction is derived from optics (see, for example [1:8]). If we consider, for ease of the following representation, a 1-D holographic aperture illuminated from a source as shown in Figure 1.4(a) the diffraction field on the α-axis is given by [1:4]

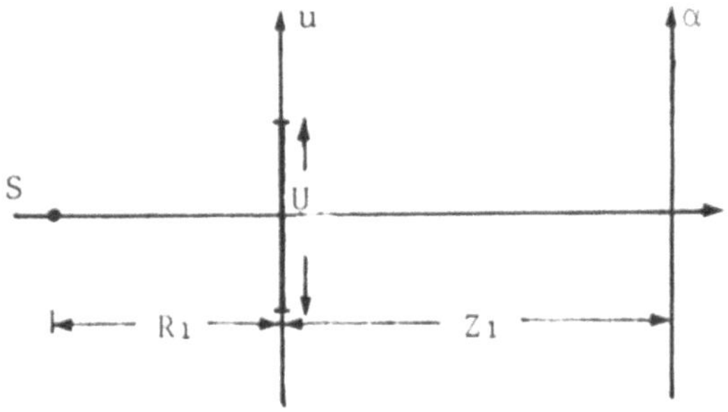

Figure 1.4(a) : One-dimensional coordinates for image reconstruction.

Figure 1.3(b) : The reference antenna in position for measurements using the OTS satellite source.

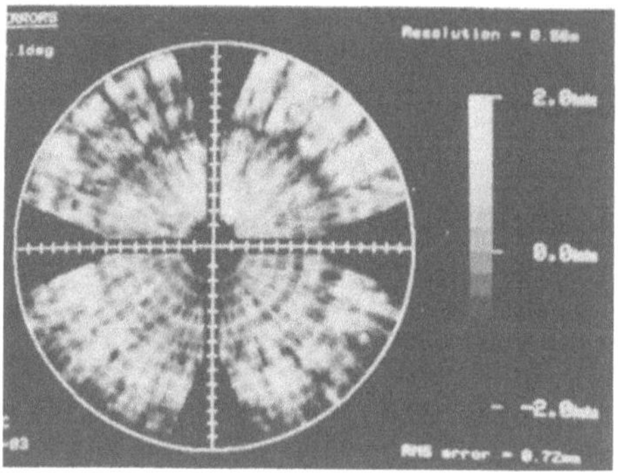

Figure 1.3(c) : The profile error map (16 levels over range ± 2.0 mm) .

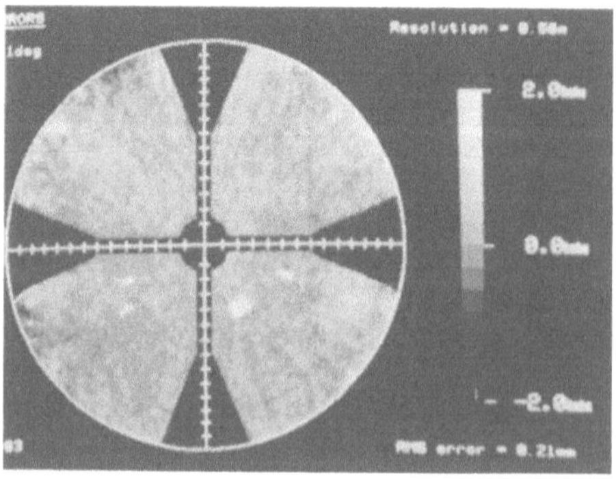

Figure 1.3(d) : Resolution of panels deliberately displaced or titled through 2.0 mm.

(Photographs reproduced by courtesy of Max-Planck Institute fur Radioastronomie & Eikontech Limited.)

$$C(\alpha) = A_1 \int_\xi \int_u D(\xi) \exp\left[j\frac{k}{2}\left(\frac{|\xi|^2}{Z_1} + \frac{|\alpha|^2}{Z_2}\right)\right]$$

$$\exp\left[-jk\,u\left(\frac{\xi}{Z_1} + \frac{\alpha}{Z_2}\right)\right]$$

$$\exp\left[-jk\,\frac{u^2}{2}\left(\frac{1}{Z_1} - \frac{1}{R_1} + \frac{1}{R_2} + \frac{1}{Z_2}\right)\right]d\xi\, du \qquad (1.1)$$

where Z_1, R_1 are the distances of the object and reference source from the hologram in the recording step and ξ is the object coordinate. The Fresnel zone quadratic approximation is removed by setting

$$\frac{1}{Z_1} - \frac{1}{R_1} + \frac{1}{R_2} + \frac{1}{Z_2} = 0$$

i.e. the focusing condition. Hence the image

$$C(\alpha) + A_2 \int_\xi D(\xi) \exp\left[j\frac{k|\xi|^2}{2Z_1}\right] \text{sinc}\, k\left(\frac{\xi}{Z_1} + \frac{\alpha}{Z_2}\right) U\, d\xi \qquad (1.2)$$

The image may be considered 'tolerably aberrated' provided $U^4 < \lambda Z_1^3$.

In digital image processing, $C(\alpha)$ may be recovered with a quadratic correction, followed by the FFT of the sampled data within U. The focal region for the case of a uniform 1-D distribution within U and zero outside is shown in Figure 1.4(b) and illustrates the poor resolution in Z. Therefore, the permissible data window size, object recording distance and resolution are limited accordingly. These limitations to the wider application of small scale long wavelength image diagnostics were the legacy of the origins in optical holography which, in our own experience, was finally shed with the introduction of inverse diffraction and multiview recording methods [1:9].

2. INVERSE DIFFRACTION 'PLANE-TO-PLANE' IMAGING TECHNIQUES

The limitations of Fresnel focusing in the near-field may be overcome by inverse diffraction (often known as backwards propagation). Any propagating field, $u(x,y,z)$, can be decomposed into an angular spectrum of plane waves [2:1], i.e. an equivalent distribution of plane waves propagating in all possible directions. If the complex field distribution recorded in a plane z_j is $u_j(x_j,y_j,z_j)$ then its angular spectrum of plane waves

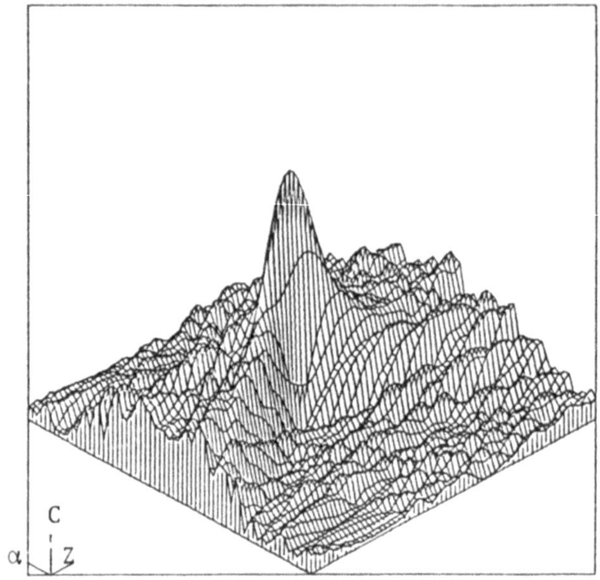

Figure 1.4(b) : Topological representation of the focal region intensity distribution.

is the 2D Fourier transform (2DFT) of u_j and is designated $\hat{U}_j(s_x,s_y;z_j)$. The value of $\hat{U}_j(s_x,s_y;z_j)$ at spatial frequency s_x,s_y is the plane wave component propagating with direction cosines $\lambda s_x,\lambda s_y$. It has been shown [2:2] that the propagation process may be regarded as a linear filter acting on this spectrum so that the spectrum of the field distribution in any other plane, e.g. z_ℓ (where $z_\ell > z_j$ and the field is propagating in the positive z direction) is given by :

$$U_\ell(s_x,s_y;z_\ell) = U_j(s_x,s_y;z_j).B_{j\ell}(s_x,s_y,z_j - z_\ell) \quad (2.1)$$

where $$B_{j\ell}(s_x,s_y;z_j - z_\ell) = \exp\{jk\,m\,(z_j - z_\ell)\} \quad (2.2)$$

and $$m = \sqrt{1 - (\lambda s_x)^2 - (\lambda s_y)^2}\ .$$

The form of $B_{j\ell}$ for a distance $z_j - z_\ell = 5\lambda$ is shown in Figure 2.1 where it is noticeable that the propagation process effectively bandlimits the field distribution to spatial frequencies within $\pm\ 1/\lambda$. The imaging process (by which the distribution is propagated back towards its source) is obtained when $z_\ell < z_j$ in eqn.(2.1). However, because the evanescent waves have been attenuated by the propagation process it is usual to assume that they cannot be recovered in the imaging stage. Therefore the filter for inverse diffraction or focusing becomes $B'_{j\ell}(s_x,s_y,z_j - z_\ell)$ where :

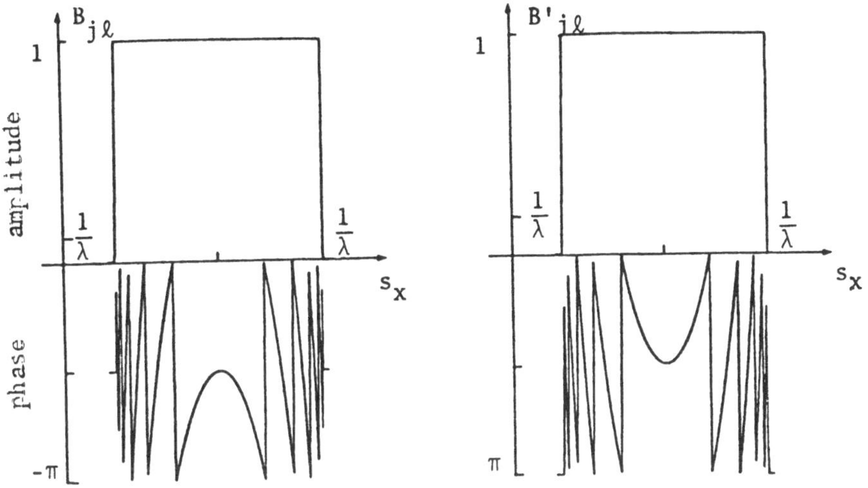

Figure 2.1(a) : Filter for forward propagation $B_{j\ell}(s_x, s_y, s_j - z_1)$ for $z_j - z_1 = 5\lambda$.
(b) : Filter for backward propagation (inverse diffraction) $B'_{j\ell}$.

$$B'_{j\ell}(s_x, s_y, z_j - z_\ell) = \exp\{-jkm|z_j - z_\ell|\} \qquad s_x^2 + s_y^2 \leqslant 1/\lambda^2$$
$$= 0 \qquad s_x^2 + s_y^2 > 1/\lambda^2 \tag{2.3}$$

The form of $B'_{j\ell}$ for $z_j - z_\ell = -5\lambda$. The final stage of the imaging process is to inverse 2DFT $\hat{U}_\ell$ to yield the field distribution in the plane z_ℓ, which is shown in Figure 2.2. When the plane z_ℓ coincides with the source of the scattered field distribution then it can be considered a diffraction-limited focused image of that source. In the context of the simple planar imaging considered in this section the source is often considered as a reflector illuminated from a fixed direction by a plane wave. However, in the practical case it is often desirable to scan the illumination source and the receiver as a unit and this is accounted for by effectively halving the wavelength in eqns. (2.1) - (2.3).

The above discussion of inverse diffraction has implied that the measured data is known continuously on an infinite aperture. In practice this is neither possible nor desirable. Since the propagation process bandlimits the data to spatial frequencies within $\pm 1/\lambda$ it is, of course, unnecessary to sample the data at

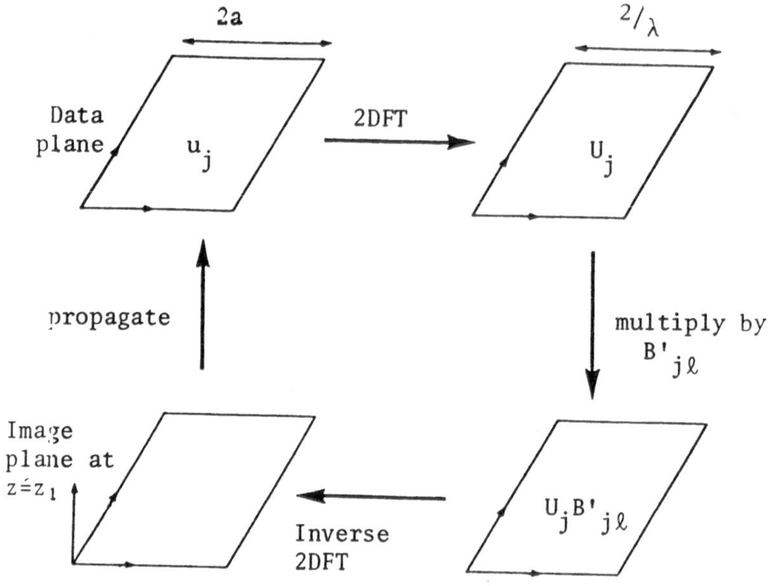

Figure 2.2 : Data processing stages.

intervals less than $\lambda/2$. In addition the finite recording aperture effectively further bandlimits the data which may allow a further increase in the sampling interval. It may be shown [2:3] that, given the point source and recording aperture shown in Figure 2.3, the recorded data will be approximately bandlimited at spacial frequencies S_c^+ and S_c^- given by

$$S_c^+ = \frac{\sin\theta^+}{\lambda} \tag{2.4}$$

$$S_c^- = \frac{-\sin\theta^-}{\lambda} \tag{2.5}$$

For a point source at the origin $\theta^+ = \theta^- = \theta = \tan^{-1}(a/z)$ and the total bandwidth, S_t, is given by :

$$S_t = |S_c^+ - S_c^-| = \frac{2\sin\theta}{\lambda} \tag{2.6}$$

The resolution for this combination of a and z will then be $1/S_t$. Also, since the data is bandlimited to S_t, the aperture sampling interval need not be less than $1/S_t$. In practice the data may be measured with a receiving sensor which is directional and may further reduce the high frequency content of the data and so relax the sampling requirements.

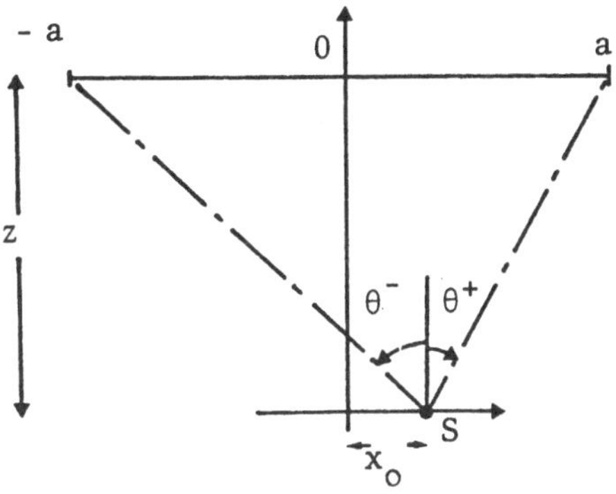

Figure 2.3 : Dimensions controlling aperture bandwidth.

It may be noted that as the aperture moves into the Fresnel region so that a << z then eqn. (2.6) becomes $S_t = 2a/z\lambda$ and the required sampling interval becomes $z\lambda/2a$ which is, of course, consistent with the requirements for data recording in section 1.

These data recording and processing techniques have been applied to a wide variety of objects. Figure 2.4 shows the microwave image of a musical instrument obtained with Q-band radiation at a recording distance of 0.6 m. The resemblance to an optical image is apparent since 'high spots' are expected from both wavelength regions for this smooth object.

A useful application of microwave imaging is the mapping of sub-surface pipes and cables [1:8]. Figure 2.5 shows some recent results indicating the presence of two pipes (one metal, one plastic) in the form of an X at a depth of 0.3 m in wet soil (∿ 25% moisture). The image was obtained at 430 MHz and represents an area of 2 m x 2 m. The plane-to-plane inverse technique is particularly useful for processing data scans close to the object region.

3. SYNTHETIC APERTURE OR DIFFRACTION TOMOGRAPHY

3.1 Introduction to 'SAT'

The major limitation of the techniques discussed in sections 1 and 2 is that objects are assumed to be simple flat reflectors and so the images generated from 2-D scans are essentially 2-D. Some progress in overcoming these limitations is made in the tomographic mode of imaging in which the ability of the radiation to penetrate the object is used to characterise its scattering structure more fully.

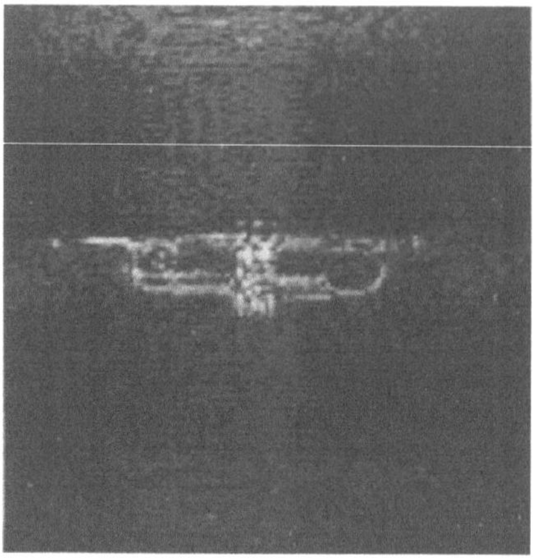

Figure 2.4 : Q-band image generated by a $B'_{j\ell}$ filter.

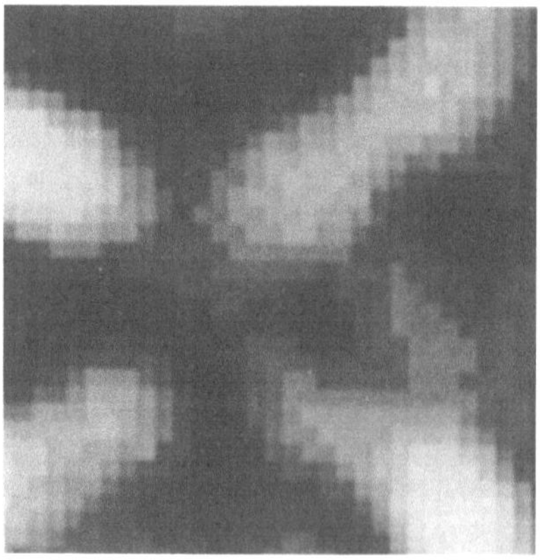

Figure 2.5 : 430MHz image of sub-surface pipes in wet soil generated by a $B'_{j\ell}$ filter.

These techniques may now be grouped under the general heading 'inverse scattering' since they generally involve, either directly or indirectly, a solution of the scalar wave equation for propagation in the medium. For a field distribution u_t and a source distribution ρ this may be written :

$$\nabla^2 u_t + \frac{\omega^2}{c^2} u_t = -\rho \qquad (3.1)$$

The source distribution ρ may be a region of physical sources (in which case the problem is considered as 'inverse source' [3:7]) or alternatively a region of inhomogeneity described by f and contained within a volume V_1. This case is the 'inverse scattering' problem and the source distribution is given by $\rho = f.u_t$. It is usual to consider that there is no closed form solution to eqn. (3.1) and so some simplifying approximations (e.g. Born, Rytov) are necessary. A solution has been attempted by workers in many fields. For example, in optics it was Wolf [3:2] who first derived a Fourier relationship between the object and the scattered field subject to the Born approximation.

In acoustics the consideration of 'inverse scattering' is becoming increasingly important [3:3]. Although the widely used 'B-scan' technique is naturally tomographic in that it provides an image of a slice of the object it is difficult to obtain quantitative information.

In electromagnetics inverse scattering might be considered to have its roots in the work on radar target shape identification [3:4] and, of course, considerable effort is still being applied to this problem [3:1]. However, in medical electromagnetic imaging there is little which may be considered as inverse scattering probably due to the severe limitations of the Born approximation in this case.

In essence 3D image reconstruction requires scans which use either multifrequency/echo-timing methods or multiview CW methods. The approach described here originates from synthetic aperture holography and combines the inverse scattering techniques introduced in section 2 with the multiview techniques which are common in X-ray CAT.

3.2 Image Space Solution

In general the solution of equation (3.1) for interaction of a scalar field u_t with a region of inhomogeneity contained within a volume V_1 can be written [3:5, 2:3] :

$$u_t(\underline{r}) = u_i(\underline{r}) + \int_{V_1} f(\underline{r_1})\, u_t(\underline{r_1})\, g\,(\underline{r},\underline{r_1})\, dV_1 \tag{3.2}$$

where as illustrated in Figure 3.1, $u_i(r)$ is the incident field and g is the ideal isotropic medium Green's function,

$$g(\underline{r},\underline{r_1}) = \frac{\exp(j\,k\,|\underline{r} - \underline{r_1}|)}{4\pi\,|\underline{r} - \underline{r_1}|} \tag{3.3}$$

and $k = 2\pi/\lambda$ is the propagation constant in the homogeneous medium and where $f(\underline{r_1})$ characterises the scattering strength of the inhomogeneities, e.g. for an object of (possibly complex) refractive index $\eta(\underline{r_1})$

$$f(\underline{r_1}) = [\eta^2(\underline{r_1}) - 1]\, k^2$$

The field u_t is measured over the plane at $z_1 = z_j$ and it will be assumed that the measurement aperture is infinite (although the effects of a finite aperture may be included as shown in section 2). The field due to scattering by the object, $u(\underline{r})$, is simply the integral part of eqn. (3.2), i.e. :

$$u(\underline{r}) = u_t(\underline{r}) - u_i(\underline{r}) \tag{3.4}$$

and on the measurement plane at $z_1 = z_j$ this field is defined as

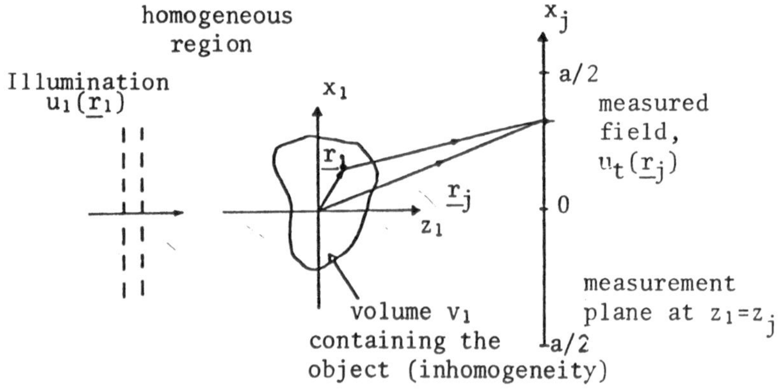

Figure 3.1 : Data acquisition scheme for one view of an object.

$u_j(x_j, y_j)$ where

$$u_j(x_j, y_j) = u(\underline{r}_j) = u(\underline{r})\big|_{z_1 = z_j} \tag{3.5}$$

The inverse scattering problem is to solve eqn. (3.2) for the scattering properties of the object, $f(r_1)$. However, no closed form solution exists so that it is necessary to adopt an approximation. The first-order approximation, often known as the Born approximation, is to assume that the object is a weak scatterer so that the field inside V_1 is given by the incident field u_i. A fictive object may now be modelled by an ensemble of point scatterers, each of which re-radiates the incident wave field modified by the scattering strength at that point, $f(r_1)$, as a diverging spherical wavefront. With this approximation the scattered field on the measurement plane becomes :

$$u_j(x_j, y_j) \simeq \int_{V_1} f(\underline{r_1}) u_i(\underline{r_1})\; g\,(\underline{r}_j, \underline{r_1})\; dV_1 \tag{3.6}$$

Our approach to the solution of eqn. (3.6) is a multiview technique analogous to the convolution back projection (CBP) process which is a well known method for the determination of an object from its projections along straight lines.

As shown in Figure 3.1 the illumination is considered to be a plane wave so that :

$$u_i(\underline{r_1}) = \exp(j\,k\,z_1) \tag{3.7}$$

The measured data or 'view' u_j may be used to generate a 3D image by a focusing or inverse diffraction technique which is equivalent to the 'back projection' process of CBP. The inverse diffraction technique is implemented as described in section 3 :- first a 2D Fourier transform (2DFT) is performed on the data :

$$\hat{U}_j(s_x, s_y) = \mathcal{F}_{2D}\,\{u_j(x_j, y_j)\} \tag{3.8}$$

which yields, combining equns. (3.6) - (3.8) :

$$u_j(s_x, s_y) = \int \hat{F}(s_x, s_y, z_1)\, \exp(j\,k\,z_1)\; G\,(s_x, s_y, z_j - z_1)\; d\,z_1 \tag{3.9}$$

where the 2DFT of the spherical wave function, g, is given by [2:1] :

$$G(s_x, s_y, z_j) = B_o \frac{\exp\{j\,k\,m(z_j - z_1)\}}{m} \tag{3.10}$$

where $m = \sqrt{1 - (\lambda s_x)^2 = (\lambda s_y)^2}$ and B_o is a frequency dependent constant and $\hat{F}$ is the 2DFT of the object, f. Secondly the inverse diffraction filter described in section 2 is applied although here it is modified to compensate both for the amplitude dependent term B_0/m in eqn.(3.10) and also for the illumination function. Thus the inverse diffraction filter B_j becomes :

$$B_j(s_x, s_y, z) = \frac{m}{B_o} \exp\{-j\,k\,m(z_j - z)\}\exp(-j\,k\,z) \tag{3.11}$$

when $s_x^2 + s_y^2 \leqslant 1/\lambda^2$ and $B_j = 0$ when $s_x^2 + s_y^2 > 1/\lambda^2$. Application of B_j has the effect of propagating the 2DFT of the measured data back towards the source. Thus in a plane z the frequency domain, $\hat{U}$, is given by eqns. (3.9) - (3.11) :

$$\hat{U}(s_x, s_y, z) = \int F(s_x, s_y, z)\exp\{-j\,k\,(z - z_1)\}\exp\{j\,k\,m(z - z_1)\}\,dz_1 \tag{3.12}$$

And, finally, on applying an inverse 2DFT to $\hat{U}$ a focused image in the plane z is achieved :

$$h(x,y,z)\Big|_z = \mathcal{F}^{-1}_{2D}\,\hat{U}(s_x, s_y, z)\} \tag{3.13}$$

Repetition of the focusing at all depth planes, z, yields a full 3D focused image of the object which, from eqns. (3.12) and (3.13) is given by a convolution integral,

$$h(x,y,z) = \iiint f(x_1, y_1, z_1)p(x - x_1, y - y_1, z - z_1)dx_1dy_1dz_1 \tag{3.14}$$

$$\text{where } p(x,y,z) = \iint \exp(-j\,k\,z)\exp(j\,k\,mz)\exp\{j2\pi(s_xX + s_yY)\}ds_xds_y \tag{3.15}$$

Thus the 3D image volume generated from one 'view' of the measured data may be considered a convolution of the object with a defocusing function, p, i.e.

$$h(x,y,z) = f(x,y,z) \circledast p(x,y,z) \tag{3.16}$$

A 2D computer simulation of the function p(x,z) is shown in Figure 3.2(a). This is a focused image of a point scatterer generated from a linescan of the field at $z_j = 16\lambda$ with aperture $a = 16\lambda$. This function is the single view imaging system point spread function (PSF).

Figure 3.2(a) illustrates that the imaging system PSF exhibits high lateral resolution but that the resolution in depth is considerably worse. This is overcome by a multiview approach in which the object is rotated by an incremental angle, $\Delta\alpha$ say, with respect to the fixed scan plane and illumination. Measurement of the scattered field is repeated for this new 'view' which is then focused to yield a 3D image volume. This is repeated for each of N views where $N = \alpha_{max}/\Delta\alpha - 1$ (ideally $\alpha_{max} = 360^o$ although in practice it may be reduced to as little as 180°). The 3D image volume for the mth view may be described by the convolution :

$$h_\alpha(x,y,z) = f(x,y,z) \circledast p_\alpha(x,y,z) \qquad (3.17)$$

where p_α is the defocusing function, p, rotated to the view angle α and $\alpha = (m-1)\Delta\alpha$. Figure 3.2(b) illustrates p_α in 2D for a view at $\alpha = 45^o$ rotation.

The next stage is the generation of an intermediate image volume by summation of the constituent images, h_α, generated at each view :

$$h_t(x,y,z) = \sum_{\alpha=0,\alpha_{max}} h_\alpha(x,y,z)$$

$$= f(x,y,z) \circledast \sum_{\alpha=0,\alpha_{max}} p_\alpha(x,y,z) \qquad (3.18)$$

The final stage is an attempt to remove the effect of the overall convolving function from eqn. (3.18). This is done by further pursuing the analogy with CBP where it is known that, in 2D, the overall convolving functions tends to a function $^1/r$ in the limit of an infinite number of views. The effect of this function can be removed by applying a 'ramp' filter in the frequency domain [3:6]. Similarly in diffraction tomography it is noted that the convolving function for one view (e.g. see Figure 3.2) may, in the region of the scatterer, be approximated by a line extending in the view direction. So that the overall function is a summation of rotated lines which, in the limit of a large number of views and in 2D, will tend to the function $^1/r$. Therefore the ramp filter will suffice as an attempt to deconvolve eqn. (3.18). Firstly a 2D Fourier transform is performed on the summed image :

$$H_t(s_x,s_z) = \mathcal{F}_{2D}\{h_t(x,y)\}$$

$$= F(s_x,s_z).\{\mathcal{F}_{2D}[\sum_{\alpha=0,\alpha_{max}} p_\alpha(x,z)]\}. \qquad (3.19)$$

H_t is then ramp filtered so that :

$$H_f(s_x,s_z) = H_t(s_x,s_z).\ A(s_x,s_z) \qquad (3.20)$$

where $A(s_x,s_z) = |\rho|$ and $\rho = \sqrt{s_x^2 + s_z^2}$ (3.21)

The final image is generated by an inverse 2D Fourier transform :

$$h_f(x,z) = \mathcal{F}_{2D}^{-1}\ \{H_f(s_x,s_z)\}. \qquad (3.22)$$

In practice it has been found necessary to apply a gaussian weighting to the ramp filter to avoid the over-emphasis of high spatial frequencies, so that :

$$A(s_x,s_z) = |\rho|.\exp(-2\pi\sigma^2\rho^2), \qquad (3.23)$$

where the parameter σ determines the resolution of the filter. If sufficient views are taken the overall resolution of the process is of the same order as the lateral resolution obtainable from one view and is equal in all directions.

3.3 Frequency Domain Interpretation of Diffraction Tomography

A further analogy with traditional projection tomography may be found in the frequency domain approach to diffraction tomography [3:3]. The frequency domain interpretation of projection tomography is based on the Projection Theory of Fourier transforms. The equivalent process for diffraction tomography may be obtained by performing a 3D Fourier transform on the focused image from one view as given by eqn. (3.16) :

$$H(s_x,s_y,s_z) = F(s_x,s_y,s_z).P(s_x,s_y,s_z) \qquad (3.24)$$

where

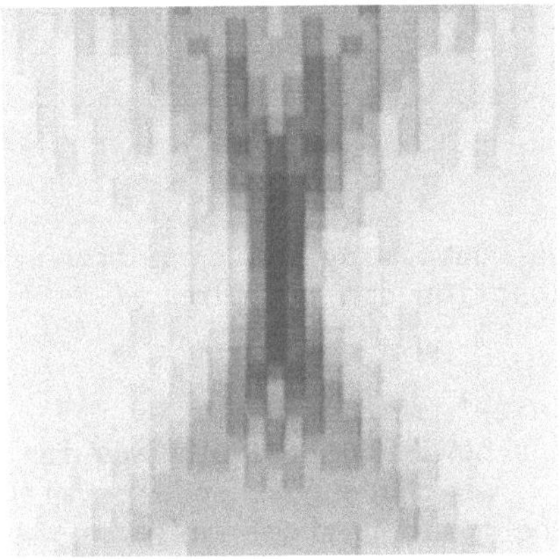

Figure 3.2(a) : Focused image of a point scatterer at the origin.

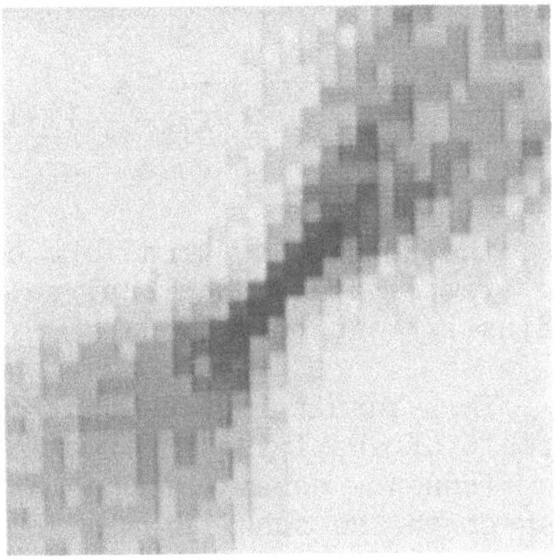

Figure 3.2(b) : Focused image of the point scatterer from a view at 45°.

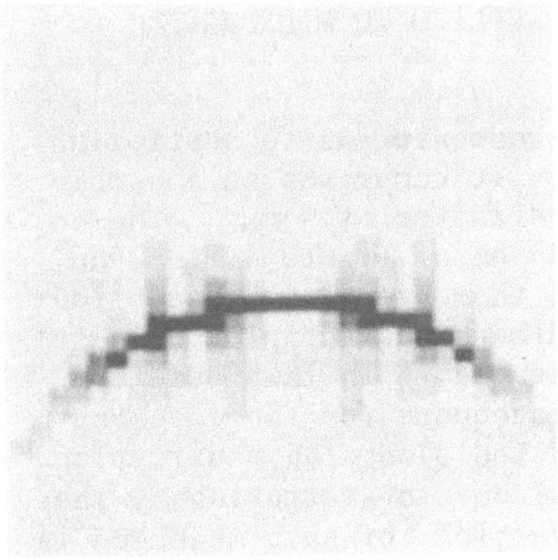

Figure 3.3(a) : 2-DFT of Figure 3.2(a).

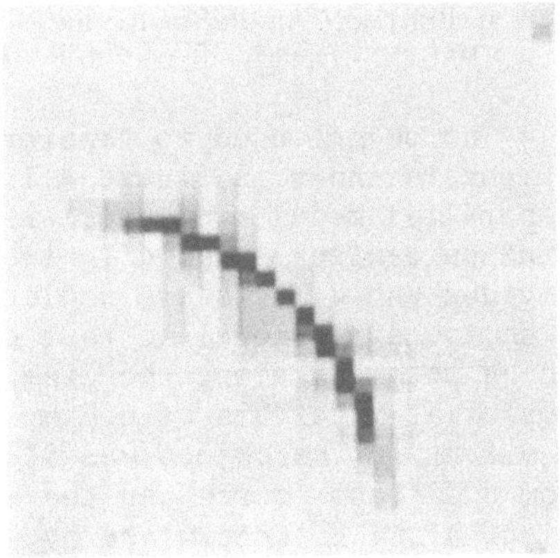

Figure 3.3(b) : 2-DFT of Figure 3.2(b).

$$P(s_x,s_y,s_z) = \mathcal{F}_{3D}\{p(x,y,z)\}$$

$$= \delta(s_z - \frac{(m-1)}{\lambda}) \qquad (3.25)$$

i.e. the 3D Fourier transform of the convolving function from one view is a hemispherical delta function distribution of radius $1/\lambda$ and centre, 0,0, $^{-1}/\lambda$.

This result implies that one 'view' of the object is capable of supplying information on a hemispherical surface in the frequency domain of the object. As the object rotates with respect to the scan/illumination, the hemispherical surface rotates with respect to the Fourier transform of the object and hence completely maps the frequency domain within the bandlimit $\rho<\sqrt{2}/\lambda$. Figure 3.3 illustrates this interesting correspondence between the image space and frequency domain approaches by display of the 2D Fourier transform of the focusing functions shown in Figure 3.2. They are semi-circular delta function distributions rotated to the appropriate view angle. Figure 3.3 also serves to illustrate the concept introduced in section 2 that the effect of a finite aperture is to attenuate the high frequency components of the measured data.

4. EXPERIMENTAL RESULTS FROM SAT APPLIED TO MICROWAVE IMAGING [4:1]

The object used to generate a response due to scattering sources is shown in Figure 4.1(a). It comprises an ensemble of co-parallel metal rods each with a diameter of 5 mm. The rotation centre is shown in the diagram of Figure 4.1(b) and, provided end effects are neglected, the object may be considered to be two dimensional so that one dimensional complex data sets can be recorded along the line indicated. In this case 60 samples at intervals of 6.5 mm are adequate for the wavelength of 9.1 mm. The object was illuminated along the z-direction from a distant source and the signal due to scattering by the object alone is determined by subtraction of data measured in an initial scan without the object. This step removes from the data that part of the wavefront not dependent on the object but the effect of the small phase curvature from the source remains. Figure 4.1(c) shows the image region obtained from the inverse diffraction operation on these data from one view. The complex superposition of 20 views of the object rotated through a total angle of 180° is seen in Figure 4.1(d). This image contains superposition artefact $\sum h_\alpha$, most of which can be removed by applying the ramp filter in the frequency domain, and the

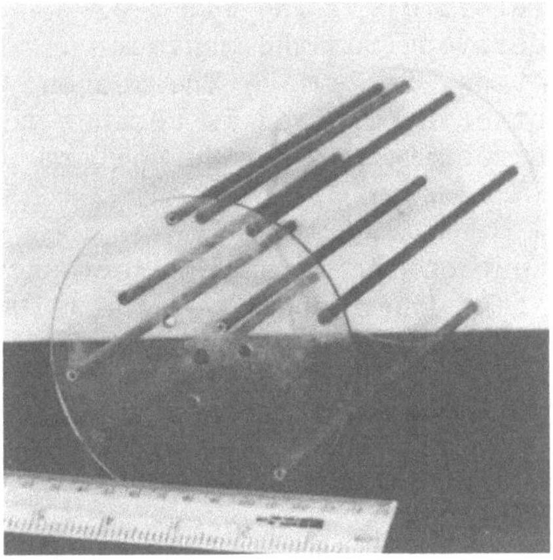

Figure 4.1(a) : Photograph of object.

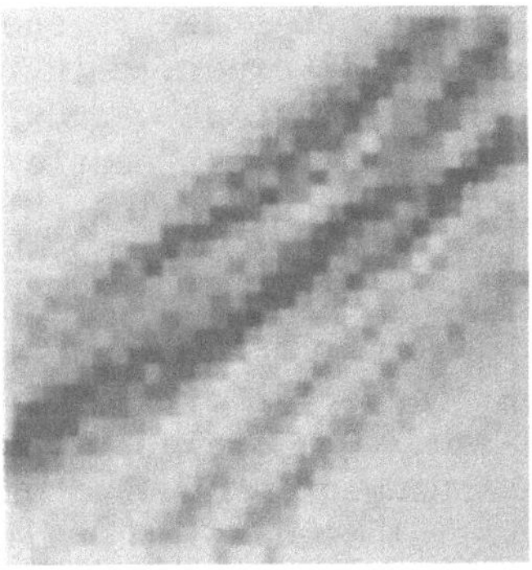

Figure 4.1(c) : Reconstruction from one view.

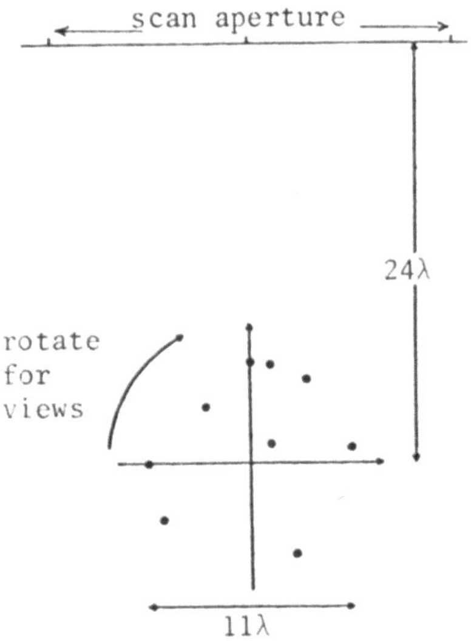

Figure 4.1(b) : Recording geometry.

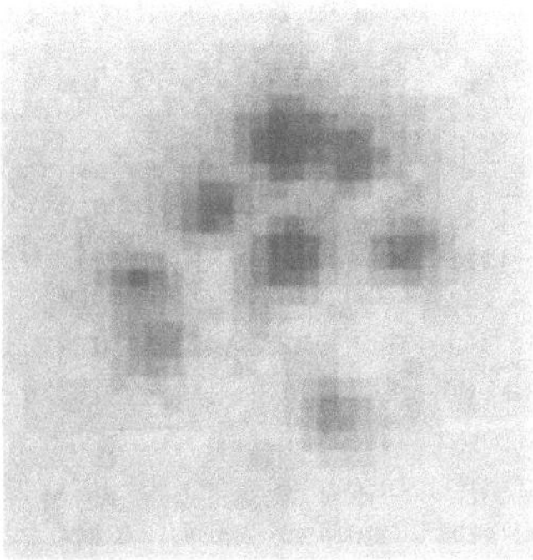

Figure 4.1(d) : Reconstruction from 20 views.

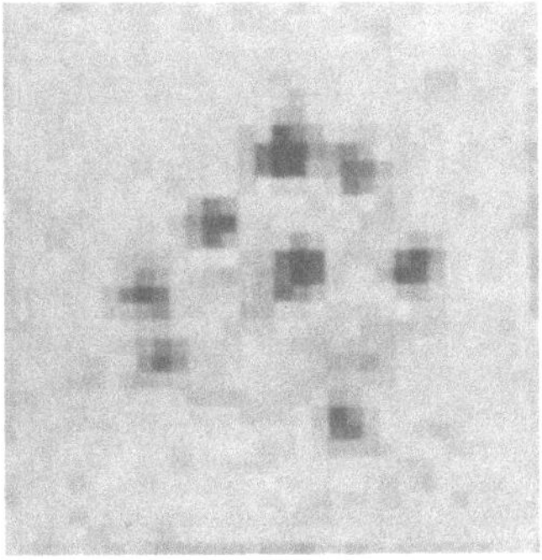

Figure 4.1(e) : Deconvolved microwave tomogram.

deconvolved image is shown in Figure 4.1(e). The limit of resolution expected in this case, derived from the maximum spatial frequency bandwidth of one view, is 7 mm. The reason that the closest scatterers have not been resolved is because the receiving antenna pattern was too directive to provide uniform synthetic aperture weighting.

Experiments were also carried out on the perspex object shown in Figure 4.2(a) below in order to test the response of the system to a dielectric scattering structure. An identical measurement procedure was used to provide a tomogram of the central slice of the object. The result is shown in Figure 4.2(b).

5. EXPERIMENTAL RESULTS FROM SAT APPLIED TO ULTRASONIC IMAGING [5:1]

Demonstration of the tomographic technique applied to ultrasonic imaging has been performed by experiments at a frequency of 1 MHz in a water tank in which the wavelength is approximately 1.5 mm. As in the previous microwave experiments, objects with only 2-D variations have been used so that 1-D scans are sufficient for recording the data, i.e. samples of complex field transmitted through the object. In this case, the object is a block of reticulated polyurethan foam of the type suggested as material for the construction of tissue-like phantoms and is shown in Figure 5.1(a). Internal structure in the form of a longitudinal hole and 4 wire scatterers was created and this structure is seen clearly in Figure 5.1(b) which also shows the

Figure 4.2(a) : Photograph of object.

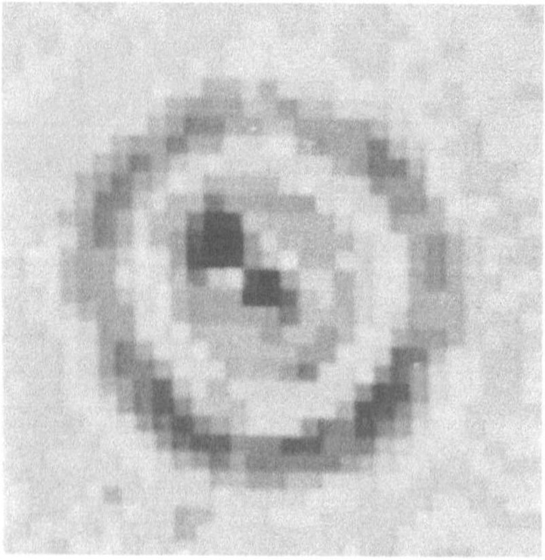

Figure 4.2(b) : Deconvolved microwave tomogram.

recording dimensions. 25 views of the object taken through a total angular range of 180° were recorded. Each view contained 64 complex data samples obtained after subtracting the unperturbed illumination field as described in the microwave experiment.

The subtracted lines of data due to the object alone were used to form the ultrasonic tomogram using the procedure described in section 3. The final deconvolved tomogram is shown in both amplitude and phase in Figure 5.1(c) and (d). It is obvious that the amplitude tomogram gives the better representation of the scatterers and that the phase tomogram gives the better representation of the hole. This property of SAT could be valuable. The foam material is known to have an attenuation of 0.5 dB cm^{-1} at 1 MHz and 1% velocity perturbation with respect to water. If attenuation may be introduced as a complex compressibility (i.e. a loss factor equivalent to its representation by complex permittivity) then the phase tomogram should represent the spatial distribution of this loss factor. This property of holographic-based tomography is examined next.

6. INVESTIGATION OF THE PHASE TOMOGRAM OBTAINED FROM SAT

6.1 Effect of a Complex Dielectric

It was evident in the results of the imaging experiment presented in section 5 that the phase tomogram, a natural result of SAT or diffraction tomography using CW illumination with complex field recording, provides a better representation of dielectric features than the amplitude image. Consider the case of an attenuating dielectric object which may be represented by :

$$f(\underline{r}_1) = k^2 \left[\frac{\varepsilon'(\underline{r}_1) - j\,\varepsilon''(\underline{r}_1) - \varepsilon_0}{\varepsilon_0}\right]$$

Provided that the Born approximation holds, the reconstituted amplitude distribution is given by :

$$f_a(r_1) = \frac{k}{\varepsilon_0}\sqrt{\{\varepsilon'(r_1) - \varepsilon_0\}^2 + \{\varepsilon''(r_1)\}^2}$$

and the phase image by :

$$f_p(r_1) = \tan^{-1}\left[\frac{-\varepsilon''(r_1)}{\varepsilon'(r_1) - \varepsilon_0}\right]$$

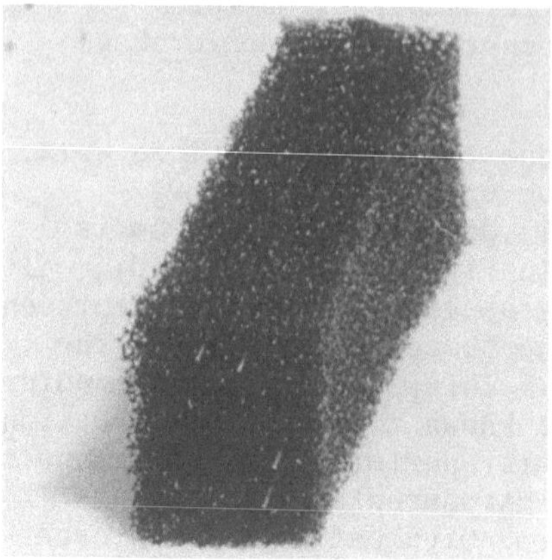

Figure 5.1(a) : Photograph of object.

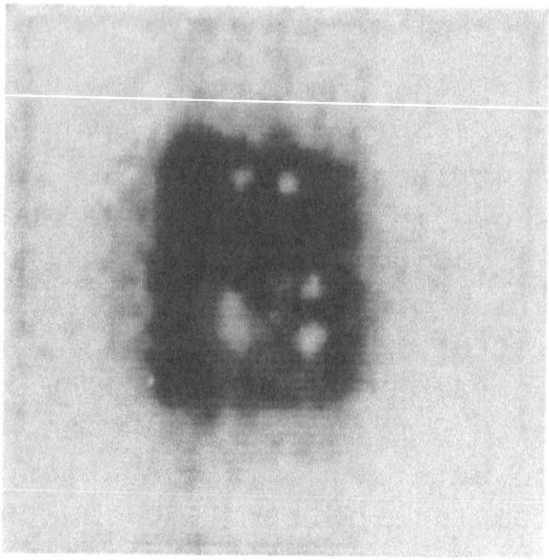

Figure 5.1(c) : Amplitude tomogram.

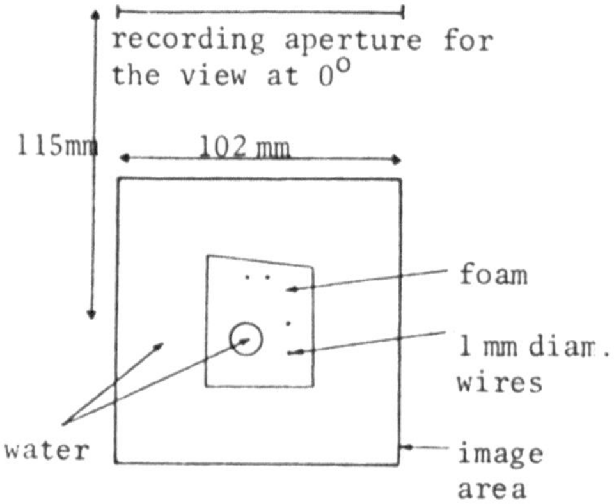

Figure 5.1(b) : Recording geometry.

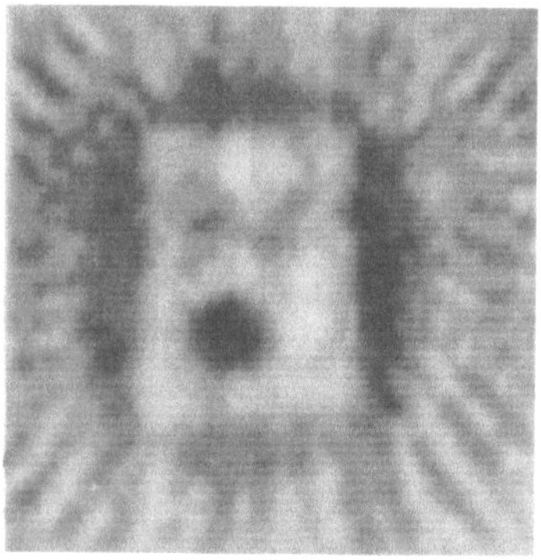

Figure 5.1(d) : Phase tomogram.

i.e. represents the distribution of loss tangent within the object. Since $\varepsilon'' > \varepsilon' - \varepsilon_0$ typically, f_p will be in the range approaching $\pi/2$ and therefore the phase image will be a sensitive function of attenuation.

6.2 Simulation of Diffraction by a Composite Dielectric Cylinder

This important property of the phase tomogram has been simulated by considering the propagation of an illuminating wavefront through a composite dielectric cylinder as indicated by the model shown in Figure 6.1. The classical approach to predicting the scattered field calculates the Hankel functions [6:1]. In order to conveniently introduce the attenuation factors, the plane-to-plane diffraction technique described earlier has been applied to this problem in a novel way [6:2]. A band-limited one-dimensional plane wave function is propagated in the z-direction toward the cylinder model using eqn.(2.1). The linear filter transfer function $B_{j\ell}$ controls the forward propagation process and may be used to increment the sampled wavefront in steps of $\lambda/2$, for example. Since this process can be implemented efficiently by the FFT algorithm it is convenient to modify $B_{j\ell}$ as it traverses the object by modifying the wave number k in each region, and by weighting it by the attenuation factor $e^{-\alpha\Delta z}$. Hence a subroutine can be developed to generate successive lines of data in x culminating in the required recording

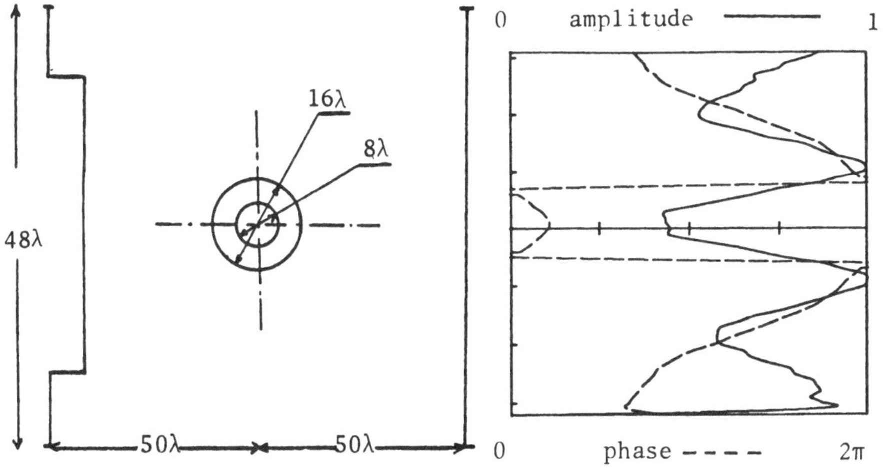

Figure 6.1(a) : Data simulation geometry.

Figure 6.1(b) : Phase and amplitude patterns for $\varepsilon_r = 1.1$

plane z_1. The simple modifications to $B_{j\ell}$ above appear to be sufficient for $f(\underline{r}_1)$ within the Born and the Rytov approximations [6:3][6:4].

The simulated scattered field due to the composite cylinder alone is obtained by subtracting the distributions obtained with and without the object. These residual data are also shown in Figure 6.1.

6.3 Simulation of the Tomograms

To simulate the amplitude and phase tomograms using the image domain approach described in section 3, the diffraction data due to the composite cylinder alone is subjected to the inverse diffraction filter $B'_{j\ell}$ for backwards propagation through the object region. In a real case the object transmission characteristics will be the unknown internal distribution required. The simulated data in this case would, following previous practice, be back-propagated with the constant wave number k_0. However, the simulated diffraction data in Figure 6 does provide the net phase shift through the object and therefore an average wave number k_{av} could be introduced in the reconstruction. The back propagated image regions for many views at different angles are superimposed, interpolated and filtered according to section 3 with all the views being the same for this model. The examples of filtered images from 36 views in Figure 6.2 (a)(b)(c)(d) show the power of the phase image to represent the change in attenuation.

7. CONCLUSIONS AND POSSIBLE FUTURE DEVELOPMENTS

This paper has shown how holographic Fourier-based inverse methods can provide 2-D and 3-D diagnostic imaging for microwaves and ultrasound. Despite their limitations, particularly in the case of low S/N data, we have found that satisfactory images can be produced in many cases using fast and convenient FFT - based algorithms, which moreover, may be amenable to hardware implementation.

The exploitation of the B-filter and B'-filter derived from the plane-to-plane diffraction and inverse diffraction processes gives a useful prediction and visual representation of the field patterns within and around an object structure. This imaging of the propagating diffraction fields may also be applied to the understanding dual reflector antenna configurations and the effects of radomes. In the former case, the diffraction field distributions indicate the shift of the position of the focal maximum from the geometric focus and the effects of blockages, for example a secondary feed or a nearby screen. A similar process is being applied to the effects of radome material on an

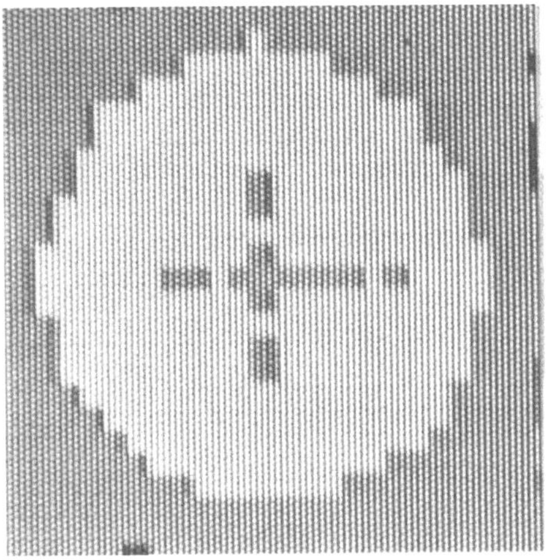

Figure 6.2(a) : Amplitude tomogram.

Figure 6.2(b) : Phase tomogram.

$\varepsilon_r = 1.1$; $\alpha_1 = \alpha_2 = 0.10$ dB m^{-1}; reconstructed with k_o

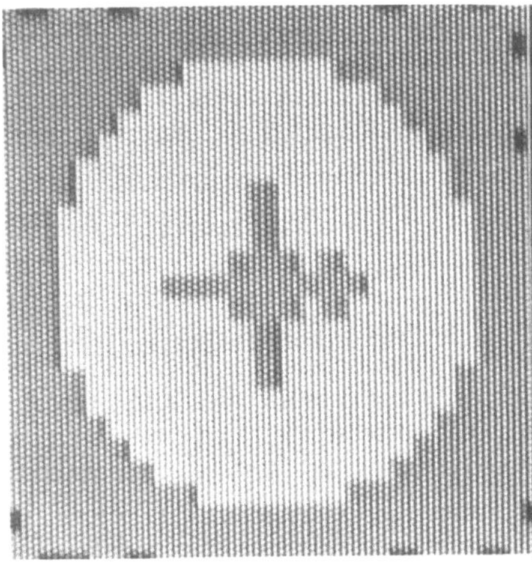

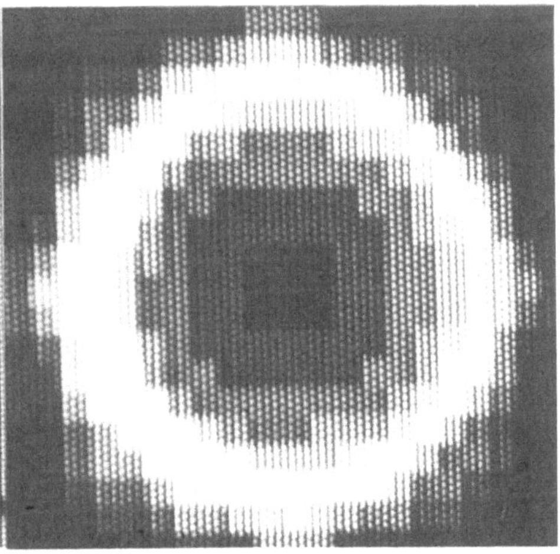

Figure 6.2(c) : Amplitude tomogram.

Figure 6.2(d) : Phase tomogram.

$\varepsilon_r = 1.1$; $\alpha_1 = 0.05$ dB m^{-1}; $\alpha_2 = 0.10$ dB m^{-1}; reconstructed with k_o

antenna where improvements in the simulation of the transfer function of the dielectric are currently being incorporated. Similar techniques may be applied to the modelling outlined in section 6.

The significance and interpretation of the phase tomogram continues to be studied. The processes outlined in this paper give insight into the diagnostic potential of this image representation. The Department of Electronic and Electrical Engineering is investigating its application to the diverse subjects of antenna metrology and the ultrasonic imaging of tissue.

REFERENCES

[1:1] Kock, W.E. and Harvey, F.K., 1951, B.S.T.J. pp.564-567

[1:2] Dooley, R.P., 1965, Proc. I.E.E.E., pp.1733-1735

[1:3] Tricoles, G. and Farhat, N.H., 1977, Proc. I.E.E.E., pp.108-121

[1:4] Anderson, A.P., 1977, I.E.E. Reviews, pp.946-962

[1:5] Bennett, J.C., et al., 1976, I.E.E.E. Trans. AP.24, pp.295-303

[1:6] Anderson, A.P., et al., 1978, I.E.E. Conf. Pub. No. 169, pp.128-131

[1:7] Godwin, M.P., et al., 1981, I.E.E. Conf. Pub. No.195, pp.232-236

[1:8] De Velis, J.B. and Reynolds, G.O., 'Theory and Applications of Holography' (Addison Wesley 1967)

[1:9] Anderson, A.P., 1979, Proc. 9th Eu. M.C. pp.64-73

[2:1] Shewell, J.R. and Wolf, E., 1968, J.O.S.A., pp.1596-1603

[2:2] Goodman, J.W., 'Introduction to Fourier Optics' (McGraw-Hill 1968)

[2:3] Adams, M.F., 1982, Ph.D. Thesis, University of Sheffield

[3:1] Boerner, W.M. Jordan, A.K. Kay, I.W., 1981, I.E.E.E. Trans. AP-29, pp.185-189

[3:2] Wolf, E., 1969, Optics Commun., pp.153-156

[3:3] Mueller, R.K. Kaveh, M. Wade, G., 1979, Proc. I.E.E.E., pp.567-587

[3:4] Lewis, R.M., 1969, I.E.E. Trans. AP-17, pp.308-314

[3:5] Morse, P.M. Ingard, K.V., 'Theoretical Acoustics' (McGraw-Hill 1968)

[3:6] Scudder, H.J., 1978, Proc. I.E.E.E., pp.628-637

[3:7] Devaney, A.J., 'Optics in Four Dimensions' 1981, Am. Inst. Phys., pp.613-626

[4:1] Adams, M.F. and Anderson, A.P., 1982, Proc. I.E.E. Pt.H, pp.83-88

[5:1] Anderson, A.P. and Adams, M.F., Acoustical Imaging Vol.12 (Prentice-Hall 1982)

[6:1] Boerner, W.M. et al., 1971, Can. J. Phys., pp.804-819

[6:2] Anderson, A.P. and Aitmehdi, R., submitted for publication

[6:3] Devaney, A.J., 1981, Opt. Lett., pp.374-376

[6:4] Kaveh, M., et al., Acoustical Imaging Vol.12 (Prentice-Hall 1982)

ACKNOWLEDGEMENTS

The authors wish to acknowledge the support of the Science & Engineering Research Council for the overall programme of long wavelength imaging research. In addition to the acknowledgements given with respect to Figures 1.3(a)-(d), the support of the subsurface imaging project (Figure 2.5) by British Telecom is also acknowledged.

V.5 (PT.4)

DIFFRACTION TOMOGRAPHY

A. J. Devaney

Schlumberger-Doll Research
P. O. Box 307, Ridgefield, CT 06877

ABSTRACT

The theoretical foundations of computed tomography using diffracting wavefields (diffraction tomography) are reviewed and contrasted with the foundations of conventional CT. The generalized projection-slice theorem is presented and shown to lead directly to inversion algorithms for diffraction tomography that are natural generalizations of the filtered backprojection algorithm of conventional CT. The inversion algorithms - called filtered back*propagation* algorithms - are developed both for the classical tomographic configuration where the detector array rotates about the object being scanned and for geophysical applications, such as offset vertical seismic profiling and well-to-well tomography, where the detector array is fixed in space. Results of computer simulations testing the algorithms are also presented.

1. INTRODUCTION

In recent years there has been considerable interest in employing computerized tomography - CT - in applications where the source of radiation is not X-rays. For example, Greenleaf [1-3] has employed CT algorithms in ultrasound tomography while Lytle and Dines [4,5] and others [6] have proposed the use of CT methods in well-to-well tomography using sound or electromagnetic radiation. Unfortunately, the quality of the tomographic reconstructions obtained with sound or electromagnetic radiation are considerably inferior to those obtained with X-rays. The principle reason for this is that, unlike X-rays,

W.-M. Boerner et al. (eds.), Inverse Methods in Electromagnetic Imaging, Part 2, 1107–1135.

sound and low frequency electromagnetic radiation diffracts and coherently scatters when passing through an object. The conventional CT reconstruction algorithms such as ART [7] and the filtered back-projection algorithm [7,8] do not account for these effects and, hence, are not suited for use with such wavefields.

The first workers who addressed the problem of deriving inversion algorithms for tomography with diffracting wavefields - now known as "diffraction tomography" - appear to be Mueller and co-workers [9,10] and Iwata and Nagata [11]. They based their work on Wolf's formulation [12] of the inverse scattering problem within the first Born approximation. Wolf showed that for an incident plane wave, the two-dimensional spatial Fourier transform of the scattered field over any plane surface lying outside the scattering volume is directly proportional to the three-dimensional spatial Fourier transform of the scattering potential (i.e., the "object") over a certain section of a spherical surface called an Ewald sphere [13,14]. Mueller et. al. [9,10] showed that an analogous relationship existed between the complex phase of the transmitted field and the scattering potential within the Rytov approximation [15,16]. The Rytov approximation is of interest in diffraction tomography since it is the generalization, to finite wavelengths, of the straight line ray approximation upon which conventional CT algorithms are based [7,8]. Consequently, reconstruction methods based on the Rytov approximation should, in principle, work at least as well as conventional CT methods. In addition, there are reasons to believe that the Rytov approximation is superior to the Born approximation for large, extended objects such as occur in most tomographic applications [16-18].

The relationship between the Fourier transform of the complex phase of the transmitted field and the Fourier transform of the scattering potential is now known as the *generalized projection-slice theorem* [9, 19-23]. It forms the basis for diffraction tomography just as the conventional projection-slice theorem [7,8] forms the basis of conventional tomography. As discussed above, the generalized projection-slice theorem is based on the Rytov approximation and just as the Rytov approximation reduces to the straight line ray approximation in the short wavelength limit, the generalized projection-slice theorem reduces to the conventional projection-slice theorem in this limit [19].

Although the generalized projection-slice theorem has been known for some time, true generalizations of conventional CT reconstruction algorithms have only recently been obtained. Virtually all reconstruction methods have, in the past, been based on interpolation methods in Fourier space ($\underline{K}$ space) where the idea is to interpolate known values of the object's Fourier transform specified on sections of Ewald spheres onto a regular cubic sampling grid. The interpolated Fourier coefficients are then transformed using the discrete Fourier transform - usually a Fast Fourier Transform (FFT) - to obtain the sought after object profile.

Interpolation methods have a number of drawbacks. Perhaps the most severe of these are the "image artifacts" that are introduced due to aliasing caused by undersampling in the spatial frequency domain. This undersampling can often not be avoided due to limits in the number of Ewald spheres over which the transform of the object profile is known. Each Ewald sphere requires a single experiment to be conducted and in many applications there are limitations to the number of experiments that can be performed.

Interpolation methods have drawbacks even in cases where a sufficient number of experiments can be performed to avoid aliasing. In particular, these methods typically require large amounts of computer memory and rather sophisticated interpolation algorithms. Finally, there is the reluctance of the CT community to accept interpolation reconstruction algorithms in general. Such algorithms have been employed in conventional CT [24] and have been found to be inferior to the "direct" reconstruction algorithms such as the filtered backprojection algorithm [25].

The most widely used conventional CT reconstruction algorithm is the well-known filtered backprojection algorithm [7,8]. This algorithm generates **partial** reconstructions of the object from data collected in separate experiments and coherently sums the partial reconstructions to obtain the final, full object reconstruction. The filtered backprojection algorithm generally yields significantly greater image quality than interpolation schemes and is conceptually simple, easy to implement, and requires relatively little on-line computer memory.

In a recent paper [19] this author derived a reconstruction algorithm for ultrasound diffraction tomography that is the natural generalization of the filtered backprojection algorithm of conventional CT. This algorithm, called the filtered back**propagation** algorithm, operates in a similar fashion to the back**projection** algorithm in that partial reconstructions are generated from data collected in single experiments and the partial reconstructions are then coherently summed to yield the final object reconstruction. The two algorithms differ principally in the replacement of the backprojection process by backpropagation. In backprojection, filtered data is **projected back** into image space along parallel straight line rays while in backpropagation filtered data is **propagated back** into image space using a Green function governing propagation in the background medium surrounding the object. Anderson and Adams [26,27] have obtained similar algorithms for tomographic reconstruction within the Born approximation.

The reconstruction algorithms discussed in [19-23,26,27] apply to the so-called classical tomographic configuration where the detector array rotates with the direction on propagation of the incident plane wave. Recently, filtered backpropagation algorithms have been derived for geophysical applications such as offset vertical profiling (offset VSP) and well-to-well tomography where the detector array is fixed in space and not all angles of plane wave illumination of the object can be employed [28,29]. Although, these latter algorithms were derived primarily for geophysical applications they can be employed also for ultrasound tomography in cases where it is infeasible to employ the classical scan configuration.

In this paper we review the underlying theory of diffraction tomography within the Rytov approximation and present the filtered backpropagation algorithms for both the classical tomographic configuration employed in most medical applications and for the case where the detector array is fixed in space as most often occurs in geophysical applications. We shall limit our treatment to applications governed by the scalar wave equation (Schrodinger equation), although most of the treatment is readily extended to the electromagnetic and elastic wave cases.

In Section 2.0 we review the theoretical foundations of the theory and present the generalized projection-slice theorem for both the classical CT configuration and the case of a fixed detector array. In this section both two and three-dimensional objects are considered. In Section 3.0 we present the appropriate filtered backpropagation algorithms for both geometries. To simplify the presentation we consider only the case of two-dimensional objects (cylindrical objects). In this section we also present the results of applying these algorithms in computer simulations of diffraction tomography within the Rytov approximation. Finally, Section 4.0 summarizes the results presented in the paper and discusses future outlooks for the field of diffraction tomography.

2. THEORETICAL FOUNDATIONS OF DIFFRACTION TOMOGRAPHY

As mentioned in the Introduction, we shall restrict our attention in this paper to applications involving the reduced scalar wave equation (Schrodinger equation)

$$\left[\nabla^2 + k^2 - k^2 O(\underline{r}, \omega)\right] U(\underline{r}, \omega) = 0. \tag{2.1}$$

In this equation $U(\underline{r}, \omega)$ is the Fourier amplitude at angular frequency ω of a time dependent wavefield $U(\underline{r}, t)$

$$U(\underline{r}, \omega) = \int_{-\infty}^{\infty} dt\, U(\underline{r}, t) e^{i\omega t}, \tag{2.2}$$

and k is the wavenumber of the field $U(\underline{r}, \omega)$ in the region of space where the "object profile" $O(\underline{r}, \omega) = 0$. The object profile will be assumed to be piecewise continuous and to be localized within a sphere of radius R_0 centered at the origin.

The goal of diffraction tomography is to employ field measurements performed at a single, fixed frequency ω to deduce the object profile $O(\underline{r}, \omega)$ at that specific frequency. In other words, diffraction tomography is developed and applied at a fixed frequency in the frequency domain. Because of this, the frequency ω plays the role of a parameter and can be removed from the arguments of all quantities without loss of generality. We shall do this here to simplify the notation. Thus, we shall denote the field $U(\underline{r}, \omega)$ by simply $U(\underline{r})$ and the object profile $O(\underline{r}, \omega)$ by $O(\underline{r})$. However, it should be kept in mind that this is done only for convenience and that, in general, both of these quantities depend explicitly on ω.

Applications governed by Eq. (2.1) include nonrelativisitic quantum mechanical scattering [30,31] and electromagnetic scattering in cases where polarization effects can be neglected such as in certain optical applications [32]. Of particular importance is acoustic wave scattering in cases where the density is constant but the velocity varies as a function of position. This later situation arises, for example, in ultrasound tomography of soft tissues[1-3,9,10,19-23]. In all of these applications the object profile $O(\underline{r})$ is related to an index of refraction profile $n(\underline{r})$ according to the equation

$$n(\underline{r}) = \frac{C_0}{C(\underline{r})} = \sqrt{1-O(\underline{r})} \tag{2.3}$$

where $C(\underline{r})$ is the (possibly complex) phase velocity of the medium at the point $\underline{r}$ and at frequency ω and C_0 is the velocity (assumed constant) at points outside the object volume (i.e., for $|\underline{r}| > R_0$). It follows from Eq. (2.3) that there is a one to one correspondence between the object profile, index of refraction and phase velocity.

We shall be concerned here with wavefields that arise from the interaction of an incident plane wave $U_0\exp(ik\underline{s}_0 \cdot \underline{r})$ with the object. We shall denote the field for such cases by $U(\underline{r}, \underline{s}_0)$ to indicate its dependence on the direction of propagation $\underline{s}_0$ of the incident plane wave. We shall also restrict our attention to cases where the Rytov approximation [15,16] can be employed. Within this approximation the field $U(\underline{r}, \underline{s}_0)$ is expressed in terms of its complex phase $\Psi(\underline{r}, \underline{s}_0)$ according to the equation

$$U(\underline{r}, \underline{s}_0) = U_0 e^{\Psi(\underline{r}, \underline{s}_0)} \tag{2.4}$$

and the complex phase $\Psi(\underline{r}, \underline{s}_0)$ is approximated by

$$\Psi(\underline{r}, \underline{s}_0) = ik\underline{s}_0 \cdot \underline{r} - \frac{k^2}{4\pi} e^{-ik\underline{s}_0 \cdot \underline{r}} \int d^3r' O(\underline{r}') e^{ik\underline{s}_0 \cdot \underline{r}'} \frac{e^{ik|\underline{r}-\underline{r}'|}}{|\underline{r}-\underline{r}'|} . \tag{2.5}$$

The Rytov approximation as embodied in Eq. (2.5) is seen to be in the form of a **linear** mapping between the object profile and the complex phase $\Psi(\underline{r}, \underline{s}_0)$ of the field. This approximation differs from the more familiar Born approximation [12] in that the Born approximation results in a linear mapping between the scattered field amplitude $U^{(s)}(\underline{r}, \underline{s}_0) \equiv U(\underline{r}, \underline{s}_0) - \exp(ik\underline{s}_0 \cdot \underline{r})$ and the object profile. Although we shall deal here only with the Rytov approximation, the inversion methods presented in the following sections can be readily applied to fields satisfying the Born approximation (see, for example, Refs. 19,20,28,29,32). We employ the Rytov approximation here since this approximation appears to be superior to the Born approximation in most diffraction tomographic applications [18].

In diffraction tomography the total field $U(\underline{r}, \underline{s}_0)$ (or, equivalently, the complex phase $\Psi(\underline{r}, \underline{s}_0)$) is measured over one or more plane surfaces lying outside the support volume of the object for one or more values of $\underline{s}_0$. The goal is then to reconstruct the object profile $O(\underline{r})$ from this information. In the classical tomographic configuration shown in Fig. 1, and almost universally employed in ultrasound tomography, a single measurement plane is employed for each $\underline{s}_0$ value. This plane is taken to be perpendicular to $\underline{s}_0$ and is placed on the side of the object opposite to the source of the incident wave so that only the forward scattered field is measured. Note that in this configuration, the measurement plane rotates around the object as different directions of propagation of the incident plane wave are employed. An alternative configuration, shown in Fig. 2 and of use both in ultrasound tomography and certain geophysical applications [28,29], employs a measurement plane whose orientation is fixed in space.

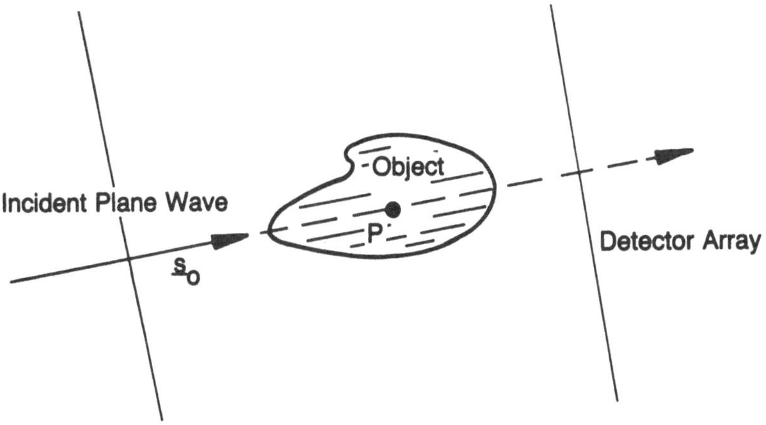

Fig. 1: Classical tomographic configuration. The detector array rotates about a point P in the object volume as different directions of propagation are employed.

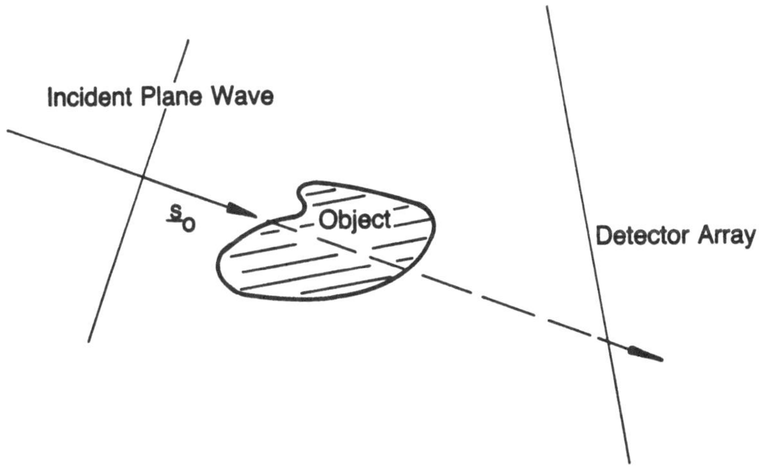

Fig. 2: Fixed detector array configuration. The detector array remains fixed in space as different directions of propagation are employed.

Within the Rytov approximation diffraction tomography reduces mathematically to "solving" the set of equations (2.5) for $O(\underline{r})$ from specification of the complex phase $\Psi(\underline{r}, \underline{s}_0)$ over the measurement planes for the set of incident unit wavevectors $\underline{s}_0$ employed in generating the data. These equations are most simply treated in spatial Fourier transform space where they reduce to simple algebraic equalities between the two-dimensional spatial Fourier transform of the data taken over the measurement plane and the three-dimensional spatial Fourier transform of the object profile evaluated over certain hemispherical surfaces in Fourier transform space. We shall obtain these relationships for the two geometries illustrated in Figs. 1 and 2 where the measurement plane either rotates with the direction of propagation of the incident wave or remains fixed in space. For the classical tomographic configuration illustrated in Fig. 1, these relationships form a natural generalization of the *projection-slice theorem* of conventional computed tomography to diffraction tomography and are sometimes referred to as the *generalized projection-slice theorem* [9,19-23]. There appears to be no analog in conventional tomography of these equalities in the case of a fixed measurement plane [28,29].

We consider first the case where the measurement plane is fixed. Without loss of generality we can take this to be the plane $z = l_0 > R_0$ in the right-handed Cartesian coordinate system (x,y,z) whose origin lies at the center of the object. We introduce the quantity

$$D(\underline{r}; \underline{s}_0) \equiv \frac{2i}{k^2}\left[\Psi(\underline{r}, \underline{s}_0) - ik\underline{s}_0 \cdot \underline{r}\right] e^{ik\underline{s}_0 \cdot \underline{r}} \tag{2.6}$$

and its two-dimensional spatial Fourier transform performed over the measurement plane $z = l_0$:

$$\tilde{D}(\underline{\kappa}, z=l_0; \underline{s}_0) \equiv \int dxdy\, D(\underline{r}; \underline{s}_0)\big|_{z=l_0} e^{-i(\kappa_x x + \kappa_y y)} . \tag{2.7}$$

$D(\underline{r}; \underline{s}_0)|_{z=l_0}$ can be regarded as being the "data" in that it is uniquely specified over the measurement plane by the unit propagation vector $\underline{s}_0$ and the complex phase $\Psi(\underline{r}, \underline{s}_0)$. We note that because evanescent waves [33] are, in practice, not measurable outside the object volume, $D(\underline{r}; \underline{s}_0)|_{z=l_0}$ is essentially bandlimited to within a circle of radius k; i.e.,

$$\tilde{D}(\underline{\kappa}, z=l_0; \underline{s}_0) \approx 0 \tag{2.8}$$

if $|\underline{\kappa}| > k$.

By employing a procedure entirely analogous to that used in Ref. 28 for the one-dimensional case it is readily shown that the two-dimensional Fourier transform over the measurement plane of both sides of Eq. (2.5) leads to the following relationship between the "data" and the object profile:

$$\tilde{D}(\underline{\kappa}, z=l_0; \underline{s}_0) = \frac{e^{i\gamma l_0}}{\gamma} \tilde{O}[k(\underline{s}-\underline{s}_0)]. \tag{2.9}$$

In Eq. (2.9) $\tilde{O}(\underline{K})$ is the three-dimensional Fourier transform of the object profile

$$\tilde{O}(\underline{K}) \equiv \int d^3r\, O(\underline{r}) e^{-i\underline{K}\cdot\underline{r}}, \tag{2.10}$$

and the unit vector $\underline{s}$ is defined in terms of the two-dimensional spatial frequency vector $\underline{\kappa}$ according to the equations (recall that $|\underline{\kappa}|<k$)

$$s_x = \frac{\kappa_x}{k}, s_y = \frac{\kappa_y}{k}, s_z = \frac{\gamma}{k}, \tag{2.11}$$

with

$$\gamma = \sqrt{k^2-\kappa^2}. \tag{2.12}$$

Eq. (2.9) relates the three-dimensional Fourier transform $\tilde{O}(\underline{K})$ of the object profile $O(\underline{r})$ to the two-dimensional transform $\tilde{D}(\underline{\kappa}, z=l_0; \underline{s}_0)$ of the "data" $D(\underline{r}; \underline{s}_0)|_{z=l_0}$. For fixed $\underline{s}_0$ (fixed incident wave) the locus of points $\underline{K} = k(\underline{s}-\underline{s}_0)$ over which $\tilde{O}(\underline{K})$ is determined by Eq. (2.9) lie on the surface of a hemisphere of radius k, centered at $\underline{K} = -k\underline{s}_0$ and over which $K_z \geq -k s_{oz}$. This last condition is a consequence of the fact that over this locus of points $K_z = ks_z - ks_{oz}$ and $ks_z \geq 0$ (see Eqs. (2.11) and (2.12)). Shown in Fig. 3 is the intersection with the K_x, K_z plane of a hemispherical surface arising from an incident plane wave whose unit propagation vector $\underline{s}_0$ lies in the x-z plane.

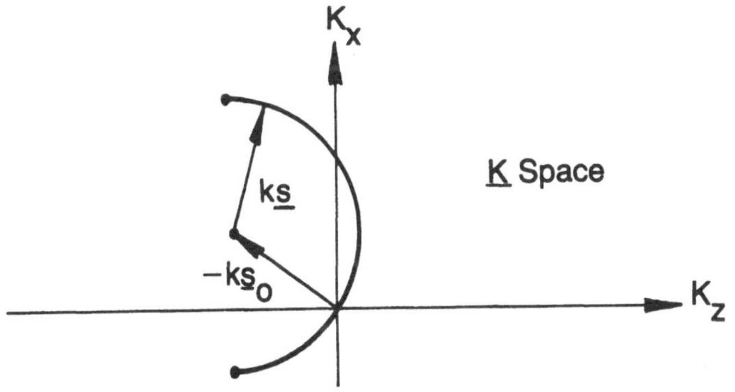

Fig. 3: Intersection of Ewald hemisphere with K_x, K_z plane for the case of a fixed detector array alligned parallel with the x, y plane. The incident unit wavevector $\underline{s}_0$ is assumed to lie in the x, z plane.

We conclude from the discussion presented above that for the case of a fixed measurement plane, diffraction tomography consists of determining the object profile $O(\underline{r})$ from specification of its Fourier transform $\tilde{O}(\underline{K})$ over the hemispherical surfaces defined above. A similar situation is encountered in the classical tomographic configuration where the measurement plane rotates with the direction of propagation of the incident plane wave. In this case the measurement plane is perpendicular to $\underline{s}_0$ and is labelled by Cartesian coordinates, say, (ξ,ρ). Eq. (2.7) then is replaced by

$$\tilde{D}(\underline{\kappa},\ \eta=l_0;\ \underline{s}_0) \equiv \int d\xi d\rho D(\underline{r};\ \underline{s}_0)\big|_{\eta=l_0} e^{-i(\kappa_\xi \xi+\kappa_\rho \rho)}, \qquad (2.13)$$

where η is the third coordinate in the Cartesian system (ξ,ρ,η) and κ_ξ,κ_ρ are the ξ,ρ components of the spatial frequency vector $\underline{\kappa}$. Eq. (2.9) is replaced by

$$\tilde{D}(\underline{\kappa},\ \eta=l_0;\ \underline{s}_0) = \frac{e^{i\gamma l_0}}{\gamma}\tilde{O}[k(\underline{s}-\underline{s}_0)], \qquad (2.14)$$

where, however, $\underline{s}$ is now given by

$$s_\xi = \frac{\kappa_\xi}{k},\ s_\rho = \frac{\kappa_\rho}{k},\ s_\eta = \frac{\gamma}{k}. \qquad (2.15)$$

Note, that for this geometry, the η axis coincides with the direction of the incident wavevector so that over the measurement plane $\underline{s}_0 \cdot \underline{r} = l_0$ and Eq. (2.6) then simplifies to become

$$D(\underline{r};\ \underline{s}_0)\big|_{\eta=l_0} \equiv \frac{2i}{k^2}[\Psi(\underline{r},\ \underline{s}_0)-ikl_0]e^{ikl_0}. \qquad (2.16)$$

As was the case where the measurement plane was fixed, the locus of points over which $\tilde{O}(\underline{K})$ is determined is a hemisphere of radius k, centered at $\underline{K} = -k\underline{s}_0$. However, for the classical configuration the **orientation** of the hemisphere changes as a function of $\underline{s}_0$. In particular, since $K_\eta \equiv ks_\eta - ks_{0\eta} \geq -ks_{0\eta}$ the axis of the hemisphere is always $-k\underline{s}_0$. This is readily apparent from Fig. 4 where we show the intersection with the K_x, K_z plane of the hemipherical surface generated by the same $\underline{s}_0$ vector employed for the case of a fixed measurement plane illustrated in Fig. 3.

As discussed in the Introduction, the hemispherical surfaces over which the Fourier transform of the object profile is specified by the data are sections of the well known *Ewald* spheres of X-ray crystallography [13,14]. Wolf [12] appears to have been the first to derive these relationships for inverse scattering from inhomogeneous objects within

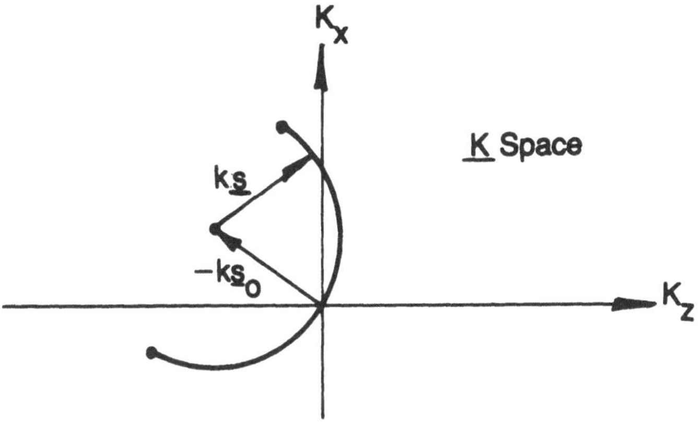

Fig. 4: Intersection of Ewald hemisphere with the K_x, K_z plane for the classical tomographic configuration. Incident unit wavevector $\underline{s}_0$ is the same as employed in Fig. 3.

the first Born approximation while Mueller et. al. [9,10] and others [11,34] extended Wolf's results to be applicable within the Rytov approximation. As pointed out by Wolf [12], these relationships place an upper limit on the resolution with which the object profile can be reconstructed from scattered field data. In particular, he showed that in the most favorable situation, where both forward and backscattered field measurements are performed, the object profile transform $\tilde{O}(\underline{K})$ can, at most, be determined only within the so-called Ewald *limiting* sphere; i.e., only for $\underline{K}$ values satisfying the inequality $|\underline{K}| \leq 2k$. In the classical tomographic configuration where only forward scatter measurements are performed, the limiting sphere has a radius of only $\sqrt{2}\,k$ [19,20]. Because of this ultimate limit in resolution it is, in principle, only possible to generate estimates of the object profile that are low-pass filtered versions of $O(\underline{r})$; i.e., these estimates will be of the form

$$\hat{O}(\underline{r}) = \frac{1}{(2\pi)^3}\int d^3K\, \tilde{F}(\underline{K})\tilde{O}(\underline{K})e^{i\underline{K}\cdot\underline{r}} \tag{2.17}$$

where $\tilde{F}(\underline{K})$ is a low pass filter. For example, for the classical tomographic configuration $\tilde{F}(\underline{K}) = 1$ if $|\underline{K}| \leq \sqrt{2}\,k$ and 0 otherwise [18,19]. In applications employing a fixed measurement plane $\tilde{F}(\underline{K})$ can be quite complicated [28] and the estimate $\hat{O}(\underline{r})$ is no longer the simple low pass filtered estimate obtained in the case of the classical tomographic configuration.

We conclude this section with a discussion of the two-dimensional case. In particular, suppose that the object profile is only a function of the two coordinates x-z and the unit propagation vector $\underline{s}_0$ lies in the x-z plane, making the angle ϕ_0, $[-\pi \leqslant \phi_0 \leqslant \pi]$, with the positive z axis. This then corresponds to the case of a cylindrical (two-dimensional) object aligned along the y axis and illuminated by plane waves having their wavefronts parallel to the y axis. For this case $D(\underline{r}, \underline{s}_0)$ and $O(\underline{r})$ are both functions only of x,z and Eq. (2.9) becomes

$$\tilde{D}(\kappa, z=l_0; \underline{s}_0) = \frac{e^{i\gamma l_0}}{\gamma}\tilde{O}[k(\underline{s}-\underline{s}_0)] \tag{2.18}$$

where now the unit vector $\underline{s}$ lies in the x-z plane and has components

$$s_x = \kappa/k\,, \tag{2.19a}$$

$$s_z = \gamma/k\,, \tag{2.19b}$$

with $\gamma = \sqrt{k^2-\kappa^2}$.

Similarly, we find that Eq. (2.14) becomes

$$\tilde{D}(\kappa, \eta=l_0; \underline{s}_0) = \frac{e^{i\gamma l_0}}{\gamma}\tilde{O}[k(\underline{s}-\underline{s}_0)]\,, \tag{2.20}$$

with

$$s_\xi = \kappa/k\,, \tag{2.21a}$$

$$s_\eta = \gamma/k\,. \tag{2.21b}$$

We note for future reference that the coordinates of a point in the x-z coordinate system are related to the coordinates of that point in the ξ, η system via the transformation

$$\xi = z \sin\phi_0 - x\cos\phi_0$$

$$\eta = z\cos\phi_0 + x\sin\phi_0\,. \tag{2.22}$$

We conclude from the above that the two-dimensional case is entirely analogous to the three-dimensional one. The principal difference between the two cases is that the loci of points in $\underline{K}$ space over which the object profile is specified from the data are hemispheres in three dimensions and semi-circles in two dimensions. Throughout the remainder of the paper we shall restrict our attention to the two-dimensional case although essentially all of our results can be readily extended to three-dimensional objects.

3. THE FILTERED BACKPROPAGATION ALGORITHMS

We showed in the preceding section that the data and object profile are related in Fourier transform space according to Eqs. (2.9) or (2.14) in the three-dimensional case, and according to Eqs. (2.16) or (2.18) in the case of two dimensions. In this section we present algorithms for inverting these relationships for an estimate $\hat{O}(\underline{r})$ of the object profile in terms of the data. These algorithms have been derived elsewhere and will not be rederived here. The reader is referred to References 19 and 28 for detailed derivations.

The algorithms presented here are completely analogous to the so-called **filtered backprojection algorithm** of conventional X-ray tomography [7,8]. The appropriate generalization of the X-ray algorithm to the case of diffraction tomography requires, however, that the process of back**projection** be replaced by back**propagation**, whereby a filtered complex phase is made to propagate backwards through the image space. For this reason the algorithms are called **filtered backpropagation algorithms**.

Just as in the X-ray case, the reconstruction of the object profile is "built up" from a sequence of partial reconstructions, each partial reconstruction being generated from data obtained for a single incident plane wave. Each such "experiment" then corresponds to a fixed view angle in the classical configuration illustrated in Fig. 1 and employed in ultrasound tomography. In this paper we shall use the term "view angle" to also denote the angle of incidence of the incident plane wave. For the two-dimensional case, the view angle is, thus, the angle ϕ_0 made by the incident unit wavevector $\underline{s}_0$ with the positive z axis. The partial reconstruction is obtained by first convolutionally filtering the measured complex phase with a filter that will, in general, depend on the angle of incidence of the incident plane wave, i.e., the view angle ϕ_0. The filtered phase is then backpropagated into the image space. This step is usually implemented with a bank of convolutional filters - one filter for each depth coordinate in image space - although other schemes are possible [21]. In the classical configuration the filtered and backpropagated phase must then be interpolated from the (ξ,η) coordinate system to the fixed (reference) coordinate system [(x, z) system]. This step is not required for the case of a fixed measurement plane. The complete object reconstruction is obtained by summing up the partial reconstructions. A block diagram of the overall process is presented in Fig. 5.

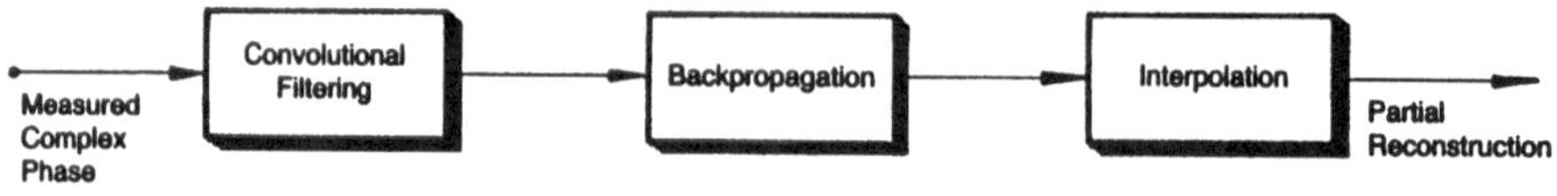

Fig. 5: Generating a partial reconstruction from a filtered backpropagation algorithm. For the case of a fixed measurement plane the convolutional filter will depend on the view angle but the interpolation step will not be required.

3.1 Classical Configuration

The filtered backprojection and filtered backpropagation algorithms are most similar for the classical configuration illustrated in Fig. 1. In this configuration, the convolutional filter employed in filtered backpropagation is identical to the convolutional filter employed in the X-ray algorithm [7,8]. In both of these cases the filtering operation is mathematically defined by

$$Q_{\phi_0}(t) \equiv \frac{k}{4\pi} e^{-ikl_0} \int_{-\infty}^{\infty} d\kappa\, \tilde{h}(\kappa) \tilde{D}(\kappa, \eta = l_0; \underline{s}_0) e^{i\kappa t}$$

$$= \frac{k}{2} e^{-ikl_0} \int_{-\infty}^{\infty} d\xi'\, D(\xi', \eta = l_0; \underline{s}_0) h(t-\xi'), \tag{3.1}$$

where $Q_{\phi_0}(t)$ denotes the filtered phase corresponding to view angle ϕ_0 and where

$$h(t) = \frac{1}{2\pi} \int_{-\infty}^{\infty} d\kappa\, \tilde{h}(\kappa) e^{i\kappa t}. \tag{3.2}$$

$h(t)$ is the convolutional filter and $\tilde{h}(\kappa)$ its Fourier transform. As discussed in Ref. 20, $\tilde{h}(\kappa)$ is usually implemented as the product of the "ideal" X-ray filter transform $|\kappa|$ with a low pass smoothing filter such as the Shepp and Logan filter [35] or a Blackman Harris filter [36].

The rational for referring to $Q_{\phi_0}(t)$ as a filtered *phase* is apparent if we express $D(\xi', \eta = l_0; \underline{s}_0)$ in terms of the *relative* phase

$$\delta\psi(\xi', \eta = l_0; \underline{s}_0) \equiv \Psi(\xi', \eta = l_0; \underline{s}_0) - ikl_0, \tag{3.3}$$

where ikl_0 is the phase of the field over the measurement plane in the absence of the object; i.e., $\delta\psi$ represents the perturbation in the phase of the incident plane wave introduced by the presence of the object.

Note that, in general, $\delta\psi$ will be complex. Upon expressing $D(\xi', \eta = l_0; \underline{s}_0)$ in Eq. (3.1) in terms of $\delta\psi$ we obtain

$$Q_{\phi_0}(t) \equiv \frac{i}{k}\int_{-\infty}^{\infty} d\xi' h(t-\xi')\delta\psi(\xi', \eta = l_0; \underline{s}_0) \,. \tag{3.4}$$

Eq. (3.4) is the most common way of expressing the filtered phase and is the one that is universally employed in X-ray CT [8]. It shows that, indeed, $Q_{\phi_0}(t)$ is a linearly filtered version of the relative phase $\delta\psi$.

The backpropagation process is defined mathematically as

$$\Pi_{\phi_0}(\xi,\eta) = \int_{-\infty}^{\infty} d\xi' Q_{\phi_0}(\xi')G(\xi-\xi', \eta-l_0) \tag{3.5}$$

where $\Pi_{\phi_0}(\xi,\eta)$ is the filtered and backpropagated phase and represents the partial reconstruction of the object profile at the view angle ϕ_0. $G(\xi,\eta)$ is the Green function governing the backpropagation of complex phase pertubations within the Rytov approximation in the background medium [19,37]. This quantity can be expressed in the following integral representation [37]

$$G(\xi - \xi', \eta - l_0) \equiv \frac{e^{-ik\underline{s}_0\cdot(\underline{r}-\underline{r}')}}{2\pi}\int_{-k}^{k} d\kappa e^{i[\kappa(\xi-\xi') + \gamma(\eta-l_0)]} \,, \tag{3.6}$$

where $\underline{r}' = (x' , z = l_0)$ denotes a point on the measurement plane. In the classical configuration $\underline{s}_0 = \hat{\underline{\eta}}$ so that $\underline{s}_0 \cdot \underline{r} = \eta$, $\underline{s}_0 \cdot \underline{r}' = l_0$ and (3.6) reduces to [19]

$$G(\xi - \xi', \eta - l_0) = \frac{1}{2\pi}\int_{-k}^{k} d\kappa e^{i[\kappa(\xi-\xi') + (\gamma-k)(\eta-l_0)]} \,, \tag{3.7}$$

which when substituted into Eq. (3.5) leads to the following computationally useful form for the filtered and backpropagated phase $\Pi_{\phi_0}(\xi,\eta)$:

$$\Pi_{\phi_0}(\xi,\eta) = \frac{1}{2\pi}\int_{-\infty}^{\infty} d\kappa\, \tilde{Q}_{\phi_0}(\kappa) e^{i(\gamma-k)(\eta-l_0)}\, e^{i\kappa\xi}, \tag{3.8}$$

where $\tilde{Q}_{\phi_0}(\kappa)$ is the Fourier transform of the filtered phase.

Eq. (3.8) is readily implemented using a Fast Fourier Transform (FFT) algorithm. In particular, by discretizing the (ξ , η) coordinate system into an N X N square array, $\Pi_{\phi_0}(\xi,\eta)$ is calculated for any fixed value of η (say, $\eta = \eta_n$, n = 1, 2, ..., N) by performing an N point FFT of the product of $\tilde{Q}_{\phi_0}(\kappa)$ with the backpropagation filter

$\exp[i(\gamma - k)(\eta_n - l_o)]$. Clearly, N, N point FFT's are required to determine $\Pi_{\phi_0}(\xi, \eta)$ at all N X N points in image space.

The final reconstruction of the object profile is obtained by interpolating each partial reconstruction $\Pi_{\phi_0}(\xi,\eta)$ onto the fixed (x,z) coordinate system and summing the interpolated partial reconstructions in a pixel by pixel fashion; i.e.,

$$\hat{O}(x,z) = \frac{1}{2\pi}\int_{-\pi}^{\pi} d\phi_0\, \Pi_{\phi_0}(z \sin\phi_0 - x\cos\phi_0,\, z\cos\phi_0 + x\sin\phi_0) \quad (3.9)$$

where $\hat{O}(x,z)$ is the estimate of the object profile generated by the algorithm and where we have made use of Eqs. (2.22) to denote the interpolation from the (ξ,η) system to the (x,z) system.

In the zero wavelength limit $(k\to\infty)$ the Green function $G(\xi,\eta)$ reduces to the Dirac delta function $\delta(\xi)$ [19] so that in this limit, $\Pi_{\phi_0}(\xi,\eta)$ is independent of η (depth) and equal to the filtered phase $Q_{\phi_0}(\xi)$. This is precisely what is done in X-ray tomography in what is known as backprojection. Thus, we have the result that for the classical configuration the filtered backpropagation algorithm reduces to the filtered backprojection algorithm in the zero wavelength limit.

Shown in Fig. 6 are reconstructions of four circular disks each having a radius "a" equal to five units. The wavelength was equal to one unit (corresponding to a ka = 31.4). Note that even at this relatively short wavelength significant focusing occurs in the backpropagation process so that reasonable reconstructions are obtained from very few view angles.

Fig. 7 shows the reconstruction of the circular Shepp and Logan head phantom for unit wavelength and one, seven, eleven and 101 views. The head phantom is 80.5 units across and is composed of circular disks ranging in radius from 1.15 to 38.41 units. (The reader is referred to Ref. 20 for further details). Fig. 8 compares the reconstructions obtained by filtered backpropagation with those obtained by filtered backprojection for wavelengths of 0.1 and 1.0 units. This comparison is of interest because, until recently, the filtered backprojection algorithm was commonly used in ultrasound tomography [2]. Note that for the $\lambda = 0.1$ case, the filtered backprojection algorithm yields a high quality reconstruction virtually indistinguishable from that obtained from filtered backpropagation. This result is a consequence of the fact that, as mentioned above, the two algorithms become identical in the very short wavelength limit.

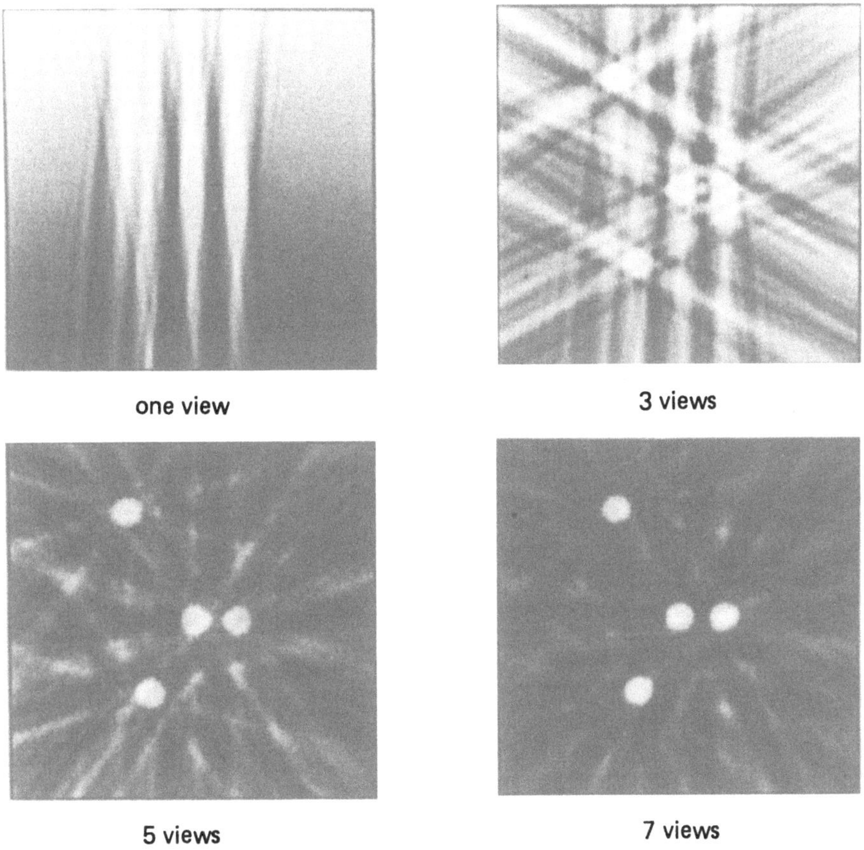

Fig. 6: Reconstructions of four circular disks for 1, 3, 5 and 7 views and a wavelength of 1 unit. Each disk has a radius of 5 units.

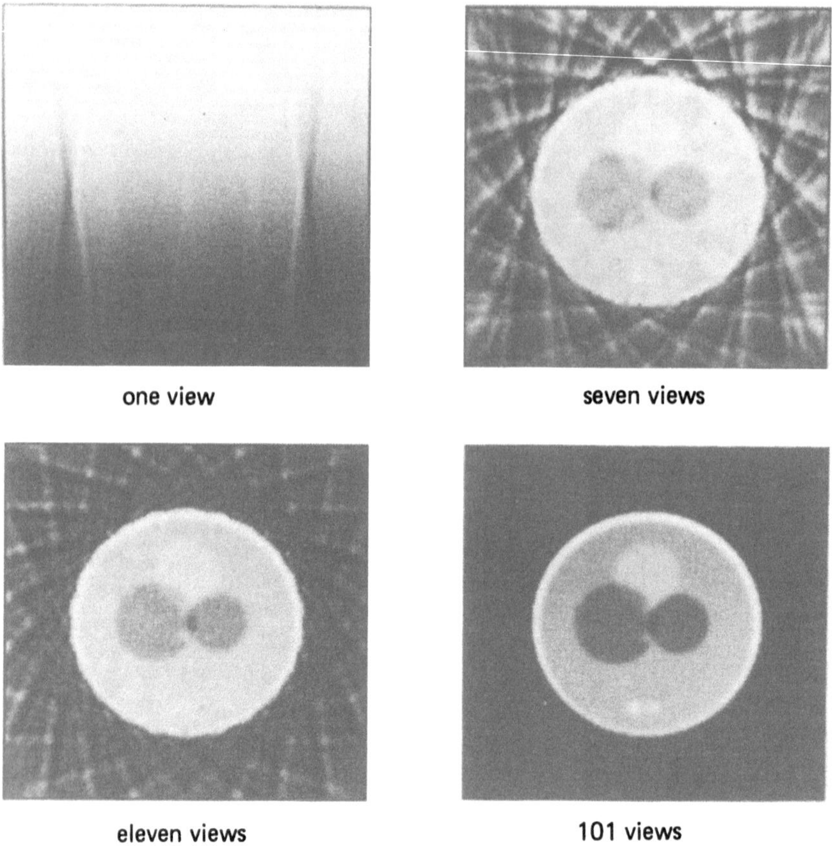

Fig. 7: Reconstructions of a circular Shepp and Logan head phantom. Wavelength was fixed at 1 unit and phantom is 80.5 units across.

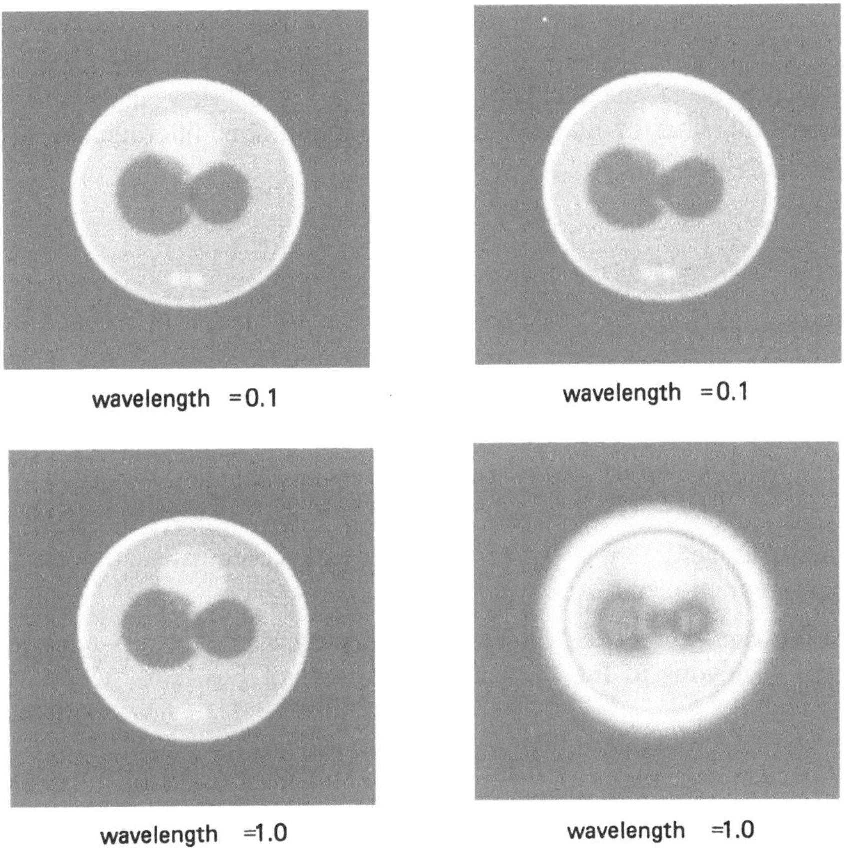

Fig. 8: Comparison of filtered backpropagation (left) with filtered back-projection (right).

The data in all of these simulations were generated using the Rytov approximation so that they can be expected to be representative only for weakly scattering objects for which this approximation holds. The question of the validity of the Rytov approximation is beyond the scope of this paper and can be found discussed in [16-18].

3.2 Fixed Measurement Plane

For most geophysical applications such as well-to-well tomography and offset vertical seismic profiling (offset VSP) a fixed measurement plane is employed. For such cases the convolutional filter in the filtered backpropagation algorithm will depend on the angle of incidence ϕ_0 of the incident plane wave - the "view angle." If we denote this filter by $h(x;\phi_0)$, then the convolutional filtering step is, in analogy with Eq. (3.4), given by

$$Q_{\phi_0}(t)=\frac{i}{k}\int_{-\infty}^{\infty} dx'\,\delta\psi(x', z=l_0; \underline{s}_0)h(t-x', \phi_0)\,. \tag{3.10}$$

Here, as in Eq. (3.4), $\delta\psi$ represents the perturbation in the phase of the incident plane wave introduced by the presence of the object. However, in the case of a fixed measurement plane the unperturbed phase $ik\underline{s}_0\cdot\underline{r}'$ is not constant over the measurement plane so that $\delta\psi$ is given by

$$\delta\psi(x', z=l_0; \underline{s}_0) \equiv \Psi(x', z=l_0; \underline{s}_0) - ik\underline{s}_0\cdot\underline{r}'\,, \tag{3.11}$$

where, as before, $\underline{r}'=(x', z=l_0)$ denotes a coordinate point on the measurement plane.

We can express the filtered phase in terms of $D(x', z=l_0;\underline{s}_0)$ in a form analogous to Eq. (3.1) by making use of Eqs.(2.6), (3.11) and the identity:

$$D(x', z=l_0;\underline{s}_0)e^{-ik\underline{s}_0\cdot\underline{r}'} \equiv \frac{e^{-iks_{oz}l_0}}{2\pi}\int_{-\infty}^{\infty} d\kappa\tilde{D}(\kappa+ks_{ox},z=l_0;\underline{s}_0)e^{i\kappa x'}. \tag{3.12}$$

We obtain:

$$Q_{\phi_0}(t) = \frac{k}{4\pi}e^{-iks_{oz}l_0}\int_{-\infty}^{\infty} d\kappa\tilde{D}(\kappa + ks_{ox}, z=l_0;\underline{s}_0)\tilde{h}(\kappa,\phi_0)e^{i\kappa t}\,, \tag{3.13}$$

where $s_{oz}=\cos\phi_0$, $s_{ox}=\sin\phi_0$. Note that when $s_{ox}=0$ then $s_{oz}=1$, and Eq. (3.13) reduces to Eq. (3.1). We also note, for future reference, that the Fourier transform of $Q_{\phi_0}(t)$ is, from Eq. (3.13), found to be

$$\tilde{Q}_{\phi_0}(\kappa) = \frac{k}{2}e^{-iks_{oz}l_0}\tilde{D}(\kappa + ks_{ox}\cdot, z=l_0;\underline{s}_0)\tilde{h}(\kappa,\phi_0)\,. \tag{3.14}$$

The precise functional form of the convolutional filter depends on the specific application and the total range of view angles employed. A detailed discussion of the appropriate filters for offset VSP is given in Ref. 28 while Ref. 29 gives the filter for well-to-well tomography. We give here the filter for cases, such as well-to-well tomography, where incident plane waves are employed whose unit wave vectors $\underline{s}_0$ make angles ϕ_0 with the $+z$ direction ranging from -90 to +90 degrees. For such cases $\tilde{h}(\kappa,\phi_0)$ is given by

$$\tilde{h}(\kappa,\phi_0) = |\kappa' s_{oz} - \gamma' s_{ox}|, \tag{3.15}$$

for $|\kappa| \leq k$ and zero otherwise. Here, $\kappa' = \kappa - ks_{ox}$ and $\gamma' = \sqrt{k^2 - \kappa'^2}$.

Following the convolutional filtering operation, the filtered phase $Q_{\phi_0}(t)$ is backpropagated into the object space which, for the case of a fixed measurement plane, is the region $z \leq l_0$. Mathematically, this operation is given by Eq. (3.5) with ξ, η replaced by x, z. The Green function $G(x-x', z-l_0)$ is given by Eq. (3.6) where, again, ξ, η are replaced by x , z . Note that for the case of a fixed measurement plane $\underline{s}_0 \cdot \underline{r}$ and $\underline{s}_0 \cdot \underline{r}'$ do not simplify as they do for the classical configuration so that Eq. (3.7) does not apply in this case. By substituting for the Green function from Eq. (3.6) into (3.12) we obtain, in analogy with Eq. (3.8), the following form for the filtered and backpropagated phase:

$$\Pi_{\phi_0}(x,z) = \frac{e^{-iks_{oz}z}}{2\pi} \int_{-\infty}^{\infty} d\kappa\, \tilde{Q}_{\phi_0}(\kappa - ks_{ox}) e^{i(\gamma - ks_{oz})(z-l_0)}\, e^{i\kappa x}. \tag{3.16}$$

Finally, substituting for $\tilde{Q}_{\phi_0}(\kappa - ks_{ox})$ from Eq. (3.14) we find that

$$\Pi_{\phi_0}(x,z) = \frac{k}{4\pi} e^{-ik\underline{s}_0 \cdot \underline{r}} \int_{-\infty}^{\infty} d\kappa\, \tilde{h}(\kappa - ks_{ox};\phi_0) \tilde{D}(\kappa, z=l_0;\underline{s}_0) e^{i[\gamma(z-l_0)+\kappa x]} \tag{3.17}$$

Note that the filter $\tilde{h}(\kappa - ks_{ox},\phi_0)$ is simply given by (3.15) with $\kappa' = \kappa$ and $\gamma' = \gamma$.

In the case of a fixed measurement plane each partial reconstruction Π_{ϕ_0} (x, z) is generated on the fixed x, z coordinate system. The final reconstruction of the object profile is obtained by simply summing the partial reconstructions in a pixel by pixel fashion over the available view angles. This then differs from what is required in the classical configuration where each partial reconstruction $\Pi_{\phi_0}(\xi,\eta)$ must first be interpolated onto the x, z coordinate system using Eq. (2.22) prior to final summation over view angles.

It is important to note that if the range of view angles is limited, as in the case of well-to-well tomography where $-90° \leq \phi_0 \leq 90°$, then the region in $\underline{K}$ space over which the transform of the object profile $\tilde{O}(\underline{K})$ is specified by the data will contain "holes"; i.e., there will be certain regions over which $\tilde{O}(\underline{K})$ will not be specified by the data. We show in Fig. 9 the region in $\underline{K}$ space over which $\tilde{O}(\underline{K})$ is specified when ϕ_0 spans the range $-90° \leq \phi_0 \leq 90°$. As discussed in the preceeding section, the object reconstruction obtained via the filtered backpropagation algorithm reduces mathematically to a convolution of the actual profile $O(\underline{r})$ with the transform of the function which is unity over the region illustrated in Fig. 9 and zero off this region [28].

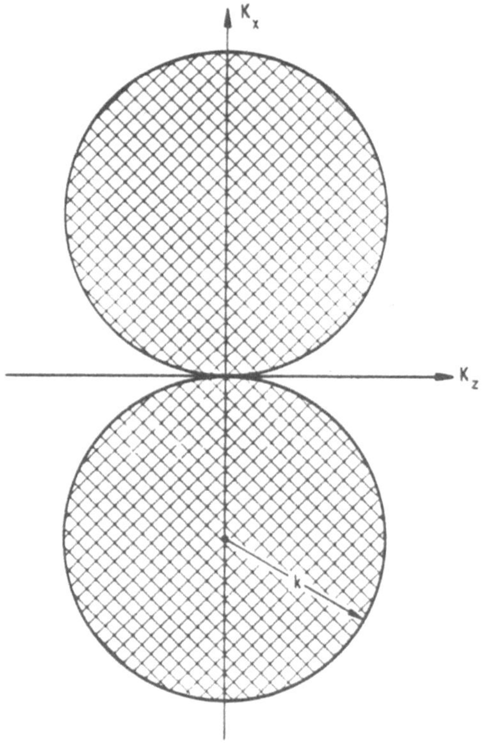

Fig. 9: Regions of coverage in Fourier space for well-to-well tomography employing plane waves forming angles ϕ_o ranging from -90 to +90 degrees.

Shown in Fig. 10 are well-to-well tomographic reconstructions of the four circular disks used in the simulation shown in Fig. 6. These reconstructions were obtained for a wavelength of five units (corresponding to a ka = 6.28) which is more representative of the structure to wavelength scales to be expected in well-to-well tomography then the ka value of 31.4 employed in Fig. 6. Note the inferiority of the reconstructions compared with the corresponding reconstructions obtained for the classical configuration. This reduction in resolution is due not only to the greater wavelength employed but also (and primarily) to incomplete coverage in Fourier transform space caused

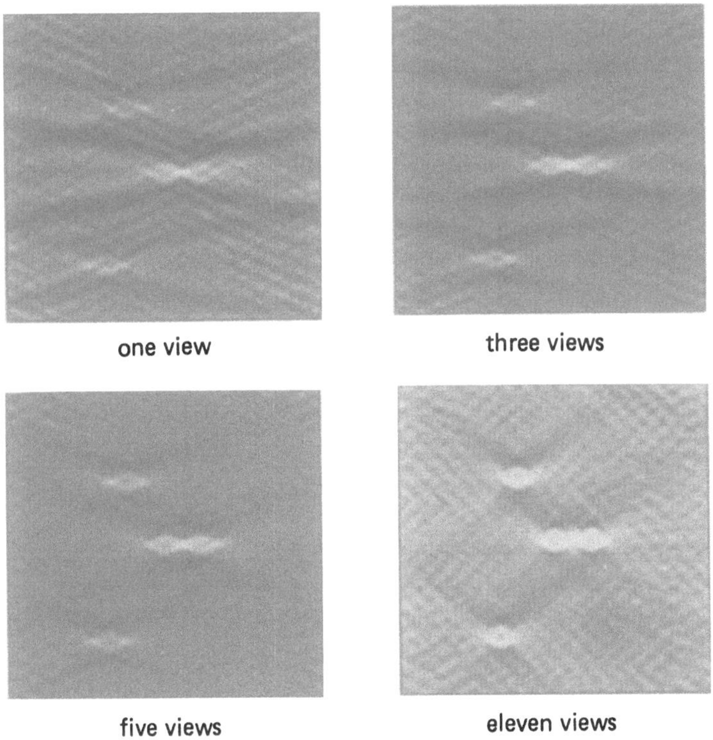

Fig. 10: Reconstructions of the four circular disks employed in Fig. 6 obtained in a well-to-well tomographic simulation. Wavelength employed was 5 units. Reduced resolution from that obtained in Fig. 6 is primarily due to incomplete coverage in Fourier space caused by limited view angle range in well-to-well tomography.

by limited view angle range imposed by the well-to-well geometry (see Fig. 9).

The reconstruction of a system of parallel layers is shown in Fig. 11. The reconstruction employed a wavelength of five units and 73 view angles (ϕ_0) equally spaced from -90 to +90 degrees. The reconstruction is essentially exact because the object contains no spatial frequencies along the K_z axis so that the "holes" in Fourier space leading to degradation in the image of the four circles shown in Fig. 10 have no effect in this case. It should be noted in this connection that for many geophysical applications the objects (geological formations) will be essentially layered so that the spatial frequency spectrum will be localized around the K_x axis so that the reconstructions can be expected to be quite good. As in the simulations performed for the classical configuration, the data for these simulations were generated using the Rytov approximation.

Fig. 11: Reconstruction of layered object (model formation) in a well-to-well tomographic simulation. The high resolution stems from the fact that the spatial frequency content of the object is concentrated about the K_x axis.

4. FUTURE DIRECTIONS

The current theory and algorithms of diffraction tomography are based on the Born or Rytov approximations. In some applications, such as ultrasound medical tomography of soft tissues, these approximations are probably adequate. In many others they will be inadequate, and the theory and algorithms presented here will no longer be valid. One of the most vital and interesting new directions for research in diffraction tomography will be to extend the theory and algorithms beyond the Born and Rytov approximations so as to have a wider domain of applicability. One possible avenue for achieving this generalization is through the use of the so-called distorted wave Born approximation DWBA [38]. The filtered backpropagation algorithms presented here are based on a certain inverse scattering identity valid within the Born and Rytov approximations [19,39,40]. This identity has recently been generalized to the DWBA [31] and it is hoped that this new identity will lead to generalized filtered backpropagation algorithms that hold within the DWBA.

A second avenue for future work will be to test the filtered backpropagation algorithms on real data. Greenleaf [3] has performed some preliminary testing on real data in ultrasound medical tomography that indicate that the algorithm performs significantly better than conventional filtered backprojection. More work, especially in geophysical diffraction tomography, is, however, required.

The filtered backpropagation algorithms presented here are for plane wave illumination - so-called "parallel beam" tomography. In many applications, such as well-to-well tomography, point source illumination is used. In order to use the filtered backpropagation algorithms in such cases it is then necessary to artificially create the plane wave response from the point source response using, for example, the method of "slant stack" [41]. An alternative, and in many ways preferable, procedure is to employ a "fan beam" algorithm that directly uses the point source response rather than the plane wave response. Fan beam filtered backpropagation algorithms have recently been developed [42] but, as yet, have not been tested either on computer generated data or real data. An important area for future work will focus on the implementation and testing of these algorithms.

Finally there is the extension of the theory and algorithms presented here to the electromagnetic and elastic wave cases. Yang and Boerner have investigated the electromagnetic case [43] while the present author has generalized the algorithms to the acoustic case where both the density and bulk modulus of the object can vary [44]. However, the full elastic wave case has, as yet, not been adequately treated and remains an area for future research.

REFERENCES

1. J. F. Greenleaf, S. A. Johnson, S. L. Lee, G. T. Herman, and E. H. Wood, "Algebraic reconstruction of spatial distributions of acoustic absorption with tissues from their two-dimensional acoustic projections," in *Acoustical Holography*, Vol.5, P. S. Green, Ed., New York: Plenum, 1974, pp. 591-603.
2. J. F. Greenleaf and Robert C. Bahn, "Clinical imaging with transmissive ultrasonic computerized tomography," *IEEE Trans. Biomed. Eng.,* Vol. BME-28, pp.177-185, 1981.
3. J. F. Greenleaf, "Computerized tomography with ultrasound," *Proc. IEEE*, Vol. 71, pp. 330-337, 1983.
4. K. A. Dines and R. J. Lytle, "Computerized geophysical tomography," *Proc. IEEE*, Vol. 67, pp. 1065-1073, 1979.
5. R.J. Lytle and K.A. Dines, "Iterative ray tracing between boreholes for underground image reconstruction," *IEEE Trans on Geoscience and Remote Sensing,* Vol. Ge-18, pp. 234-240, 1980.
6. E. J. Witterholt, J. L. Kretzsehmar and K. L. Jay, "The application of crosshole electromagnetic wave measurements to mapping of a steam flood," *Proceedings of the Petroleum Society of CIM,* 1982.
7. W. S. Swindell and H. H. Barrett, "Computerized tomography: taking sectional X-rays," *Physics Today,* pp. 32-41, Dec. 1977.
8. A. C. Kak, "Computerized tomography with x-ray emission and ultrasound sources," *Proc. IEEE,fR Vol. 67, pp. 1245-1272, 1979.*
9. *R. K. Mueller, M. Kaveh, and G. Wade, "Reconstructive tomography and applications to ultrasonics," Proc. IEEE,* Vol. 67, pp. 567-587, 1979.
10. R. K. Mueller, M. Kaveh, and R. D. Inverson, "A new approach to acoustic tomography using diffraction techniques," in *Acoustical Imaging*, Vol. 8, A. F. Metherell, Ed. New York: Plenum, 1980, pp. 615-628.

11. K. Iwata and R. Nagata, "Calculation of refractive index distribution from interferograms using the Born and Rytov's approximation," *Japan J. Appl. Phys.*, Vol. 14 Suppl. 14-1, pp. 379-383, 1974.

12. E. Wolf, "Three-dimensional structure determination of semitransparent objects from holographic data," *Optics Commun.*, Vol. 1, pp. 153-156, 1969.

13. H. Lipson and W. Cochran, *The determination of crystal structures,* Ithaca, New York: Cornell University Press, 1966.

14. B. K. Vainshtein, *Diffraction of X-rays by chain molecules,* New York: John Wiley, 1974.

15. V.I Tatarski, *Wave propagation in a turbulent medium* New York: McGraw-Hill, 1961, Chap. 7.

16. A.J. Devaney, "Inverse scattering within the Rytov approximation," *Opts. Letts.* Vol. 6, pp. 374-376, 1981.

17. J.B. Keller, "Accuracy and validity of the Born and Rytov approximations," *J. Opt. Soc. Am.*, Vol. 59, pp. 1003-1004, 1969.

18. M. Kaveh, M. Soumekh and R.K. Mueller, "A comparison of Born and Rytov approximations," in *Acoustical Imaging,* Vol. 11, J. Powers, Ed. New York: Plenum, 1981, pp. 325-335.

19. A.J. Devaney, "A filtered backpropagation algorithm for diffraction tomography," *Ultrasonic Imaging,* Vol. 4, pp. 336-350, 1982.

20. A.J. Devaney, "A computer simulation study of diffraction tomography," *IEEE Trans. Biomed. Eng.,* Vol. BME-30, pp. 377-386, 1983.

21. M.L. Oristaglio, "A geometric approach to the filtered backpropagation algorithm," *Ultrasonic Imaging,* Vol. 5, pp. 30-37, 1983.

22. S.X. Pan and A.C. Kak, "A computational study of reconstruction algorithms for diffraction tomography," *IEEE Trans. Acoustics, Speech and Signal Processing* (to appear).

23. Malcolm Slaney and A.C. Kak, "Diffraction tomography," *Inverse Optics*, ed. A.J. Devaney, Proc. SPIE Vol. 413, pp. 2-19, 1983.

24. R. M. Mersereau and A. V. Oppenheim, "Digital reconstruction of multidimensional signals from their projections," *Proc. IEEE,* Vol. 62, pp. 210-229, 1974.

25. H. Stark and I. Paul, "An Investigation of Computerized Tomography by Direct Fourier Inversion and Optimum Interpolation," *IEEE Trans. Biomedical Eng.,* Vol. BME-28, pp. 496-505, 1981.

26. M.F. Adams and A.P. Anderson, "Synthetic aperture tomographic (SAT) imaging for microwave diagnostics," *Proc. IEE,* pt.H, Vol. 129, pp. 83-88, 1982.

27. A.P. Anderson and M.F. Adams, "Synthetic aperture tomographic imaging for ultrasonic diagnostics," in *Acoustical Imaging,* Vol. 12, E.A. Ash, Ed., New York: Plenum, 1982, pp. 565-578.

28. A.J. Devaney, "Geophysical diffraction tomography," *IEEE trans. Geoscience and Remote Sensing* - special issue on electromagnetic methods in applied geophysics - (to appear Jan. 84).

29. A.J. Devaney, "Well-to-well tomography using diffracting wavefields," presented at the 1983 geoscience and remote sensing conference "IGARSS'83". To be submitted for publication.

30. A.J. Devaney and E. Wolf, "A new perturbation expansion for inverse scattering from three-dimensional finite-range potentials," *Phys. Letts.,* Vol. 89A, pp. 269-272, 1982.

31. A.J. Devaney and M.L. Oristaglio, "Inversion procedure for inverse scattering within the distorted-wave Born approximation," *Phys. Rev. Letts.,* Vol. 51, pp. 237-240, 1983.

32. A.J. Devaney "Inverse scattering as a form of computed tomography," in *Applications of mathematics in modern optics*, ed. H. Carter, Proc. SPIE, Vol. 358, pp. 10-16, 1982.

33. J. W. Goodman, *Introduction to Fourier Optics*, New York: McGraw-Hill, Sec.3-7, 1968.

34. F. Stenger and S. A. Johnson, "Ultrasonic transmission tomography based on the inversion of the Helmholtz equation for plane and spherical wave insonification," *Appl. Math. Notes,* Vol. 4, pp. 102-127, 1979.

35. L.A. Shepp and B.F. Logan, "The Fourier reconstruction of a head section," *IEEE Trans. on Nuclear Sci.,* Vol. NS-21, pp. 21-43, 1974.

36. F. J. Harris, "On the use of windows for harmonic analysis with the discrete Fourier transform," *Proc. IEEE,* Vol. 66, pp. 51-83, 1978.

37. A.J. Devaney, H. Liff, and S. Apsell, "Spectral representations for free space propagation of complex phase perturbations of optical fields," *Opt. Commun.*, Vol. 15, pp. 1-5, 1975.

38. R.J. Taylor, *Scattering Theory*, New York: Wiley, 1972.

39. A.J. Devaney, "Inversion formula for inverse scattering within the Born approximation," *Opt. Letts.*, Vol. 7, pp. 111-112, 1982.

40. G. Beylkin, "The fundamental identity for iterated spherical means and the inversion formula for diffraction tomography and inverse scattering," *J. Math. Phys.*, Vol. 24, pp. 1399-1400, 1983.

41. R. A. Phinney, K. Kay Chowdhury and L. Neil Frazer, "Transformation and analysis of record sections," *Jour. Geophys. Res.*, vol. 86, pp. 359-377, 1981.

42. A.J. Devaney and G. Beylkin, "A filtered backpropagation algorithm for fan beam diffraction tomography," in *Acoustical Imaging*, Vol. 13 (Plenum Press, New York) to be published.

43. Chau-Wing Yang and W. Boerner, "Vector diffraction tomography," see paper this proceedings.

44. A.J. Devaney, "Inverse source and scattering problems in ultrasonics," *IEEE trans. Sonics and Ultrasonics* (to appear Nov. 1983).

V.6 (PT.3)

ALGORITHMS AND ERROR ANALYSIS FOR DIFFRACTION TOMOGRAPHY USING THE BORN AND RYTOV APPROXIMATIONS

M. Kaveh and M. Soumekh

Department of Electrical Engineering
University of Minnesota
Minneapolis, MN 55455

ABSTRACT

In this paper, diffraction tomography is presented as an inverse problem of an inhomogenous differential equation obtained from the Helmholtz equation in terms of a general transformation of the propogating wave. Born and Rytov transformations are shown to be special cases of this general form. Frequency and spatial domain algorithms are given for the reconstruction of a spatially varying index of refraction that is under study. Errors in the images based on Born and Rytov's approximations are compared and numerical and experimental results on these techniques are presented.

Introduction

The objective of diffraction tomography is to reconstruct, from the observed scattered field, the spatial distribution of appropriate parameters in a test object. The data is normally obtained from measurements of the scattered field from a number of object orientations. The mathematical nature of the reconstruction is then one of solving an inverse problem of the appropriate equation of motion. The challenge, of course, is in obtaining a solution that is computationally efficient and subject to reasonable interpretation. This in general is not possible. In many cases of interest, however, the relatively small perturbation of object parameters compared to the surrounding medium give the possibility of using approximations to the equation of motion based on perturbation techniques. These techniques substantially simplify the inversion problem and lend themselves to computational advantages of fast Fourier transform processing.

W.-M. Boerner et al. (eds.), Inverse Methods in Electromagnetic Imaging, Part 2, 1137–1146.

In its more common form, then, diffraction tomography is formulated as an approximate inversion of the Helmholtz equation with a spatially varying coefficient. This inversion is accomplished in practice by first transforming the Helmholtz equation into an inhomogeneous one with constant coefficients. Two transformations leading to two different approximate inversions have been most commonly used. These are the Born and Rytov transformations. The appealing feature of these methods is the fact that the inversion algorithms are non-iterativew and computationally efficient. Computational efficiency and algorithmic simplicity, however, is accompanied by limitations on the ranges of validity for these methods.

In this paper we first introduce the Born and Rytov approximations as special cases resulting from a generalized transformation of the Helmholtz equation. We then review methods of practical image reconstruction and comment on some of the signal processing issues of concern to the Rytov transformation. This discussion is then followed by numerical examples that treat the consequences of both the approximations in the Born and Rytov method and practical signal processing algorithms on the reconstructed images. Finally some experimental results, using these techniques are presented.

The Wave Equation and a Generalized Transformation

In the following we consider the excitation of a soft isotropic scattering medium with a single frequency plane-wave. For a two-dimensional geometry, the equation describing the propagating wave is assumed to be given by

$$\nabla^2\psi + k^2\psi = 0 \tag{1}$$

where ∇^2 is the Laplacian operator, $k(x,y)$ is a complex wave number including velocity of propagation changes and possible attenuation in the medium under study and $\psi(x,y)$ is the propagating wave. We model $k^2(x,y)$ as:

$$\begin{aligned} k^2(x,y) &= k_0^2[1 + f(x,y)] \\ k_0 &= \frac{\omega_0}{C_0} + i\varepsilon, \ \varepsilon << 1, \ \left| f(x,y) \right| << 1 \end{aligned} \tag{2}$$

k_0 is the complex (small attenuation) wave number of the homogeneous medium surrounding the object and $f(x,y)$ is the complex function describing the object that is to be imaged. The imaging procedure is to estimate $f(x,y)$ from measurements of $\psi(x_0,y)$ along a linear array at x_0, external to the object, for different angular orientations of the object [1-3]. This can be accomplished, approximately, by a preliminary transformation on $\psi(x,y)$.

The aim of expressing equation (1) in terms of a function of ψ is to approximate the resulting equation as a linear constant coefficient system relating the resulting function of the measurement to f(x,y), this in turn making computationally efficient reconstruction possible. Thus, let $\psi_0 = A_0 e^{ik_0 x}$ be the incident wave. Define χ as a differentiable scalar function of $\alpha = \frac{\psi}{\psi_0}$. Thus let

$$\chi = g(\alpha) \text{ and } \chi_1 = \psi_0 g(\alpha) \tag{3}$$

Expressing equation (1) in terms of χ_1, α and g(•) one obtains:

$$\nabla^2 \chi_1 + k_0^2 \chi_1 = -\psi_0 \frac{dg(\alpha)}{d\alpha} \alpha k_0^2 f + \psi_0 \frac{d^2 g(\alpha)}{d\alpha^2} (\nabla\alpha)^2 = -\psi_0 h \tag{4}$$

The operator on left hand side of equation (4) is a familiar one. This is the system used in the inversion based on Born and Rytov approximations. Indeed appropriate choices of the transformation g(•) do lead to these two formulations. The general form of equation (4), however, gives one a chance of investigating other transformations as well. This, without computational concerns about the actual inversion algorithm beyond the initial transformation.

An interesting feature of equation (4) is the manner is which the function of interest, f(x,y) and other extraneous terms appear. Following the approach in references [1-4], the ideal equation that results in a distortionless reconstruction of f(x,y) is of the form

$$\nabla^2 \chi_1 + k_0^2 \chi_1 = -b\psi_0 f \tag{5}$$

where b is a constant. Therefore, for a given g(•), the terms appearing on the right hand side of (4) that are extra to that in (5) will be considered as distortion terms. A good transformation would minimize these extra terms. In any case the actual reconstructed image is that of h rather than the ideal f.

We will now reexamine the Rytov and Born approximations as results of special forms of g(•). A class of transformations which includes the former two as limiting cases is given in [5].

i) Rytov transformtion. One obtains this form by noting from equations (4) and (5) that for b=1

$$\frac{dg(\alpha)}{d\alpha} \alpha = 1 \text{ and } g(\alpha) = \ell n a \tag{6}$$

It is important to note the very desirable property that this transformation removes the modulation of f(x,y) by a spatially dependent function that depends on the scattered wave. The distortion term in this approximation results from $D_r = -(\nabla\alpha)^2/\alpha^2\psi_0$.

ii) The Born transformation. In this case $g(\alpha) = \alpha - 1 = \frac{\psi - \psi_0}{\psi_0}$ and the distortion term is due to $D_b = -(\psi - \psi_0)k_0 f$ resulting in the very limited range of validity of this approximation [3].

Inversion

In this section we briefly discuss the reconstruction algorithm for h(x,y) from measurements of $\psi(x_0,y)$ for different orientation of f(x,y) given by rotation angles $\{\theta_i\}_i$. Thus let $\underline{\Theta}$ denote the usual rotation operator by an angle θ and denote various functions obtained for an object rotation θ by superscript θ. The general equation (4) can now be expressed as

$$(\nabla^2 + k_0^2)\chi_1^{(\theta)} = -\psi_0 h^{(\theta)} \quad , \theta = \theta_1, \ldots, \theta_\kappa \tag{7}$$

Thus, irrespective of the transformation g(•) used, equations (7) can be inverted using a general formulation as shown below. The quality of the actual reconstructed image h (i.e. its proximity to $k_0 f$) depends of course, on the transformation used.

Let $[\mu,\upsilon]$ represent the spatial frequency vector and '~' denote the spatial Fourier tranform of the function over which it appears. Then, it can be shown [4] that an equation of the form (7) can be inverted in the frequency domain as

$$\tilde{h}([\mu_0,\upsilon]\underline{\Theta}^{-1}) \propto \tilde{T}(\Theta,\upsilon)\exp(-j\mu_0 x_0)\sqrt{k_0^2-\upsilon^2} \tag{8}$$

where

$$\mu_0 = -k_0 + \sqrt{k_0^2-\upsilon^2} \tag{9}$$

$$\tilde{T}(\Theta,\upsilon) = F.T._{(y)}\,[\chi^{(\theta)}(x_0,y)] \tag{10}$$

Note that for values of $|\upsilon| < k_0$, the pair (μ_0,υ) trace a line contour in the spatial frequency domain. From (9), it can be seen that this contour is in form of a semicircle that passes through the origin. We call the contours defined by

$$[\mu,\lambda] \equiv [\mu_0,\upsilon]\,\underline{\Theta}^{-1}. \tag{11}$$

i.e. the rotated versions of $[\mu_0, \upsilon]$ contour, the object reconstruction contours (RCONs) in the spatial frequency domain.

We associate a small imaginary part, ε_0, with k_0 to represent loss in the homogeneous medium. Denote the Laplace transform of $h(x,y)$ by $L(s_1,s_2)$. The inversion equation (8) can be rewritten as:

$$L(s_1,s_2) = \tilde{U}(\theta,\upsilon) \tag{12}$$

where

$$U(\theta,\upsilon) \equiv T(\theta,\upsilon)\ \sqrt{k_0^2 - \upsilon^2}\ e^{-s_0 x_0} \tag{13}$$

and

$$s_0 = j(-k_0 + \sqrt{k_0^2 - \upsilon^2})$$

$$\begin{bmatrix} s_1 \\ s_2 \end{bmatrix} = \begin{bmatrix} \gamma_1 + j\mu_1 \\ \gamma_2 + j\lambda_1 \end{bmatrix} \equiv \underline{\Theta}^{-1} \begin{bmatrix} s_0 \\ j_\upsilon \end{bmatrix} = \begin{bmatrix} \cos\theta & -\sin\theta \\ \sin\theta & \cos\theta \end{bmatrix} \begin{bmatrix} s_0 \\ j\upsilon \end{bmatrix} \tag{14}$$

The Jacobian of the transformation from (s_1,s_2) to (θ,υ) can be found from (14) to be given by:

$$\frac{\partial(s_1,s_2)}{\partial(\theta,\upsilon)} = \tilde{J}(\upsilon,\theta) = \frac{k_0|\upsilon|}{\sqrt{k_0^2 - \upsilon^2}} \tag{15}$$

From the inverse Laplace transform integral equations, we have:

$$h(x,y) = \iint L(s_1,s_2)\exp(s_1 x + s_2 y)\ ds_1 ds_2 \tag{16}$$

With help of (12) and (13), we make variable tranformations in (16) to obtain

$$h(x,y) = \iint J(\theta,\upsilon)\ U(\tilde{\theta},\upsilon)\exp[(\alpha_1 + j\mu_1)x + (\alpha_2 + j\lambda_1)y]\ d\upsilon d\theta \tag{17}$$

Equation (17) is the basis of the interpolation algorithms that follow.

i) Spatial domain interpolation (backpropogation [7])

This method, developed by Devaney [7], can be derived by rewriting (17) as:

$$h(x,y) = \iint \tilde{J}(\theta,\upsilon)\ \tilde{U}(\theta,\upsilon)\ \exp\{[x,y] \begin{bmatrix} s_1 \\ s_2 \end{bmatrix}\}\ d\upsilon d\theta \tag{18}$$

Denote $[x',y'] \equiv [x,y]\ \underline{\Theta}^{-1}$. From (18) we have:

$$h(x,y) = \int [\int J(\theta,\upsilon)\, U(\theta,\upsilon)\exp(s_0 x' + j\upsilon y')d\upsilon]\, d\theta \quad (19)$$

Introduce a new two-dimensional function I_Θ by:

$$I_\Theta(x',y') = F.T.^{-1}(\upsilon)[\tilde{J}(\theta,\upsilon)\, \tilde{U}(\theta,\upsilon)\exp(s_0 x')] \quad (20)$$

Using (20) in (19) gives:

$$h(x,y) = \int I_\theta(x',y')\, d\theta$$

or

$$h(x,y) = \int I_\Theta(x\cos\theta + y\sin\theta, -x\sin\theta + y\cos\theta)\, d\theta \quad (21)$$

Equation (21) is approximated to interpolate $f(x,y)$ in a spatial grid [7]. Assume K profiles in an object grid of N x N is used. Then it can be shown that this algorithm requires (K.N) N-point FFT and assignment of $K \cdot N^2$ points in (x,y) domain [6].

ii) Frequency domain interpolation

Taking the Fourier transform of equation (21) leads to:

$$\tilde{h}(\mu,\lambda) = \int\int J(\theta,\upsilon)U(\theta,\upsilon)H(\mu,\mu_1,\gamma_1)\cdot H(\lambda,\lambda_1,\gamma_2)\, d\upsilon d\theta \quad (22)$$

where

$$H(\eta,\beta,\gamma) \equiv \int e^{(\gamma+j\beta-j\eta)x}dx$$

Equation (22) is approximated to interpolate $h(\mu,\lambda)$ in a spatial frequency grid. To obtain $h(x,y)$, a two-dimensional inverse FFT is performed on h. This algorithm requires (2.N) N-point FFT assignment of K.N points in (μ,λ) domain. Various versions of this approach are discussed in [6].

It is important to consider an important practical problem that arises with the use of Rytov's approximation. This is the problem of determining the unwrapped phase of the normalized field in conjunction with the natural logarithm of a complex function, α, along y. Let $\phi(\Theta,y)$ denote the phsae of $\alpha^{(\theta)}(x_0,y)$. We derive $\phi(\theta,y)$ from its principal value, $[\phi(\theta,y)]$, by using the following reasonable physical constraints that

i) Observations are made in a homogeneous medium. Therefore, ϕ is a continuous function of y.

ii) ϕ is also a continuous function of θ.

iii) On the edges of the reciever line, i.e. $y=\pm a_0$, $\psi^{(\theta)} \approx \psi_0$ which implies that $\phi(\theta,\pm a_0) << \pi$. As a result we have: $\phi(\theta,\pm a_0)=[\phi(\theta,\pm a_0)]$.

To utilize the above constraints for the available finite and discrete data base, we restate (i) and (ii) as follows:

i) $|\phi(\theta_i,y_j) - \phi(\theta_{i,j-1})| < \pi$

ii) $|\phi(\theta_i,y_j) - \phi(\theta_{i-1},y_j)| < \pi$

On the basis of the above constraints, we outline the following phaseunwrapping algorithm to obtain ϕ from $[\phi]$:

1) Consider the ith profile. From (iii) we can set $\phi(\theta_1,-a_0)=[\phi(\theta_1,-a_0)]$

2) Set $\phi(\theta_i, y_j) = [\phi(\theta_i, y_j)] + 2\pi\ell$ where ℓ is an integer that ensures (i).
Repeat step (2) for all values of j starting with the sample adjacent to $y=-a_0$.
3) Test the unwrapped value of $\phi(\theta, a_0)$ for (iii). If it satisfies the constraint, the ith profile is properly unwrapped. Go to step (6).
Otherwise,
4) Set $\phi(\theta_i, y_j) = [\phi(\theta_i, y_j)] + 2\pi\ell$ where ℓ is an integer that ensures (ii).
Repeat step (4) for all values of j.
5) Lowpass-filter the unwrapped values of the ith profile that were obtained in step (4) in the spatial frequency domain of y.
6) Increment i and go to step (1).
This procedure has yielded accurate phase determination in several experimental situations[5].

Numerical and Experimental Examples

In this section we present some examples comparing the Born and Rytov approximations in diffraction tomography. The comparative examples between Born and Rytov methods are conclusive. Theoretical analysis of the approximation errors of the two method as they pertain to diffraction tomography can be found in reference [6]. The first example clearly demonstrates the breakdown of Born's approximation, due to the distortion term which directly depends on the scattered waves from a cylindrical complex f that were numerically generated. Figure 1 shows the Born and Rytov reconstructions for this simulation. The distortion terms discussed earlier are prominent here.

The next example is a tomogram of an experimental phantom. This phantom is a 60λ diameter gelatine cylinder with two 8λ holes in water. 64 samples/profile at 2λ spacing and 64 profiles were used. The disturbance imaged by the Rytov method (Figures 2(a) and (b)) is mainly real, indicating the difference in the velocity of propagation. The tomogram based on Born's method (Figures 2(c) and (d)) is again in error, with the real part washed out and the imaginary part incorrecly showing an outline of the object. These examples show the uselessness of the Born transformation in most tomographic applications involving a large amount of forward scattering.

Figure 3 shows reconstructed profiles of a two dimensional Gaussian object using the spatial (S) and Fourier domain (F) interpolations. The reconstructions are the result of using the values of the object Fourier transform as the input on the RCONS. The Fourier domain interpolation, clearly shows a better reconstruction for this type of an object. This is perhaps not surprising, in that f is a low-pass function and in the lower frequency region the Fourier method performs well.

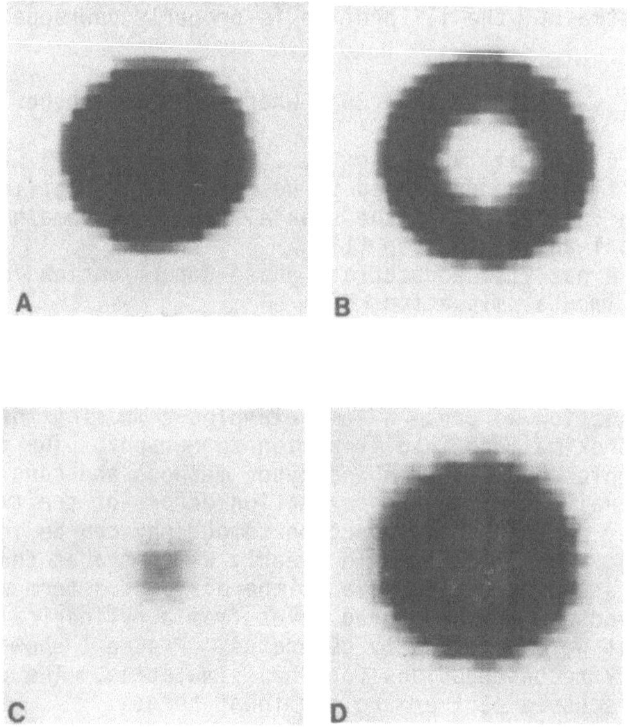

Figure 1. Complex annular cylinder. Real [f(r)]= -0.1, r<2λ; Im[f(r)] = 0.19, λ<r<2λ. Rytov: (a) Real, (b) Imaginary; Born: (c) Real, (d) Imaginary.

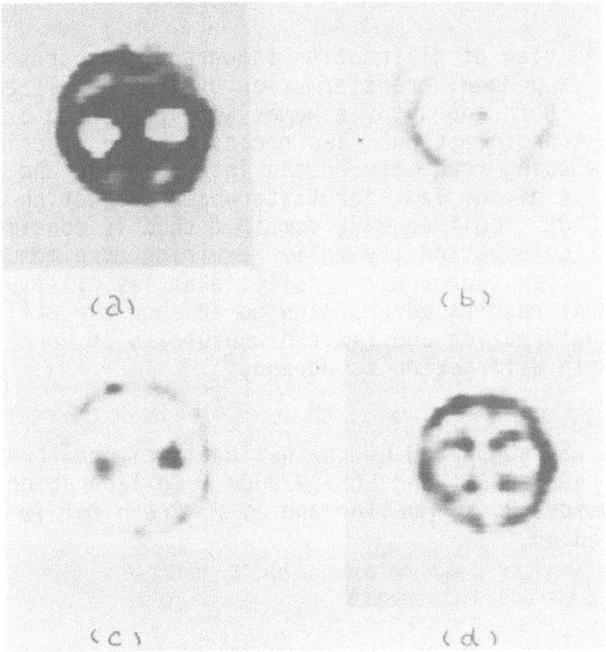

Figure 2. 60λ diameter gelatine cylinder with 8λ holes.
Rytov: (a) Real, (b) Imaginary; Born: (c) Real, (d) Imaginary.

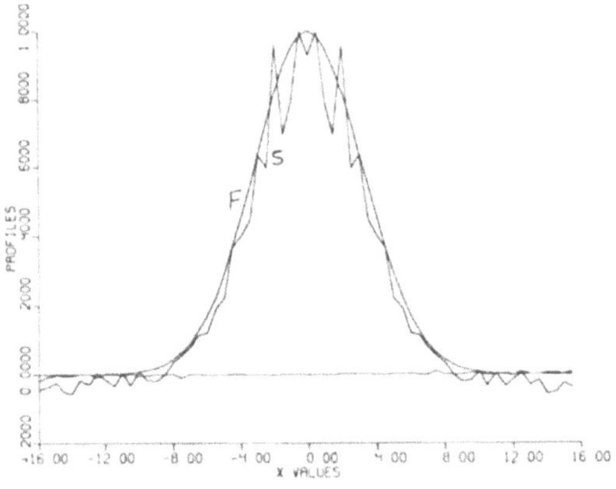

Figure 3. Cuts of spatial and Fourier domain reconstructions of a Gaussian object function.

Conclusion

A different view of diffraction tomography was presented as a consequence of a general transformation of the normalized measured wave. Born and Rytov schemes were presented as special cases of this transformation. Two reconstruction algorithms were presented, one using frequency domain interpolation and the other a generalization of the familiar filter-back projection method developed in [7]. The frequency domain method is considerably more efficient computationally while requiring more memory than the backpropagation approach. Finally, examples of simulation and experimental results were presented to show the utility of the Rytov transformation and limited usefulness of Born's approximation in diffraction tomography.

Acknowledgement

This work was supported by the National Science Foundation under Grants ENG76-84521 and ECS-7926008. Collaboration on this work by Professors R. K. Mueller and J. F. Greenleaf is gratefully acknowledged.

References

1. M. Kaveh, R. K. Mueller and R. D. Iverson, Ultrasonic tomography based on perturbation solutions of the wave equation, Computer Graphics and Image Processing, 9-1:105 (1979).

2. R. K. Mueller, M. Kaveh and G. Wade, Reconstructive tomography and applications to ultrasonics, IEEE Proceedings, 67:567 (1979).

3. M. Kaveh, M. Soumekh and R. K. Mueller, A comparison of Born and Rytov approximations in acoustic tomography, in: "Acoutical Imaging, Vol. 10," J. Powers, ed., Plenum, New York (1981).

4. M. Kaveh, M. Soumekh and R. K. Mueller, ICASSP '82 Proceedings, Paris (1982).

5. Z. Q. Lu, Internal Memo, Department of Electrical Engineering, University of Minnesota (1982)

6. M. Soumekh, Ph.D. Dissertation, University of Minnesota (1983).

7. A. Devaney, A filtered backpropagation algorithm for diffraction tomography, Ultrasonic Imaging, Vol. 4 (1982).

V.7 (PT.5)

EXTENSION OF SCALAR TO VECTOR PROPAGATION TOMOGRAPHY - A COMPUTER NUMERICAL APPROACH

Brian D. James, Chau-Wing Yang and Wolfgang-M. Boerner

Communications Laboratory, Electromagnetic Imaging Div.,
Department of Electrical Engineering & Computer Science,
M/C 154, University of Illinois at Chicago
851 S. Morgan St., P.O. Box 4348, Chicago, IL 60680, USA

Abstract: The basis for developing projection tomographic reconstruction algorithms has been the assumption of straight-line ray path propagation. But, in the case in which propagation occurs within discretely inhomogeneous media at wavelengths of the order of the size of the scatterer, phenomena such as refraction, reflection and diffraction can no longer be neglected and a straight -line projection tomographic approach fails. This is especially evident when a large difference in refractive index occurs, such as that encountered with electromagnetic dm-to-mm wave propagation in inhomogeneous media, representing hydrometeorite distributions, the marine ocean boundary layer, the ground surface underburden [23,25], or bone and soft layers within soft tissue [7,8]. In an effort to decrease the amount of energy imparted to the object, a long wavelength analysis was developed and applied to ultrasound tomography and diffraction tomography most recently. These extensions to ultrasonic projection tomography represented scalar diffraction along spherical wavefronts using Green's functions. In an effort to increase the applicability of these methods to the electromagnetic case, it is natural to extend this scalar diffraction approach to the more general vector diffraction approach. An exact solution for the general vector scattering case which strictly requires a polarimetric radiative transfer approach, is not available. However, in this paper, the assumption is made that the medium is weakly diffracting so that an iteration scheme can be introduced which will converge in one iteration using the Born/Rytov first approximation. To compare with existing methods in diffraction tomography, we shall use a linear iteration scheme involving Green's dyadic functions.

W.-M. Boerner et al. (eds.), Inverse Methods in Electromagnetic Imaging, Part 2, 1147–1163.

1. STATEMENT OF THE PROBLEMS

1.1 The Diffraction Problem:

In this paper, we are investigating the interaction of electromagnetic waves with a geometrically fixed material body which may be considered as a "polarization-sensitive scatterer with spatial and temporal resonance features". We shall assume that the object and clutter (object interior) are at rest so that temporal and doppler information will be excluded from this specific analysis. Within the dm-to-mm wavelength region obstacle sizes within the clutter become of the order of the wavelength, and thus, strong depolarization mechanisms must be incorporated into the wave-target interrogation formalism. This, in turn, requires a complete polarimetric radiative transport approach to the propagation description. However, no solution for this sophisticated vector scattering problem exists, even for the direct problem. Therefore, we introduce an iterative approach with the intent to uniformly converge to the measured propagated electromagnetic wave. For comparison with existing work on scalar diffraction tomography, we assume Born and Rytov approximations that are used as initial values in the iteration process.

In Section 2, we develop a linear iteration equation that is used to solve the (vector) nonhomogeneous-Helmholtz equation for the measured electric field $\underline{E}(\underline{r})$. This nonhomogeneous-Helmholtz equation is a result of separating Maxwell's equations. For comparative purposes, we choose the standard linear iteration which has a transform kernel in the form of a space-invariant field propagator $\bar{\bar{g}}(\underline{r})$ (often chosen to be a series of Green's dyadic functions), multiplied by the characteristic matrix $\bar{\bar{f}}(\underline{r})$ (which is associated with the object profile function $O(\underline{r})$). In choosing a kernel of this form, the field propagator must satisfy the Helmholtz equation for an impulse response. The iteration transform, denoted $I\{..\}$, and its inverse, denoted $I^{-1}\{..\}$, can be described by:

$$(1)\quad I\underline{E}(\underline{r}) = \int d\underline{r}'(\bar{\bar{g}}(\underline{r}-\underline{r}') \cdot \bar{\bar{f}}(\underline{r}')) \cdot \underline{E}(\underline{r}')$$

$$(2)\quad I^{-1}\underline{E}(\underline{r}) = F^{-1}F^{-1}\,\{F[\bar{\bar{g}}(\underline{\mu})^{-1} \cdot FI\underline{E}(\underline{\mu})]\,(\underline{\nu}) \cdot FF\bar{\bar{f}}(\underline{\nu})^{-1}\}\,(\underline{r})$$

where the Fourier transform and its inverse are denoted as $F\{..\}$ and $F^{-1}\{..\}$, respectively, and $\underline{\mu},\underline{\nu}$ are Fourier coordinates.

Then when $\underline{E}(\underline{r})$ is the initial approximation to the propagated field, we apply N applications of $I\{..\}$ to $\underline{E}(\underline{r})$ until convergence is obtained for N sufficiently large.

1.2 The Reconstruction Problem:

The basis for developing tomographic reconstruction algorithms has been the usage of Radon's transform and its associated inverse transform to reconstruct object profile information from limited samples outside the object (inside the support of the profile function). Radon's transform has been extensively surveyed for problems where straight-line propagation of the electromagnetic field is appropriate [Dean(1983)]. Radon's transform has also been applied to problems where scalar-diffraction along spherical wavefronts applies [Pan(1983);Slaney(1984);Devaney(1983)]. It is our intent to model the vector-diffraction problem by either a linear iteration procedure involving the Green's dyadic functions, or by a tensorial scattering procedure. In either case, it is necessary to reconstruct the electric field (in its vector form) from samples of the field outside the object. To this end, we define Radon's transform of a vector as the Radon transform of each of its components.

The Radon transform has been developed (Radon(1917)[21]) and is stated in parametric form here. The parametric form of the Radon transforms is explicitly shown as the sampling along 1-dimensional measurement lines of functions defined in a 2-dimensional space. In general, the sampling along (n-1)-dimensional hyper-planes of functions is defined in a n-dimensional space. Software implementation is simplified by using the parametric form since indicies to arrays can be easily renormalized. In addition, other experimental configurations can be easily defined. Denoting the Radon transform as R{..} and its inverse as $R^{-1}\{..\}$, we write:

$$(3)\quad R\underline{E}(|\underline{p}|,\hat{p}) = \int d\underline{r}\underline{E}(\underline{r})\ \delta(\underline{r}\cdot\hat{p}-|\underline{p}|) \quad \text{where } \underline{p} = \nabla_{\underline{t}}\ \underline{r}'(\underline{t})$$

$$(4)\quad R^{-1}\underline{E}(\underline{r}) = 4(2\pi)^{1-n}(-1)^{(1-n)/2}\Delta_{\underline{r}}^{\ (n-1)/2} \int d\hat{p}\ \underline{E}(|\underline{p}|,\hat{p})$$

where the gradient and the fractional laplacian with respect to $\underline{r}'$ are denoted by $\nabla_{\underline{r}'}\{..\}$, and $\Delta_{\underline{r}'}\{..\}$, respectively.

1.3 Definitions on Selected Tomography Configurations:

In an effort to define the terms: straight-line propagation tomography, (scalar) diffraction tomography, and vector-diffraction tomography in the framework of this study, we proceed with the following definitions:

1.3.1 Straight-line propagation tomography:

α. first order approximation: $I^n\{..\} = I^1\{..\} = E^s$.

β. isotropic approximation: $\underline{E}$=E, $\bar{\bar{f}}$=f, etc.

γ. straight-line propagation approximation:

$$(5)\quad g(\underline{r}-\underline{r}') = \delta(\underline{r}\cdot\hat{r}' - |\underline{r}'|)$$

δ. uniform initial approximation to field: $E(\underline{r})=E$.
ε. linear identification between object profile function and the characteristic function: $O(\underline{r})=f(\underline{r})$.

1.3.2 Scalar-diffraction tomography:
α. first order approximation: $I^n\{..\} = I^1\{..\} = E^s(\underline{r})$.
β. isotropic approximation $\underline{E}=E$, $\bar{\bar{f}}=f$, etc.
γ. single point source spherical scalar-diffraction approximation:

$$(6)\quad g(\underline{r}-\underline{r}') = \frac{(\exp j|\underline{k}||\underline{r}-\underline{r}'|)}{|\underline{r}-\underline{r}|}$$

δ. either Born or Rytov approximation:

$$(7)\quad E(\underline{r}) = (\exp j\underline{k}\cdot\underline{r}), \text{ or } E(\underline{r}) = (\exp j\underline{k}\cdot\underline{r})\,(\exp j\phi(\underline{r}))$$

ε. linear identification between profile and characteristic function: $O(\underline{r})=f(\underline{r})$

1.3.3 Vector-diffraction tomography:
α. first order approximation: $I^n\{..\} = I^1\{..\} = \underline{E}^s(\underline{r})$.
β. do not make isotropic approximation
γ. m-point source spherical vector-diffraction approximation:

$$(8)\quad \bar{\bar{g}}(\underline{r}-\underline{r}') = \frac{\Sigma_i^m \bar{\bar{I}}(\exp j|\underline{k}||\underline{r}-\underline{r}'- z_i|)}{|(\underline{r}-\underline{r}')-\underline{z}_i|}$$

δ. extended Born or extended Rytov approximation:

$$(9)\quad \underline{E}(\underline{r}) = \underline{a}(\exp j\underline{k}\cdot\underline{r}), \text{ or } \underline{E}(r) = \underline{a}(\exp j\underline{k}\cdot\underline{r})(\exp j\phi(\underline{r}))$$

ε. tensorial identification between the object profile function in terms of the span of the characteristic matrix: $O(\underline{r}) = \text{span}\{\bar{\bar{f}}(\underline{r})\}$.

1.4 Objectives:
α. Suggest an expression for the object profile function in terms of the span of the characteristic matrix:

$$(10)\quad O(\underline{r})=\text{span}\{\bar{\bar{f}}(\underline{r})\}.$$

β. To develop an expression for the characteristic matrix in terms of the measured propagated electric field and the incident electric field:

i. The expression must be in a form that is amenable to direct computation so that a "work bench" of computer programs can be developed.

ii. The expression must be in the form of an n'th order iteration to take into account nontrivial convergence.

γ. Compare with existing methods used in diffraction and straight-line tomography.

δ. Examine tensorial-diffraction; an alternate vector diffraction approach.

ε. Formulate several computational algorithms using Radon theory to reconstruct functions in laboratory space.

2. ANALYTICAL APPROACHES:

This section addresses the problem of object profile reconstruction by identifying the profile function with the electromagnetic characteristics of the object. The electromagnetic characteristics of the object are contained in the permittivity ($\bar{\varepsilon}$), permeability ($\bar{\mu}$), conductivity ($\bar{\sigma}$), and coupling matrices ($\bar{\gamma}$). The matrices are lumped together into a characteristic matrix which is assumed to be related to the object profile function. Because the object profile function is a scalar (rank-0 invariant), and the span of the characteristic matrix is a scalar, we suggest a direct relation between the object profile and the span of the characteristic matrix.

The analytical approaches are divided as follows:

Use Maxwell's equations to describe the interaction between the media and the incident electric field. Formulate these equations into the Helmholtz equation so that the object's description can be lumped into one matrix that we shall call the characteristic matrix.

Because the Helmholtz equation is a nonhomogeneous equation, we must use an iteration procedure to solve this equation for the characteristic matrix. Although there are several techniques to solve inhomogeneous partial differential equations, we shall adopt a linear iteration scheme. The linear scheme has the advantage of guaranteed uniform convergence, and guaranteed inversion. In addition, using the linear iteration, we can make direct comparisons with existing work.

We present a general inversion procedure that can be used to invert an iteration to solve for the characteristic matrix. Fourier inversion (a series method mentioned in 2.3) is used to invert the iteration equation. Comparisons are made with some of the existing methods.

An alternate scattering approach is presented that might yield an accurate first estimate to the characteristic matrix when the electric field has undergone strong repolarization.

2.1 Interfacing Maxwell's Equation

Maxwell's equations for a linear, stationary medium (although anisotropic and nonhomogeneous) can be "decoupled" into a non-homogeneous-Helmholtz equation:

$$(11) \quad (\nabla \cdot \nabla)\underline{E}(\underline{r}) + a^2\underline{E}(\underline{r}) = \bar{\bar{f}}(\underline{r})\cdot\underline{E}(\underline{r})$$

where the coupling matrix, $\bar{\bar{\gamma}}$ is defined by:

$$(12) \quad \nabla\ X(\bar{\bar{\mu}}(\underline{r} \cdot \underline{H}(\underline{r})) = \bar{\bar{\gamma}}(\underline{r}) \cdot \bar{\bar{\mu}}(\underline{r}) \cdot \underline{H}(\underline{r})$$

and we define the characteristic matrix, $\bar{\bar{f}}$ by:

$$(13) \quad \bar{\bar{f}}(\underline{r}) = \bar{\bar{\gamma}}(\underline{r}) \cdot \bar{\bar{\mu}}(\underline{r}) \cdot \left(\omega^2\bar{\bar{\varepsilon}}(\underline{r}) - j\omega\bar{\bar{\sigma}}(\underline{r})\right)$$

and the operator a^2 as:

$$(14) \quad a^2\ \underline{E}(\underline{r}) = -\nabla(\nabla \cdot \underline{E}(\underline{r}))$$

We have assumed that the electric field and the magnetic field are of the form $\underline{E}(\underline{r})(\exp j\omega t)$ and $\underline{H}(\underline{r})(\exp j\omega t)$, respectively.

2.2 Development of Helmholtz Equation:

The Helmholtz equation can be cast into the (homogeneous) Fredholm equation [27] using logical induction (this is analytically simpler and seems no less general than applying some variant of scalar diffraction theory [16] in generalized vector form [11]). We make the hypothesis of the Fredholm equation which we use as a linear iteration:

$$(15) \quad I^{i+1}\underline{E}(\underline{r}) = \int d\underline{r}[\bar{\bar{g}}(\underline{u}-\underline{r}) \cdot \bar{\bar{f}}(\underline{u})] \cdot I^{i}\underline{E}(\underline{u})$$

In this manner, we should interpret $I^{i}\underline{E}$ to be the i'th approximation to the field at any point, and interpret $\underline{E}$ to be the initial approximation (possibily the transmitted field) to the field at every point.

It is possible to show [27] that, for arbitrary n, $I^{i+n}\underline{E}$ is related to $I^{i+n-1}\underline{E}$ by a Fredholm equation. Thus, we need only show that this is true for some i. Since we want this to be true for arbitrary i, we pose the initial constraint on $\bar{\bar{g}}$ by substituting the Fredholm equation into the Helmholtz equation. This poses the constraint:

$$(16)\quad (\nabla \cdot \nabla)\ \bar{\bar{g}}(\underline{u}-\underline{r}) + a^2\ \bar{\bar{g}}(\underline{u}-\underline{r}) = \bar{\bar{I}}\delta(\underline{u}-\underline{r})$$

It is well known (and can be verified by direct substitution) that a sufficient solution of this equation for the propagator $\bar{\bar{g}}$ can be expressed as a sum of Green's dyadic functions:

$$(17)\quad \bar{\bar{g}}(\underline{u}-\underline{r}) = \Sigma_i \frac{\bar{\bar{I}}\ (\exp -j|\underline{k}||\underline{u}-\underline{r}-\underline{z}_i|)}{|\underline{u}-\underline{r}-\underline{z}_i|}$$

where $\underline{z}_i$ is the location of the i'th "independent" scattering center within the object's interior.

2.3 <u>A General Inversion Procedure for Linear Iterations:</u>

Inversion of the Helmholtz equation uses an iterative procedure. We have developed a linear iteration procedure for comparitive purposes. At any stage of the iteration, any number of methods can be used to invert this integral equation. Of course, before any process can be implemented on computing machines, we must first discretize the problem. This essentially involves converting functions with variables to arrays with subscripts, and converting integrals to summations.

It has been shown [27] that the Fredholm equation will converge uniformly to a unique solution under the following reasonable conditions:

α. Each element in the propagation matrix $\bar{\bar{g}}$ can be expanded in a Taylor series around any $\underline{r}$. This condition is also called continuity, or infinite differentiability.

β. If a redundant solution (one such that IE(r)=f(r)+λE) occurs, then f(r) must be orthogonal to E($\underline{r}$):

$$(18)\quad \int d\underline{r}\ f(\underline{r})\ E(\underline{r}) = 0$$

The first condition is reasonable from a physical viewpoint and is also satisfied by our choice for the propagator. The second condition need only be tested if we wish to prevent "infinite looping" of the general inversion procedure.

We present a general inversion algorithm, with the intent of demonstrating the purpose of Fourier transforming, so often employed in tomography.

A) Initialize old guess
B) Invert Fredholm equation for new guess using old guess and newly iterated data. There are several methods commonly in use:
   a) matrix inversion techniques using P-simplex method [23], algebraic reconstruction techniques (ART) [25,5], or maximum entropy techniques [4].
   b) or, transform the Fredholm equation into a new vector space, invert the new equation and take the results back into the original vector space. Some commonly used series expansions with associated inverses are:
      i) Fourier and inverse Fourier using FFT algorithm [20, 15].
      ii) Laurent and Hilbert transform [13].
      iii) Base-2 and Fermat number theoretic transform [6].
   c) or, construct graphs of the original matrix and identity matrix. Then use the graph-search procedure (algorithm A*) to produce an inverse graph by finding an isomorphism between the original graph and the identity graph [17,22].
   d) or, successive approximation from Neumann series approximation [1].
C) Reset the old guess to newly computed guess.
D) Go to step B.

Often, only first approximations are necessary; therefore, the repetition of steps B and C are not necessary. For example [2,3,9] use a first Born/Rytov approximation when inverting the Fredholm equation. Although the matrix inversion approach is tempting, it often becomes numerically unstable, particularly when the matrix is Hilbertian. Graph construction is often difficult to implement in software. The potential advantage of graph construction would be to handle large but sparse matrices. Unfortunately, the image matrix in tomography is seldom sparse. Furthermore, the graph isomorphism algorithm is not polynomially bounded in time [17].

The Hilbert transform technique suffers due to lack of an efficient computational algorithm. The number-theoretic transform, which has the main advantage of converting a floating point problem into a corresponding fixed point problem [6], usually requires very large integer sizes, and would require a reformulation of the physics involved for best implementation. Finally, this leaves us with the Fourier transform, which has been implemented using the Fast Fourier transform technique in one and more dimensions [20]. In tomography, we are often able to approximate the scattering involved by Green's functions; these functions are similar to the Fourier kernels, and therefore suggest possible simplifications.

2.4 Characteristic Matrix From the Iteration Equation:

At this point we focus our attention on inverting the linear iteration equation for the characteristic matrix $\bar{\bar{f}}(\underline{r})$:

$$(19) \quad I^{i+1}\underline{E}(\underline{r}) = \int d\underline{u}[\bar{\bar{g}}(\underline{u}-\underline{r}) \cdot \bar{\bar{f}}(\underline{u})] \cdot I^{i}\underline{E}(\underline{u})$$

As was pointed out in the general inversion algorithm described in Section 2.3, there are several known ways to invert this equation. We shall adopt, for comparative purposes, the Fourier transform technique, whereby the Fredholm equation is cast into Fourier space by expanding each function in its Fourier series. The orthogonality of the Fourier system is then employed, which leads to:

$$(20) \quad F\, I^{i+1}\underline{E}(\underline{\mu}) = F\bar{\bar{g}}(\underline{\mu}) \cdot \int d\underline{\nu}\; F\bar{\bar{f}}(\underline{\mu}-\underline{\nu}) \cdot FI^{i}\underline{E}(\underline{\nu})$$

The convolution is eliminated by taking a second Fourier transform, giving:

$$(21) \quad F\{\, F\bar{\bar{g}}(\underline{\mu})^{-1} \cdot FI^{i+1}\underline{E}(\underline{\mu})\}(\underline{\nu}) = FF\bar{\bar{f}}(\underline{\nu}) \cdot FFI^{i}\underline{E}(\underline{\nu})$$

We rewrite the above equation with obvious notational changes as:

$$(22) \quad \underline{e}(\underline{\mu}) = \bar{\bar{f}}(\underline{\mu}) \cdot \underline{E}(\underline{\mu})$$

Now, the intent is to solve this equation for the matrix $\bar{\bar{f}}$. This can be done by diagonalizing the matrix in a new coordinate system ν rotated with respect to the μ-system via a tensorial transformation [14], denoted as T{..}. To perform this transform we expand the functions of (22) in the $\underline{\nu}$-system and employ the orthogonality of the new system to yield:

$$(23) \quad T\underline{e}(\underline{\nu}) = T\bar{\bar{f}}(\underline{\nu}) \cdot T\underline{E}(\underline{\nu})$$

The matrix $T\bar{\bar{f}}$ has been diagonalized, and therefore:

$$(24) \quad T\bar{\bar{f}}(\underline{\mu}) = \{\bar{\bar{I}}T\underline{e}(\underline{\nu})/T\underline{E}(\underline{\nu})\}$$

where $\{\bar{\bar{I}}T\underline{e}(\underline{\nu})/T\underline{E}(\underline{\nu})\}$ should be interpreted as component-wise division with component-wise multiplication along the diagonal of the identity matrix.

To reconstruct $\bar{\bar{f}}$, substitute for the notational changes made in going from (21) to (22), rotate back to the old $\underline{\mu}$-system, and inverse Fourier transform from the $\underline{\mu}$-system to the $\underline{r}$-system:

$$(25) \quad \bar{\bar{f}}(\underline{r}) = F^{-1}F^{-1}T^{-1}\{\bar{\bar{I}}\, \frac{TF\{F\bar{\bar{g}}(\underline{\mu})^{-1} \cdot FI^{i+1}\underline{E}(\underline{\mu})\}(\underline{\nu})}{TFFI^{i}\underline{E}(\underline{\nu})}\}\,(\underline{r})$$

2.5 Comparisons with Existing Methods:

For comparisons with existing methods, (25) is simplified to the forms found in [2,3,9]. These results are not as general as that of (25) since approximations such as first order Born/Rytov diffraction, scalar-diffraction, or straight-line propagation are often employed.

2.5.1 (Scalar) Diffraction Tomography:

We can recover the results of scalar diffraction tomography by substituting the approximations made in 1.3.2 into (25) (see [18] for details). Note that, the tensor operation T{..} becomes a null operation, the matrix inverse becomes a reciprocal, and recall [15] that:

$$(26) \quad F\bar{\bar{g}}(\underline{\sigma}) = \frac{1}{(|\underline{\sigma}|^2-|\underline{k}|^2)}$$

Then:

$$O(\underline{r}) = \int d\underline{\mu}(\exp -j\underline{r}\cdot\underline{\mu}) \int d\underline{\nu}(\exp -j\underline{\mu}\cdot\underline{\nu}) \frac{\int d\underline{\sigma}(\exp j\ \underline{\nu}\cdot\underline{\sigma})(|\underline{\sigma}|^2-|\underline{k}|^2)FE^s(\underline{\sigma})}{(\exp j\underline{\nu}\cdot\underline{k})}$$

$$(27) \quad = \int d\underline{\sigma}\ (|\underline{\sigma}|^2 - |\underline{k}|^2)\ FE^s(\underline{\sigma}) \int d\underline{\mu}(\exp -j\underline{r}\cdot\underline{\mu}) \int d\underline{\nu}\ (\exp j\underline{\nu}\cdot(\underline{\sigma}-\underline{k}-\underline{\mu}))$$

$$= \int d\underline{\sigma}\ (|\underline{\sigma}|^2 - |\underline{k}|^2)\ FE^s(\underline{\sigma})(\exp -j\underline{r}\cdot(\underline{\sigma}-\underline{k}))$$

We find agreement with the results presented in [2,3,9] when we take the Fourier transform of the above results:

$$(28) \quad E(\underline{r}) = \int d\underline{\mu}\ \frac{FO(\underline{\mu}-\underline{k})}{|\underline{\mu}|^2-|\underline{k}|^2}\ (\exp -j\underline{\mu}\cdot\underline{r})$$

Modified Filtered Backprojection:

We may fill the Fourier space of $FE^s(\underline{\sigma})$ by interpolating in Fourier space the Fourier transform along lines outside the object with position vector $\underline{b}$. By taking the "Fourier weighted" Radon transform of the electric field in (28), we may write:

$$(29) \quad \int d\underline{r}\ E(\underline{r})\ (\exp j\underline{\alpha}\cdot\underline{r})\ \delta(b-\hat{b}\cdot\underline{r})$$

$$= R\{E(\underline{r})\ (\exp j\underline{\alpha}\cdot\underline{r})\}\ (b,\hat{b})$$

$$= \int d\underline{\mu}\ \frac{FO(\underline{\mu}-\underline{k})}{|\underline{\mu}|^2-|\underline{k}|^2} \int d\underline{r}\ (\exp j\underline{r}\cdot(\underline{\alpha}-\underline{\mu}))\ \delta(b-\hat{b}\cdot\underline{r})$$

$$= \int d\underline{\nu}\ \frac{FO(\underline{\mu}(\underline{\nu})-\underline{k})}{(|\underline{\mu}(\underline{\nu})|^2-|\underline{k}|^2)}\ (\exp j(b/\hat{b}_1)(\alpha_1-\mu_1(\underline{\nu})))\delta(\underline{\nu}-\underline{\sigma})$$

$$= FO(\underline{\alpha}-\underline{k}) \frac{e^{-j\underline{\alpha}\cdot\underline{b}}}{\underline{\alpha}\cdot\hat{b}} ;$$

where $|\underline{\alpha}|^2 = |\underline{k}|^2$ and $\underline{\nu} = \{-\hat{b}_1\mu_2, \hat{b}_2\mu_1\}$.

For comparison with [2,3], pick $\underline{b} = \underline{k} = \{0,k\}$. Then we may express the above results in the form:

$$(30) \quad \int dr_1\, E(r_1, r_2) = \frac{FO(\alpha_1, \alpha_2-k)}{\alpha_2} (\exp -j\alpha_2 k)$$

where

$$(31) \quad \alpha_2 = \sqrt{k^2 - \alpha_1^2}$$

2.5.2 Straight-line Propagation Tomography:
In order to compare with existing methods, we make the first-approximation, uniform initial approximation, and isotropic approximation as was done in Section 1.3.1. Next, we formulate straight-line diffraction by chosing $\bar{\bar{g}}(\underline{r}-\underline{u}) = \delta(\underline{r}\cdot\underline{u}-|\underline{u}|)$ which results in a Radon transform that can be inverted by any of the methods presented in Section 4.

2.6. Tensorial Scattering -- An Alternate Approach:
We shall model the scattering process in two equations to be held true simultaneously. Both equations account for scalar diffraction and depolarization (vector diffraction).

Scalar and vector diffraction can be modeled with the tensorial-diffraction equation involving the scattering matrix $\bar{\bar{t}}(\underline{r}_{n+1}, \underline{r}_{n-1})$ from the point $\underline{r}_{n-1}$ to the point $\underline{r}_{n+1}$. The tensorial scattering operation will be denoted by H{..}:

$$(32) \quad H\underline{E}(\underline{r}_{n+1}) = \bar{\bar{t}}(\underline{r}_{n+1}, \underline{r}_{n-1}) \cdot \underline{E}(\underline{r}_{n-1})$$

Scalar and vector diffraction may also be modeled with the linear iteration equation:

$$(33) \quad I\underline{E}(\underline{r}) = \int d\underline{u}\, (\bar{\bar{g}}(\underline{u}-\underline{r}) \cdot \bar{\bar{f}}(\underline{u})) \cdot \underline{E}(\underline{u})$$

The objective is to find the scattering matrix at every point $\underline{r}$ and use that information to reconstruct the object profile at every point.

2.6.1. Recursion Relation For The Scattering Matrices:
Using (32), we have (see [18] for more details):

$$(34) \quad H\underline{E}(\underline{r}_{n+1}) = \bar{\bar{s}}(\underline{r}_{n+1}, \underline{r}_0) \cdot \underline{E}(\underline{r}_0)$$

where we define the net scattering matrix $\bar{\bar{s}}$ as:

$$(35)\quad \bar{\bar{s}}(\underline{r}_{n+1}, \underline{r}_0) = \bar{\bar{t}}(\underline{r}_n, \underline{r}_{n-1}) \cdot \ldots \cdot \bar{\bar{t}}(\underline{r}_1, \underline{r}_0)$$

From the definition of the net scattering matrix, we may write a recursion relation that can be inverted to find the scattering matrix at any point:

$$(36)\quad \bar{\bar{t}}(\underline{r}_n, \underline{r}_{n+1}) = \bar{\bar{s}}(\underline{r}_{n+1}, \underline{r}_0) \cdot \bar{\bar{s}}^{-1}(\underline{r}_n, \underline{r}_0)$$

2.6.2 Determination of the Net Scattering Matrix:
Since we are interested in finding the scattering matrix over the object, we will relate the net scattering matrix with the field $\underline{E}$ at some point on the object.

The net scattering matrix can be found by performing two independent tests. The first test transmits $\underline{E}$ in the "r_1-polarized state" (1,0); the second test transmits $\underline{E}$ in the "r_2-polarized state" (0,1).

Using (34), and transmitting in the r_1/r_2-polarized states, we can recover the net scattering matrix at $\underline{r}_n$ in terms of the field $\underline{E}$ at $\underline{r}_n$:

$$(37)\quad \underline{t}_1(\underline{r}_n, \underline{r}_0) = \underline{E}(\underline{r}_n) \text{ when } \underline{E}(\underline{r}_0) \text{ in } r_1\text{-polarized state } (1,0)$$

$$\underline{t}_1(\underline{r}_n, \underline{r}_0) = \underline{E}(\underline{r}_n) \text{ when } \underline{E}(\underline{r}_0) \text{ in } r_2\text{-polarized state } (1,0)$$

So that a compact expression can be written for the scattering matrix, we write the matrix $\bar{\bar{s}}$ as a dyad:

$$(38)\quad \bar{\bar{s}}(\underline{r}_n, r_0) = \underline{t}_1(\underline{r}_n, \underline{r}_0) \;\; | \;\; \underline{t}_2(\underline{r}_n, \underline{r}_0)$$

and hence the scattering matrix $\bar{\bar{t}}$ is:

$$(39)\quad \bar{\bar{t}}(\underline{r}_n, \underline{r}_0) = \bar{\bar{s}}(\underline{r}_{n+1}, \underline{r}_0) \cdot \bar{\bar{s}}^{-1}(\underline{r}_n, \underline{r}_0)$$

2.6.3. Relation Between Scattering Matrix And Object Profile:
As presented previously, we write the tensorial scattering equation from the points of field transmission $\underline{r}'$ to the points of field reception $\underline{r}$:

$$(40)\quad \underline{E}(\underline{r}) = \bar{\bar{s}}(\underline{r}, \underline{r}') \cdot \underline{E}(\underline{r}')$$

Then using the expression for the net scattering matrix (38), and interfacing equation (40) with the characteristic matrix by substituting for the electric field the expression given in (33), gives:

$$(41)\quad \bar{\bar{s}}(\underline{r})=\left(\int d\underline{u}\ [\bar{\bar{g}}(\underline{u}-\underline{r}).\bar{\bar{f}}(\underline{u}).\{1,0\}]\right)\ \Big|\ \left(\int d\underline{u}\ [\bar{\bar{g}}(\underline{u}-\underline{r}).\bar{\bar{f}}(\underline{u}).\{0,1\}]\right)$$

$$= \int d\underline{u}\ [\bar{\bar{g}}(\underline{u}-\underline{r}) \cdot \bar{\bar{f}}^{L}(\underline{u})]\ |\ [\bar{\bar{g}}(\underline{u}-\underline{r}) \cdot \bar{\bar{f}}^{R}(\underline{u})]$$

$$= \int d\underline{u}\ \ \bar{\bar{g}}(\underline{u}-\underline{r}) \cdot [\bar{\bar{f}}^{L}(\underline{u}))\ |\ \bar{\bar{f}}^{R}(\underline{u})]$$

$$\bar{\bar{s}}(\underline{r})= \int d\underline{u}\ \bar{\bar{g}}(\underline{u}-\underline{r}) \cdot \bar{\bar{f}}(\underline{u})$$

where we have denoted $\{..\}^{L}$ and $\{..\}^{R}$ as the left/right half, respectively, of the enclosed matrix.

Equation (41) may be inverted for the characteristic matrix by the usual Fourier transform procedure giving:

$$(42)\quad \bar{\bar{f}}(\underline{r}) = F^{-1}\{F\bar{\bar{g}}(\underline{\mu})^{-1}\cdot F\bar{\bar{s}}(\underline{\mu})\}(\underline{r})$$

By comparing this inversion result with that of Section 2.4 equation (25), we observe a substantially simplified expression that can be more rapidly computed. Thus, since the elements of the scattering matrix can be easily measured, and because of the overall simplicity of the inversion result, an implementation of this method may be highly desired. Observe that this equation can be simplified further if we invoke the Kirchhoff scalar diffraction approximation, giving:

$$(43)\quad \bar{\bar{f}}(\underline{r}) = F^{-1}\{(|\underline{\mu}|^2-|\underline{k}|^2)F\bar{\bar{s}}(\underline{\mu})\}(\underline{r})$$

3. COMPUTER-NUMERICAL APPROACHES

In practice it is not possible to measure $\underline{E}$ for all $\underline{r}$, nor $F\underline{E}(\underline{\nu})$ for all ν. By the very nature of the problem, a probe cannot be inserted into the object, nor can we construct an infinite number of channels to measure the Fourier spectrum. Often a finite number of frequency channels lead to acceptable results.

For notational simplicity, consider finding the Fourier transform of any element $E(\underline{r})$ in $\underline{E}(\underline{r})$. We follow the development as presented in [25], which interfaces the Fourier and Radon transform theories.

Consider the Radon transform of a function $E(\underline{r})$ which we define as $RE(p,\hat{p})$ and is measured along a line perpendicular to the unit direction vector $\hat{p}$:

$$(44)\quad RE(p,\hat{p}) = \int d\underline{r}\ E(\underline{r})\ \delta(p - \underline{r}\cdot\hat{p})$$

We discuss the mathematical forms of each method; and their scope of implementation is what follows. Observe (for computer implementation) that the integrals over $\{\hat{p}\}$ should be replaced with integrals over spatial angles, and that the function $\hat{p}$ should be replaced with functions of the spatial angles. The integral $1/2\int d\hat{p}$ is referred to as the backprojection operation.

(i) Fractional Laplacian

$$(45)\quad E(\underline{r}) = (2\pi j)^{\frac{1-n}{2}}\,\Delta^{(n-1)/2} \int d\hat{p}\; RE(p\hat{p})$$

Although this inversion procedure is simple in mathematical form, it could be numerically too unstable and exhaustive to compute the fractional Laplacian.

(ii) Backprojection of gradient (odd dimensional space)

$$(46)\quad E(\underline{r}) = 2(\pi j)2/2 \int d\hat{p}\; \nabla_p \nabla_p \Big|_{p=\hat{p}\cdot\underline{r}} RE(p,\hat{p})$$

Although the arrays might be excessively large, the Laplacian operator is fairly stable numerically. This procedure has been investigated using NMR[19], and for the far scattered field case when a unit ramp field is incident [29,28].

(iii) Backprojection of Hilbert transform (even 2-dimensional space):

$$(47)\quad E(\underline{r}) = (2\pi j)/2 \int d\hat{p} \int dp\, \frac{\nabla_p E(p,\hat{p})}{p - t}\Big|_{t=\underline{r}\cdot\hat{p}}$$

This method shows some promise for implementation. Note that the gradient becomes a first derivative for 2-dimensional reconstruction, and that first derivatives are fairly stable. Unfortunately, the computationally expensive Hilbert transform appears.

(iv) 2-dimensional Fourier inversion from interpolations:

$$(48)\quad \underline{E}(\underline{r}) = \int d\underline{k}\; (\exp -j2\pi\underline{r}\cdot\underline{k}) \underset{\text{over } \hat{p}}{\text{interpolate}} \{\int dp(\exp jsp) RE(p,\hat{p})\; |\hat{p}s\}$$

There are several computationally attractive properties of this inverse method. The 1-dim/2-dim Fourier transforms are reasonably numerically stable (if aliasing errors can be neglected), and the interpolation procedure is usually simple to perform. This method is often called the direct Fourier inversion method and has been used extensively [3,26].

(v) Backprojection and 2-dimensional Fourier Transforms:

$$(49)\quad E(\underline{r})= \int d\underline{k}(\exp -j2\pi \underline{r}\cdot\underline{k})|\underline{k}| \int d\hat{p}(\exp j2\pi\underline{r}\cdot\underline{k})\ 2 \int dp \quad RE(p,\hat{p})$$

Again, this method can be implemented using the fast Fourier transform routines that are stable and quick. However, the storage space could easily become excessive because of the number of 2-dimensional arrays involved. This method has its main advantage in implementing the Fourier transforms using an optical lens [16,10].

(vi) Backprojection and 1-dimensional Fourier Transforms:

$$(50)\quad E(\underline{r})= 2\int d\hat{p} \int dk(\exp -j2\pi sk)\ |k| \int dp(\exp j2\pi pk)RE(p,\hat{p})\Big|_{s=\underline{r}\cdot\hat{p}}$$

This method shares the same numerical stability as with method (v), and also has the advantage of working with only 1-dimensional arrays. For these reasons this method has been extensively implemented [10].

5. CONCLUSIONS AND FUTURE RESEARCH NEEDS

A closed form expression has been proposed for the object profile function in terms of the (n+1)th and n'th approximation to the received electric field. This expression is independent of a particular initial approximation and explicitly shows the need for convergence on the first iteration. Hence, the first approximation (whether it be Born, Rytov, or some other) should be as close as possible to the actual electric field. Further research is needed to derive an expression in terms of the (n+1)th and the first approximation.

We have suggested a direct relationship between the object profile function and the span of the characteristic matrix. We believe that computational experimentation is needed to determine the effects and applicability of incorporating the coupling matrix and Green's dyadic functions to account for vector-diffraction in tomography.

Connections and comparisons have been made between the objectives and results between vector-diffraction tomography (using the modified-filtered backpropagation technique) and straight-line propagation tomography. Further interface is needed with the filtered backpropagation technique [9] and Rytov first approximation technique used in scalar-diffraction tomography.

5. ACKNOWLEDGEMENTS

This research was supported, in part, by the U.S. Office of Naval Research, Contract No. US ONR-N0001480-C-00708, and we wish to express our sincere gratitude to Drs. Stuart L. Brodsky, and Charles J. Holland for their continued interest in our research. Here we also wish to thank Dr. Anthony J. Devaney for his continued interest and constructive criticisms. The authors wish to express special thanks to Mr. Richard W. and Mrs. Deborah A. Foster for the efficient typing of the manuscript.

6. REFERENCES

[1] F.G. Tricomi, "Integral Equations," Interscience, New York, 1957.

[2] M. Slaney, A.C. Kak, E. Larsen, "Limitations of Imaging with first order Diffraction Tomography," IEEE Trans. on Microwave Theory and Techniques, July 1984. (to appear)

[3] S.X. Pan, A.C. Kak, "A Computational Study of Reconstruction Algorithms for Diffraction Tomography," IEEE Trans. on Acoustics Speech and Signal Processing, Vol. 31, No. 5, Oct. 1983, pp. 1262-1275.

[4] S.J. Wernecke, L.R. D'Addario, "Maximum Entropy Image Reconstruction", IEEE, Trans. on Computers, Vol. 26, April 1977, pp. 351-364.

[5] R. Gordon, G.T. Herman, "Reconstruction of pictures from their projections", Commun. ACM. Vol. 14, 1971, pp. 759-768.

[6] R.C. Agarwal, C.B. Burrus, "Fast Convolution Using Fermat Number Transforms with Applications to Digital Filtering," IEEE Trans. on Acoustics and Signal Processing, Vol. 22, No. 2, Apr. 1974, pp. 87-97.

[7] L.E. Larsen, J.H. Jocobi, "Microwave Interrogation of Dielectric Targets," Medical Physics, Vol. 5, No. 6, Nov./Dec. 1978, pp. 500-508.

[8] J.H. Jocobi, L.E. Larsen, C.T. Hast, "Water-Immersed Microwave Antennas and Their Applications to Microwave Interrogation of Biological Targets", IEEE Trans. on Microwave Theory and Techniques, Vol. 27, No. 1, Jan. 1979, pp. 70-78.

[9] A.J. Devaney, "Inverse Source and Scattering Problems in Ultrasonics. IEEE Trans. on Sonics and Ultrasonics," Vol. 30, No. 6, Nov. 1983, pp. 355-364.

[10] R.H.T. Bates, T.M. Peters, "Towards Improvments in Tomography," New Zealand Journal of Science, Vol. 14, pp. 883-896.

[11] A.T. Friberg, E. Wolf, "Angular Spectrum Representation of Scattered Electromagnetic Fields," Journal Optical Society of America, Vol. 73, No. 1, Jan. 1983, pp. 26-32.

[12] M.M. Syslo, N. Deo, J.S. Kowalk, Discrete Optimization Algorithms, Prentice-Hall, Englewood Cliffs, NJ., 1983.

[13] M.R. Spiegel, Complex Variables, McGraw-Hill, New York, 1964.

[14] I.S. Sakolnikoff, Tensor Analysis, John Wiley & Sons, New York, 1964.

[15] M.R. Spiegel, Fourier Analysis, McGraw-Hill, New York, 1974.

[16] J.W. Goodman, Introduction to Fourier Optics, McGraw-Hill, New York, 1968.

[17] N. Deo, Graph Theory with Applications to Engineering and Computer Science, Englewood Cliffs, N.J. Prentice-Hall, 1974.

[18] B.D. James, "An Approach to Electromagnetic Propagation Tomography", M.Sc. Thesis, Communications Laboratory, Electrical Engineering & Computer Science Department, University of Illinois at Chicago, December 1984.

[19] L.A. Shepp, "Computerized Tomography and Nuclear Magnetic Resonance", J. Computer Ass. Tomog., Vol. 4, pp. 94-107.

[20] P. Corsini, G. Frosini, "Properties of the Multidimensional Generalized Discrete Fourier Transform," IEEE Trans. on Computers, Vol. 28, No. 11, Nov. 1979, pp. 819-830.

[21] S. Helgason, The Radon Transform, Birkhauser, Boston, 1980 (Ed. J.Coates, S.Helgason).

[22] J.R. Ullman, "An Algorithm for Subgraph Isomorphism," J. Ass. Comput. Mach., Vol. 23, No. 1, Jan. 1976.

[23] R.J. Lyth, "Iterative Ray Tracing Between Boreholes for Underground Imaging Reconstruction", IEEE Trans. on Geoscience and Remote Sensing, Vol. 18, No. 3, July 1980.

[24] S.K. Kenue, J.F. Greenleaf, "Limited Angle Multifrequency Diffraction Tomography," IEEE Trans. on Sonics and Ultrasonics, Vol. 29, Jul. 1982, pp. 213-217.

[25] S.R. Deans, The Radon Transform and Some of Its Applications, John Wiley & Sons, New York, 1983.

[27] J.W. Dettman, Mathematical Methods in Physics and Engineering, McGraw-Hill, New York, 1962.

[28] W.-M. Boerner, C.-M. Ho, "Analysis of Physical Optics Far Field Inverse Scattering for the Limited Data Case Using Radon Theory and Polarization Information," Wave Motion, Vol. 2, Sep. 1980 (North Holland)

[29] C.L. Bennett, "Theoretical Methods for Determining the Interactions of Electromagnetic Waves with Structures," (Ed. J.Skwirzynski) Sijthoff & Nordhoff, 1981.

V.8 (OI.4)

IMAGE PROCESSING: ANALYSIS BEYOND MATCHED FILTERING

Harold H. Szu
Optical Sciences Division
Naval Research Laboratory, Washington DC 20375

Matched filters are revisited from the viewpoint of regulations of the ill-posed problem and the super-resolution. Object-matched minimum mean-square image spread (MMIS) filters are derived from a general contrained moment expansion method. An inverse problem of image processing is to determine an optimum trade-off potential $U(\vec{f})$ or $V(\vec{f})$ from the knowledge of the restored object $Q(\vec{f})$ or the output transfer function $\Psi(\vec{f})$ as well as the knowledge of the optical transfer function (OTF) $S(\vec{f})$. Analytic examples and exact solutions are given for the coherent and incoherent image restorations. A MMIS filter is found to yield an identical minimum mean square estimation-error (MMSE) filter of Wiener in the special case, similar to the visual MTF or contrast sensitivity function, peaked at the middle band of spatial frequency channels. The direct problem for the linear and noisy motion-blurred image restoration is solved. By means of the Gel'fand-Levitan inverse transform technique for the reference potential, the inverse problem is solved and the design of the energy constrained MMIS filter follows. Finally, both the direct and inverse problems are exemplified with the dc-incision, namely the central dark field method of Abbe and Zernike, useful for a systematical design of both the imaging OTF and image restoration filters under noise. The simple rule of thumb is given for man-made sciences.

1. INTRODUCTION

The direct problem is, given some trade-off potential, to derive the output transfer function Ψ from which an energy constrained minimum mean-square image sread(MMIS) filter follows

W.-M. Boerner et al. (eds.), Inverse Methods in Electromagnetic Imaging, Part 2, 1165–1187.

$$M(\vec{f}) = S_o(\vec{f})^{-1}\ \Psi(\vec{f}). \tag{1}$$

In other words, the noise-to-signal trade-off potential is known

$$V_o(\vec{f}) = \frac{\langle |N|^2 \rangle}{|S_o|^2} \tag{2}$$

in terms of the ratio of the power spectral densities of noise and of the optical transfer function (OTF) $S_o(\vec{f})$. The solution of the direct problem is obtained by solving the following Schrodinger equation:

$$-\nabla^2\Psi + V(\vec{f})\Psi = E\Psi \tag{3}$$

which balances the noise-to-signal energy ratio with the image spread in terms of the radius of gyration and $\sqrt{E}$ is the coordinate shift of the restored image. The solution is called an apodization function Ψ used to soften the exact inverse filter Eq(1). Also, it is called the output transfer function, in the 2-D spatial frequency domain $\vec{f} = (\nu,\mu) = (f_x,f_y)$ because it transfers the object O to the restored object Q,

$$Q(\vec{f}) = M(\vec{f})\ I(\vec{f}) = \Psi(\vec{f})\ O(\vec{f}) + M(\vec{f})\ N(\vec{f}), \tag{4}$$

if one assumes the linear and shift invariant image equation.

$$I(\vec{f}) = S_o(\vec{f})\ O(\vec{f}) + N(\vec{f}) \tag{5}$$

An object-matched MMIS filter is derived

$$-\nabla^2\bar{Q} + U(\vec{f})\bar{Q} = E\,\bar{Q} \tag{6}$$

where the object-dependent U is the deviation of Wiener's Ψ_W from unity,

$$U(\vec{f}) = 1 - \Psi_W \tag{7}$$

The inverse problem is to derive the trade off potential V knowing the output transfer Ψ. The necessary and sufficient condition of the unique inverse solution (from E, Ψ to solve V) has been generally known for the Sturm-Liouville equation of which Schrodinger equation is a special case.[1] It is known in the quantum mechanical potential scattering problem[2] as the Gel'fand-Levitan inverse (scattering) transform technique, which will be applied to the design of imaging and restoration filters for the first time.[3,4,5]

Firstly, we compare Wiener filtering with dynamic filtering. Can the energy-constrained dynamic filter be identical to the least squares filter of Wiener? What is the imaging OTF $S(\vec{f})$ that fulfills both the Least Squares (LS) and the constrained LS? To answer both, an equivalence differential equation is derived for arbitrary object spectrum $P = \langle |O|^2 \rangle$. Two specific questions will be answered. The first compares Wiener filtering with dynamic filtering. The second studies the noisy image restoration. What is the trade-off potential that includes the ideal restoration $\Psi(\vec{f}) = 1$ as its eigenfunctions? Since the exact restoration Q=0 implies no shift $\Psi=1$ and the ideal inverse filtering

$$\Psi = \exp(i\sqrt{E}\,\vec{f}) \to 1, \quad \text{as} \quad E = 0$$

$$M_i = S_o^{-1}\Psi = S_o^{-1}$$

then it is desirable to design the OTF for the exact restoration. But the presence of noise can be overly amplified by any zero of the OTF S_o. As a result, the exact inverse filtering S_o^{-1} must be apodized by the output transfer function $\Psi(\vec{f})$. Therefore, the second question amounts to find a new OTF S which admits $\Psi=1$ and a new potential V by means of the Gel'fand-Levitan (GL) inverse transform. The physical meaning of these solutions will be discussed, and an experiment modifying the imaging condition will be shown.

2. MATCHED FILTERS

We would like to implement a spatial filter $m(\vec{r})$ which can take into account a set of constraints based on a priori knowledge. For example, a useful constraint is a constant output noise from a processor when the input noise power spectrum is filtered by $m(\vec{r})$. Since a positive real function can be identified as an energy if it is derived from the squared modulus of another function, then the spatial integrated noise power spectrum will be referred to as total noise energy. A technique which applies the constraint of constant output noise energy to the method of minimum radius of gyration will be presented. It will be shown to be a special case of a general constrained moment method when used to measure the image spread. Thereby, general applications to image restoration and pattern recognition become possible. The performance criterion of the technique will be identified as MMIS as opposed to the conventional MMSE.[6,7,8,9,10] Since the present criterion does not depend on the unknown estimation errors, it is convenient for image restoration. The resulting filter is a bandpass filter which is analytic in the spatial frequency domain. It will be referred to as an energy-constrained filter because the total energy is traded between the output noise energy and the energy associated with the radius of gyration. Such an energy constrained filter that belongs to the object-independent class of constrained least squares filters[11,12] can not be entirely new[13,14], but potential applications to image restoration and pattern recognition problems[15,16] have not been explored. Hence, a general method, a constrained moment method, will be presented to construct a regulated inverse filter for image restoration. An object-matched MMIS filter is derived for the first time.

(i) Imaging Model: Since a spatial filter is usually compared to the Wiener matched filter used commonly for both image restorations and traditional pattern recognitions, a Wiener model[17,18,19], as used by Helstrom, for imaging through a random medium is adopted here for the necessary comparison. A stochastic object $o(\vec{r})$ is imaged with a lens point spread function (psf) denoted by $s_o(\vec{r})$ and an independent additive Gaussian white noise of zero mean denoted by $n(\vec{r})$. Under the assumption of linear, incoherent and space-invariant imaging, the noisy image is given by (8) in the coordinate domain

$$i(\vec{r}) = s_o(\vec{r}) * o(\vec{r}) + n(\vec{r}), \tag{8}$$

where the asterisk stands for the convolution integral $s_o(\vec{r}) * o(\vec{r}) = \int s_o(\vec{r}-\vec{r}')\, o(\vec{r}')d\vec{r}'$. By the assumption of statistical independence and zero mean $\langle n(\vec{r})\, o(\vec{r}')\rangle = 0$; $\langle n(\vec{r})\rangle = 0$ where angular brackets specify an ensemble average. For simplicity of presentation, the discussion will be limited to a deterministic psf. For this $n(\vec{r})$, we would like to construct a deterministic spatial filter $m(\vec{r})$ which when convolved with a stochastic image $i(\vec{r})$ will produce a stochastic object estimator $q(\vec{r})$ in the form

$$q(\vec{r}) = m(\vec{r}) * i(\vec{r}). \tag{9}$$

However, such a simple deconvolution of (8) can be ill-conditioned. As a matter of fact, the exact inverse filter $m_i(\vec{r})$ indicated by the subscript i and used for deconvolution as shown by (9) is ill-conditioned for two reasons: it does not exist everywhere and it produces noise amplification. Since both are important aspects for subsequent development, we shall briefly review the divergent nature of ill-posed problems[20] and then describe what may be called a Wiener's method of regulation. Moreover, in order to reject back-ground clutter, we will apply the Gel'fand-Leviten (GL) transform between two potentials to fabricate analytically an overall psf $s(\vec{r}) = s_o(\vec{r}) * s_p(\vec{r})$, where $s_p(\vec{r})$ is a pre-imaging filter for example the incision of dc spectrum to get rid of the frame aperture background .

$$s_p(\vec{r}) = s_o(\vec{r}) * s(\vec{r}) = s(\vec{r}) - \int s(\vec{r})d\vec{r}.$$

(ii) Image Restoration: We begin with the definition of exact inverse filter $m_i(r)$ to indicate reasons why it is ill-conditioned. Then from the regulation viewpoint of the ill-conditioned deconvolution, Wiener's performance criterion: minimum mean-square estimation errors, $(q-o)^2$, is presented. As a result, an alternative and practical performance criterion namely a minimum mean-square image-spread follows. Such an output energy constrained criterion, related to a human visual performance will be modelled by a composite filter $M(\vec{f})$ namely the (high-pass) inverse filter $S_o(\vec{f})^{-1}$ being apodized by a (Gaussian) regulation function $\Psi(\vec{f})$. Here and henceforth upper case letters denote Fourier transforms in 2-D spatial frequency domain. And the Fourier transform of the point spread function $s_o(\vec{r})$ is commonly called as the (OTF) $S_o(\vec{f})$. Since the imaging OTF $S_o(\vec{f})$ has the effect of low-pass filter, then the inverse OTF, $S_o^{-1}(\vec{f})$ has the effect of high-pass filter such that their product is the unity, all-pass filter.

(a) Exact Inverse Filter: An exact inverse filter $m_I(\vec{r})$ may be defined as one with the property of no image spread

$$m_I(\vec{r}) * s_0(\vec{r}) = \delta(\vec{r}). \tag{10}$$

Postponing for the moment consideration of the existence of m_I, one can substitute definition (8) into (9) and use (10) to obtain

$$q(\vec{r}) = o(\vec{r}) + m_I(\vec{r}) * n(\vec{r})$$

Taking the ensemble average of q, one would then obtain a perfect restoration of no image spread $\langle q(\vec{r})\rangle = \langle o(\vec{r})\rangle$. However, due to the convolution theorem, the Fourier transform of definition (10) is $M_I(\vec{f})\ S_0(\vec{f}) = 1$, so that $M_I(\vec{f}) = 1/S_0(\vec{f})$. Consequently, the function $M_I(\vec{f})$ becomes divergent at the zeros and outside the band-limits of the OTF $S_0(\vec{f})$. A simple 1-D example shows an ad hoc regulation method as follows.

(b) Ad hoc Regulation. Since the blur psf s_0 = rect (x) has the OTF $S_0 = \sin(\pi\nu)/\nu\pi$ associated with an infinite set of zeros where sine function vanishes, the inverse filter becomes divergent. But it has been fabricated by Tsujichu's technique of clipping peaks, namely a valid approximation for a band-limited restoration. Thus, the first problem of ill-conditioning that concerns the divergence of m_I has been piecewise regulated in an ad hoc fashion. Nevertheless, more serious is the noise amplification problem due to the high pass character of the inverse filter. From the Fourier transform of (10) and the result of (5) follow

$$Q(\vec{f}) = O(\vec{f}) + N(\vec{f})/S_0(\vec{f}) \tag{11}$$

Thus, the ever decreasing signal-to-noise ratio at higher and higher spatial frequencies, prevents the exact inverse filter from giving a faithful restoration.

(c) Analytic Regulation. A simple way to insure the overall existence of $m_I(\vec{r})$ as well as regulate the divergence of output noise $m_I(\vec{r}) * n(\vec{r})$ is to allow the replacement of the ideal image-spread δ-function in (10) with a broader regulation function $\psi(\vec{r})$,

$$m(\vec{r}) * s_0(\vec{r}) = \psi(\vec{r}) \tag{12}$$

Our problem remains to implement $\Psi(\vec{f})$ in an analytic fashion for an arbitrary noise $N(\vec{f})$.

(iii) Wiener's Regulation: To illustrate this regulation concept (12) Wiener's filtering of white noise $n(\vec{r})$ is re-interpretated as follows. Given the Wiener filter M_w, to be derived later, it follows readily, by factorization, the Wiener's regulation function, Ψ appended by the subscript w, is

$$M_w = S_0^{-1}\ \Psi_w \tag{13a}$$

$$\Psi_w(\vec{f}) = (1 + R(\vec{f}))^{-1} \tag{13b}$$

where R stands for the ratio of noise-to-signal

$$R(\vec{f}) = \langle|N|^2\rangle/(\langle|O(\vec{f})|^2\rangle|S_o(\vec{f})|^2) \qquad (14)$$

The real, positive and monotonously decreasing function Ψ_W can also be called Wiener's apodization function in the context of imaging. The amount of broadening between (10) and (13) may be measured with their difference $B_W(f)$ in the Fourier domain defined by $B_W = 1 - \Psi_W$ which, using (13) equals to $B_W = R(1+R)^{-1}$. or after a proper normalization, $B_W/\Psi_W = R$.

(a) Vander Lugt Filter: Evidently two limiting situations can happen simultaneously, i.e. $R(\vec{f}_\ell) \ll 1$ and $R(\vec{f}_h) \gg 1$; where ℓ and h denote respectively the low and the high spatial frequencies. Because the optical imaging has the effect of a low-pass filter, a relative small noise $R(\vec{f}_\ell)$ occurs over a dominant object spectrum at a low frequency region, while a large $R(f_h)$ appears at a high frequency region. As a result, using binomial expansions of (13) in these limits

$$\Psi_W(\vec{f}_\ell) \approx 1 - R(\vec{f}_\ell), \qquad R\ll1, \qquad (15a)$$

$$\Psi_W(\vec{f}_h) \sim 1/R(\vec{f}_h), \qquad R\gg1, \qquad (15b)$$

an approximate broadening near the low frequency,

$$M_I(\vec{f}_\ell) - M_W(\vec{f}_\ell) = B_W S_o^{-1} \approx (\langle|N|^2\rangle/\langle|O|^2\rangle)|S_o|^{-3}, \qquad (16)$$

is inversely proportional to the cubic power of the OTF, while in the high frequency limit (15) the Wiener filter is identical to the (North or Van Vleck and Middelton) matched filter implemented by Vander Lugt for optical character recognition,

$$M_W(\vec{f}_h) = (\langle|O|^2\rangle/\langle|N|^2\rangle)\, S_o^*(\vec{f}_h) \qquad (17)$$

(b) Absence of Super-resolution: If use is made of the inverse identity to replace Ψ_W,

$$\Psi_W = (1+R)^{-1} = 1 - R(1+R)^{-1}. \qquad (18)$$

then, using (18), from (9)(13) follow

$$Q = O - R(1+R)^{-1}\, O + M_W N, \qquad (19)$$

Thus, the object spectrum O will always be substracted off so long as $R \neq 0$. Furthermore, the object spectrum can never be restored in the limit $S_o(f) \to 0$ when $R \to \infty$. The last fact can be restated as that a Wiener filter can never be able to have a super-resolution beyond the diffraction limit where the OTF $S_o(\vec{f}) = 0$. But, as a sufficient result of no super-resolution, a Wiener filter can suppress the output noise divergence in the mean square sense,

$$\langle|M_W N|^2\rangle = (1+R)^{-2}\, R\, \langle|O|^2\rangle, \qquad (20)$$

which vanishes in the same limit $S_o(f) = 0$ that $R = \infty$. However, while these sufficient proofs are valid for all frequencies we left with the suspicion whether a global optimization procedure for all frequency is necessary, and the possibility that Wiener might have indiscriminately and overly killed the ill-conditioned deconvolution for all frequencies. This viewpoint will be elaborated as follows.

(c) MMSE Wiener Filter. The Wiener filter results from the global performance criterion of minimum mean-square estimation-error, $(q(\vec{r}) - o(\vec{r}))^2$,

$$\int \langle (q - o)^2 \rangle \, d\vec{r} = \text{minimum}, \tag{21}$$

or in the Fourier domain, using Parseval formula,

$$\int \langle |Q-O|^2 \rangle \, d\vec{f} = \text{minimum} \tag{22}$$

The standard variation of (22) with respect to M yields the Wiener filter

$$M_W = \langle I^* O \rangle \, / \, \langle I^* I \rangle \tag{23}$$

(d) Orthogonal Projection: The result (23) can be directly obtained from definition (4) after averaging it with $I^*(\vec{f}')$

$$\langle I^*(\vec{f}') \, Q(\vec{f}) \rangle = M(\vec{f}) \, \langle I^*(\vec{f}') \, I(\vec{f}) \rangle \tag{24}$$

and replacing $\langle I^*Q \rangle$ with $\langle I^*O \rangle$ by invoking the extremum principle of orthogonal projection, expressed in the inner product bracket notation (,) as

$$\langle (I, Q-O) \rangle = 0 \tag{25}$$

i.e. the estimation error, Q-O, is an extremum if Q-O becomes orthogonal to the input data I alone.

Obviously, there exists, other than the input data I alone, viz. a priori knowledge (about the object denoted by $(O' < O)$ that is the subset of O which yields $(I' < I)$ such that the principle of orthogonal projection might work better than (25)

$$\langle (I', Q-O') \rangle = 0$$

Also the correlation of neighborhood frequency $\vec{f}'$ with $\vec{f}$ is not used in (24) because, according to the Wiener-Khinchine theorem, no mode coupling, $\delta(\vec{f}'-\vec{f})$, is possible for the stationary correlation $\langle I^*(\vec{f}')I(\vec{f}) \rangle$. Instead, the unknown object statistics $\langle O^*(\vec{f}')O(\vec{f}) \rangle$ is needed, but is impractical to be ascertained. Based on these criticisms for image applications, we would like to point out the possibility that, instead of using these unknown covariance alone, a priori knowledge, such as a local spatial frequency coupling through its curvature effects may produce in some localized image cases better restoration.

(iv) A Constrained Moment Method: A real function $g(\vec{r})$ to be employed as a general performance measure for a constrained moment method will be introduced. Thus, in viewing various possibilities, an arbitrary regulation function $g(\vec{r})$ is conveniently treated in the linear vector space and then specialized to $\psi(\vec{r})$ later. Since noise is generally characterized by its real and positive power spectral density, the noise suppression to be traded with object distortion requires the square of a bipolar $g(\vec{r})$ as a real and positive energy density $\rho(\vec{r}) = g(\vec{r})^2 \geq 0$. Since g is square-integrable, then a constant total energy follows

$$\int \rho(\vec{r}) \, d\vec{r} = \int g(\vec{r})^2 \, d\vec{r} = \text{constant}$$

Since only lower orders of image moments are usually measured, then only lower orders of spatial frequency derivatives are known through Parseval's formula,

$$\int |2\pi\vec{r}|^{2n} g(\vec{r})^2 d\vec{r} = (-)^n \int G^*(\vec{f}) \nabla_f^{2n} G(\vec{f}) d\vec{f}$$

Furthermore, the estimated function $g(\vec{r})$ based on the partial series must be consistent with the constraints imposed upon the function $g(\vec{r})$ itself. Thus, the purpose of constrained moment method is to intersect the hyper-volume representing the partial series with the hyper-surfaces representing those constraints, so that the estimated uncertainty may be minimized in Hilbert space. This is analytically fulfilled using the minimum energy concept of Hamiltonian system. The partial series is associated with the kinetic energy while the constraints is casted in the form of potential energy.

There are two kinds of constraints. The first kind is intrinsic to the image. The second kind is external and pertinent to the imaging distortion and restoration system. The first kind is concerned about the symmetry, size, positivity and other interesting features of the image known a priori. The second kind is generally known and related to the system that forms and processes the imagery. Both are useful for the 2-D information retrieval. The essence of a constrained moment method is to provide a convenient framework that can incorporate these constraints.

(a) <u>Object Independent MMIS Filters For Image Restorations:</u> An alternative performance criterion is the following MMIS $\Psi = MS_0$, which is to control mathematically the broadened regulation function that will allow us to restore imagery without relying implicitly upon the unknown object. This will lead to a local mode coupling theory in the spatial frequency domain as follows. Since the convenient performance criterion that replaces the estimation error $(Q-0)^2$ with the image-spread $\Psi = MS_0$ is no longer depending on the unknown object 0, then we can choose, for the sake of reasoning, a test point object $\delta(\vec{r})$. Since $\langle o\rangle = \delta$, then one expects $\langle q\rangle \approx \delta$. Therefore one defines $\langle q\rangle = m*i = m * s_0 * \langle o\rangle = \psi * \delta = \psi$ where, due to the low-pass imaging lens s_0 and arbitrary noise n, the ideal image δ is replaced by $\psi = m * s_0$. Therefore, the spot size of Ψ is conveniently measured by the second moment or the so-called radius of gyration given by:

$$\int |2\pi\vec{r}|^2 (m * s_0)^2 d\vec{r} = \text{minimum} \tag{26}$$

which we demand to be as small as possible. Such a MMIS performance criterion must be constrained by a constant output noise in order to regulate the divergence problem (11) (20)

$$\langle \int (m * n)^2 d\vec{r}\rangle = \text{constant} \tag{27}$$

and, moreover constrained by a constant total image intensity throughput in order to be meaningful

$$\int (m * s_0)^2 \, d\vec{r} = \text{constant} \tag{28}$$

This set of constrained least squares equations can be implemented in a slightly different and perhaps more flexible manner than those constrained least squares filters implemented in the coordinate domain. One obtains the result (3) as follows. The Laplacian operation, $\nabla^2 = |d/d\vec{f}|^2$, in the Fourier domain is derived from (26) by means of Parseval's formula and integration by parts,

$$4\pi^2 \int |\vec{r}|^2 \, |\psi(\vec{r})|^2 \, d\vec{r} = - \int \Psi^*(\vec{f}) \, \nabla^2 \, \Psi(\vec{f}) \, d\vec{f} \tag{29}$$

The potential energy, V(f), is derived from the constant of output noise (27) by definition $\Psi = M \, S_0$,

$$\int |M|^2 \, \langle |N|^2 \rangle \, d\vec{f} = \int \Psi^* \, \frac{\langle |N|^2 \rangle}{|S_0|^2} \, \Psi \, d\vec{f} \; . \tag{30}$$

The total energy E is the Lagrangian multiplier that multiplies the quadratic normalization (28)

$$\int \psi^2 \, d\vec{r} = \int \Psi^* \, \Psi \, d\vec{f} = 1 \quad . \tag{31}$$

Since all integrands of (29-31) are real and homogeneously quadratic, then the Lagrangian equation associated with the standard Hamilton variation principle requires that the sum of integrands must vanish (3). Formally we can rewrite (3) as a Hamiltonian system $H \, \Psi_n = E_n \, \Psi_n$ where eigenvalues E_n and eigenfunctions Ψ_n are self-consistently determined by the condition of high frequency boundedness. If a proper choise of total output intensity E_n is given, then eignefunction Ψ_n will yield a dynamic filter $M_n = \Psi_n \, S_0^{-1}$. It is clear that the ground state Ψ_0 is associated with the minimum image-spread provided that $\sqrt{E_0}$ is related to the output image location.

Solutions of Schrodinger equation are abundant in quantum mechanics and well documented using analytic approximations and numerical methods for arbitrary potentials.[2] Thus an analytic formula may be available for the trade-off between the relative noise energy $V(\vec{f})$ and the image-spread function, $-\nabla^2\Psi$. For example, such a resulting dynamic filter may behave like a high pass inverse filter, $S_0(\vec{f})^{-1}$, multiplied by a Gaussian apodization function, $\Psi(\vec{f})$ being a typical ground state of Schrodinger equation (3), $M(\vec{f}) = (a + b \, |\vec{f}|^2) \exp(-c|\vec{f}|^2)$ Since it is similar to a model of band-pass filter for visual perception, it is tempting to conjecture that this MMIS, might be actually corresponding to a physiological model of visual processing. Although any further study[21] would require the material beyond the present scope, we would like to support the conjecture with some observations in image processing[22] which show empirical tests using a similar functional form[23] giving better results[24] than those of Wiener's.

(b) Object-Matched MMIS Filters For Pattern Recognitions: In order to be useful for pattern recognitions, the knowledge of the object to be recognized must be incorporated. Thus, the contrained moment method can derive an object-matched MMIS filter. Instead of minimizing the image spread of the output transfer function ψ due to a point object, we can minimize the spread of the restored object $q(\vec{x})$. Although two are identical for a point object, for an extended object the generalization allows us to derive a new object-matched dynamic filter as follows. Parseval's formula gives

$$4\pi^2 \int |\vec{x}|^2 q^2(\vec{x})d\vec{x} = \int |\nabla Q(\vec{f})|^2 d\vec{f} = -\int Q^*(\vec{f})\, \nabla^2 Q(\vec{f})\, d\vec{f} \quad (32)$$

We will assume a piecewise space invariant image for the general (object) space variant image. This will allow us to use Fourier deconvolution of image equation within the large piecewise region.

We minimize the spread of the restored object under two constraints: (1) a constant output noise energy

$$\int \langle \hat{n}(\vec{x}) \rangle d\vec{x} = \int \langle |\hat{N}(\vec{f})|^2 \rangle d\vec{f} = \int \langle Q^* \frac{|N|^2}{|SO+N|^2} Q \rangle d\vec{f} \quad (33)$$

where use is made of

$$\hat{N} \equiv MN \equiv QN/I$$

and (2) a constant restored energy, using Lagrangian multiplier E,

$$\int \langle q^2(\vec{x}) \rangle d\vec{x} = \int \langle |Q(\vec{f})|^2 \rangle d\vec{f} \quad (34)$$

Then we use the identical technique in obtaining Schrodinger's equation of one realization of the restored object ensemble in Fourier domain denoted by the overhead bar in (6). The Wiener output transfer function Ψ_W in (15) enters naturally into (33) and becomes the difference of Schrodinger's potential from the unit output noise energy

$$U(f) = \langle \frac{|N|^2}{|SO+N|^2} \rangle = \frac{R}{1+R} \equiv 1 - \Psi_W \quad (35)$$

we arrive at the object matched dynamic filter

$$M(\vec{x}) = \int \bar{Q}(\vec{f})/\bar{I}(\vec{f}) \exp(i2\pi \vec{f}\cdot\vec{x})\, d\vec{f} \quad (36)$$

(c) Discussion: Since a blurred object tends to be diffusively vanishing over a larger spatial frequency domain, then the scheme of minimizing the radius of gyration of the object in the coordinate domain will produce in the spatial frequency domain the desired frequency spread under a constant output noise energy. It reduces for a point object, 0 = 1 to the previously derived Schrodinger's equation of the output transfer function.

$$Q(\vec{f}) = \Psi(\vec{f}) + MN \simeq \Psi(\vec{f}),\ R(\vec{f}) = V(\vec{f}),\ U(\vec{f}) = V(\vec{f})(1 + V(\vec{f})) \simeq V(\vec{f})$$

We expect that the object-matched MMIS filter should work locally better than the previous one, because the a priori knowledge of the expected object spectral density $P=\langle|O|^2\rangle$ has been included in the optimization scheme (32). We shall discuss the implications in Section 5, where the Abraham-Moses application of the Gel'fand-Levitan transformation is used for the reconstruction of difference potentials. The coherent and incoherent restoration will be shown.

3. EXACT SOLUTION OF INVERSE PROBLEM

It is interesting to find a consistent trade-off potential between two approaches, namely the LS method of Wiener and the energy-constrained LS method. The possibility of identical Wiener and dynamic filter is a specific inverse problem that can be answered with a straightforward algebraic approach. Wiener filter has been casted as the apodization of inverse filtering (13) where the Wiener apodization function is re-written for arbitrary object $P = \langle|O|^2\rangle$ in terms of the trade-off potential $V(\vec{f})$.

$$\Psi_W(\vec{f}) = \left(1 + \frac{V(\vec{f})}{P(\vec{f})}\right)^{-1} \tag{37}$$

On the other hand, the formal inverse solution of Schrodinger's equation in (3) may be written as

$$V(\vec{f}) = \frac{\nabla^2\Psi}{\Psi} + E \tag{38}$$

If $\Psi = \Psi_W$, then solving (37) for V,

$$V = (\Psi(\vec{f})^{-1} - 1)\ P(f)\ . \tag{39}$$

Equating (38) with (39), we obtain the differential equation

$$\nabla^2\Psi + (P+E)\ \Psi = P \tag{40}$$

with a simple solution

$$\Psi = [\nabla^2 + (P+E)]^{-1}\ P \tag{41}$$

The one dimensional real and symmetric result is shown for $P = 1$.

$$\Psi(\nu) = \frac{1}{1+E} - \cos(\sqrt{E+1}\ \nu) \tag{42}$$

The trade-off potential is found from (39) to be

$$V(\nu) = \left[\frac{E}{1+E} + \cos(\sqrt{E+1}\ \nu)\right]\left[\frac{1}{1+E} - \cos(\sqrt{E+1}\ \nu)\right]^{-1} \tag{43}$$

and the optical transfer function is, when chosen to be real,

$$S(\nu) = \left[\frac{1}{1+E} - \cos(\sqrt{1+E}\ \nu)\right]^{1/2}\left[\frac{E}{1+E} + \cos(\sqrt{1+E}\ \nu)\right]^{-1/2} \tag{44}$$

Then, by definition (1), the identical filter for both performance criteria follows:

$$M(\nu) = [\frac{1}{1+E} - \cos(\sqrt{1+E}\ \nu)]^{1/2} [\frac{E}{1+E} + \cos(\sqrt{1+E}\ \nu)]^{1/2} \quad (45)$$

We make three comments as follows.

(a) For the unbounded domain, the complex apodization function that satisfies the equivalent equation (40) is

$$\Psi(\nu) = \frac{1}{1+E} + e^{i\sqrt{1+E}\ \nu}$$

(b) for arbitrary object spectral density the equivalent differential (40) can be expressed in terms of the net broadening, $\Psi-1$,

$$\nabla^2(\Psi-1) + (E+P)(\Psi-1) = -E$$

which is a Schrodinger equation of a constant inhomogeneity $-E$.

(c) It is interesting to observe that the matched filtering is similar to the visual contrast sensitivity function. Note that the optical transfer function (44) is a high pass filter which is unusual in optical imaging, but possible using computer generated holograms. The equivalence differential (40) makes the analytical comparison between the two types of filters possible.

4. MOTION-BLURRED NOISE IMAGE RESTORATION: PHYSICAL MEANING OF E

The concept of reference filters[4,5], is closely related to the reference potential that is the main concept of the inverse problem of quantum mechanics. The exact relationship is made possible by generalizing MMSE filter of Wiener's to the energy-constrained the MMIS filter. Then the apodization function of the inverse filter satisfies a Schrodinger-like equation in the spatial frequency ν domain, rather then in the spatial x domain. To appreciate the importance of this fact, we consider a special case of noiseless imaging. Since the tradeoff potential is zero in the case of no noise, then the output transfer function Ψ of the Schrodinger equation is integrated to give the phase factor

$$\Psi(\nu) = \exp(ik\nu); \quad k = (E)^{1/2}. \quad (46)$$

The restoring output image becomes, by definition (4), $Q=\Psi O$.

$$Q(\nu) = \exp(ik\nu)\ O(\nu) \quad (47a)$$

The important "gauge" freedom means the invariant intensity.

$$|Q(\nu)|^2 = |O(\nu)|^2 \quad (47b)$$

and implies that the physical meaning of E is, in the spatial domain, a shifted but perfect restoration of the object

$$q(x) = o(x + (k/2\pi)) = o(x + \sqrt{E}/2\pi) \quad (48)$$

The further significance of such a solution can be best seen in that it unifies results of Harris[25] and Swindell[26] for the case of linear motion blur and consequently a new "de-ghost" filter is obtained as follows. The psf is

$$s(x) = \mathrm{rect}(x/2L)/2L \quad (49)$$

which gives, in Fourier domain, the OTF, with $K \equiv 2\pi L$,

$$S(\nu) = \sin(K\nu)/K\nu \quad (50)$$

Accordingly, the restoring processor filter, using (46) in $M \equiv \Psi/S$, has the real and imaginary parts

$$M(\nu) = K\nu \cos k\nu/\sin K\nu + iK\nu \sin k\nu/\sin K\nu \tag{51}$$

The real part gives Swindell's result while the imaginary part gives Harris' result. In the spatial domain, one can Fourier transform the solution (51) and therefore derive the general solution given by:

$$m(x) = L\frac{d}{dx}\left[\sum_{n=0}^{\infty} \delta(x-(2n+1)L + (k/2\pi))\right] - L\frac{d}{dx}\left[\sum_{n=0}^{\infty} \delta(x+(2n+1)L + (k/2\pi))\right] \tag{52}$$

where use is made of the straightforward long division $1/(\exp(ik\nu) - \exp(-ik\nu))$ producing the expansion

$$1/\sin K\nu = 2\sum_{n=0}^{\infty} \sin((2n+1)K\nu) \tag{53}$$

Thus, the Swindell's result is derived from (52) in the case the ground state ($k = \sqrt{E} = 0$) output transfer function $\Psi = 1$, while the Harris' result follows the central portion ($|x|<2L$) of the excited state ($k = K$) output transfer function (i.e. the shifted version of the ideal inverse filter.) If in the spatial frequency domain $k = K$, then the first term $M(\nu)$ of (51) is the "de-ghost" filter which cancels the goast image produced by the second term: $iK\nu \to L\, d/dx\, \delta(x)$ namely the spatial differential operator of Harris.

For white noise, the trade-off potential may be approximated by a crystal lattice (Kronig-Penny model)[27] where $a=1/2L$

$$V(\nu) = \left(\frac{\nu}{\sin\nu}\right)^2 \langle|N|^2\rangle \simeq \sum_{n=-\infty}^{\infty} a\delta(\nu-na) - a\delta(\nu) \tag{54}$$

The void "impurity" at the origin is due to $S(0) \neq 0$ in (50) and consequently $V(0) \neq \delta(\nu)$in (54). Using the Green's function,

$$G(\nu,\zeta) = \frac{i}{2k}\exp(ik|\nu-\zeta|) \tag{55}$$

and the Bloch theorem $\psi(\nu + na) = \exp(i\mu na)\,\psi(\nu)$, the output transfer function is known

$$\psi(\nu) = \frac{a}{2k}\psi(0)\exp(i\mu(\nu-\nu'))\,\frac{\exp(i\mu a)\sin k\nu' - \sin k(\nu'-a)}{\cos\mu a - \cos ka} \tag{56}$$

where $\nu' = \nu - a \cdot$ integral part of $[\nu/a]$. Then, the standard perturbation approach follows from Saxon and Hutner's work[28] and will not repeat it here.

5. DIRECT AND INVERSE PROBLEMS OF COHERENT IMAGE RESTORATION

The OTF of coherent (laser) imaging is represented by

$$S_c(\nu) = \mathrm{rect}(\nu/\pi) \tag{57}$$

which is 1 if $|\nu| < \pi/2$ and zero outside the range. Four special cases are considered as follows.

(i) No Noise: In the first case of negligible phase (speckle) noise, we solve the equivalent Schrodinger equation, $H_0 = -d^2/d\nu^2 + V_c$ for a particle in a box, where $V_c \simeq 0$ in the zero-th order of approximation and obtain

$$-\frac{d^2}{d\nu^2}\cos(n\nu) = n^2\cos(n\nu) \tag{58}$$

Thus the even output transfer function is

$$\Psi_n(\nu) = \left(\frac{2}{\pi}\right)^{1/2}\cos(n\nu) \tag{59}$$

which vanishes at the infinite potential well at $\nu = \pm\pi/2$ that

$$\int_{-\infty}^{\infty}\Psi_n(\nu)^2\,d\nu = C_n = 1; \qquad E_n = n^2$$

Similarly the odd eigenfunction follows

$$\Psi_n(\nu) = \left(\frac{2}{\pi}\right)^{1/2}\sin(n\nu)\,. \tag{60}$$

Thus, the associated restoration filters become $M_n(\nu) = \Psi_n(\nu)S_c^{-1}$ that are first derived by Papoulis[13] for noiseless coherent imagery.

(ii) Arbitrary Noise Perturbation: Due to the Smith's formulation[14] we can go beyond Papoulis by including the arbitrary speckle noise power spectral density as the potential

$$V_c = \frac{\langle|N|^2\rangle}{|S_c|^2} \tag{61}$$

Then in case of arbitrary speckle noise power spectrun the conventional perturbation method can be used to calculate the eigenfunction Ψ_c of $H_c = H_0 + V_c$ and the constant eigenvalue E_c. In particular, if $\langle N^2\rangle \simeq \varepsilon_0$ is a piecewise constant <1, then a fast covergent series solution follows

$$\Psi_{cn} = \Psi_n + \varepsilon_0 \sum_{m \neq \nu} \Psi_m/(n^2 - m^2) \tag{62}$$

$$E_{cn} = n^2 + \varepsilon_0 \tag{63}$$

Contrary to positive ψ_1, Ψ_{c1} becomes bipolar cancellation of noise.

(iii) Gel'Fand-Levitan Method: In order to solve the inverse problem of imaging we recapitulate the Gel'fand-Levitan (GL) integral transform[2] for the 1-D Spatial frequency domain

$$\Phi(\nu) = \Psi(\nu) + \int_{-\infty}^{\nu} K(\nu, \mu)\, \Psi(\mu)\, d\mu \tag{64}$$

The kernel of GL transform itself satisfies the Fredholm integral equation of the second kind, at a fixed $\nu > \mu > 0$,

$$K(\nu,\mu) = -\Omega(\nu,\mu) - \int_{-\infty}^{\nu} K(\nu,\nu')\, \Omega(\nu,\mu)\, d\nu' \tag{65}$$

where Ω is experimental data. Solving K from (65) the tradeoff potential follows

$$V(\nu) = 2\, d\, K(\nu,\nu)/d\nu \tag{66}$$

Discrete states and real constants b_n or c_n are assumed.

$$\Omega_2(\nu,\nu') = \sum_n \Phi_n \Phi_n'/b_n - \sum_n \Psi_n \Psi_n'/c_n \tag{67}$$

Solving K_2 from (68), the tradeoff potentials V_2 follows.

$$K_2(\nu,\mu) = -\Omega_2(\nu,\mu) - \int_{-\infty} K_2(\nu,\nu')\, \Omega_2(\nu,\mu)\, d\nu' \tag{68}$$

$$V_2(\nu) = 2\, d\, K_2(\nu,\nu)/d\nu = V - V_1 \tag{69}$$

where V_1 is associated with Ψ while V is with Φ.

(iv) Add On Potential Problem: We shall demand the extra ground state $n = 0$ of the form (57). Since we are not going to willfully alter the other eigenfunctions Ψ_n, unless they are necessarily adjusted due to the required new potential $V = V_1 + V_2$, then we assume tacitly an identical set of formal eigenfunctions denoted by the lower case ϕ_n,

$$\phi_n = \Psi_n; \quad b_n = c_n \; ; \; n = 1, 2, 3, \ldots \tag{70}$$

$$\phi_0(\nu) = S_c(\nu) \tag{71}$$

Consequently we have arrived at a single factorized term

$$\Omega_2(\nu,\nu') = \sum_{n=0} \frac{\phi_n \phi_n'}{b_n} - \sum_{n=1} \frac{\Psi_n \Psi_n'}{c_n} = \frac{\phi_o \phi_o'}{b_o} \tag{72}$$

Solving the degenerated Fredholm integral equation, we obtain

$$K_2(\nu, \mu) = \frac{-\phi_0(\nu)\,\phi_0(\mu)}{b_0 + \int_{-\infty}^{\nu} \phi_0^2(\nu')d\nu'} \tag{73}$$

The actual eigenfunctions denoted by the upper case Φ_n are

$$\Phi_n(\nu) = \Psi_n(\nu) + \int_{-\infty}^{\nu} K_2(\nu,\mu)\, \Psi_n(\mu)\, d\mu \tag{74}$$

due to the add-on (difference) potential

$$V_2(\nu) = 2\, dK_2(\nu,\nu)/d\nu \tag{75}$$

Given

$$\Omega_2(\nu,\mu) = \phi_0(\nu)\ \phi_0(\mu)/b_0 \tag{76}$$

The solution $K_2(\nu, \mu)$ when evaluated at $\mu = \nu$ has a simple form

$$K_2(\nu,\nu) = -\frac{d}{d\nu}\ \ell n\ (b_0 + \int_{-\infty}^{\nu} \phi_0{}^2(\nu')d\nu') \tag{77}$$

and then the solution of the difference potential becomes

$$V_2(\nu) = -\ 2\,\frac{d^2}{d\nu^2}\ \ell n(b_0 + \int_{-\infty}^{\nu} \phi_0{}^2(\nu')d\nu') \tag{78}$$

If we consider to add rect(ν/π) to the first case of coherent imaging, viz. a particle in a box, we obtain from substituting (71) into the above formula (77)

$$K_2(\nu,\mu) = \frac{-\ \text{rect}(\nu/\pi)\ \text{rect}(\mu/\pi)}{b_0 + \frac{\pi}{2} + \nu} = -\ \Phi_0(\nu)\ \phi_0(\mu), \tag{79}$$

where, as obviously identified from (74, 68, 76),

$$\Phi_0(\nu) = \text{rect}(\nu/\pi)/\ (b_0 + \frac{\pi}{2} + \nu) \tag{80}$$

Consequently, we can determine the corresponding imaging OTF (61) from the solution (78) of potential

$$V_2(\nu) = -\ 2(b_0 + \frac{\pi}{2} + \nu)^{-2} \tag{81}$$

where the normalization constant $b_0 = [(\pi^2 + 4\pi)^{1/2} - \pi]/2$

(v) Deletion Potential Problem: If we consider to delete the ground state Ψ_1 then we obtain, instead of Eqs (76, 77, 78),

$$\Omega_2(v,\ \mu) = -\ \Psi_1(\nu)\ \Psi_1(\mu)/C_0 \tag{82}$$

$$K_2(\nu,\nu) = -\frac{d}{d\nu}\ \ell n(C_0 - \int_{-\infty}^{\nu} \Psi_1{}^2(\nu')\ \ d\nu'\ ) \tag{83}$$

$$V_2(\nu) = -\frac{d^2}{d\nu^2}\ \ \ell n(C_0 - \int_{-\infty}^{\nu} \Psi_1{}^2(\nu')\ \ d\nu'\ ) \tag{84}$$

For coherent imaging, we delete the Papoulis ground state

$$\Psi_1 = \left(\frac{2}{\pi}\right)^{1/2} \cos\nu; \qquad c_1 = 1 \tag{85}$$

we obtain the following potential slightly different from the result of Abraham and Moses[29] derived for different problem.

$$V_2(\nu) = \frac{8 \sin 2\nu + 32(\cos \nu)}{(2\nu - \pi + \sin 2\nu)^2} \tag{86}$$

6. DIRECT AND INVERSE PROBLEMS OF INCOHERENT IMAGING

The incoherent imaging point spread function $s_0(x)$ is the intensity of the diffraction pattern $s_i(x) = (\sin \pi x/\pi x)^2$, behind the ideal shuttle aperture, e.g. 1-D slit rect(ν) = 1 if $|\nu| <$ 1/2. Then the OTF is the roof top function due to the Fourier transform (FT) convolution theorem, Fig.1

$$S_i(\nu) = FT\{s_i(x)\} = \mathrm{rect}(\nu) * \mathrm{rect}(\nu) \equiv \Lambda(\nu) = 1 - |\nu| \tag{87}$$

Thus, the tradeoff potential is a hyperbolic well shown in Fig. 2

$$V_i(\nu) = \frac{<|N|^2>}{|S_i|^2} = \frac{1}{SNR(1-|\nu|)^2} \tag{88}$$

The eigenfunctions are shown in Fig. 3, filters shown in Fig. 4.

Since it is desirable to incise the dc spectrum from the OTF $S_i(\nu)$, then the new OTF $S(\nu)$, without the subscript i, gives the new tradeoff potential, where S is to be determined.

$$V(\nu) = \frac{<N^2>}{|S|^2} \simeq \frac{1}{SNR} \left[\frac{1}{(1 - |\nu|)^2} + \delta(\nu) \right] \tag{89}$$

We solve, in case of weak SNR = 1/2, the ground state E = 0,

$$\left[-\frac{d^2}{d\nu^2} + \frac{2}{(1 - |\nu|)^2} + 2\delta(\nu) \right] \left[\frac{1}{1 - |\nu|} \right] = E \left[\frac{1}{1-|\nu|} \right] = 0 \tag{90}$$

Thus, we obtain exactly the ground state apodization function

$$\Phi_0(\nu) = \frac{1}{1 - |\nu|} = \frac{1}{S_i(\nu)} \tag{91}$$

which turns out to be the ideal inverse filter of the incoherent imaging. Consequently, the dynamic filter for the dc-incision incoherent imaging follows

$$M_0 = \Phi_0 S^{-1} = (S_i S)^{-1} \tag{92}$$

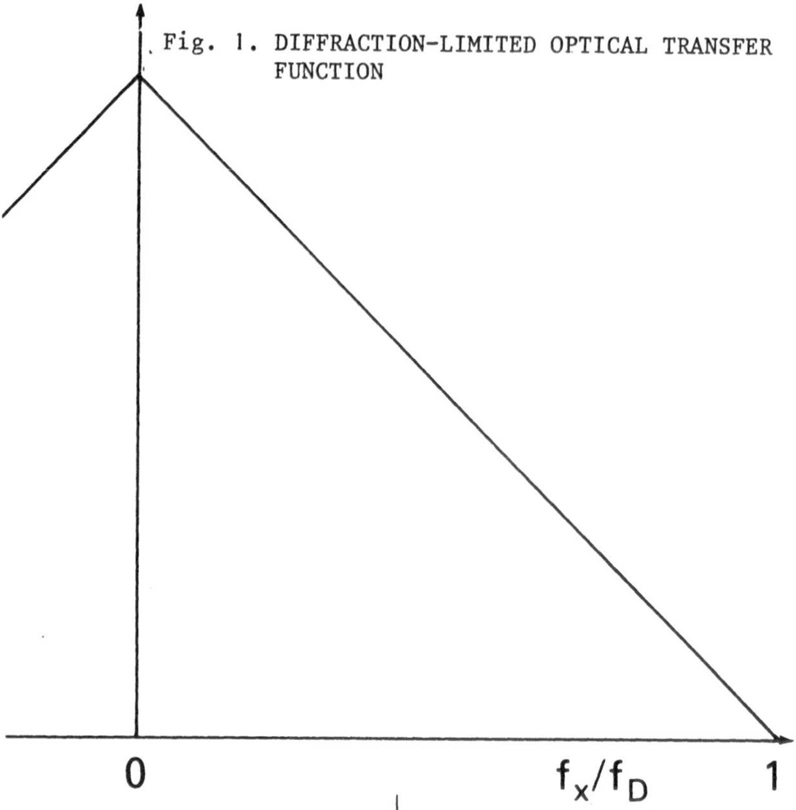

Fig. 1. DIFFRACTION-LIMITED OPTICAL TRANSFER FUNCTION

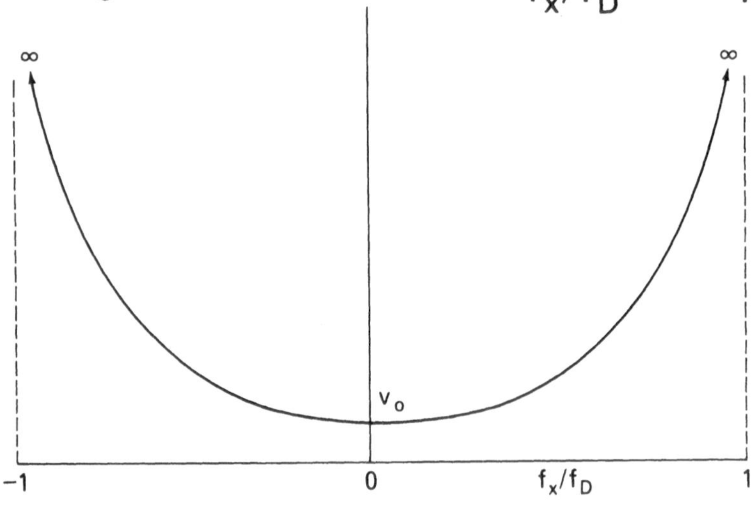

Fig. 2. TRADE-OFF POTENTIAL $V_D(f_x) = \frac{\langle |N(f_x)|^2 \rangle}{\langle |S(f_x)|^2 \rangle}$

Fig. 3. EIGEN SOLUTIONS

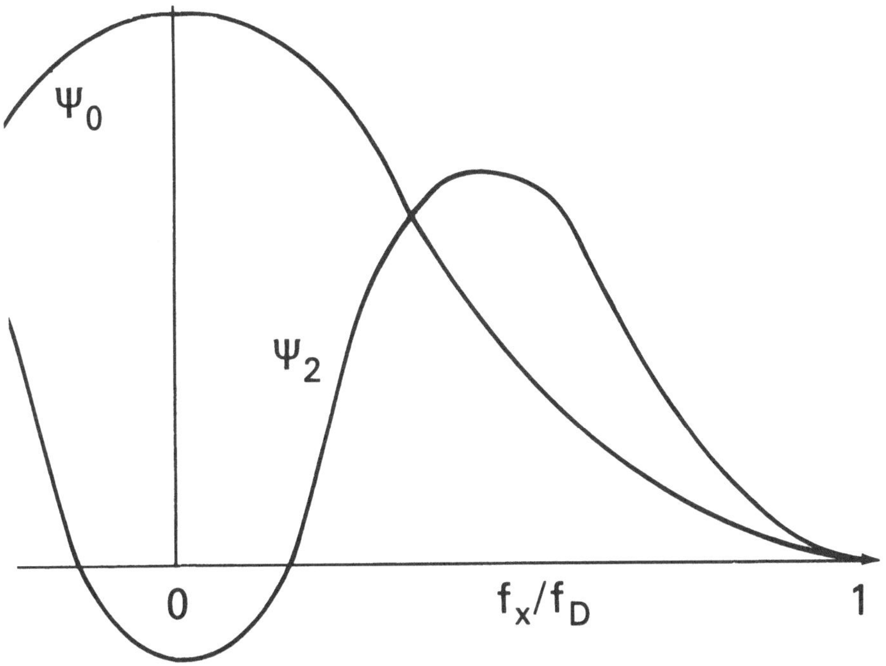

Fig. 4. IMAGE FILTERS

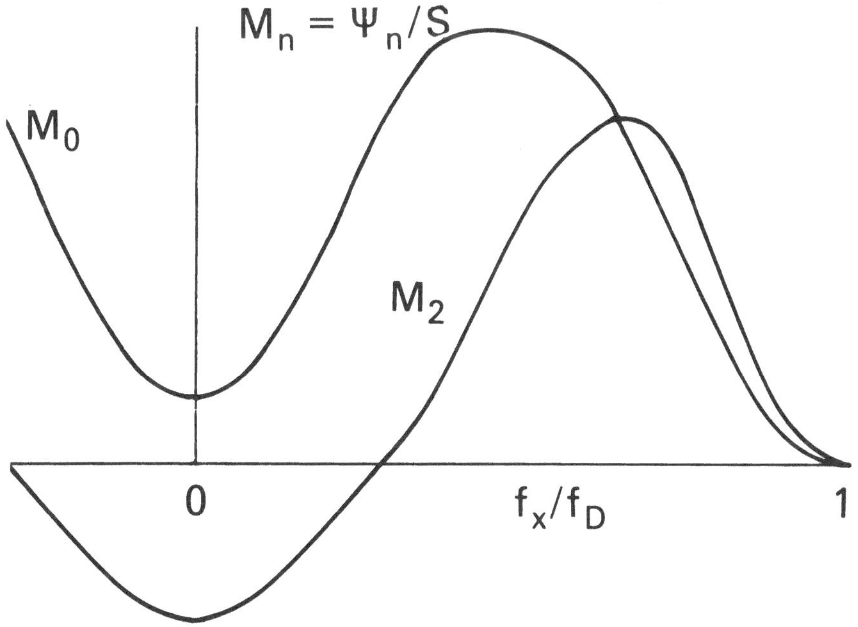

Since we use the principle value (P) to represent $\delta(\nu)$ in (89).

$$\delta(\nu) = \text{Real Part}\left[\frac{1}{-i\pi}\left(P\frac{1}{\nu} - \frac{1}{\nu - ia}\right)\right] = \frac{a}{\pi(\nu^2 + a^2)} \quad (93)$$

then we obtain

$$\frac{1}{|S(\nu)|^2} = \frac{1}{(1-|\nu|)^2} + \delta(\nu) = \frac{(1+\frac{a}{\pi})\nu^2 - \frac{2a}{\pi}|\nu| + a^2 + \frac{a}{\pi}}{(1-|\nu|)^2(\nu^2 + a^2)} \quad (94)$$

Solving the singular algebraic equation, we find

$$S(\nu) = (1 - |\nu|)\left(\frac{\nu + ia}{\nu - c}\right) \quad (95)$$

where $c = \frac{a}{\pi + a} + i\frac{\sqrt{\pi a}}{\pi + a}(1 + a\pi + a^2)^{1/2}$

The OTF at zero frequency in the limit a = 0 is a phase plate.

$$S(0) = -i \lim_{a\to 0} \frac{a}{c} = -i\pi , \quad (96)$$

and the real part of S(0) vanishes due to the incision of dc. Therefore, we conclude by the exact analysis that a new incoherent imaging (95), with the suppression of dc similar to the dark field electron microscope, can be restored by means of the ideal inverse filter $|S_i|^{-2}$

$$M_0 = (1 - |\nu|)^{-2}\left(\frac{\nu - c}{\nu + ia}\right) \quad (97)$$

where the extra factor is the complex phase factor given by (96).

7. CONCLUSION

In conclusion, we plot various filters in Fig. 5. The simple rule of thumb is the following: derive an appropriate Sturm-Liouville differential equation by the constrained moment expansion Sect. 2 and then apply the Gel'fand-Levitan inverse tansform to both the direct and the inverse problems Sect. 4. In the image processing, the output transfer function $\Psi(\vec{f})$ is governed to the lowest order moments by a Schrodinger-like equation, rather than that of the image filter $M(\vec{f})$ itself. Then it is straightforward to apply the Gel'fand-Levitan inverse transform to synthesize both designs of the imaging OTF $S(\vec{f})$ and the noisy image restoration filters $M(\vec{f})$. If someone argues that one shall not squeeze the physics into the mathematics between the direct and the inverse problems, the author agrees in principle but disagrees in the present case, because both the image and the signals are man-made and therefore man can create them with specific processing in mind. If we properly formulate the problem with a solution in mind, then the systematic approach is bound to work and so to speak pass the utility filter of man-made sciences. In order to be both useful and efficient, the present GL approach provides a complete system design of the pre-imaging system and the post-imaging restoration. This viewpoint reminisces the signal spectrum pre-shaping before the sending channel and then follows a systematic restorative decoding. The analysis beyond matched filtering opens up a new prospect in image and signal processing for the first time.

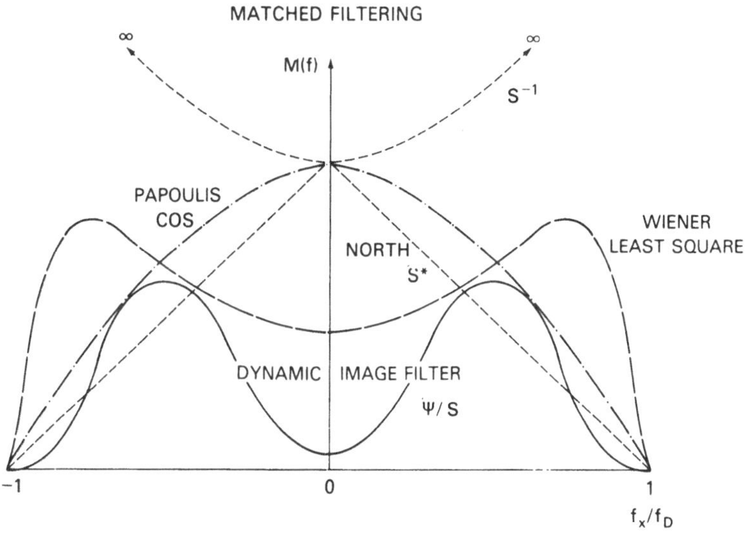

Fig. 5. Comparison among various matched filters. Papoulis' cosine filter[13] was derived for a smooth object in noise-free case. The energy-constrained MMIS filter is the quotient of the restoration output transfer function $\Psi(\vec{f})$ and the optical transfer function $S(\vec{f})$.

REFERENCES

1. H. H. Szu, C. E. Carroll, C. C. Yang, and S. Ahn, J. Math. Phys. 17, 1236 (1976).
2. K. Chadan and P. C. Sabatier, "Inverse Problems in Quantum Scattering Theory", (Spring-Verlag, New York, 1977), Ch. 3.
3. L. D. Faddeyev, J. Math. Phys. 4, 72 (1963)
4. F. Dyson, J. Opt. Soc. Am. 65, 551 (1975).
5. M. Kanal, H. E. Moses, S. K. Mitter, Transp. Theo. and Stat. Phys. (USA) 8, 163 (1979).
6. N. Wiener, "Extrapolation, Interpolation and Smoothing of Stationary Time Series," (John Wiley & Sons, New York, 1949) p.59.
   A. Kolmogoroff, "Interpolation and Extrapolation von Stationaren zufaligen Folgen", Bulletin de lacademie des science de U.R.S.S. ser. Math. 5, pp. 3-14 (1941).
7. D. O. North, "Analysis of the Factors which determine signal-to-noise Discrimination in Radar", RCA labs., Princeton N.J., Classified Tech. Rept. PTR-6C June 1943.
   D. O. North, Proc. IEEE 51, 1016 (1963).
8. J. H. Van Vleck and David Middleton, J. Appl. Phys. 17, 940 (1946).
9. "Special Issue on Matched Filters", IRE Trans. Inf. Theo IT-6, 310-408 (June 1960).
10. "Benchmark Papers in Electrical Engineering and Computer Science", Series Editor J. B. Thomas, "Detection and Estimation: Applications to Radar", editor S. S. Haykin (IEEE Inc. 1976).
11. J. Herrmann, J. Opt. Soc. Am. 70, 28 (1980).
12. A. B. Vander Lugt, IEEE Trans. Inf. Theo. IT-10, 139 (1964).
    J. W. Goodman, "Introduction to Fourier Optics", (McGraw-Hill, San Francisco, 1968) Ch. 7.
13. A. Papoulis, J. Opt. Soc. Am. 62, 1423 (1972).
14. H. A. Smith, J. SIAM Appl. Math. 14, 23 (1966).
15. H. J. Caulfield, R. Haimes, and D. Casasent, Opt. Eng. 19, 152 (1980).
16. A. A. Q. Kadkly, Opt. Acta 26, 461 (1979).
17. B. Roy Frieden, J. Opt. Soc. Am. 64, 682 (1974).
    B. Roy Frieden, Comp. Graphics Image Proc. 12, 40 (1980).
18. B. E. A. Saleh, Appl. Opt. 13, 1833 (1974).
19. M. J. Lahart, J. Opt. Soc. Am. 69, 1333 (1979).
20. H. H. Szu, "The foundation of single frame image processing," In: Optics in 4-dimension", Proceeding Am. Inst. Phys.65, (Int. Com. Opt., Mexico, August 1980).
21. R. A. Messner and H. H. Szu, Proc. Computer Vision and Pattern Recognition (IEEE Computer Soc. 1983) p. 522.
22. B. R. Hunt, Proc. IEEE 63, 693 (1975).
23. B. R. Hunt, IEEE Trans. AU-20, 94 (1972).
24. B. R. Hunt, IEEE Trans. C-22, 805 (1973).

25. J. L. Harris, J. Opt. Soc. Am. 56, 569 (1966).
26. W. Swindell, Appl. Opt. 9, 2459 (1970).
27. R. de L. Kronig and W. G. Penny, Proc. Roy. Soc. A130, 499 (1931).
28. D. S. Saxon and R. A. Hunter, Philips Res. Rept. 4, 81 (1949).
29. P. B. Abraham and H. E. Mose, Phys. Rev. A22, 1333 (1980).

V.9 (OI.3)

MODIFICATIONS OF THE GERCHBERG METHOD APPLIED TO ELECTROMAGNETIC IMAGING.

F.Gori and S.Wabnitz

Dipartimento di Fisica - P.le A.Moro, 2 - Roma 00185
Italy

Abstract

A general eigenfunction analysis of the iterative Gerchberg method (GM) and related procedures for the restoration of images of finite extent objects in coherent or incoherent illumination is given. Noise sensitivity considerations will lead to a new iterative algorithm fully exploiting the information content of the image. The discretization of the GM is then examined in view of a minimization of the number of significant samples. A new procedure is also described to give exact extrapolation of the image spectrum over a finite band.

1. INTRODUCTION

In 1974 Gerchberg proposed an iterative method [1] for increasing the resolution in the image of a coherent finite object assuming that the optical imaging system had a rectangularly shaped, low-pass pupil. The explanation of the method as given by Gerchberg rested on the analyticity properties of the object spectrum and on the Parseval theorem. An analysis of the Gerchberg method (GM from now on) by means of prolate spheroidal wave-functions was given, nearly simultaneously, in Refs. [2] and [3] . Furthermore, the method was extended to images formed through any positive pupil both one and two-dimensional so as to include, e.g., incoherent imaging [4]. These works showed that the GM could be usefully applied to the case of small space-bandwidth products (SBP), whereas even for moderately large values of the SBP (e.g.10) both the noise sensiti-

W.-M. Boerner et al. (eds.), Inverse Methods in Electromagnetic Imaging, Part 2, 1189–1203.

and the slowness of convergence made the method unfeasible.

Since then, the GM has attracted the interest of a large number of researchers. Being a method for spectrum extrapolation, it is by no means limited to optics and was indeed applied to several other fields. Here, we would hardly discuss a complete list of references. Nevertheless, we can quote several papers that improved on and extended the previous works in some sense or another. Among the various results, we shall quote analog implementation of the GM by optical processors with feedback [5,6], Wiener filtered [7] and regularized version of the method [8], non-iterative modified methods [8,9] ,increased convergence rates [10], generalizations of other kinds [11-13].

In this paper, a brief eigenfunction analysis will be given of both the GM and its main extensions. The relationship between the GM and the Van Cittert method will be evidentiated. In particular, it will be shown why the GM is effective only for small values of the space-bandwidth product (Sec.2). The GM uses only that part of the image which corresponds to the object region (geometrical image). It will be pointed out that the outher part of the image contains information as well and a modified method using such information will be presented (Sec.3). It will be shown that a further modification leads to use the external and internal information on an equal footing in order to reduce the noise sensitivity of the method (Sec.4). The original proposal of the GM referred to the continuous case. When passing to a discrete implementation a careful analysis is required. It will be shown that the discrete filters to be used differ from the sets of samples of the filter which would be used in the continuous case. This is to be taken into account expecially when trying to minimize the number of image and spectrum samples (Sec.5). Furthermore, the original GM would require managing infinite bandwidths. Truncating the bandwidth to finite values introduces errors. A modification of the method will be given which allows the correct spectrum extrapolation to be obtained on a finite band (Sec.6).

2. ANALYSIS OF THE GERCHBERG METHOD AND ITS EXTENSIONS

We shall analyze the GM in a generalized form so as to encompass several cases of interest. Let us suppose that a set of object functions O(x) is mapped into a set of image functions I(x). Although we use a one-dimensional notation for the x-coordinate, two- or three-dimensional cases are included by thinking of x as a vectorial coordinate. The operator leading from O(x) to I(x) is denoted by $\underset{\sim}{S}$, or

$$I(x) = \underset{\sim}{S}\, O(x). \qquad (2.1)$$

For the moment we do not specify the mathematical character of $\underset{\sim}{S}$

except that it is a linear integral operator. As a general input-output relationship, eq.(2.1) applies to many different optical situations like image formation with both coherent or incoherent light either through space-variant or space-invariant optical systems, near-zone or far-zone diffraction, light scattering and so on.

In order to analyze the relationship between the object and the image we consider the following eigenvalue equation

$$\underset{\sim}{S}\Phi(\eta,x) = \mu(\eta)\ \Phi(\eta,x). \tag{2.2}$$

Here, $\Phi(\eta,x)$ is an eigenfunction of $\underset{\sim}{S}$ and η is to be thought of as a parameter whose values specifies a particular eigenfunction within the whole set of eigenfunctions. Depending on the mathematical structure of $\underset{\sim}{S}$, η can vary either in a continuous or in a discrete (finite or denumerably infinite) set. The possibly complex number $\mu(\eta)$ is the eigenvalue corresponding to $\Phi(\eta,x)$. The set of the eigenfunctions may be or may not be complete with respect to the set of object functions, but it is generally complete with respect to image functions. Accordingly, we shall assume that the following expansions hold

$$O(x) = \int o(\eta)\ \Phi(\eta,x)d\eta + R(\eta)\ , \tag{2.3}$$

$$I(x) = \int i(\eta)\ \Phi(\eta,x)d\eta\ , \tag{2.4}$$

where the integration domain depends on the structure of $\underset{\sim}{S}$. In particular, the integrals of eqs.(2.3), (2.4) become sums or series if η is a discrete variable. The function $R(x)$, which is non-vanishing whenever the eigenfunction set is not complete with respect to the object functions, represents that part of the object which does not contribute to the image. It is, in fact, such that

$$\underset{\sim}{S}\ R(x) = 0, \tag{2.5}$$

and can be thought of as an eigenfunction with zero eigenvalue. We shall refer to the other part of the object, namely the one expressed as an integral in eq.(2.3), as the transmittable part. The functions $o(\eta)$ and $i(\eta)$ can be called the Φ-transforms of $O(x)$ and $I(x)$, respectively. The object to image relationship is then established in terms of Φ-transforms. On using eqs.(2.2) through (2.5) in eq.(2.1) we obtain

$$i(\eta) = \mu(\eta)\ o(\eta). \tag{2.6}$$

We shall assume $\mu(\eta)\neq 0$ as the possibly zero eigenvalues are accounted for by the presence of $R(x)$ in eq.(2.3). Then, in principle, $o(\eta)$ can be recovered from $i(\eta)$ and the transmittable part of the object can be reconstructed by the superposition integral of eq.(2.3). Note that in order to perform this procedure both the eigenfunctions and the eigenvalues of $\underset{\sim}{S}$ must be known. This is a strong requirement in the general case. On the contrary, the iterative method which we are going to discuss does not require any knowledge of eigenfunctions and eigenvalues. Let us introduce the following operator

$$\underset{\sim}{K} = \underset{\sim}{\mathcal{I}} - \underset{\sim}{S} , \qquad (2.7)$$

where $\underset{\sim}{\mathcal{I}}$ is the identity operator. Now, consider the iterative algorithm

$$O_N(x) = I(x) + \underset{\sim}{K}\, O_{N-1}(x), \qquad (2.8)$$

with the initial estimate $O_1(x)= I(x)$. On using eqs.(2.2),(2.4) and (2.7) we see that eq.(2.8) can be written

$$\begin{aligned} O_N(x) &= \int i(\eta)\{1+[1-\mu(\eta)] +\ldots+[1-\mu(\eta)]^{N-1}\}\Phi(\eta,x)d\eta = \\ &= \int i(\eta) \frac{\{1-[1-\mu(\eta)]\}^N}{\mu(\eta)} \Phi(\eta,x)\, d\eta = \\ &= \int o(\eta)\{1-[1-\mu(\eta)]^N\} \Phi(\eta,x)d\eta , \end{aligned} \qquad (2.9)$$

where eqs.(2.4) and (2.6) have been used.
Suppose that the following condition holds for any η

$$|1 - \mu(\eta)| < 1 . \qquad (2.10)$$

Then, it is seen that when $N \to \infty$ eq.(2.9) leads to

$$O_\infty(x)= \int o(\eta)\, \Phi(\eta,x)d\eta = O(x) - R(x), \qquad (2.11)$$

where use has been made of eq.(2.3). Thus, the transmittable part of the object has been recovered.

Let us now examine some particular cases in order to appreciate both the potentialities and the limits of this method. As a first case, suppose $\underset{\sim}{S}$ is a space-invariant convolution operator so that eq.(2.1) becomes the familiar equation

$$I(x) = \int_{-\infty}^{\infty} O(y)\, S(x-y)\, dy . \qquad (2.12)$$

In this case, eq.(2.2) is a Fredholm homogeneous integral equation on an infinite domain. Its eigenfunctions form a continuous set being in fact given by the Fourier exponentials

$$\Phi(\eta,x) = \exp(2\pi i\eta x) , \qquad (2.13)$$

whereas the continuous set of the eigenvalues coincides with the set of values of the transfer function, say $\hat{S}(\eta)$

$$\mu(\eta) = \hat{S}(\eta) = \int_{-\infty}^{\infty} S(x)\exp(-2\pi i\eta x)dx \;. \quad (2.14)$$

The Φ-transforms of O(x) and I(x) are simply Fourier transforms. The previous iterative procedure was devised by Van Cittert as early as 1931 [14] and leads to what is commonly known as inverse filtering [15] . This is seen from eqs.(2.11) and (2.14). It is also evident that the iterative method fails to converge when the transfer function takes on negative values. Furthermore, when noise and round-off errors are taken into account the result of the method at frequencies where $S(\eta)$ is very near to zero will be mainly amplified noise. In view of these limitations and because of the availability of fast inverse filtering methods based on the Fast Fourier Transform, the Van Cittert method is mainly of historical interest. Nevertheless when the method is supplemented by the "a priori" knowledge that the object has finite extent the results change in a drastical way. Suppose in fact that the object function vanishes outside a finite domain A. Then, the operator $\underset{\sim}{S}$ may be assumed to imply truncation to the basic domain in addition to a superposition operation. In other words, eq.(2.1) becomes

$$I(x) = \int O(y)S(x,y)dy \;. \quad (2.15)$$

The eigenvalue equation for $\underset{\sim}{S}$ is a Fredholm homogeneous integral equation. Both the eigenfunction and the eigenvalue sets are discrete. We shall accordingly denote by $\Phi_n(x)$ and μ_n the eigenfunctions and eigenvalues, respectively. The integrals in eq.(2.3), (2.4), (2.9) and (2.11) are replaced by sums or series and the Φ-transforms of O(x) and I(x) are given by the sets of coefficients of the pertaining series expansions. The kernel S(x,y) represents the impulse response of the physical device leading from the object to the image. For the sake of simplicity we shall assume that it is hermitian. It can be either space-variant or convolution-like, i.e., space invariant (within A). The set of the eigenfunctions can be finite, e.g. for imaging through arrays of point-like elements [16]; it can be infinite but not complete in the object space, e.g. for a single light scattering experiment [17]; finally it can be complete in the object space as it happens, e.g., for both coherent and incoherent imaging through clear pupils [18]. It is in fact one of the latter cases, namely coherent imaging through perfect low-pass pupils, the one considered by Gerchberg in its paper of 1974. It is to be mentioned however that the mathematical iterative technique for solving eq.(2.15) had been proposed by Landweber [19] and that a similar iterative scheme had been used by Landau and Miranker [20] for recovering band-limited signal distorted by companding operations.

The iterative method can be applied in the cases mentioned above provided that condition (2.10) is met. This requires the (real) non zero eigenvalues to be positive and less than two.

As far as passive optics is concerned, energy considerations lead to eigenvalues with less than unity modulus. Positiveness of the eigenvalues is a stronger requirement which can be violated even for common impulse responses like those pertaining to defocussed optical systems. However, for images formed through space-invariant systems a suitable filtering can invert the negative lobes of the transfer function which are responsible for negative eigenvalues. Furthermore, for several cases of interest, e.g. imaging through clear pupils, the positiveness of the eigenvalues is warranted[18]. If this is the case the iterative method should lead to a perfect recovery of the transmittable part of the object.

In order to outline the behaviour of the iterative method we shall refer to the case considered by Gerchberg. A one-dimensional coherent object vanishing outside a single interval (-a/2, a/2) is imaged through a space-invariant system with a low-pass pupil extending from $-\nu_M$ to ν_M.

We introduce the following two operators. $\underset{\sim}{T}$ truncates any function to the basic interval, i.e.

$$\underset{\sim}{T}\, f(x) = f(x)\ \mathrm{rect}\left(\frac{x}{a}\right) , \tag{2.16}$$

where, as usual, rect(x)=1 for $|x|\leq 1/2$; rect(x)=0 for $|x|> 1/2$. $\underset{\sim}{B}$ is a band-limiting operator whose action is expressed by

$$\underset{\sim}{B}\, f(x) = \int f(\xi)\ \frac{\sin(2\pi\nu_M(x-\xi))}{\pi(x-\xi)}\, d\xi \quad . \tag{2.17}$$

In our present hypothesis, the object to image relationship becomes

$$I(x) = \underset{\sim}{B}\,\underset{\sim}{T}\, O(x), \tag{2.18}$$

and the eigenfunctions satisfying eq.(2.2)

$$\underset{\sim}{B}\,\underset{\sim}{T}\,\Phi_n(x) = \mu_n \Phi_n(x), \qquad (n=0,1,2,\dots) , \tag{2.19}$$

are the prolate spheroidal wavefunctions [21].

They form a complete set in $L^2(-a/2, a/2)$ so that the transmittable part of the object coincides with the object itself. The eigenvalues are positive and less that one. As it is well known, they exhibit a peculiar behaviour being very close to one for the indexes less than the SBP, i.e. $2\nu_M a$, and reaching very small values for greater indexes. The range of indexes in which the transition between nearly unity to nearly zero values occurs grows like $\log(2\nu_M a)$.

As can be seen from eqs.(2.4), (2.6) and (2.9) the N-th iterate has the same form as the initial image but the eigenvalues μ_n are replaced by the factors $[1-(1-\mu_n)^N]$. These factors tend to one when $N\to\infty$ so that the object is progressively recovered. The speed of convergence depends upon μ_n. In particular when

μ_n is much less than one we can write $[1-(1-\mu_n)^N] \approx N\mu_n$. Therefore, the number of iterations required to replace a small μ_n with a nearly unity factor is of the order of $1/\mu_n$. This can be a very high number. As an example, when the space-bandwidth product is 10.19 the eigenvalue μ_{30} is of the order of 10^{-28} [22]. This makes it clear that the method can recover only eigenvalues not too small. In practice the recovery can take place for indexes not beyond the strip of transition from near to one to near to zero eigenvalues. This has a small effect when the SBP is large because the width of the transition strip is small compared to the SBP. On the contrary, the effect can be quite relevant when the SBP is of the order of unity. In this case, the overall effect can easily be the same as doubling or triplicating the SBP [2,4] The limitations examined so far were connected to the number of iterations which can sensibly be used in the method. When noise (including round-off errors) is taken into account a more fundamental limitation occurs. Referring to the literature for a detailed analysis [8,9], we simply observe that too small eigenvalues make the corresponding information be completely masked by noise.

Summing up, the GM works for small values of the SBP and for signal to noise ratios large enough. The usual approach is to push up the number of iterations until noise effects are exhibited. Such an intuitive and interactive approach can be avoided when an "a priori" estimate of the signal to noise ratio is available by exploiting some kind of regularization [7-8].

We finally mention that the GM can be extended to non-hermitian operators. Basically, this can be obtained by transforming eq.(2.1) into the equation

$$\underset{\sim}{S}^{\dagger} I(x) = \underset{\sim}{S}^{\dagger}\underset{\sim}{S}\, O(x) \,, \tag{2.20}$$

where $\underset{\sim}{S}^{\dagger}$ is the adjoint of $\underset{\sim}{S}$. The operator $\underset{\sim}{S}^{\dagger}\underset{\sim}{S}$ being always non-negative the GM can be applied.

3) A MODIFIED GM USING THE OUTER PART OF THE IMAGE.

We have seen in the preceding section that the GM uses only the truncated image discarding the outer part. However, the outer part of the image contains information as well. We can even show that the outer part alone allows us to reconstruct the object. In order to see this, let us write the image in the form

$$I(x)= \sum \mu_n [\underset{\sim}{T}\Phi_n(x) + \overline{\underset{\sim}{T}}\Phi_n(x)]\; o_n \,, \tag{3.1}$$

where $\overline{\underset{\sim}{T}} = \underset{\sim}{\delta} - \underset{\sim}{T}$. Taking the outer part of I and band limiting it we obtain

$$\underset{\sim}{B}\,\overline{\underset{\sim}{T}}\, I(x) = \sum \mu_n o_n \underset{\sim}{B}[\Phi_n(x) - \underset{\sim}{T}\Phi_n(x)] =$$

$$=\sum \mu_n o_n (1-\mu_n)[\underset{\sim}{T}\,\Phi_n(x) + \overline{\underset{\sim}{T}}\,\Phi_n(x)] \ . \tag{3.2}$$

Equation (3.2) can be read as follows. On applying the operator $\underset{\sim}{B}$ to the outer part of the image we obtain the same result which we would obtain by imaging the object with a system described by the operator

$$\underset{\sim}{B}\,\underset{\sim}{T} - (\underset{\sim\sim}{BT})^2, \tag{3.3}$$

whose eigenvalues are $\mu_n(1-\mu_n)$. The operator (3.3) being positive definite we can apply the GM in the form

$$O_N(x) = J(x) + [\underset{\sim}{\delta} - \underset{\sim\sim}{BT} + (\underset{\sim\sim}{BT})^2]O_{N-1}(x) \ , \tag{3.4}$$

where $J(x)=\underset{\sim\sim}{B\overline{T}}\, I(x)$ and the initial estimate is $O_1(x)=J(x)$.
It is easily seen that the N-th iterate takes the form

$$O_N(x)= \sum o_n\{1-[1-\mu_n+\mu_n^2]^N\}\ \Phi_n(x). \tag{3.5}$$

The recovery of the object coefficients takes place at a different speed for every index. The fastest recovery occurs for $\mu_n \simeq 0.5$ i.e. for n nearly equal to the SBP. The present algorithm can find application when the inner part of the image is damaged for some reason or when the noise affects more the inner than the outer part. The main interest of the preceding argument, however, is to point out that information in the outer part of the image can be used. In this respect a method which uses mainly the outer part of the image without discarding the inner part can be devised. To this end, let us introduce the function, say $O^{(b)}(x)$, which coincides with $O(x)$ inside the basic domain and is band-limited. It can be constructed in the following way

$$O^{(b)}(x)= \sum o_n \underset{\sim}{T}\,\Phi_n(x) + \sum o_n \overline{\underset{\sim}{T}}\,\Phi_n(x), \tag{3.6}$$

where

$$o_n = \int_{-a/2}^{a/2} O(x)\ \Phi_n(x)dx. \tag{3.7}$$

It is obvious that

$$I(x)= \underset{\sim\sim}{BT}\ O(x) = \underset{\sim\sim}{BT}\ O^{(b)}(x) \tag{3.8}$$

whereas

$$O^{(b)}(x) = \underset{\sim}{B}\ O^{(b)}(x). \tag{3.9}$$

Furthermore, we can write

$$O^{(b)}(x) = \underset{\sim}{T}\, O^{(b)}(x) + \underset{\sim}{\bar{T}}\, O^{(b)}(x) \qquad (3.10)$$

On inserting from eq.(3.10) into eq.(3.9) and using eq.(3.8) we obtain

$$O^{(b)}(x) = I + \underset{\sim\sim}{B\bar{T}}\, O^{(b)}(x) \qquad (3.11)$$

Equation (3.11) can be solved through the algorithm

$$O_N^{(b)}(x) = I + \underset{\sim\sim}{B\bar{T}}\, O_{N-1}^{(b)} \quad , \qquad (3.12)$$

with the initial estimate $O_1^{(b)}(x) = I(x)$. The N-th iterate has the series expansion

$$O_N^{(b)}(x) = \Sigma \;\; o_n[1-(1-\mu_n)^N] \;\; \Phi_n(x). \qquad (3.13)$$

It is seen that in this case the recovery of the object coefficients takes place at the same speed as in the original GM. Nevertheless at each iteration the outer part of the preceding estimate is utilized (because of the presence of $\bar{T}$ in (3.12)). As far as the result is concerned, the main difference between the present method and the GM is that the, so to say, allied band-limited object (3.6) instead of the original object is recovered.

4) A MODIFIED GM UTILIZING BOTH PARTS OF THE IMAGE.

Whenever the whole image is affected by the same kind of noise it may be advisable to use in the same manner both the inner and the outer part of the image. This can be easily obtained in the following way. Let us write the allied band-limited object in the form

$$O^{(b)}(x) = \frac{1}{2}\, O^{(b)}(x) + \frac{1}{2}\, \underset{\sim\sim}{BT}\, O^{(b)}(x) + \frac{1}{2}\underset{\sim\sim}{B\bar{T}}\, O^{(b)}(x) =$$

$$= \frac{1}{2} O^{(b)}(x) + \underset{\sim\sim}{BT} O^{(b)}(x) - \frac{1}{2}\underset{\sim\sim}{BT} O^{(b)}(x) + \frac{1}{2}\, \underset{\sim\sim}{B\bar{T}} O^{(b)}(x) = \qquad (4.1)$$

$$= BT\; O^{(b)}(x) + \frac{1}{2}\, \underset{\sim}{T} O^{(b)}(x) - \frac{1}{2}\, \underset{\sim\sim}{BT} O^{(b)}(x) + \frac{1}{2}\underset{\sim}{\bar{T}} O^{(b)}(x) + \frac{1}{2}\underset{\sim\sim}{B\bar{T}} O^{(b)}(x),$$

or

$$O^{(b)}(x) = I(x) + [\frac{1}{2}(\underset{\sim}{\delta} - \underset{\sim}{B})\underset{\sim}{T} + \frac{1}{2}(\underset{\sim}{\delta} + \underset{\sim}{B})\underset{\sim}{\bar{T}}]\, O^{(b)}(x). \qquad (4.2)$$

Equation (4.2)can be solved iteratively by the algorithm

$$O_N^{(b)}(x) = I(x) + [\frac{1}{2}(\underset{\sim}{\delta} - \underset{\sim}{B})\underset{\sim}{T} + \frac{1}{2}(\underset{\sim}{\delta} + \underset{\sim}{B})\underset{\sim}{\bar{T}}]\, O_{N-1}^{(b)}(x), \qquad (4.3)$$

with the initial estimate $O_1^{(b)}(x) = I(x)$. It is easily seen that the following relations hold

$$\left[\frac{1}{2}(\underset{\sim}{\delta}-\underset{\sim}{B})\underset{\sim}{T}+\frac{1}{2}(\underset{\sim}{\delta}-\underset{\sim}{B})\underset{\sim}{\bar{T}}\right]\underset{\sim}{T}\,\Phi_n(x) = \frac{1}{2}(1-\mu_n)\underset{\sim}{T}\,\Phi_n(x) - \frac{1}{2}\mu_n\underset{\sim}{\bar{T}}\,\Phi_n(x), \tag{4.4}$$

$$\left[\frac{1}{2}(\underset{\sim}{\delta}-\underset{\sim}{B})\underset{\sim}{T}+\frac{1}{2}(\underset{\sim}{\delta}+\underset{\sim}{B})\underset{\sim}{\bar{T}}\right]\underset{\sim}{\bar{T}}\,\Phi_n(x) = \frac{1}{2}(1-\mu_n)\underset{\sim}{T}\,\Phi_n(x) + \frac{2-\mu_n}{2}\underset{\sim}{\bar{T}}\,\Phi_n(x). \tag{4.5}$$

It follows from eqs.(4.4), (4.5)

$$\left[\frac{1}{2}(\underset{\sim}{\delta}-\underset{\sim}{B})\underset{\sim}{T}+\frac{1}{2}(\underset{\sim}{\delta}+\underset{\sim}{B})\underset{\sim}{\bar{T}}\right]\Phi_n(x) = (1-\mu_n)\,\Phi_n(x), \tag{4.6}$$

so that the N-th iterate has again the series expansion

$$O_N^{(b)}(x) = \Sigma\; o_n\left[1-(1-\mu_n)^N\right]\Phi_n(x). \tag{4.7}$$

The important point to be made is that at each iteration a transfer of information occurs between the outer and the inner part of the image and viceversa as evidentiated by eqs.(4.4) and (4.5). This is of benefit to increase the noise immunity of the method.

5) DISCRETIZATION OF THE METHOD

The original GM as well as the extensions discussed so far refer to the continuous case. For numerical implementation, however, discretized versions are needed. In fact, the methods are generally implemented by means of the Fast Fourier Transform. Although a brute force approach can be used by sampling both the image and the spectrum at a rate high enough to obtain good approximation to the continuous case, it is both useful and instructive to inquiry about the steps that can lead to discretization. We shall refer to the original GM where only the inner part of the image is to be considered. Then, the problem to be solved is the inversion of the following equation

$$I(x) = \int_{-a/2}^{a/2} O(y)\,\frac{\sin[2\pi\nu_M(x-y)]}{\pi(x-y)}\,dy, \quad \left(|x| \le \frac{a}{2}\right), \tag{5.1}$$

We can observe that because of the limitation ($|x| \leq a/2$), only the values of the impulse response within the interval (-a,a) are of interest. The external values do not affect the image within $(-\frac{a}{2}, \frac{a}{2})$. Therefore, if we know the image only within the interval (-a/2,a/2) we can choose arbitrarily the form of the impulse response for $|x| > a$. This means that the transfer function is not defined by eq.(5.1) alone. This is quite interesting in itself because it shows that there exist an infinite number of systems which give rise to one and the same inner part of the image of a finite extent object. Then the image of eq.(5.1) could be produced by a system with a transfer function different from the familiar low-pass filter. In fact, as we shall see shortly it could be produced even by a transfer function transmitting arbitrarily high frequencies. This can give an intuitive insight into the fundamental reason why a band-unlimited version of the object can be recovered. There is an infinite number of ways for extending the impulse response for $|x| > a$ and we first choose the simplest, choosing S(x) to be zero for $|x| > a$, i.e.

$$S(x) = \mathrm{rect}\left(\frac{x}{2a}\right) \frac{\sin(2\pi \nu_M x)}{\pi x} . \tag{5.2}$$

The corresponding transfer function is

$$\hat{S}(\nu) = \frac{\sin(2\pi a\nu)}{\pi\nu} * \mathrm{rect}\left(\frac{\nu}{2\nu_M}\right), \tag{5.3}$$

where * stands for convolution. It will be noted that the difference between $\hat{S}(\nu)$ and the original transfer function $\mathrm{rect}(\nu/(2\nu_M))$ is particularly significant for small space-bandwidth product values, the same values in fact where the GM is most likely to work.

Furthermore, $\hat{S}(\nu)$ does not vanish identically in any finite interval of the ν-axis so that it transmits arbitrarily high frequencies although with progressively lower weight.

In order to approach discretization, let us now imagine a new transfer function, say $S_1(x)$, given by a periodic repetition of the right-hand side of eq.(5.2) or

$$\begin{aligned} S_1(x) &= \sum_{-\infty}^{\infty}{}_n S(x-2na) = \\ &= \sum_{-\infty}^{\infty}{}_n \mathrm{rect}\left(\frac{x-2na}{2a}\right) \frac{\sin[2\pi\nu_M(x-2na)}{\pi(x-2na)} . \end{aligned} \tag{5.4}$$

The new transfer function is an array of Dirac δ-functions or

$$\hat{S}_1(\nu) = \frac{1}{2a} \sum_{-\infty}^{\infty} {}_n \hat{S}\left(\frac{n}{2a}\right) \delta\left(\nu - \frac{n}{2a}\right). \quad (5.5)$$

We have reached a discrete transfer function which can be represented by an array of samples. Notice, however, that the pertinent samples refer to eq.(5.3) not to the original rect($\nu/(2\nu_M)$) transfer function. This is particularly relevant for small values of the space-bandwidth product. For numerical implementation of the GM the denumerably infinite set of samples (5.5) has of course to be replaced by some finite set. Here, some approximation is unavoidable. Once $\hat{S}(\nu)$ has been computed, one can approximate the right-hand side of eq.(5.5) by truncating the series to an index M such that for $n > M$ the samples $\hat{S}(\pm n/(2a))$ are sufficiently small. This, in turn, determines the distance a/M between adjacent samples to be used in the object and image domain.

6) SPECTRUM EXTRAPOLATION ON A FINITE BAND

Let us consider again the GM in its original version and let us observe that in principle, during each iteration an infinite band of frequencies should be handled. At first sight, it seems that the obvious existence of a cut-off frequency in any practical implementation of the GM should simply limit the spectrum extrapolation to be achieved exactly only below such a frequency. This is not the case, as we shall presently show, in that the extrapolated spectrum obtained by this method up to the cut-off frequency does not coincide with the true spectrum.

Let us denote by $\underset{\sim}{B}_1$ the band-limiting operator (so far denoted by B) accounting for the low-pass character of the imaging system. Further, let $\underset{\sim}{B}_2$ be the operator relating to the feinite band of frequencies to be managed in the practical implementation of the GM. We now substitute for the original Gerchberg algorithm the following one

$$Q_N(x) = I_1(x) + (\underset{\sim}{B}_2 - \underset{\sim}{B}_1)\underset{\sim}{T}\, Q_{N-1}(x), \quad (6.1)$$

with the initial estimate $Q_1 = I_1(x) = \underset{\sim}{B}_1 \underset{\sim}{T} O(x)$. One might be led to conclude that the sequence of $Q_N(x)$ converges to the image which would be formed through $\underset{\sim}{B}_2$ namely $I_2(x) = \underset{\sim}{B}_2 \underset{\sim}{T} O(x)$. We can show that this is not the case in the following way.

Let us consider the band-pass operator $(\underset{\sim}{B}_2 - \underset{\sim}{B}_1)\underset{\sim}{T}$. This is easily shown to be a positive definite operator. Denoting by $\Psi_n(x)$ its eigenfunctions, they satisfy the equation

$$(\underset{\sim}{B}_2-\underset{\sim}{B}_1)\underset{\sim}{T}\,\Psi_n(x)=\lambda_n\Psi_n(x)\ ,\quad (|x|\leq\frac{a}{2}\ ;\ n=0,1\ldots), \tag{6.2}$$

where λ_n are the (strictly positive) eigenvalues. Scarce information is available about the eigenfunctions and the eigenvalues. For our purposes, however, it is sufficient to know that the eigenfunction set is complete in $L^2(-a/2,a/2)$ and that the eigenvalues are less than unity. We now expand the truncated image $\underset{\sim}{T}\,I_1(x)$ into a series of the $\Psi_n(x)$

$$\underset{\sim}{T}\,I_1(x)=\sum_{n=0}^{\infty}\beta_n\underset{\sim}{T}\,\Psi_n(x). \tag{6.3}$$

By using eqs.(6.1), (6.2) and (6.3) it is easily shown that the truncated iterates tend to

$$\underset{\sim}{T}\,Q_\infty(x)=\sum_{n=0}^{\infty}\frac{\beta_n}{1-\lambda_n}\underset{\sim}{T}\,\Psi_n(x)\ . \tag{6.4}$$

We want to show that such an asymptotic expression differs from $I_2(x)$. To this end, let

$$\underset{\sim}{T}\,O(x)=\sum_{n=0}^{\infty}\alpha_n\underset{\sim}{T}\,\Psi_n(x)\ , \tag{6.5}$$

be the expansion of the object into a series of $\Psi_n(x)$. The follo̲wing expression can be given for $I_1(x)$ and $I_2(x)$

$$I_1(x)=\underset{\sim}{B}_1\underset{\sim}{T}O(x)=\sum_{n=0}^{\infty}\alpha_n\underset{\sim}{B}_1\underset{\sim}{T}\,\Psi_n(x), \tag{6.6}$$

$$I_2(x)=\underset{\sim}{B}_2\underset{\sim}{T}O(x)=\sum_{n=0}^{\infty}\alpha_n\underset{\sim}{B}_2\underset{\sim}{T}\Psi_n(x). \tag{6.7}$$

We now subtract eq.(6.6) from eq.(6.7) and use eq.(6.2). The result is as follows

$$I_2(x)-I_1(x)=\sum_{n=0}^{\infty}\alpha_n\lambda_n\Psi_n(x). \tag{6.8}$$

On using eqs.(6.8) and (6.3) we obtain

$$\underset{\sim}{T}\,I_2(x)=\sum_{n=0}^{\infty}(\beta_n+\alpha_n\lambda_n)\underset{\sim}{T}\,\Psi_n(x). \tag{6.9}$$

Let us compare eqs.(6.4) and (6.9). Were the two series identical, the following relations should hold

$$\alpha_n = \frac{\beta_n}{1-\lambda_n}\,, \qquad (n = 0,1,\ldots). \tag{6.10}$$

But this would entail (see eq.(6.5)) $\mathbf{T}Q_\infty(x) \equiv \mathbf{T}O(x)$, and this cannot be true unless $\mathbf{B}_2 = \boldsymbol{\delta}$.

It is difficult to estimate the difference between eqs.(6.4) and (6.9) and of course it can be argued that the error can be kept small provided that the bandwidth pertaining to $\mathbf{B}_2$ is sufficiently larger than the one pertaining to $\mathbf{B}_1$. However the following observation can be made. Because of the features of the GM we can attain spectrum extrapolation only for small values of the SBP and, even in this case, it is hopeless to attempt extrapolation over a bandwidth greater than, say, two or three times the original one. So, it is useful to devise a method allowing spectrum extrapolation on a finite interval regardless of the actual extension of such an interval and, in particular, valid for small intervals.

This can be reached in the following way. Consider the simple identities

$$\mathbf{B}_2\mathbf{T}O(x) = \mathbf{B}_2\mathbf{T}I_1(x) + \mathbf{B}_2\mathbf{T}O(x) + \mathbf{B}_2\mathbf{T}\mathbf{B}_1\mathbf{T}\,O(x), \tag{6.11}$$

$$\mathbf{B}_1\mathbf{T}\,O(x) = \mathbf{B}_1\mathbf{T}\,I_1(x) + \mathbf{B}_1\mathbf{T}\,O(x) - \mathbf{B}_1\mathbf{T}\,\mathbf{B}_1\mathbf{T}\,O(x). \tag{6.12}$$

From eq.(6.11) we obtain the iterative algorithm

$$P_N(x) = \mathbf{B}_2\mathbf{T}\,I_1(x) + P_{N-1}(x) - \mathbf{B}_2\mathbf{T}\,V_{N-1}(x)\,, \tag{6.13}$$

where $P_N(x) = \mathbf{B}_2\mathbf{T}\,O_N(x)$ and $V_N(x) = \mathbf{B}_1\mathbf{T}\,O_N(x)$. Such an algorithm requires at each step both $\mathbf{B}_2\mathbf{T}O_{N-1}(x)$ and $\mathbf{B}_1\mathbf{T}\,O_{N-1}(x)$. The latter can also be iteratively obtained from (see eq.(6.12))

$$V_N(x) = \mathbf{B}_1\mathbf{T}\,I_1(x) + V_{N-1}(x) - \mathbf{B}_1\mathbf{T}\,V_{N-1}(x). \tag{6.14}$$

Convergence of the sequences (6.13) and (6.14) follows from the analogous property of the original GM sequence taking into account the continuity of the operators $\mathbf{B}_1\mathbf{T}$ and $\mathbf{B}_2\mathbf{T}$. In conclusion, the required image $I_2(x)$ is obtained through the combined use of the sequences (6.13) and (6.14). It is apparent that the computations in the frequency domain involve only data below the cut-off frequency pertaining to $\mathbf{B}_2$.

REFERENCES

1. R.W.Gerchberg, Opt.Acta, 21, 709 (1974)
2. P.De Santis and F.Gori, Opt.Acta, 22, 691 (1975)
3. A.Papoulis, IEEE Trans.Circuits Syst.,CAS-22, 735 (1975)
4. F.Gori, Optical Computing Conference, Washington D.C., IEEE Catalog No.75 CH0941-SC,p.137 (1975)
5. P.De Santis, F.Gori, G.Guattari and C.Palma, Opt.Acta, 23, 505 (1976)
6. R.J.Marks II, Appl.Optics, 19, 1670 (1980)
7. G.Cesini, G.Guattari, G.Lucarini and C.Palma, Opt.Acta, 25, 501 (1978)
8. J.B.Abbiss, C.De Mol and H.S.Dhadwal, Opt.Acta, 30, 107 (1983)
9. J.A.Cadzow, IEEE Trans.Acoust.Speech, Signal Processing, ASSP-27, 4 (1979)
10. H.Maitre, Opt.Acta, 28, 973 (1981)
11. D.C.Youla, IEEE Trans.Circuits Syst., CAS-22, 735 (1978)
12. C.K.Rushforth and R.L.Frost, J.Opt.Soc.Am.70, 1539 (1980)
13. R.W.Schafer, R.M.Mersereau and M.A.Richards, Proc.IEEE, 69, 432 (1981)
14. P.H.Van Cittert, Z.Physik 69, 298 (1931)
15. B.R.Frieden, "Picture Processing and Digital Filtering", Ed. T.S.Huang, Springer-Verlag, Berlin 1975
16. F.Gori and G.Guattari, J.Opt.Soc.Am., 64, 453 (1974)
17. F.Gori and L.Ronchi, J.Opt.Soc.Am., 71, 250 (1981)
18. F.Gori, J.Opt.Soc.Am.64, 1237 (1974)
19. L.Landweber, Am.J.Math., 73, 615 (1951)
20. H.J.Landau and W.L.Miranker, J.Math.An.Appl.2, 97 (1961)
21. D.Slepian and H.O.Pollack, Bell Syst.Tech.J. 40, 43 (1961)
22. D.Slepian and E.Sonnenblick, Bell Syst.Tech.J. 44, 1745 (1965)

V.10 (MI.4)

PHASE COMPARISON MONOPULSE SIDE SCAN RADAR

Peter T. Gough

University of Canterbury

Department of Electrical and Electronic Engineering

1. INTRODUCTION

Monopulse radar, both in amplitude- and phase-comparison configurations, has been in service for many years as has the terrain mapping side scan or side looking airborne radar.(1) The latter system as a coherent synthetic aperture radar (SAR) has produced microwave maps of photographic resolution. Under what circumstance can we combine a monopulse system (explained in section 2) with a noncoherent side scan radar (section 3) or a coherent synthetic aperture radar (section 4)? To justify some of the claims an experimental air sonar was built to model a side looking radar using a monopulse configuration and the results of these experiments (shown in section 6) justify a more detailed investigation.

2. STATIONARY MONOPULSE RADAR

The feature that distinguishes a monopulse system from other types of radar is the use of a pair of closely spaced receiving antennas with overlapping beam patterns. Each receiver records the normal pulse echo return as a function of time/range from a single (mono) pulse radiated; however, by comparing the information in each receiver, an estimate of the target's asimuth or bearing may be obtained.

Amplitude-comparison monopulse compares the sum and difference of the two received signals to generate an estimate of the azimuth. Phase-comparison (or phase difference) monopulse compares the phases of the two received signals (echoes) at

W.-M. Boerner et al. (eds.), Inverse Methods in Electromagnetic Imaging, Part 2, 1205–1221.

any one range to estimate the target's asimuth.

In both amplitude- and phase-comparison monopulse, the usual limitation is that there is only one target in any one range cell (that distance below the range resolution capability of the system). This restriction is necessary to prevent misleading nonreal "targets" being indicated and to ensure that the detected video of both receivers exceeds the threshold (the voltage above which a target is said to exist) for the same target. In an air surveillance radar, this restriction is not normally a limitation but it may prove troublesome in side scan radar looking towards sea- or land-clutter which may have many specular glints in one range cell.

In phase-comparison monopulse, a second limitation arises from the fact that the receiver system is based on a two element interferometer and, like all interferometers, has multiple grating lobes. Consequently, the azimuth angle computer from the phase differences between the two receivers is ambiguous since the same phase difference could be measured from a single target at several other angles. If θ is the phase difference between the antennas, then

$$\phi = \sin^{-1} \frac{\theta\lambda}{2\pi d} \tag{1}$$

where d is the separation and ϕ the azimuth angle. It can be easily seen that many different ϕ can result from the same phase difference θ depending on d/λ. Although this ambiguity can be resolved by using multiple wavelengths or a multiplicity of antennas, no ambiguity results if the acquisition area covers only the central lobe of the interferometer. In this case, since the angular separation of the grating lobes is inversely proportional to the physical separation of the receiving antennas, the size of the individual antennas and their separation are not independent design variables if a unique azimuth angle is required. Two identical antennas of linear dimension D must just touch (d equal to D) to produce unambiguous results.

Irrespective of their configuration, if the antennas just touch the angular resolving ability of the monopulse system is still no better than a single antenna with a linear dimension of 2D and a corresponding narrower beamwidth.

An advantage of the monopulse system is that it can interrogate the entire acquisiton beamwidth (that angular area of beam overlap) with a single pulse whereas the narrow beam antenna of linear dimension greater than D must be rotated between pulses and takes several pulses to interrogate the entire acquisition beam width (see Figure 1). A disadvantage of the monopulse system is that the signal to noise performance of a single pulse is much worse if the noise is clutter rather than pre-amplifier noise since the individual beamwidths are much broader than its single beam equivalent. Consequently, many integrated pulses may be necessary to detect a target at a particular range and azimuth; a target which would be detected by a single pulse from a narrow beam single antenna when it points in the correct direction.

The detection of a single target in Gaussian noise is based on the probability that the video signal from the target plus noise exceeds a certain predetermined threshold voltage.

The threshold voltage decided on is determined by the noise, the type of target and the type of detector. For simplicity, assume constant cross section, non-fluctuating target and a zero mean, unit variance, Gaussian white noise as input to the narrow band width IF amplifier and filter. The output voltage of the IF section is subjected to a nonlinearity and averaged by the detector. The output voltage from the detector is random and has a probability density function pdf.

The pdf of the signal + noise using an envelope detector is

$$p(V_E) = V_E e^{-(V_E^2+A^2)/2} \quad I_o\,(V_E A) \tag{2}$$

where V_E is the output voltage of the envelope (linear) detector,
A is the aplitude of the signal,
and I_o is a modified Bessel function.

The equivalent pdf using a square law detector is

$$p(V_S) = \frac{1}{2}\; e^{-(V_s+A^2)/2} \quad I_o(A\sqrt{V_s}) \tag{3}$$

where V_S is the output voltage of the square law detector.

Note that in both cases as A tends to zero, the pdf becomes that of pure noise.

The voltage threshold is set so there is a small probability of the noise alone exceeding the threshold but there is a high probability of the signal + noise exceeding that same threshold. Although there are no fixed rules about the absolute value of the probability of detection, p(det), and the probability of a false alarm, p(fa), a reasonable system might have a p(det) of 0.8 and a p(fa) of 0.001.

For this particular example, using a single receiver with a linear detector, and a single pulse the SNR required would be 10.5 dB.

In a monopulse system, the SNR required by a single pulse is usually higher than that required by its single antenna equivalent. Both receivers must now detect the same target if the correct azimuth angle is to be generated. The signals from the target in the two receivers are almost always correlated so that if the signal voltage (i.e. in the absence of noise) in one receiver exceeds the threshold then, perforce, so must the other. However, if the two pdf's of the signal + noise are independent (almost always), there is a finite probability that the addition of antiphase noise takes one of the signals + noise below threshold.

The overall p(fa) is set so that the noise voltage in the two receivers do not exceed the threshold at exactly the same time. Two distinct cases now occur. The first is when the noise between the two receivers is uncorrelated (i.e. system or pre-amplifier noise), then the p(fa)s of each receiver are multiplied to determine the overall p(fa). The second case is when the "noise" is clutter, then the "noise" voltage in each receiver may be correlated (it depends on d) and so the overall p(fa) may be the same as the individual receiver's p(fa). However, the signal + clutter is also correlated so the overall signal to clutter ratio (SCR) of a monopulse system is usually the same as that of a single antenna radar of the same beamwidths.

2.1 The Probability Density Function of Random Narrow Band Noise in a Phase Comparison Monopulse Radar

Looking at Figure 1, we see that the beamwidth of the antennas covers many cells of resolvable azimuth for any one range. When there is no target present, the Gaussian, zero mean, unit variance, white noise is filtered by the IF amplifier detected and averaged. If the detector uses a piecewise linear device, the pdf of the normalised noise voltage is a Rayleigh distribution (equation 2 with A zero). Usually however the detector uses a square law device (optimum for a low signal to noise environment), the pdf of the normalised output voltage is a negative exponential most apyly described by a Chi squared distribution with 2 degrees of freedom (equation 3 with A zero). This pdf has an expected value E(V) of 1.0. Using either type of detector, the phase difference between the two receivers at any range is an equiprobable distribution between $+\pi$ and $-\pi$ rads. As this phase difference has an equiprobably distribution, the noise voltage at any range may be placed equiprobably in <u>any</u> one of the M azimuth cells. Thus for each range cell (which we can treat quite independently of any other range cell), we have

M possible choices of equiprobably azimuth cells so that the probability that the noise voltage at one particular range goes into any one azimuth cell is (1/M).

After N pulses, the total integrated noise voltage (V_T) has a pdf given by

$$p(V_T)\Big|_N = \sum_{k=o}^{N} p(V,k)\ p(k) \tag{4}$$

where p(k) is the probability that k noise voltages out of a possible of N have gone into one azimuth cell,
and p(V,k) is the pdf of k added noise voltage.

Now p(k) is given by the Binomial expansion and is

$$p(k) = \binom{N}{k} \left[1-\frac{1}{M}\right]^{N} \left[\frac{1}{M}\right]^{N-k} \tag{5}$$

Also since we have specified a square law detector,

$$p(V,k) = \frac{V^{k-1}\ e^{-\frac{V}{2}}}{2^k\ \Gamma\ (k)} \tag{6}$$

where Γ is the Gamma function and so p(V,k) a Chi squared distribution with 2k degrees of freedom.

Note that

$$\sum_{k=o}^{N} p(k) = 1$$

$$\int_{0}^{\infty} p(V,k)\ dV = 1 \tag{7}$$

and $\int_{0}^{\infty} p(V_T)\ dV_T = 1$

As we are looking for signals buried/surrounded by noise the expected value of V_T is of interest

$$E\{V_T\} = \int_{0}^{\infty} V_T \sum_{k=o}^{N} p(V,k)\ p(k)\ dV_T \tag{8}$$

and by changing the order of integration and summation

$$E\{V_T\} = \sum_{k=o}^{N} \int_{o}^{\infty} V_p\,(V,k)\; p(k)\; dV$$

$$= \sum_{k=o}^{N} p(k) \int_{o}^{\infty} V_p(V,k)\; dV$$

$$= \sum_{k=o}^{N} p(k)\; E\{p(V,k)\} \tag{9}$$

where

$E\{p(V,k)\}$ is the expected value of a Chi squared distribution with 2k degrees of freedom.

Taking an example, say there are 20 possible azimuth cells (M) at every range and we integrate over 10 pulses (N) then

$$p(V_T) = 0.599\;\delta V + 0.3151\; e^{-\frac{V}{2}}$$

$$+ 0.075\,\frac{V\, e^{-\frac{V}{2}}}{4\;\Gamma(2)} + 0.011\,\frac{V^2\, e^{-\frac{V}{2}}}{8\;\Gamma(3)} + + \tag{10}$$

and

$$E\{V_T\} = 0.62 \tag{11}$$

For a comparison, take a non-monopulse radar with the same beamwidth as the individual antennas. Since a non-monopulse radar has only one azimuth cell for every range cell, M is unity and the pdf is a Chi squared with 2N degrees of freedom, given by

$$q(V_T) = \frac{V_T^{\,N-1}\; e^{-\frac{V_T}{2}}}{2^N\;\Gamma(N)} \tag{12}$$

where $q(V_T)$ is the pdf and so after 10 pulses (N)

$$q(V_T) = \frac{V_T^{\,9}\; e^{-\frac{V_T}{2}}}{2^{10}\;\Gamma(10)} \tag{13}$$

The expected value of this pdf

$$E\{V_T\} = 9.6 \qquad (14)$$

i.e. about 10 times the expected value of the noise voltage one range cell.

Consequently, the expected value of the integrated noise in an N pulse monopulse radar is much lower than that of a single antenna radar.

To complete this section, consider a monopulse radar interrogating an area in which there is a single target surrounded by noise. The SNR must be sufficient so that the threshold is exceeded at least once in N pulses otherwise on average no target will be detected at all. A more usual criterion is that for a single pulse, p(det) be greater than say 0.80. So that, after N pulses, the expected number of times the target would be detected is 0.8N. This then determines the integration improvement factor of the signal since after N pulses, 0.8N echoes have placed a target in the <u>same</u> azimuth cell at the same range. Thus the integrated voltage from the detector (proportional to signal power) is increased 0.8N times whereas the integrated noise voltage in the surrounding cells has not increased by the same factor since it is spread (equiprobably) over M cells. In our example, the number of pulses integrated N is ten so the signal improvement is 8 times and the expected value of the integrated noise is actually 1.6 times less than that in any one range cell in any one pulse echo return. Thus the integration improvement is 12.8 times.

In summary, because the noise detected at any one range is spread over M equiprobable azimuth cells whereas the signal (on average) is placed in the same azimuth cell, the signal to noise improvement factor in a phase-comparison monopulse is proportional to both the number of independent pulses N <u>and</u> the number of available azimuth cells at any one range M.

3. NON COHERENT SIDE SCAN RADAR

The simplest form of noncoherent (sometimes called incoherent) side scan radar is to transmit and receive a narrow beam and move the single antenna sideways between pulses. The Cartesian coordinates of the map plotted out become those of <u>Track</u> (along the direction of travel) and <u>Cross Track</u>. In a narrow beam system, Cross Track and Range can be used interchangeably but in a wide beam system they have a distinct difference as Figure 2 shows. Depending on the velocity of the antenna along the Track, the beamwidth and the pulse repetition frequency it may be possible to use post-detection integration on sequential pulses to improve the SNR. However, to do this the target must remain in the beam for more than one pulse. Certainly, if only a few pulses of relatively strong targets are integrated, integration efficiencies close to that of perfect pre-detection integration (3 dB per doubling) might be expected.

As an alternative to electronic integration, if the map is permanently recorded on paper (sonar mostly) or presented to an operator as a simple B scan on a long persistance CRT, sufficient visual integration may occur to approach 2.4 dB per doubling once again coming very close to that achievable with coherent pre-detection integration for a small number of pulses. Once the number of pulses on target gets above 10, then the performance of an operator using a B scan drops off in accordance with that predicted.(1)

Using a monopulse system in a non-coherent moving side-scan radar requires the post-detection integration to be done electronically and a simple B scan of "plot and forget" is untenable. Since the number of lines of Cross Track information stored at any one time may be large, a simple delay line type of integration is inadequate and a two dimension computer memory is required. The memory is used to store and spatially integrate over many pulses and many positions of track.

The simplest implementation of the memory management is to have a 16 bit pointer to an address in a 2-dimensional memory where each line in memory corresponds to a position of Track and each 8 bit storage element in that line, a range cell in Cross Track. The pointer then indicates the current position of the vehicle and, as the vehicle moves, the pointer scrolls through the memory. When a line is so far behind the point that it is no longer addressable, it is printed on some permanent recording material. As well, the memory could be mapped onto a CRT screen and the present position of the vehicle defined as midscreen. Thus any targets forward of the vehicle would be plotted in the upper half of the screen and as they move down the screen, gradually intensify and pass off the bottom of the screen to be printed out on a traditional B scan display.

An advantage of the moving side scan monopulse over a stationary side scan monopulse radar is that the restriction of only a single target in any one range cell is greatly reduced. In a stationary monopulse radar, two targets in the same range cell but different azimuth produce a confusing amplitude and phase response due to their intermodulation and, if the targets are of comparable strengths, the phase comparison monopulse system creates a single artifact at the mean azimuth and plots the two correct targets at much lower target strengths. In stationary monopulse radar, this artifact is only bothersome if the targets are also stationary in which case they are almost always eliminated by standard moving target/clutter rejection techniques. In a moving sidescan monopulse system, two stationary targets in one range cell but at different azimuths also produce an artifact, but unless the two targets are in close proximity, they appear in different range cells as soon as the radar moves to another position of Track and so they no longer interfere.

Perhaps the easiest way to compare the relative (de)merits of a non-coherent side scan monopulse viz a viz a traditional single beam side scan a 3dB beam width of 6.5^{o} at a mean frequency of 3 GHz with maximum and minimum ranges of 370 km (200 n miles) and 46 km (25 n miles) respectively. A traditional single beam side scan antenna needs a minimum linear dimension of 80 cm to produce the required beamwidth and has a 12 dB gain over an isotronic antenna. An equivalent monopulse side scan would have two adjoining antennas 40 cm long each with a beamwidth of 13° and a gain of 9.3 dB. The pulse repetition rate of 500 Hz is determined by the maximum range and we assume a 2.5dB per doubling for the integration process. Consequently, it comes as no real surprise to find that for a fixed vehicle velocity, the SNR, as well as the eventual angular resolution, for a noncoherent monopulse side scan is quite close to its single beam equivalent.

When we integrate over several pulse echo returns in a moving side scan radar using phase comparison to produce an azimuth estimate, the integrated noise follows the same pdf as that calculated for a stationary monopulse radar provided that the beamwidths of the individual antennas are not extreme (i.e. less than 2° or greater than 70°). However, the maximum number of returns that can be integrated (N in Section 2.1) depends on the ratio of the vehicular velocity to pulse repetition frequency and it is equivalent to the number of hits on a single target. This does add an extra complicating factor as N is now a function of range since a target close to the track of the vehicle will have fewer hits than for a target on the extreme limit of range. In cases where the "noise" is clutter this increase in N with range works to our advantage as the "noise" also increases with range.

3.1 Threshold in a Noncoherent Side Scan Radar

When we combine several pulse echo returns into a two dimensional Track/Cross Track terrain map, we must define two different threshold decisions. The first decision is based on a single pulse echo return and is the usual threshold decision most radar systems refer to. We call this the Range/Amplitude (RA) threshold decision. When the single pulse echo returns are integrated over many pulses and positions of Track, they form a two dimensional terrain map in which the signal is surrounded/ immersed in noise which may or may not be clutter.

If the integrated value of signal plus noise exceeds a certain threshold, a target is said to exist. This threshold decision we call the Track/Cross Track threshold decision (TCT). Once again the TCT threshold must be set so that the p(det) is high and the p(fa) from the surrounding noise is low.

It is not at all clear if there is any theoretical advantage in having a non-zero RA threshold since a target present/absent decision can be made solely from the TCT threshold system. Since the azimuth angle must be calculated for every target deemed present if it exceeds RA threshold, a considerable reduction in computing may result from setting the RA threshold at a non-zero value.

In our air model experiments described in section 5, two separate thresholds were set to reduce the computing load. In a real-time system, a CFAR type adjustment of the RA threshold would ensure that the system computational capability was fully utilised . Our RA threshold was set so that p(det) on a single pulse was 0.8 and our p(fa) was less than 0.001. After spatial and temporal integration into a two dimensional terrain map, our TCT threshold was set so that p(det) was 0.9 and p(fa) was 0.1. This TCT threshold was determined by the dynamic range and quantisation levels of the contour plotting routine rather than by a conscious choice on our part.

4. COHERENT SIDE SCAN RADAR: SAR

Now well entrenched in the remote sensing discipline, basic SAR principles need no explanation. (1,2) Suffice to say that both the amplitude and the phase (relative to that of the transmitted pulse) of the targets' echo are recorded and a series of one dimensional Fourier or Fresnel transforms are performed on each column of Cross Track. Most processing is now digital and so it is common to see coherent apertures formed from the data collected over 64, 128 or any convenient power of two positions of track. One of the essential requirements of SAR is that the compensated track of the vehicle must be linear to one quarter of a wavelength. This restriction has limited the implementation of undersea synthetic aperture sonar (SAS) since the difficulties in computing a compensated linear track are compounded by the very slow speed of propagation of sound in water (1500 m/sec), the more erratic movement of the loat/tow fish and the inherent instability of the medium. (3,4)

Consider now a monopulse SAR where instead of measuring the phase of the echo relative to that of the transmitted pulse, we measure the phase of the echo detected by receiver one relative to that detected by receiver two. (An analogous modification of PSK is the differential PSK used in many communication systems). Now let us move antenna one to a new position along the track and antenna two to the old position of antenna one. Once again we measure the phase of the echo in one relative to the echo in two. Consequently, as the vehicle moves for every position of track we record a history of the amplitude of a pulse echo returned from the target and its phase differential. Note that this differential is independent of most vehicular random motion except yaw and is also free of errors introduced by fluctuating path length variations provided they affect both channels equally.

To calculate the phase required to form a synthetic aperture, it is sufficient to pick a position of Track in the centre of the aperture, set that phase to zero and integrate the phase differentials out to the edges of the aperture. Once the amplitudes and the phases are known, it follows normal SAR techniques to reconstruct a terrain map with the greatly improved cross track resolution expected.(5)

This resolution is now solely dependent on the size of the synthetic aperture which itself is dependent on how many phase differentials can be integrated before the phase noise in each differential renders the phase integration uncertain. Other more noise resistant integration techniques exist based on phase unwrapping but they usually are restricted to phase differentials from very simple targets.(6)

5. AIR SONARS AS EXPERIMENTAL MODELS FOR RADAR SYSTEMS

Since most radar apertures and their targets are many wavelengths in diameter, the scalar approximation is usually valid so, provided cognizance is made of the different wavelengths and velocities of propagation, physically small models can be used to great effect. One property of sound in air that might initially seem to be a limitation, but is in fact a benefit, is the rapid attenuation along the path of propagation - some dB per meter. This rapid attenuation does mean that the targets must be close (10 m or less) to the transducers to produce a reasonable signal to noise ratio but it also means closely controlled experiments inside shielded environments are very easy. Almost any large empty room with minimal wall coverings is suitable.

A second advantage in using air sonars to model radar systems is that for a particular physical wavelength corresponding to common radar frequencies, the transmitted sonar frequencies are usually down in the 50 to 200 kHz range and so phase comparisons can be made without heterodyning the RF down to a more tractable IF. The simplicity of the analogue hardware is easily seen in the block diagram drawn in Figure 3.

6. EXPERIMENTAL RESULTS USING THE AIR SONAR MODELS

The five test targets were placed in the area of the test range covered by the swathe (see Figure 2). Their sonar cross section was 90×10^{-4} m^2 each where the background clutter at the grazing angle used was 14×10^{-4} per m^2. Since the range cell x the acquisition beam width covered an area of 0.5 m^2, the ratio of signal to clutter was 11dB. Thus for a single pulse, the threshold could be set so that p(det) was 0.8 and p(fa) was 0.001.

Figure 4 shows the terrain map plotted out if one receiver was turned off and the sonar operated as a simple B scan. The points to note there are that the targets can be roughly located by the position where they are closest to the Track. The second point is that target B has multiple glints within one range cell at almost all positions of Track resulting in self interference. The third point is that despite only a 11 dB SNR, the effect of the RA threshold followed by a TCT threshold is such that the overall effect of the noise is minimal.

Figure 5 shows the result of simple non-coherent monopulse integration as explained in section 3. Here because targets B1 and B2 are not well separated, at almost all positions of Track they appear in the same azimuth cell and so an artifact is generated. True images of B1 and B2 are generated only when the sonar is at the extremities of the Track and the pulse length is sufficiently short to place them in different range cells.

Finally Figures 6 and 7 show the images generated by a full SAR type processing where the phase required for the Fourier/ Fresnel transformation is generated by the integration of the measured phase differences out from the centre of the synthetic aperture selected. In Figure 6 there are 10 synthetic apertures each of length 16 cm with centres separated by 20 cm, and the terrain map generated by any one synthetic aperture is added non-coherently to the other nine. In Figure 7, there are 10 synthetic apertures each of length 32 cm with centres separated by 20 cm. In this situation, some phase difference measurements contribute to two adjacent synthetic apertures (i.e. the synthetic apertures overlap). Note that now the along Track resolution and

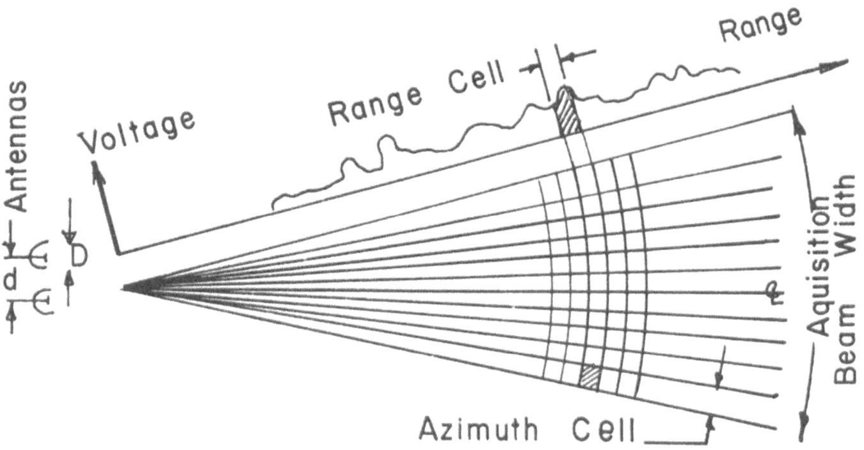

Figure 1. Monopulse Radar with M azimuth cells.

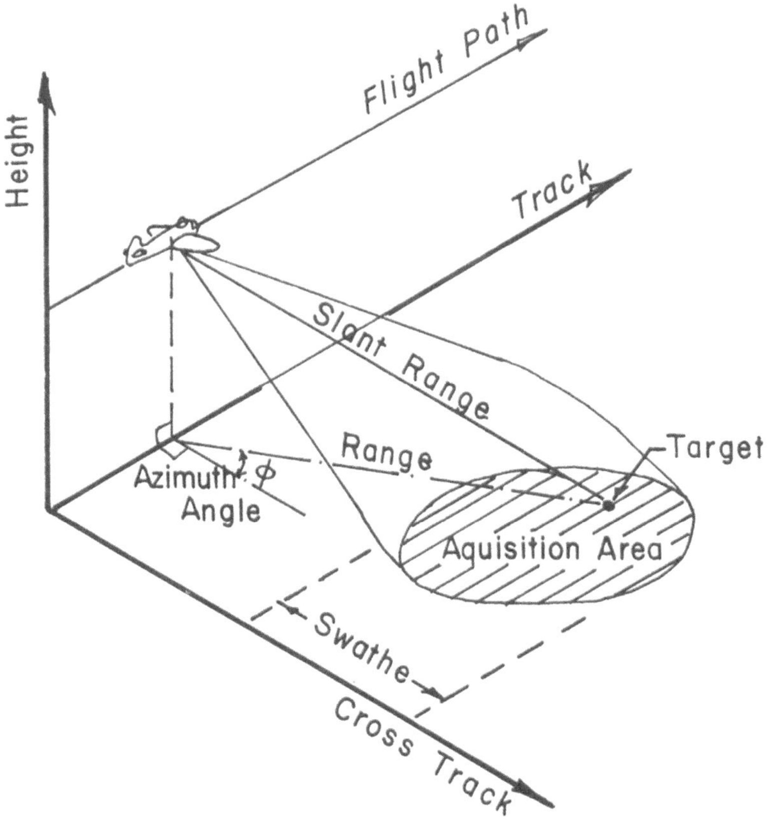

Figure 2. SLAR geometry showing commonly used terms.

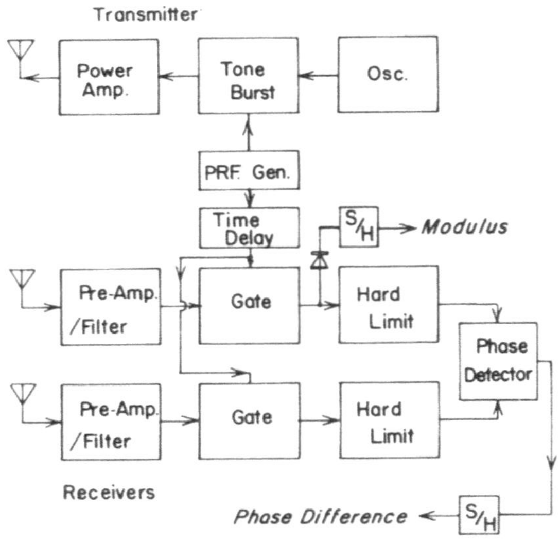

Figure 3. Block diagram of electronic system used in air sonar model.

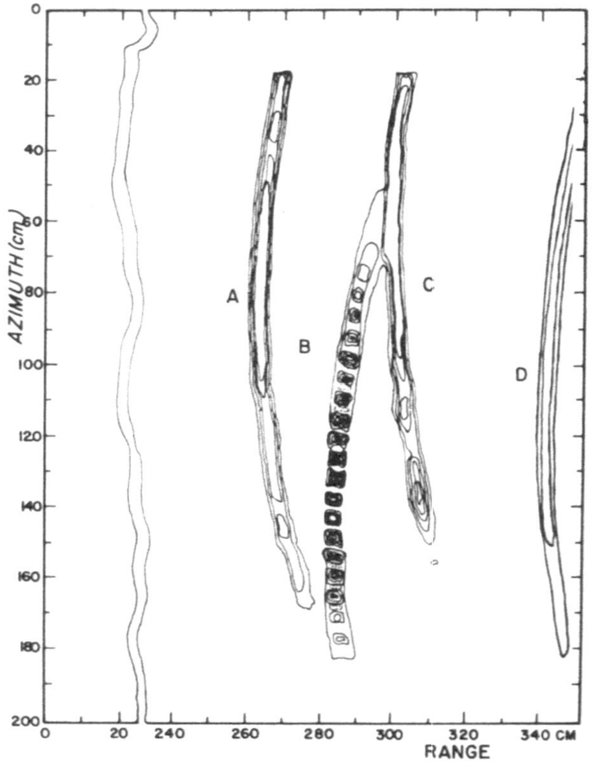

Figure 4. Single antenna simple B scan image.

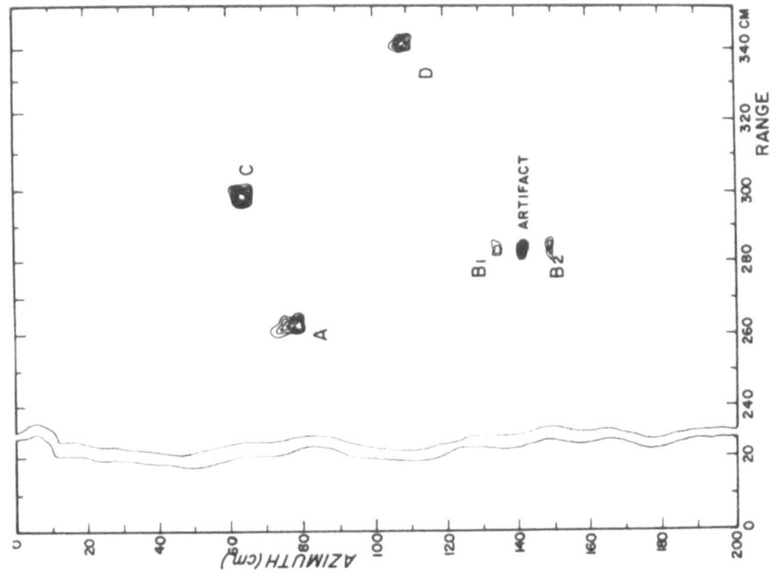

Figure 5. Non-coherent side scan image.

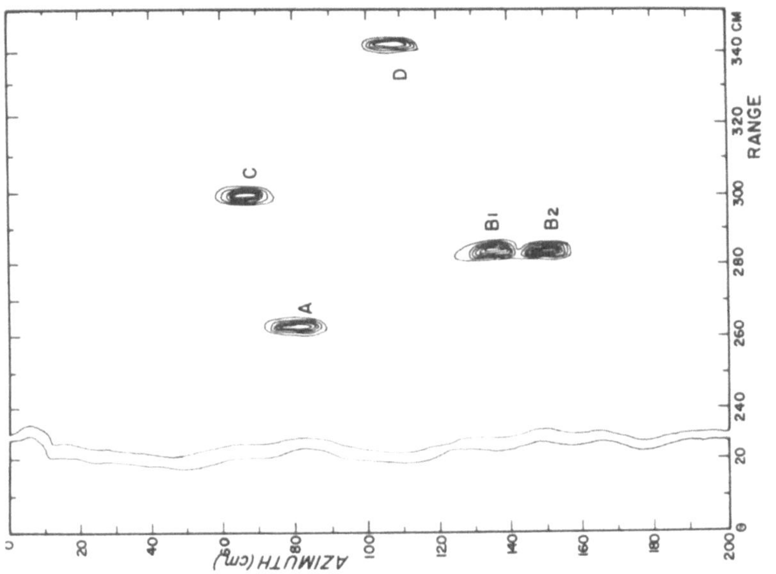

Figure 6. Image formed by the average of 10 independent synthetic apertures each 16 cm long.

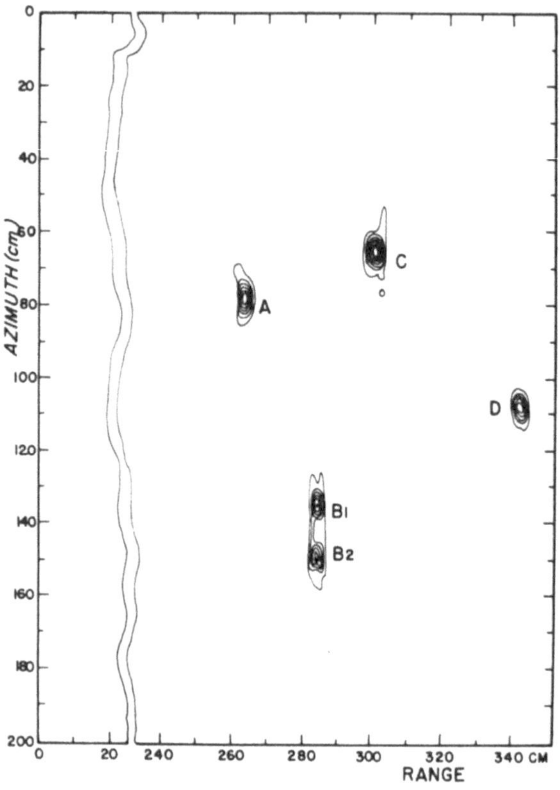

Figure 7. Image formed by the average of 10 independent synthetic apertures each 32 cm long.

the Cross Track resolution are equal but that the noise has become slightly worse. Were we to make the synthetic aperture say 64 cm, the Track resolution would be better than the Cross Track resolution and the noise would be considerably worse.

7. CONCLUSIONS

Although monopulse radar has been traditionally restricted single target surveillance, the technique does have its uses in an SLAR where it is searching for targets against background clutter. Should the targets be sparse, simplified non-coherent processing can produce usable images. It is also possible to use full SAR processing provided the true phase can be calculated from the measured phase differences.

REFERENCES

(1) Skolnik, M.I., "Introduction to Radar Systems", McGraw-Hill, NY, 2nd Edition, 1980.

(2) Kovaly, J.J., "Synthetic Aperture Radar", Artech House Inc., Dedham, Mass., 1976.

(3) Cutrona, L. J., "Comparison of Sonar system performance achievable using synthetic-aperture techniques with performance achievable by more conventional means", J. Acoust. Soc. A., 58, 1975, pp. 336-348.

(4) Williams, R.E., "Creating an acoustic synthetic aperture in the ocean', J. Acoust. Soc. Am., 60, 1976, pp.60-73.

(5) Gough, P.T., "Side-looking Sonar or Radar using phase difference monopulse technqiues: coherent and non-coherent applications", submitted to IEE, part F.

(6) Gough, P.T., "A particular example of phase unwrapping using noisy experimental data", accepted for publication IEEE Trans. ASSP.

V.11 (MM.4)

DIAGNOSTIC MEASUREMENTS AND ANALYSIS OF WAVE MECHANISMS IN RADOMES

G. Tricoles, E. L. Rope and R. A. Hayward

General Dynamics Electronics Division

Wavefronts in radome-bounded regions are related to boresight error; in particular, wavefront tilts are proportional to boresight error. The wavefronts and therefore boresight error depend strongly on wave polarization direction. A method for probing wavefronts was described. This diagnostic method was applied to specific, thin wall radome; it identified a wave mechanism, guided waves, which are omitted from approximate analytical methods that locally approximate curved radomes by assemblies of flat dielectric sheets. The diagnostic method has been applied to improve other radomes; it located reflecting regions.

W.-M. Boerner et al. (eds.), Inverse Methods in Electromagnetic Imaging, Part 2, 1223–1233.

INTRODUCTION

A microwave antenna at the nose of a missile or supersonic aircraft is usually enclosed by a pointed dielectric shell called a radome. Radomes protect antennas and give low aerodynamic drag. Although radomes are necessary, they cause errors in measuring the direction of a source or reflecting object; this error is called radome boresight error. Radomes also can have low transmittance, especially at near grazing incidence on pointed shapes.

The analysis of pointed radomes can be complicated. Boresight error and transmittance depend on several variables, and many radomes have shapes that do not fit separable co-ordinate surfaces. For a specific radome, boresight error and transmittance are derived from computed farfield patterns of the enclosed antenna.

The theory for pattern calculations is usually approximate. A common approximation describes propagation through the radome by using transmittances of flat sheets that locally approximate the curved shell.[1] This approximation is questionable for regions near the tip of a pointed radome, where the surface normal varies rapidly in the circumferential direction so that the polarization changes rapidly between parallel and perpendicular. The flat sheet approximation also omits guided waves. Some analytical methods omit reflections that arise from one side of a radome when the antenna's main beam is pointed toward the opposite side. Another approximation is in describing the antenna's aperture distribution.

To evaluate approximations, we have used an experimental diagnostic method, which is sketched in Fig. 1. Plane waves illuminate the outside of a radome, and a small receiving antenna scans the radome-bounded region. A network analyzer measures the phase and intensity of the received field. The measured values depend on the gimbal angle, which is the angle between the radome axis and the incident wave normal. In addition, the measured values depend strongly on polarization direction. Two special cases are common. In one the electric field is in the plane containing the radome axis and the wave normal; this is the E-plane. The other case has the electric field orthogonal to the plane containing the radome axis and the wave normal; this is the H-plane.

The purpose of the measurements is to identify wave mechanisms in radomes. As suggested in Figure 2, the field in the radome-bounded region consists of several constituents, which are directly transmitted, reflected, and guided waves. For coherent sources these waves interfere to produce a distorted wavefront, rather than a plane wave. A distorted wavefront would affect the boresight error and the received intensity for an aperture antenna enclosed by a radome. However, the analysis of a distorted wavefront from observation with an aperture antenna is indirect because an aperture antenna integrates the field over its aperture. Therefore, in diagnostic measurements we use a small probe to resolve variations in the interference patterns and thus to identify the constituent waves.

The probing technique was originally developed to evaluate theoretical approximations, but it has also been used to improve radomes, for example by locating reflecting regions and modifying them to reduce reflections.[2]

MEASURED RADOME BORESIGHT ERROR FOR AN APERTURE ANTENNA

Boresight error was measured for an aperture antenna in an axially symmetric radome. The radome had length 51 cm and base diameter 20 cm. It was relatively thin; thickness varied from 0.33 cm at the base to 0.43 at the tip. Dielectric constant was 2.1. The antenna was an array of four rectangular horns, dimensions of each were 4.3 cm in the E-plane and 5.3 cm in the H-plane; the array was covered by a dielectric polarizer. The horns were closely packed and the array was rotated to place the horn diagonals vertical or horizontal. Frequency was 10 GHz.

Figures 3 and 4 show E and H-plane boresight error for frequency 10 GHz. E-plane results are for the electric field horizontal; H-plane results are for the electric field vertical. It is clear that boresight error depends on polarization direction. This dependence complicates electronic compensation of boresight error with stored, measured values. For example, the difference in algebraic signs of the E and H-plane errors near $\pm 5^{\circ}$ gimbal angle means that the two planes require distinct correction.

WAVEFRONT PROBING

The apparatus in Figure 1 was used to measure the phase and intensity of the wavefront in the region bounded by the radome that was described in the above section. The probe was 3.1 cm square. The aperture antenna was absent. Figures 5 and 6 show power transmittance for the E and H-plane respectively. Figures 7 and 8 show insertion phase delay, which is the change in phase, from the phase of the incident field.

In Figures 5 and 6, we see very low transmittance values, below -8 dB, for the probe toward the positive side. For some angles and some probe positions the field exceeds that of the incident field. It is clear that intensity depends on polarization, for Figures 5 and 6 differ.

For Figures 7 and 8, we see that wavefront tilts depend on polarization. In Figure 7, for the E-plane, the tilts are upward for the smaller gimbal angles, like 4° through 8°, but for large angles the tilts are downward. This difference in behavior corresponds to the sign reversal of boresight error in Figure 3. In Figure 8 the similar and correspond to the monotonic central portion of the boresight error curved in Figure 4.

WAVEFRONT ANALYSIS

To explain the wavefront measurements, we computed phase and intensity in the radome bounded region. The field at a point was calculated by integrating over a portion of a Huygens surface and modifying each term by a plane sheet transmittance. This computational procedure is called the surface integration method.[3]

For gimbal angle 15° and horizontal polarization, Figure 9 shows computed phase and intensity on the horizontal diameter. Figure 9 also shows values measured with the probe antenna. Frequency was 10 GHz. The measured and computed results agree better for negative Y_A than for positive.

To explain the discrepancies between measured and computed values, we first formed the difference $E_m - E_c$, where E_c is the computed field component and E_m is measured. The values $E_m - E_c$ are shown in Figure 10.

We than hypothesized that the difference $E_m - E_c$ was caused by guided wave. The model is shown in Figure 11. To test this hypothesis we calculated the propagation and attenuation constants (k_g and ν) for a slab-guided wave with frequency 10 GHz.[4] For slab-thickness 0.31 cm and 0.38 cm, Figure 10 shows the amplitude and phase a guided wave would cause on the probing paths in Figure 11. The amplitude was normalized at Y_A equal 5 cm to the measured value.

The measured differences $E_m - E_c$ and the calculated curve agree, confirming the model of a guided wave. The guided wave magnitude is 0.26 near the radome wall, for a unit amplitude incident wave. This amplitude, combined with the plane sheet amplitude, transmittance of approximately 0.8 give the -8 dB power transmittance in Figure 5.

REFERENCES

1. G. Tricoles, "Boresight Error of a Microwave Antenna Enclosed by an Axially Symmetric Shell, Jour. Opt. Soc. Am., Vol. 54, pp 1094-1101 (1964).

2. G. Tricoles and N. H. Farhat, "Microwave Holography", IEEE Proc., Vol. 65, pp 108-120 (1977).

3. G. Tricoles, E. L. Rope, and R. A. Hayward, "Accuracy of Two Methods for Radome Analysis", Digest IEEE A.P.S. Symposium, pp 598-601 (1979), IEEE Catalog No. 79 CH 1456-3 AP.

4. R. F. Harrington, Time Harmonic Electromagnetic Fields, pp 163-168, McGraw-Hill (1961).

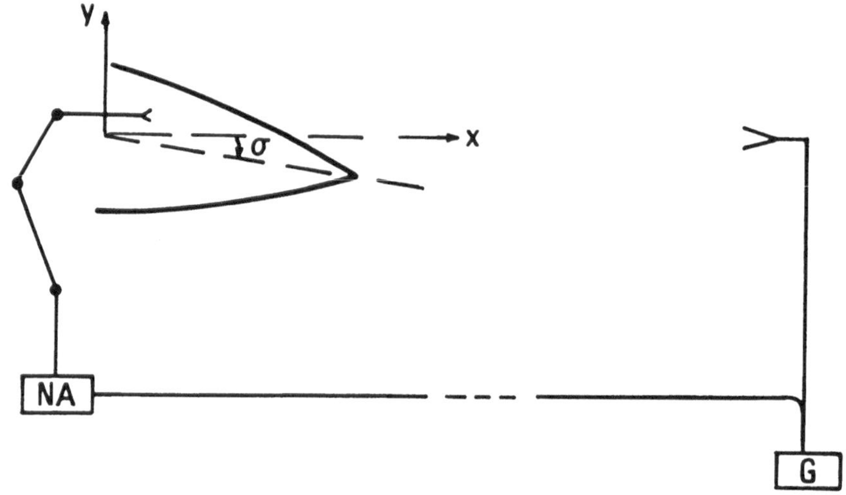

Fig. 1 Apparatus for measuring a field distribution in a radome-bounded region. G is a signal generator, which is connected to a network analyzer.

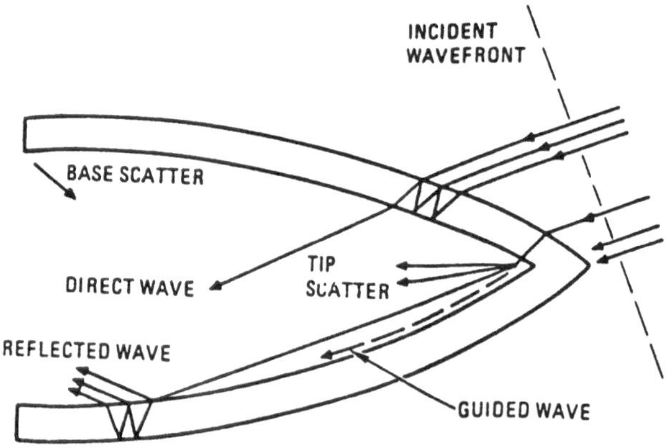

Fig. 2 Wave mechanisms in hollow dielectric shell. The fields near a radome consists of constituent waves.

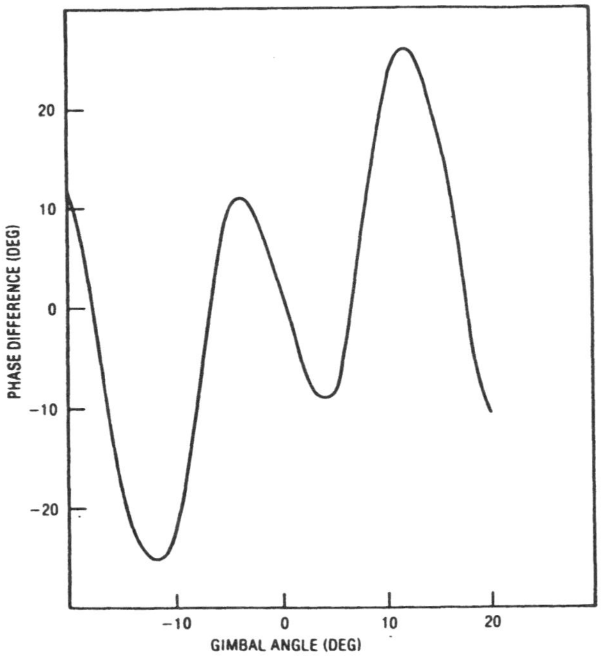

Fig. 3 E-Plane phase difference between two Antennas; Frequency 10 GHz

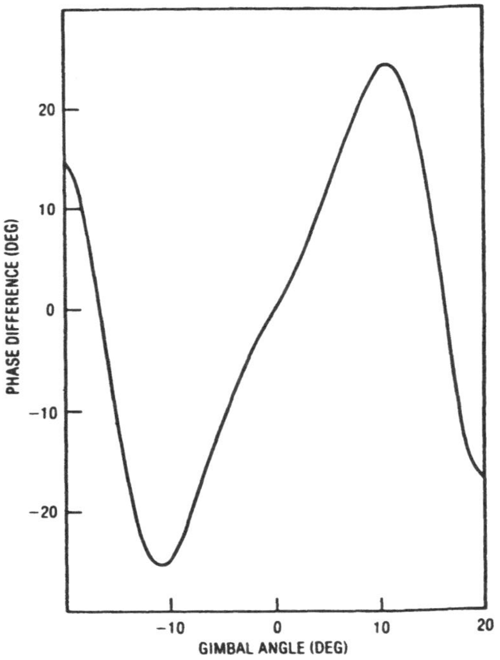

Fig. 4. H-plane phase difference between two horn antennas; Frequency 10 GHz

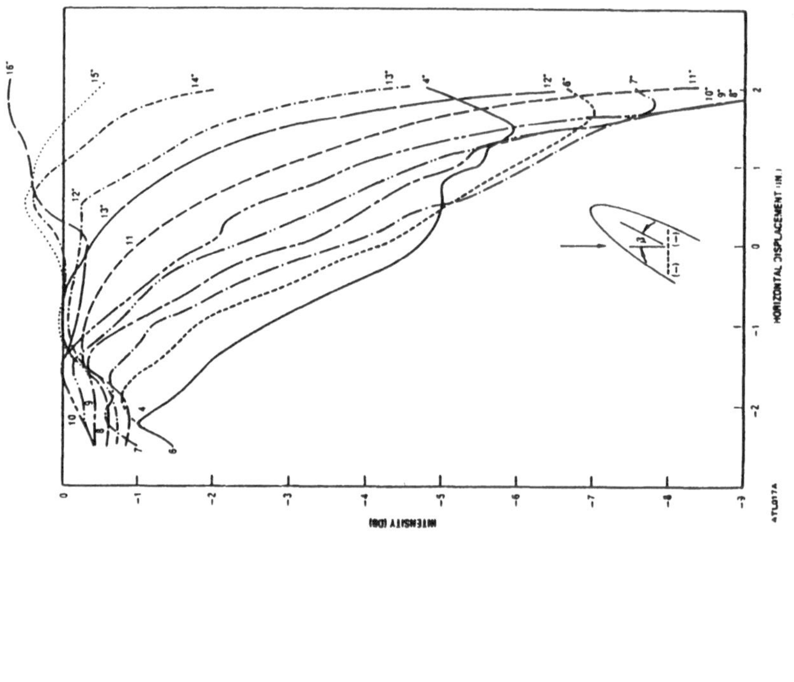

Fig. 5. Intensity Transmittance of Axially Symmetric Shell. The Inset Shows the Probing Path as a Dotted Line, the Graphs are Labeled with Values of β, Horizontal Polarization

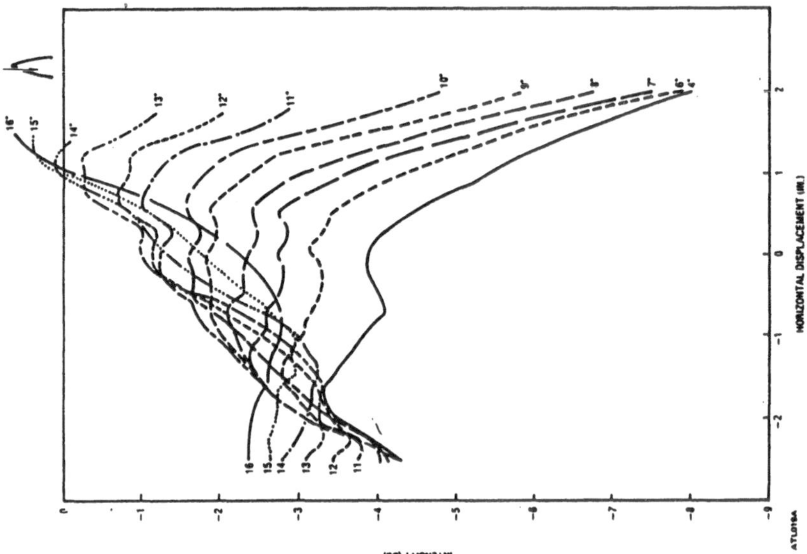

Fig. 6. Intensity Transmittance of Axially Symmetric Shell, Vertical Polarization

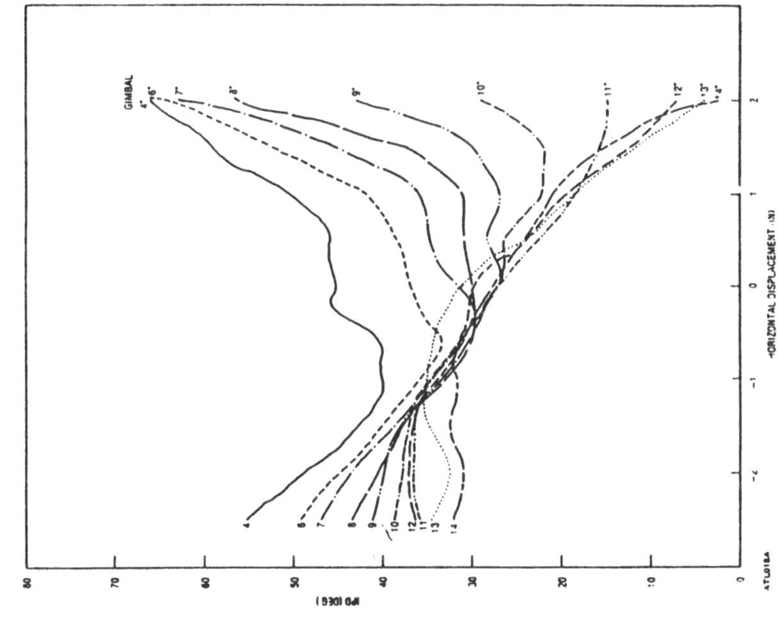

Fig. 7. Phase Delay (IPD) of Axially Symmetric Shell for Horizontal Polarization

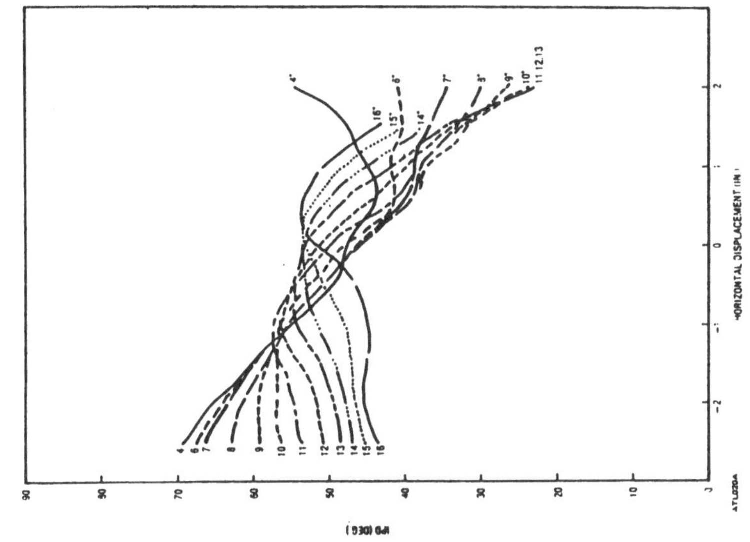

Fig. 8. Phase Delay (IPD) for Shell, Vertical Polarization

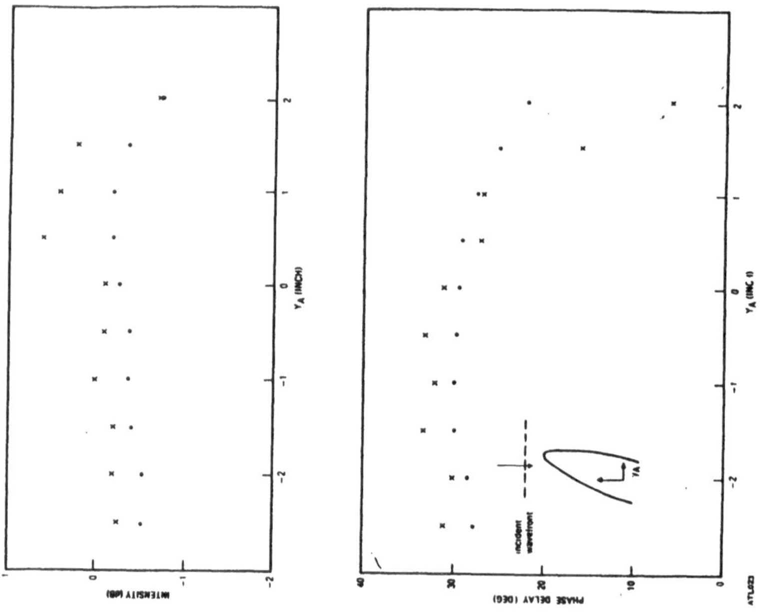

Fig. 9. Intensity and Phase Behind Axially Symmetric Shell for $\beta = 15°$; E-Plane. Measured with 1.25 in. Square Antenna (x). Calculated with Surface Integration (·) for $\ell = 40\lambda$; Frequency 10 GHz.

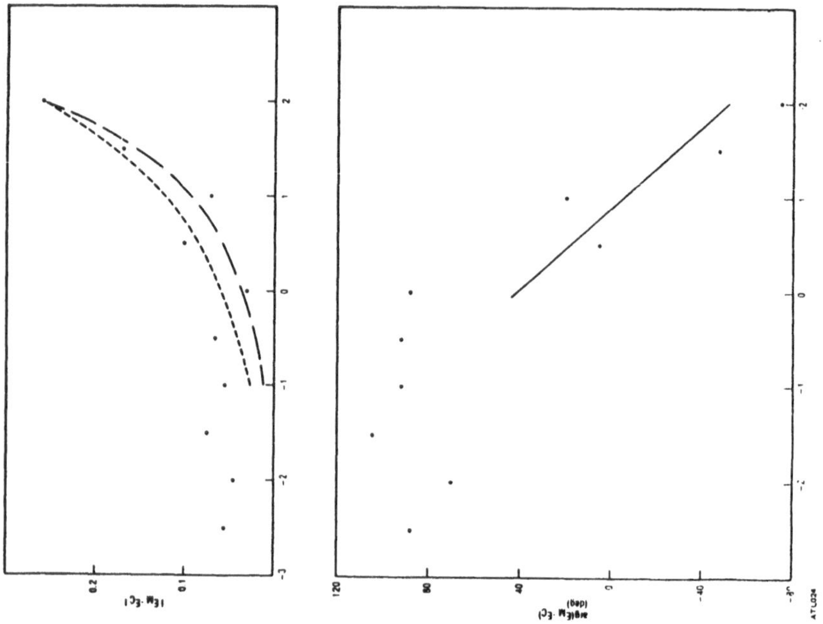

Fig. 10. The Magnitude and Phase of $E_m - E_c$ (·). Computed Values of 0.260 exp $(-v\,|y_A - 2|)$; for Thickness 0.125 in.(- - - -); for Thickness 0.150 in (— —).

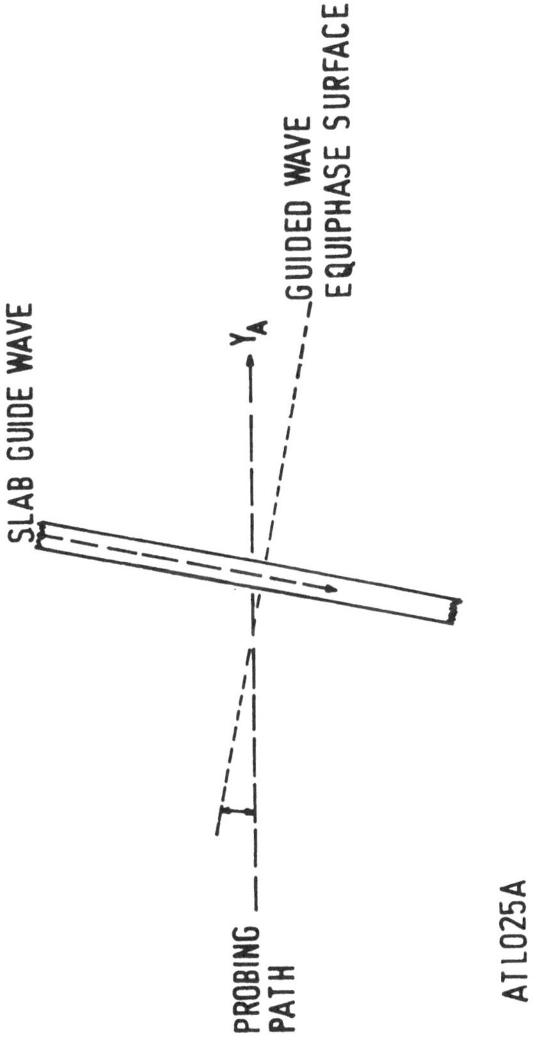

Fig. 11. This Figure Suggests a Guided Wave on a Slab that Approximates the Side of the Radome near Y_A equal 2 in. in Figure 3-6.

V.12 (OI.2)

SYNTHESIS AND DETECTION OF ELECTROMAGNETIC AUTHENTICITY FEATURES

H.P. Baltes* and A.M.J. Huiser**

Department of Electrical Engineering
University of Alberta
Edmonton, Canada T6G 2E1

We review selected results of the authors' recent work on the construction and retrieval of electromagnetic authenticity features. Our contribution starts with a system view of inverse scattering and a summary of the corresponding retrieval and synthesis problems. As a first specific question we consider the construction of grating groove profiles from desired electromagnetic diffraction efficiencies. Next we describe a method to detect diffractors that are hidden behind a diffuser. Finally, the electromagnetic scattering by random rough surfaces in the resonance region is revisited.

1. INTRODUCTION

Modern electronic signal processing is a powerful tool for measurement and control provided that the system in question is equipped with

(i) sensors or input transducers (detectors) that provide the (usually noisy and incomplete) data, and
(ii) smart recovery algorithms (based on inverse methods) to program the signal processor for optimal use of the data.

The latter component, sometimes overlooked, is the topic of the present conference.

Our study of inverse electromagnetic scattering problems was motivated by the quest for encoding information in diffracting

*formerly with Landis & Gyr Zug, Switzerland,
**formerly with EPF Lausanne, Switzerland, where most of the research presented here was done.

W.-M. Boerner et al. (eds.), Inverse Methods in Electromagnetic Imaging, Part 2, 1235–1244.

or scattering structures, reading such information using optical or electromagnetic sources and detectors, and retrieving structural information obscured by the presence of random scatterers. The goal was the automatic optical checking of identity or authenticity, e.g., of a voucher card. One example of a system based on such procedures is the PHONOCARD currently manufactured by Sodeco-Saia, Switzerland, a company of the Landis & Gyr group. The PHONOCARD is a pay phone operated by prepaid cards (see Fig. 1) with optically stored value units.

In connection with optically coded voucher or identity cards two different types of problems arise, namely, the question of devising appropriate new authenticity features. These are examples of the two principal classes of inverse scattering problems:

(i) The <u>retrieval</u> problem, i.e., the reconstruction or estimate of existing scattering objects from the measured data.

(ii) The <u>synthesis</u> problem, i.e., to devise and manufacture scattering objects tailored to yield a desired scattering behaviour.

It is well known that both types of inverse problems lead to the question of <u>uniqueness</u> and <u>stability</u> of the solution; in the case of the synthesis problems, a solution may even <u>not exist</u> at all. These questions and a number of specific inverse optical and electromagnetic problems are reviewed in two recent books [1,2].

CONSTRUCTION OF GRATING GROOVE PROFILES FROM KNOWN OR DESIRED ELECTROMAGNETIC DIFFRACTION EFFICIENCIES

It is obvious that the (re)construction of grating profiles from diffraction efficiencies is not unique in the case of Fourier optics (wavelength λ much smaller than the grating constant d), since only the absolute value of a small number of Fourier coefficients of the profile is known. The corresponding electromagnetic inverse problem (in the resonance region, where λ and d are of the same order) was broached only recently [3,4, 5]. As an example, let us briefly describe the two-step reconstruction procedure devised by Huiser, et al. [5].

We denote the incident field by $\underline{E}^{inc}(x,z)$ and the scattered field by $\underline{E}^{scat}(x,z)$; the unknown grating profile is described by the function $z = h(x)$. Invariance is assumed in the y direction. Under the assumption of perfectly conducting grating material, the profile $z = h(x)$ has to be determined such that the boundary condition

$$[\underline{E}^{inc}(x,z) + \underline{E}^{scat}(x,z)]_{z = h(x)} = 0 \tag{1}$$

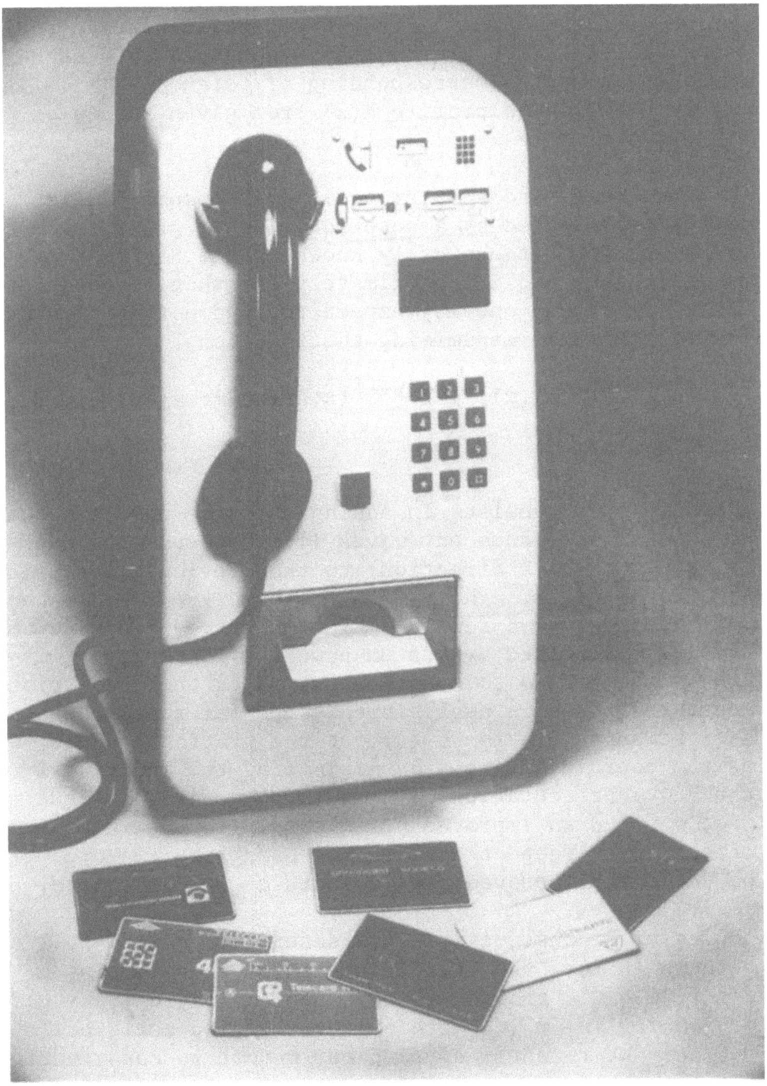

Figure 1. PHONOCARD station and card (by courtesy of Sodeco-Saia)

is fulfilled. Assuming an incident plane wave of monochromatic radiation polarized parallel to the y direction, the scattered field can be shown to have the same polarisation. Far away from the grating, the scattered field can be represented as a superposition of a number N of plane waves travelling in different directions according to Bragg's law, with coefficients c_n, n = 1, ..., N. The relative amount of energy propagated into each direction is given by the corresponding *efficiency* $|c_n|^2$. The problem is to determine a profile h(x) from given values of $|c_n|^2$.

In a first step, the scattered field is assumed to be approximately represented by a superposition of plane waves also close to the grating, (technically known as the Rayleigh hypothesis). Hence equ. (1) should hold approximately for that representation of $\underline{E}^{scat}(x,z)$ also, and one may expect to find a first approximation $h_1(x)$ of h(x) from minimizing the functional

$$F \equiv \int d^2s\,|\underline{E}^{inc}(x,h_1(x)) + \underline{E}^{scatt}(x,h_1(x))|^2 \qquad (2)$$

with $d^2s = d^2x + d^2h_1$.

The second step consists in adding a finite number M (say M = 20) of so-called evanescent waves (i.e., waves with exponential damping in the z direction) to the previous expression for the scattered field. Their wavelength is λ and their decrement is of the order of a few λ^{-1}. The relative importance of these waves is controlled by yet unknown coefficients a^1_m m = 1, ..., M. We are entitled to add these waves as their contribution to the scattered field is negligible in the far zone. Using the previous first approximation $h_1(x)$ for the grating profile, we determine the coefficients a^1_m by minimizing the functional (2) within the new representation of $\underline{E}^{scat}(x,z)$. With a^1_m thus determined, we find an improved profile $h_2(x)$ by again minimizing the functional with respect to h(x) and using the improved profile $h_2(x)$, we determine improved coefficients a^2_m and so forth.

Examples of reconstructions are described in Ref. 5. Notably, these include the difficult task of constructing a *beamsplitter*, i.e., a grating with, say, $|c_1|^2 = |c_{-1}|^2 = 0.5$ and $|c_n|^2 = 0$ for $|n| \neq 1$. With 6 iteration, we achieved $|c_1|^2 = |c_{-1}|^2 = 0.46$. We recently applied our method to constructing beamsplitters with larger grating constants in order to check the small wavelength limit. Reconstructed groove profiles are shown in Fig. 2 for the case $d = 3\sqrt{2}\,\lambda$. The reconstruction is not unique: both profiles obtained (Fig. 2) yield 0.48 efficiency in the +1 and -1 order. Apparently, the upper profile is close to the triangle-shaped profile that one would expect from geometrical optics.

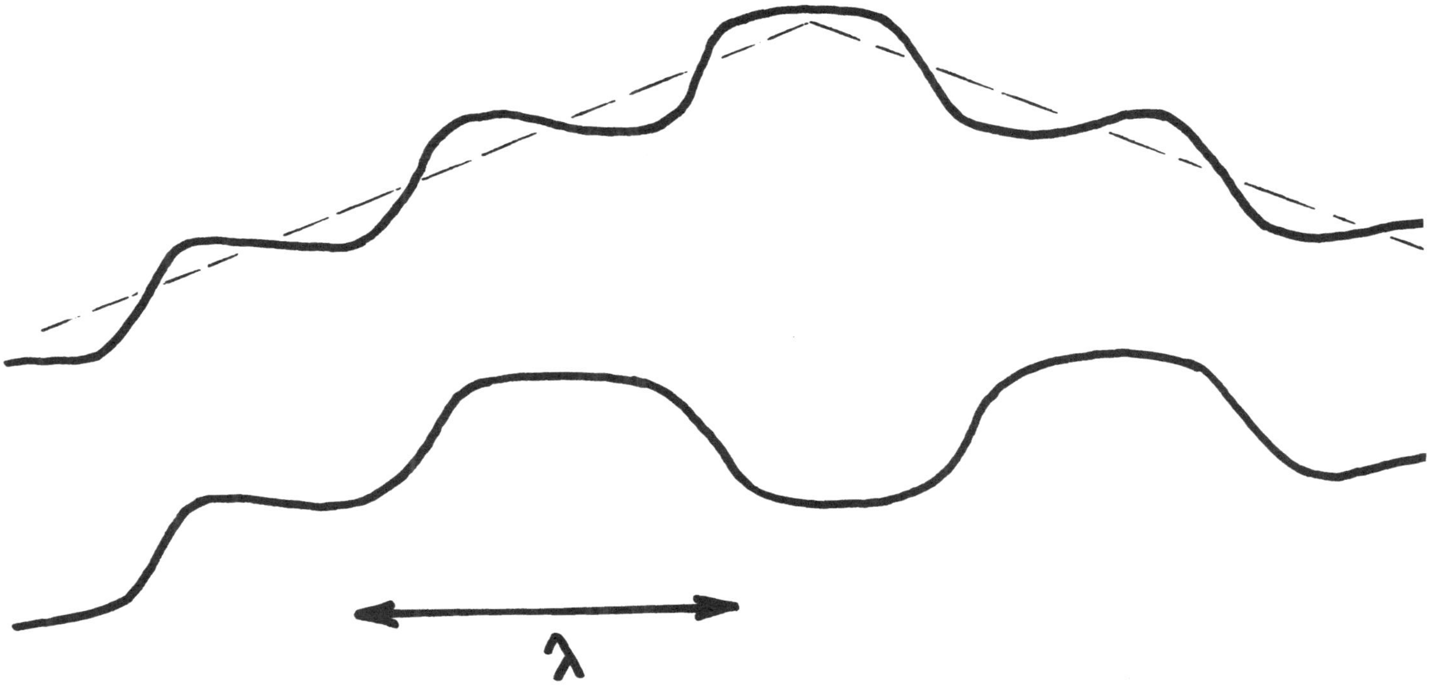

Figure 2. Reconstructed profiles of a *beamsplitter* grating with $d = 3\sqrt{2}\ \lambda$. Each profile yields $|c_1|^2 \approx 0.48$.

3. DETECTION OF DIFFRACTORS HIDDEN BEHIND A DIFFUSER

The structure of a source or scatterer manifests itself not only in the far-zone distribution of the energy of the scattered radiation, but also in the far-zone degree of coherence μ, whose absolute value $|\mu|$ is the fringe contrast that the scattered radiation produces if subject to appropriate interferometric experiments. We recall that completely incoherent radiation leads to zero fringe contrast, fully coherent radiation to maximum contrast, and partially coherent radiation to something in between.

By virtue of the Van Cittert-Zernike (VCZ) theorem, the far-zone coherence function of a quasi-monochromatic incoherent source is given by the Fourier transform of the intensity distribution in the source plane. For a compound secondary source formed by the incoherent illumination of an amplitude grating, straightforward application of the VCZ theorem leads to the prediction of a multi-peaked far-zone degree of coherence: Thus the presence of the grating can manifest itself in interferometric experiments.

For a phase grating, however, the above theorem would predict only the trivial central peak of the far-zone degree of coherence. In contrast to this classical VCZ result, the following coherence effect was predicted recently [6] and checked experimentally [7]: Even phase gratings under diffuse illumination can produce a multi-peaked far-zone degree of coherence, provided that the correlation length ℓ characterizing the diffuser is not zero, but larger than at least d/20, where d denotes the grating constant. (The classical VCZ results is, of course, reproduced in the limit $\ell/d \to 0$). Notably, there exists a regime of source correlation lengths ℓ, where the diffraction orders of the grating disappear completely in a single broad far-zone intensity lobe $I(\theta)$, but where the grating still manifests itself in the side peaks of the degree of coherence $|\mu(\theta,-\theta)|$, see Fig. 3.

The curves in Fig. 3 are obtained by evaluating the source-model cross-spectral density

$$W(x_1,x_2) = R(x_1)\ R^*(x_2)\ I(x_1)\ I(x_2),\ \mu(x_1,x_2) \tag{3}$$

with x_1, x_2 denoting position in the source plane. The function

$$R(x) = \sum_n c_n \exp(2\pi ix/d) \tag{4}$$

describes the underlying phase grating,

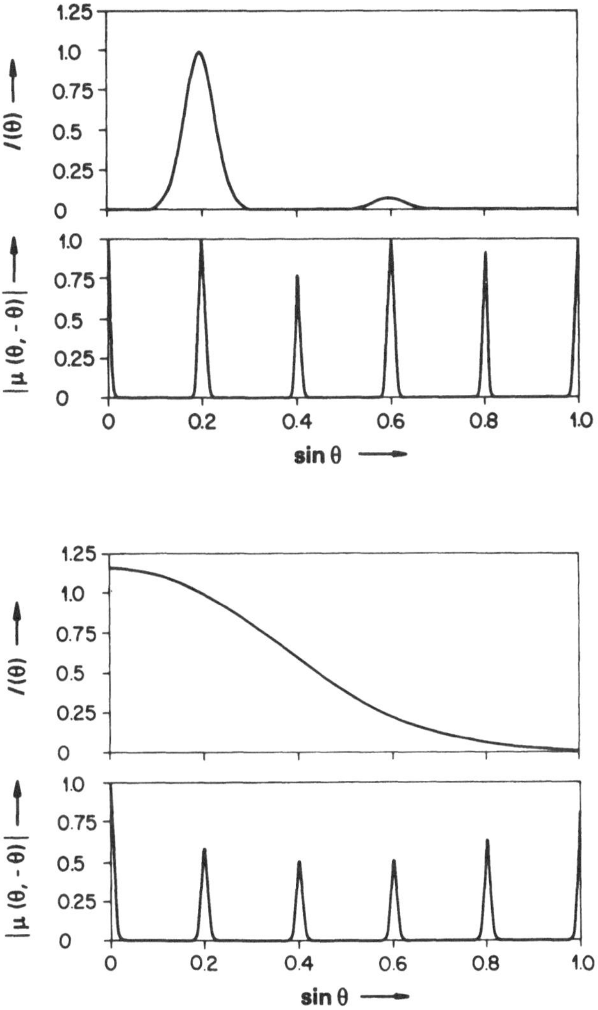

Figure 3. Angular distribution of far-zone intensity $I(\theta)$ and degree of coherence $|\mu(\theta, -\theta)|$ of radiation emanating from a lamellar phase grating hidden behind a diffuser. Upper: $\ell = d$, lower: $\ell = d/8$. [After Optica Acta 29 (1982) 169].

$$I(x) = \exp(-x^2/2a^2), \quad a \gg d, \quad a \gg \ell \tag{5}$$

the (large) overall aperture, and

$$\mu(x_1,x_2) = \exp[-(x_1 - x_2)^2/2\ell^2] \tag{6}$$

the degree of coherence in the source plane (as determined by the diffuser correlation).

The situation described above (Fig. 3, lower) can be exploited by *hiding* a grating (or some other difractor) behind a diffuser such as a rough surface or a layer with randomly varying refraction index and detecting it by using interferometric or photon correlation equipment. (The sample is moved in order to average over speckle.)

4. SCATTERERS WITH RANDOM CORRUGATIONS

In the previous section we described how judiciously chosen rough surfaces (to be prepared e.g., by some spray or by registering speckle in photo-resist) can be used to hide diffraction feature, which, however, are retrieved by more sophisticated methods. A further step would be the direct use of random (or seemingly random) surfaces as authenticity features: The authenticity could then be checked by detecting certain moments of the underlying statistics or some residual (e.g., non-Gaussian, non-monotonically decreasing) correlation of the roughness. Such samples could be produced using e.g., electron-beam lithography.

Before ideas of the kind outlined above can be realized, a better understanding of the scattering by rough surfaces in the resonant region (linear dimensions of the corrugation of the order of λ) is needed. The problem in question was revisited only recently [9,10] in terms of a perturbation expansion starting from the Ewald-Oseen extinction theorem. The expansion clearly shows the inadequacy of the Kirchhoff approximation (as one would expect from the corresponding results for gratings, see also [10]). However, a radius of convergence of the expansion could not be established.

We therefore devised an alternative perturbation expansion [11] based on the Fredholm-type integral equations for the fields and their normal derivatives. The source functions of the scattered electric and magnetic fields are developed into power series. The expansion parameter α is defined by the

$$h(x) = (\alpha/k)\, h_o(x), \quad |h_o(x)| \leq 1, \quad k = 2\pi/\lambda \tag{7}$$

for the corrugation height $z = h(x)$: The various terms of the source function expansions are obtained from recursion formulae.

In order to test our perturbation expansion we apply it to

the case of sinusoidal grating corrugation,

$$h_o = \sin ax, \; a = k/\sqrt{12} \approx 1.8 \; \lambda \qquad (8)$$

up to terms of the order α^4, and calculate the corresponding diffraction efficiencies. Agreement with the efficiencies obtained using known numerical methods is obtained for $\alpha < 1$. For $\alpha > 1$, further terms in the expansion have to be accounted for. An upper limit for the radius of convergence is $\alpha \leq k/a$ because of branch points in the kernels of the integral equations.

By the above result we feel encouraged to apply the perturbation expansion to an ensemble of rough surfaces with α not too large. In order to calculate the far-zone angular intensity and degree of coherence, we expand the underlying correlation functions in powers of α, e.g.,

$$\langle E(\theta) \; E^*(\theta)\rangle = \sum_n \alpha^n \; \Gamma_n^E(\theta,\theta') \qquad (9)$$

for the electrical field component E. Under the assumption of incident plane waves, $\langle h(x)\rangle = 0$, Gaussian and spatial stationary roughness statistics, we obtain zero correlation for odd n. For Γ_0 and Γ_2 we find results consistent with [9]. In particular, we corroborate the obliquity factor $(1 + \cos^2\theta)/2$ of the second-order far-zone intensity $[\Gamma_2^H(\theta,\theta) + \Gamma_2^E(\theta,\theta)]/2$ for unpolarized radiation.

References

(1) Baltes, H.P., editor, 1978, *Inverse Source Problems in Optics* (Springer, New York).

(2) Baltes, H.P., editor, 1980, *Inverse Scattering Problems in Optics* (Springer, New York).

(3) Roger, A., and Breidne, M., 1980, Optics Comm. 35, p. 298.

(4) Rieder, K.H., García, N., and Celli, V., 1981, Surface Science 108, p. 169.

(5) Huiser, A.M.J., Quattropani, A., and Baltes, H.P., 1982, Optics Comm. 41, pp. 149-153.

(6) Glass, A.S., and Baltes, H.P., 1982, Optica Acta 29, pp. 168-185, and references therein.

(7) Jauch, K.M., and Baltes, H.P., 1982, Optics Letts. 7, pp. 127-129, and references therein.

(8) Nieto-Vesperinas, M., and García, N., 1981, Optica Acta 28, pp. 1651-1672, and references therein.

(9) Nieto-Vesperinas, M., 1982, Optica Acta 29, pp. 961-971.

(10) Huiser, A.M.J., and Baltes, H.P., 1981, Optica Acta 40, pp. 1-4.

(11) Baltes, H.P., and Huiser, A.M.J., 1983, *A new perturbation expansion for electromagnetic scattering by stochastic surfaces*, Proc. U.R.S.I. Symposium on Wave Propagation and Remote Sensing, 9-15 June 1983, Louvain-la-Neuve, Belgium, Session 8.

V.13 (SR.5)

IMPROVEMENT OF IMAGE FIDELITY IN MICROWAVE IMAGING THROUGH DIVERSITY TECHNIQUES

H. Gniss, K. Magura

Forschungsinstitut für Hochfrequenzphysik (FHP/FGAN)
Königstr. 2
D 5307 Wachtberg-Werthhoven FRG

ABSTRACT. This paper discusses some problems concerning the recognition of man-made targets from microwave imagery. Thereby, of special interest is the dependence of the image fidelity on the lateral resolution, on the target orientation, and on the depolarization properties of the target. An improvement of the image fidelity is obtained through incoherent superposition of independent partial images using aspect and polarization diversity.

1. INTRODUCTION

Generally, a radar target can be described in the high-frequency approximation as a collection of scattering centers which can move, disappear, change their reflection coefficients, and depolarize depending on the target itself and on the illuminating electromagnetic field /1/. Therefore, in addition to the resolving power of a mapping radar system, the fidelity of a microwave (mw-) image is strongly affected by the object structure, the aspect angle, the transmit/receive polarizations, and the wavelength. Furthermore, of course, the background noise may lead to serious image distortions, too.

The coherent mw-imaging process is essentially the convolution of the reflectivity distribution on the object with the focused diffraction pattern (point scatterer response) of the imaging aperture. The principal result remains the same whether we use a direct imaging method (lens, reflector or phased array) or a two-stage wavefront reconstruction method. Much reference material concerning the latter method can be found in /2,3/.

W.-M. Boerner et al. (eds.), Inverse Methods in Electromagnetic Imaging, Part 2, 1245–1254.

A mw-image will be understood in the following context as a more or less resolved two-dimensional crossrange representation of an object similar to a photograph. The principal possibility to obtain additional resolution in the third dimension with aid of high bandwidth waveforms will not be further discussed here. Yet, it will be assumed that enough bandwidth is available for isolation of the object from the background.

The mw-images which will be presented in this paper have been obtained with a wavefront reconstruction method utilizing a quasi-monostatic cylindrical scanning technique at 36 GHz /4/. The reconstructions have been done computer-aided.

2. INFLUENCE OF RESOLUTION AND OF OBJECT STRUCTURE ON MW-IMAGE FIDELITY

Since the mw-image acuity is decisiveley influenced by the resolving power of the imaging system the question arises: Which is the necessary resolution-cell size δ in relation to the target length L, and which target discrimination level should be required? Concerning the latter of both questions, in many cases an observer would already be satisfied with a higher recognition level, i.e., the mw-imaging system should enable him not only to discern, for example, a ground vehicle, but also to discriminate between a tank, a truck, or a private car.

Besides the resolving power of the mw-imaging system the structure and the orientation of an extended target itself are of great influence on the mw-image quality. This can be easily realized at two extreme examples: If the target consists of isolated prominent scattering centers which can be individually resolved, there should be no serious problems to recognize the target, or even to identify certain target features under favourable conditions. However, if a complicated object shows a large quantity of scattering centers such that there can several interfere within a resolution cell, the mw-image exhibits a bothering speckle-like phenomenon, which is known as "coherent breakup" /5/.

The influences of the resolution and of the orientation on the recognition of two objects with entirely different structures from their mw-images are shown in Figures 1 and 2. Parameters are the aspect angle ϕ and the target length $L_0 = L(\phi = 0^0)$ (at broadside aspect) relative to the resolution cell size δ. As can be seen from the photos in the left columns, the target in Figure 1 (subsequently called TRUCK) has a relatively simple structure with large smooth surface parts and with only very few dominant scattering centers, whereas the target in Figure 2 (subsequently called TANK) shows a much more complicated structure. At the

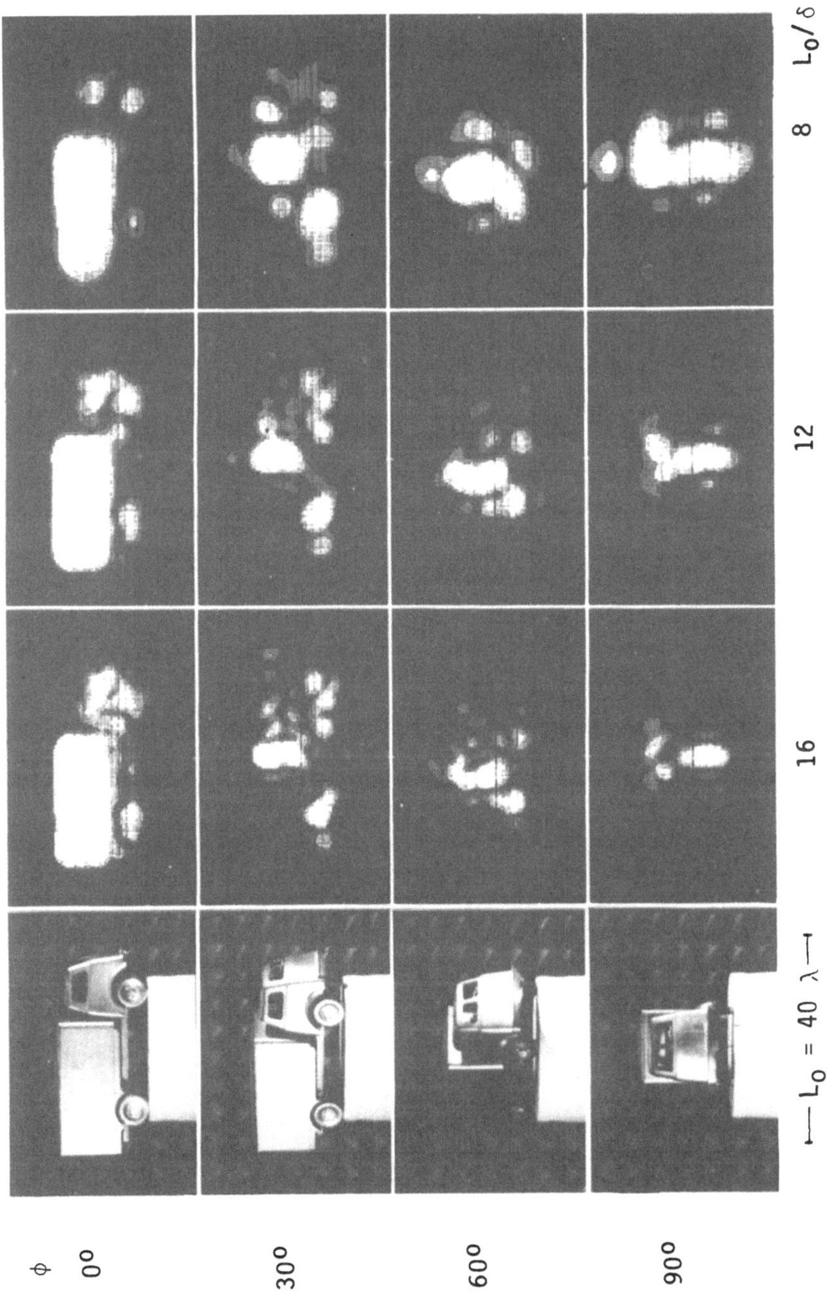

Figure 1. MW-images of the target TRUCK at various orientations and resolutions; target length L_0 = 40 λ, Pol. VV, f = 36 GHz, intensity quantisation = 4 dB

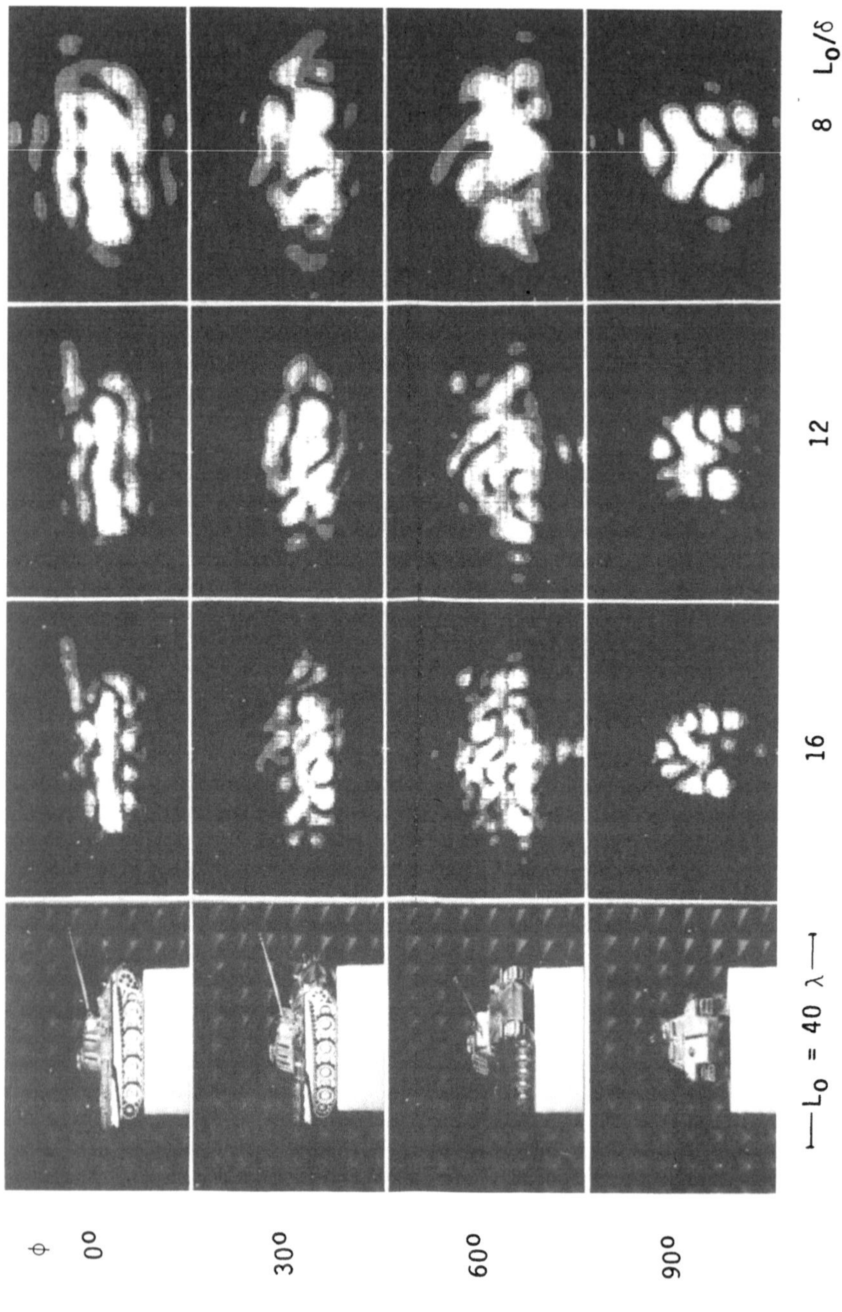

Figure 2. MW-images of the target TANK (see Figure 1)

cardinal broadside aspect both targets can be recognized definitely for $L_0 \geq 12\ \delta$ because the most prominent features - the gun barrel of the TANK and the rectangular flat sides of the TRUCK - are reflecting both specularly into the imaging aperture. But only a slight aspect change out of this optimum specular direction causes the disappearence of these outstanding features, which renders the target recognition at least more difficult if not almost impossible. This applies obviously to smooth targets like the TRUCK, where the remaining few visible scattering centers are not sufficient to recognize the target at a single view. Somewhat more promising is the situation at the more complex target TANK, where enough scattering centers are distributed within the target contour. Here we can recognize the target silhouette and estimate the target dimensions for all four aspect angles shown. A trained interpreter would with relative ease be in the position to recognize such prominent features like the turret or wheels.

To briefly summarize the foregoing: a target with a complicated structure (many scatterers) should be recognizable if $L_0 \geq 12\ \delta$ or even slightly less. This is in accordance with simulation results in /5/, where a symmetrical cross with independent scatterers on a rectangular grid was used.

3. POLARIZATION DEPENDENCE OF MW-IMAGE FIDELITY

Depolarization effects in mw-imagery may only be neglected at targets having perfectly conducting, smooth surfaces with radii of curvature large in relation to the wavelength. But generally, man-made targets of interest most likely show polarization - sensitive parts, as for example, edges, wedges, corners, and strongly curved sections. This means that the complete polarization scattering matrix has to be measured for all mw-image elements /3,4/. The polarization behavior of many simple elementary targets is extensively treated in /1/, and the available results may be applied directly to interpret mw-imagery if the polarization sensitive scatterers on a target can be resolved individually.

Much information upon a target is often irrecoverably lost if the co-polarized component of the scattered field is received only. This is clearly demonstrated in Figure 3 at the self-explanatory mw-imagery of a simple test target consisting of a flat plate and of two dihedral corners. Using linear vertical transmit polarization, the 45^{0} inclined dihedral corner remains completely invisible if the co-polarized scattered field is measured only. I.e., at least the cross-polarized component, too, must be received in order to obtain a more complete mw-image of the target. If moreover, the observer wants to identify the types (plate or dihedral corner) of the two outer scatterers, he must measure and

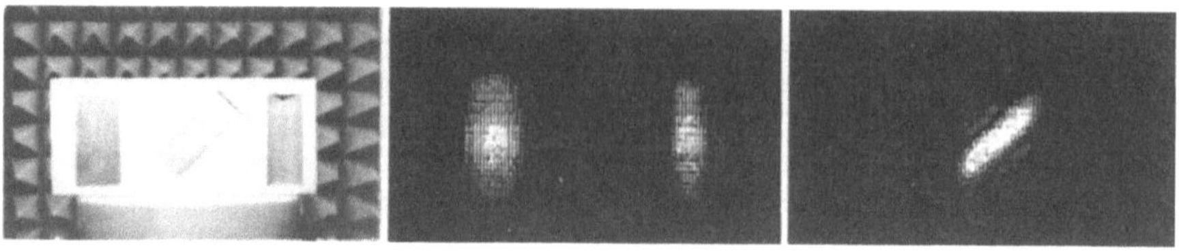

Figure 3. Polarization effects on the mw-imagery of a simple test target (from left to right: flat plate, two dihedral corners), f = 36 GHz, $\delta = 2.3\ \lambda = 20$ mm

evaluate the complete coherent polarization scattering matrix /3/. The necessity to measure the co- and cross-pol elements of the polarization scattering matrix will also be demonstrated for the more complicated target TANK. Figure 4 shows the mw-images of the TANK for the linear transmit/receive polarizations VV, HH and HV.

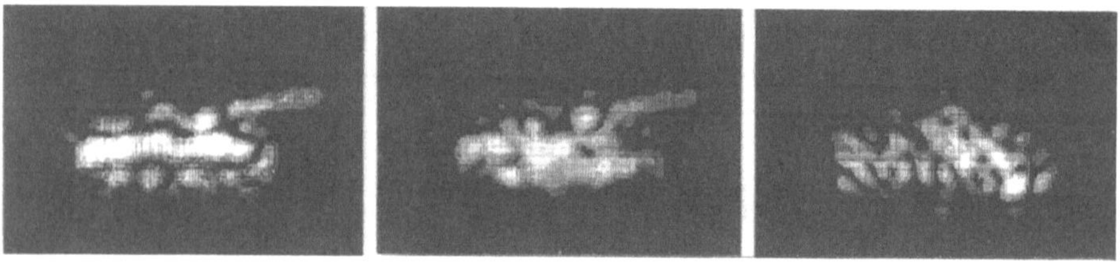

Figure 4. Polarization effects on the mw-images of the TANK, f = 36 GHz, $L_0 = 16\ \delta$

Both similar looking co-pol mw-images (VV and HH) clearly show those specular reflecting target features like the gun barrel, parts of the wheel frame and the wheels. In the cross-pol mw-image (HV) the slope of the fringes indicate the presence of prominent depolarizing parts like inclined edges, wedges, and dihedral corners.

4. MW-IMAGE ENHANCEMENT THROUGH DIVERSITY

The presence of the granularity in the mw-images of complicated targets turns out to be an unpleasant hindrance to the recognition of objects from their silhouettes. Fortunately, this fluctuation problem can be alleviated by utilizing diversity which means that several uncorrelated mw-images of a target are superposed incoherently /4-7/. Uncorrelated component images can be obtained from a given target by means of frequency, aspect or polarization diversity.

The fluctuations of the mw-image intensity I(x,y) which is measured within the target contour can be estimated with aid of the contrast C defined in /7/ as

$$C = \sigma/\bar{I} \tag{1}$$

where $\bar{I}$ is the mean value, and σ is the standard deviation of I. In the following it will be shown for two important target types which reduction of C can be expected from the superposition of p uncorrelated mw-images on an intensity basis.

Diffuse type target (Rayleigh statistics)

If a target exhibits in the resolution cell many independent randomly phased scatterers, no one of which is dominant, the intensity fluctuations of the mw-image can be statistically described in terms of a negative exponential or Rayleigh distribution /5-7/. Since a Rayleigh-distributed intensity has the well known feature $\sigma = \bar{I}$, the contrast C = 1 results. In /6,7/ it is shown that an addition of p uncorrelated images on an intensity basis reduces the contrast of the resultant image according to

$$C_p = \sigma_p/\bar{I}_p = 1/\sqrt{p}. \tag{2}$$

Specular type target (log-normal statistics)

Targets where one or a few specular reflecting surface parts dominate all other scatteres in a resolution cell often obey log-normal intensity statistics /5/. It has been shown in /8/ that the sum of log-normal variates can be statistically approximated very accurately by another log-normal distribution. If, for convenience, identical log-normal distributions are assumed for the component image intensities I_i the superposition of p uncorrelated mw-images reduces the contrast C_p of the resultant intensitiy I_p to /8/

$$C_p = \sqrt{\exp(\sigma_p^2) - 1} = \sqrt{\{\exp(\sigma^2) - 1\}/p} \tag{3}$$

where σ and σ_p denote the standard deviations of the logarithmic variables $\ln I_i$ and $\ln I_p$ in Neperian units. The transformation to the decibel base is linear according to $\sigma_{dB} = 4.343\,\sigma$.

Equations (2) and (3) show that for both target types diversity reduces the contrast C_p in accordance with a $1/\sqrt{p}$ - law. This relationship implies that the first few superpositions (p = 4 or 5) already cause a noticeable contrast reduction, whereas a further decrease would be recognizable only at the expense of large p.

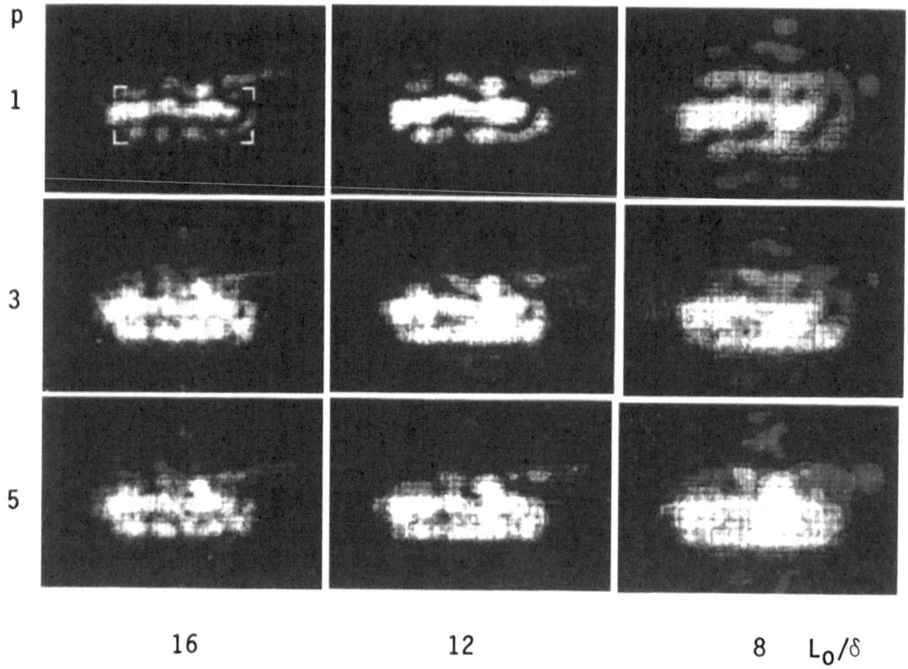

Figure 5. Aspect diversity by superposition of p = 1, 3 and 5 component images, aspect-angle increment = 5°, Pol. VV

This statement will now be checked by means of some experimental results where the TANK turns out to be a suitable target. Figures 5 and 6 show the applications of aspect and polarization diversity respectively; the expected reduction of the contrast within the target silhouette is evident.

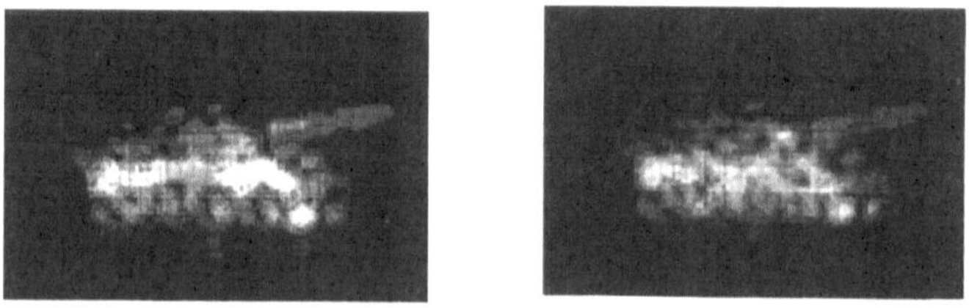

Figure 6. Polarization diversity by superposition of two component images, $L_o = 16\ \delta$

Notice the strong similarity of both smoothed mw-images in Figure 6. This implies in this special case that an incoherent superposition of all component mw-images wouldn't decrease the contrast further /4/.

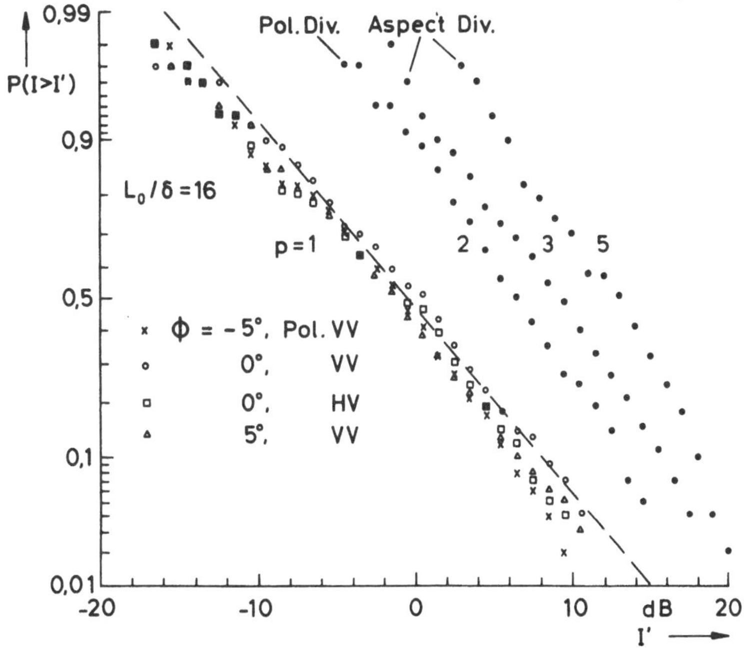

Figure 7. Distribution functions of the intensities of p incoherently superposed mw-images of the test target TANK; at p = 5 two further images with $\phi = \pm 10^0$ are added

To obtain some quantitative measure of the image enhancement, the image intensity distributions for $L_0 = 16\ \delta$ have been evaluated within the rectangular area as indicated in Figure 5. The (complements of the) distribution functions are plotted in Figure 7 over 10 log I for p = 1, 2, 3 and 5. It can be realized frome Figure 7 that the component image intensities (p = 1) exhibit nearly indentical statistics which follow the log-normal law with a fair approximation. This remains valid for p = 2, 3 and 5 where the slopes of the distribution functions are now steeper.

In Table 1 the standard deviations are given which can be obtained from fitting straight lines to the distribution functions in Figure 7. With these values the contrast C_{pmeas} can be evaluated by Equation (3); the expected contrast C_{pcalc} is in satisfactory agreement with C_{pmeas}.

Finally, it should be noted that the superposition of all copol and cross-pol mw-images on an intensity basis is also

p	1	2	3	5
σ_p/dB	6.6	5.3	5.3	4.7
C_{pmeas}	3.0	1.9	1.9	1.5
C_{pcalc}	3.0	2.1	1.7	1.3

Table 1 Contrast reduction in the mw-images of the TANK through diversity, $L_0 = 16\ \delta$

suggested by the well-known fact that the trace of the power scattering matrix remains invariant to the choice of any orthogonal polarization pair (e.g. linear or circular) /3/.

REFERENCES

1. G.T. Ruck, D.E. Barrick, W.D. Stuart, and C.K. Krichbaum, Radar Cross Section Handbook, Vol. 1 and 2, Plenum Press, New York, 1970

2. G. Tricoles, N.B. Farhat, Microwave Holography, Applications and Techniques, Proc. IEEE 65 (1977), pp. 108-121

3. W.M. Boerner, Polarization Microwave Holography: An Extension of Scalar to Vector Holography, 1980 Int. Opt. Comp. Conf., Washington, D.C., USA, Paper No. 231-23

4. H. Gniss, K. Magura, Millimeterwellenabbildung von bodengebundenen Objekten, FHP-Forschungsbericht Nr. 1-78, Wachtberg-Werthhoven, FRG, 1978

5. R. L. Mitchell, Models of Extended Targets and their Coherent Radar Images, Proc. IEEE 62 (1974), pp. 754-758

6. W.A. Penn, Signal Fidelity in Radar Processing, IRE Trans. MIL-6 (1962), pp. 204-218

7. J. W. Goodman, Some Fundamental Properties of Speckle, J.Opt. Soc.Am. 66 (1976), pp. 1145-1150

8. R.L. Mitchell, Permanence of the Log-Normal Distribution, J.Opt. Soc. Am. 58 (1968), pp. 1267-1272

V.14 (SP.6*)

FAR-FIELD TO NEAR-FIELD TRANSFORMS IN SPHERICAL COORDINATES

Doren W. Hess

Atlanta Instrumentation Division
Scientific-Atlanta, Inc.
Atlanta, Georgia U.S.A.

ABSTRACT

Experimental results are presented which demonstrate that the spherical near-field to far-field transform may be used in its inverse mode to determine the radiating near-field of a source distribution from measurements made of its far field.

INTRODUCTION

The purpose of this paper is to present a set of experimental comparison results to demonstrate the capability afforded by spherical near-field measurement techniques. Data representing the response of a probe antenna, as it is scanned over a near-field surface surrounding a radiating source antenna, may be transformed to equivalent data representing the far field of the source antenna. Similarly far-field data may be inverted to give the near field of the source on a close-in spherical surface. Comparison of the results from both these procedures against data directly measured provides convincing evidence that the theoretical basis of the transformation is correct, accurate and computationally sound.

The basis for the conversion transform which permits the mathematical computations alluded to above is the spherical near-field transmission equation. This transmission equation describes the excitation transfer between the transmitting antenna and receiving probe.

W.-M. Boerner et al. (eds.), Inverse Methods in Electromagnetic Imaging, Part 2, 1255–1266.

In the paper which follows the transmission equation is derived and results are shown which illustrate its application.

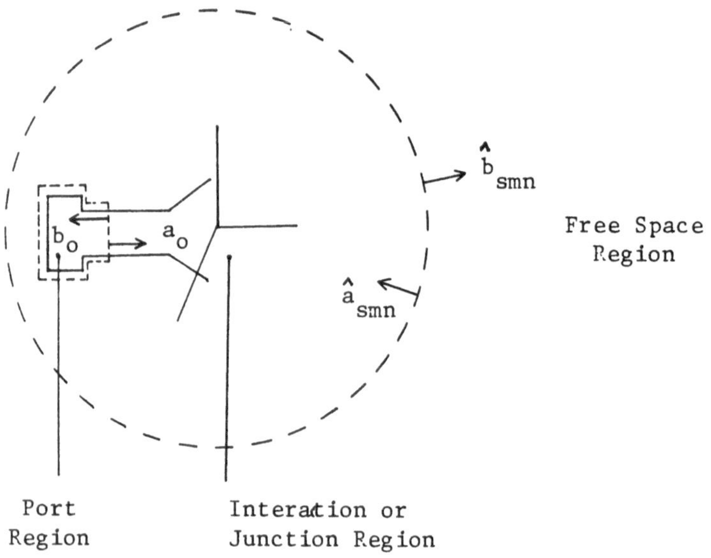

Figure 1. Schematic Diagram of the Scattering Matrix Description of an Antenna

THE SCATTERING MATRIX DESCRIPTION

The microwave antenna is described by a scattering matrix as a generalized microwave junction in which excitation is converted between free spherical modes.

Spherical coordinates are especially useful for describing the radiation of antennas because the two coordinates θ and ϕ - the polar and azimuthal angles - readily describe direction. The free space modes of the electromagnetic field in spherical coordinates are labelled by a set of three indices s,m,n. The n and m indices of the spherical harmonic functions label the rotation properties of the modes; n describes the group representation and m the row of the representation. The s index is a two valued index labelling TE or TM and pertains to the parity of the mode under parity. The wave number k and principal index n also label the representation on row of the model function's tranlations group property. Rotation and translation are essential to scanning theory.

The principal index n takes on only integer values between +1 and n_{max}, where n_{max} is given approximately by $\pi D/\lambda$. Here λ is the wavelength and D the diameter of the minimum sphere enclosing the antenna. The azimuthal index m is restricted to $|m| < n$. In spherical modal expansions one may always truncate the series at a large but nevertheless finite number of terms.

Each spatial mode of the microwave antenna's free space region can contain excitation propagating both toward and away from the junction region of the antenna. The scattering matrix connects the outward propagating modes to the ingoing excitation by establishing the linear relationship between the modal coefficients. Thus

$$\begin{vmatrix} b_o \\ \hat{b}_{s'm'n'} \\ \; \end{vmatrix} = \begin{vmatrix} \Gamma_o & R_{s'm'n'} \\ & \\ T_{smn} & S_{s'm'n',smn} \end{vmatrix} \times \begin{vmatrix} a_o \\ \hat{a}_{smn} \\ \; \end{vmatrix}$$

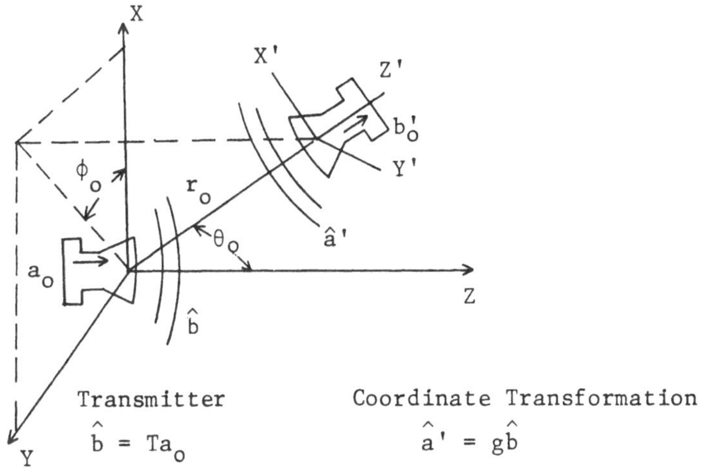

Figure 2. Schematic Diagram of the Coupled Two-Antenna System Used in Spherical Near Field Scanning

In this equation the port of the antenna is assumed to support only one mode labelled o; the free space region supports a large set of modes labelled smn. In spherical near-field theory only the port reflection coefficient Γ_o and the receiving and transmitting partitions, $R_{s'm'n'}$ or T_{smn}, of the scattering matrix are of interest.

THE SPHERICAL NEAR-FIELD TRANSMISSION EQUATION

The near-field transmission equation gives the port-to-port transfer of excitation between the transmitting antenna under test and the receiving probe.

To derive the spherical near-field transmission equation first write the equations that express the scattering mechanism in terms of the transmitting characteristic of the transmitting antenna and the receiving characteristic of the receiving antenna. The port excitation a_o is converted into free space b_{smn} excitation via the relation

$$b_{smn} = T_{smn} a_o \quad .$$

The free space excitation $a_{\sigma\mu\nu}$ at the receiver is converted to port excitation b'_o by summation of modal amplitudes and receiving coefficients

$$b'_o = \sum R'_{\sigma\mu\nu} \, a'_{\sigma\mu\nu} \quad .$$

These two equations are written in separate coordinate systems related to each other by the scanning geometry of spherical coordinates. Now convert each equation into an intermediate coordinate system. This intermediate coordinate system is one that is rotated relative to the coordinates fixed to the antenna being measured; but, its origin coincides with the test antenna's coordinate origin. The intermediate coordinate system is related to the coordinate system of the scanning probe by a translation along its z axis.

The outgoing wave coefficients of the intermediate frame $b_{s\mu n}$ are related to the outgoing wave coefficients of the test antenna frame by a linear rotation matrix $D_{\mu m}$ which parametrically depends on the Euler angles ϕ_o, θ_o, χ_o through which the rotations are made:

$$b_{s\mu n} = \sum_{m} D_{\mu m} (\phi_o \theta_o \chi_o)\, b_{smn} \quad .$$

Furthermore the ingoing wave coefficients $a_{\sigma\mu\nu}$ in the probe antenna frame are related to the outgoing wave coefficients is the intermediate frame by a linear translation matrix which parametrically depends on the wave number k and the scanning radius r:

$$a'_{\sigma\mu\nu} = \sum_{\sigma,\nu} C_{\sigma\mu\lambda,s\mu n} (kr\hat{z})\, b_{s\mu n}$$

[The convention is adapted that the probe modal labels are written in Greek letters $\sigma\mu\nu$ to correspond to the test antenna labels smn.) When combined these equations become the transmission equation

$$\frac{b'_o}{a_o} = \sum_{\substack{\sigma\mu\nu \\ smn}} R'_{\sigma\mu\nu}\, C_{\sigma\mu\nu,s\mu n} (kr\hat{z})\, D^{(n)}_{\mu m} (\phi_o \theta_o \chi_o)\, T_{smn}$$

THE FREE SPACE WAVEGUIDE

The solution of the Maxwell equations in free space is very similar to the solution in a waveguide. In the waveguide one is concerned with propagation along the guide axis. In free space using spherical coordinates one is concerned with propagation along radial lines and normal to a spherical surface which separates the source region from the rest of space.

The concept of the radial waveguide of free space gives rise to the notion that one may compute the field at any radial distance (along the guide) from data pertaining to any other distance. Whereas one is used to thinking of measuring a near field distribution and transmitting to a far-field radius, it is equally possible to measure the radiating far field and deduce from that the radiating near field.

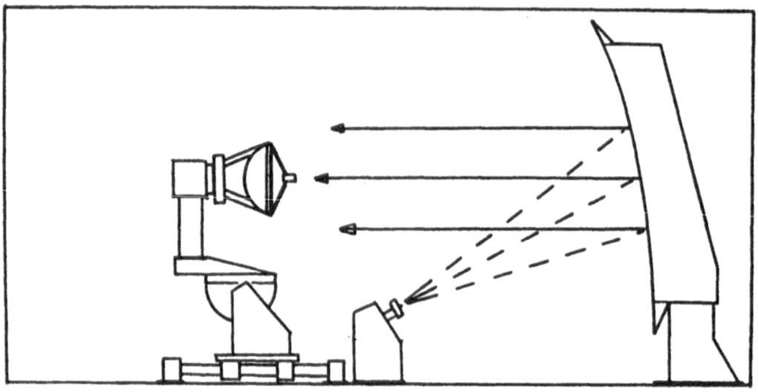

Figure 3. Schematic Diagram of the Compact Range

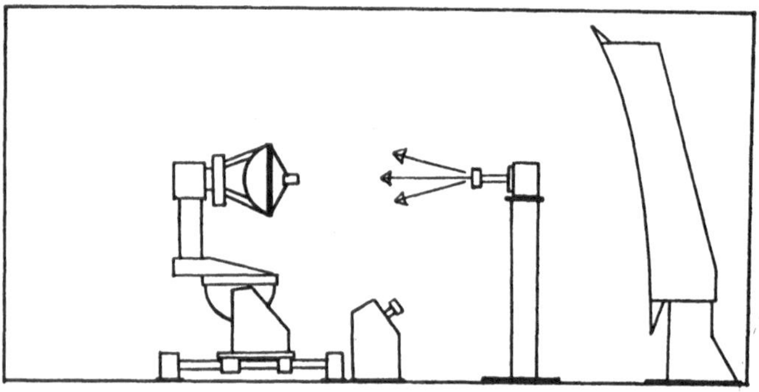

Figure 4. Schematic Diagram of the Spherical Near-Field Range

SPHERICAL NEAR-FIELD VERSUS COMPACT RANGE MEASUREMENTS

Measurements were made for both the near-field and far-field cases on a 24 inch paraboloidal dish at 13 GHz. The results demonstrated that the spherical-near-field-derived far-field pattern agreed with the far-field pattern measured directly under plane-wave illumination. The plane-wave illumination was achieved by use of a compact range. A comparison is shown in Figure 7 between the principal plane pattern cuts obtained directly on the compact range and obtained by transforming near-field measured data.

The compact range method corresponds precisely to the conventional far-field technique of pattern measurement. The antenna under test is placed on a two axis positioner and the response of the antenna is recorded as it is rotated. The rotation changes the direction of arrival of the illuminating plane wave and the response then is the receive pattern of the antenna and may be thought of as far-field scanning in spherical coordinates. See Figure 3.

The spherical near-field method consists of making spherical coordinate scans at reduced distances with a point source of radiation that give spherically curved phase fronts illuminating the test antenna. See Figure 4. The spherical near-field range consisted of a source tower erected in the compact range facility to support the point source illuminator, a broad beam round waveguide scalar feed horn.

For the spherical near-field the method of illumination was changed but the test antenna was mounted on the same positioner as for the compact range test and the same electronic hardware was used to record the data. The near-field to far-field transform was performed on the same minicomputer that was used to acquire the data. Both the compact range and the spherical near-field data could be plotted overlaid on the same graphical output to make the near-field to compact range comparison.

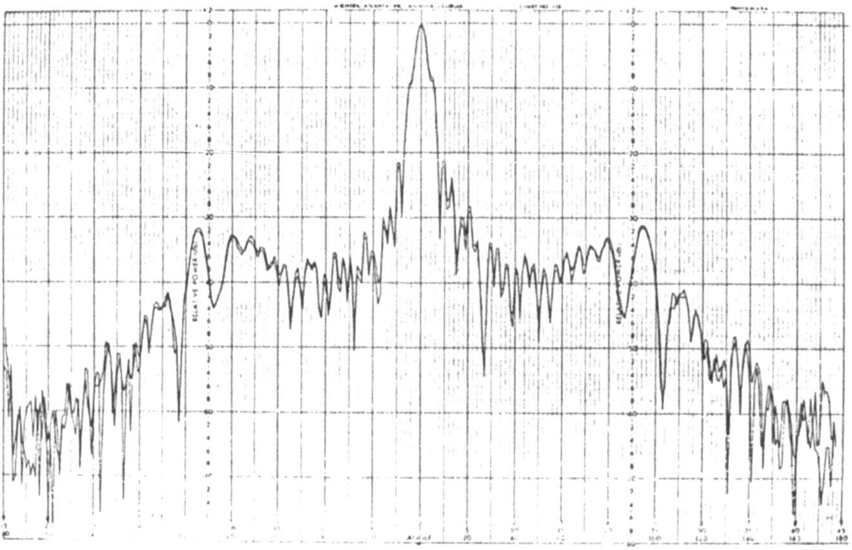

Figure 5. Comparison of R_2 = 299 cm of Measured Data versus Data Transformed from R_1 = 193 cm

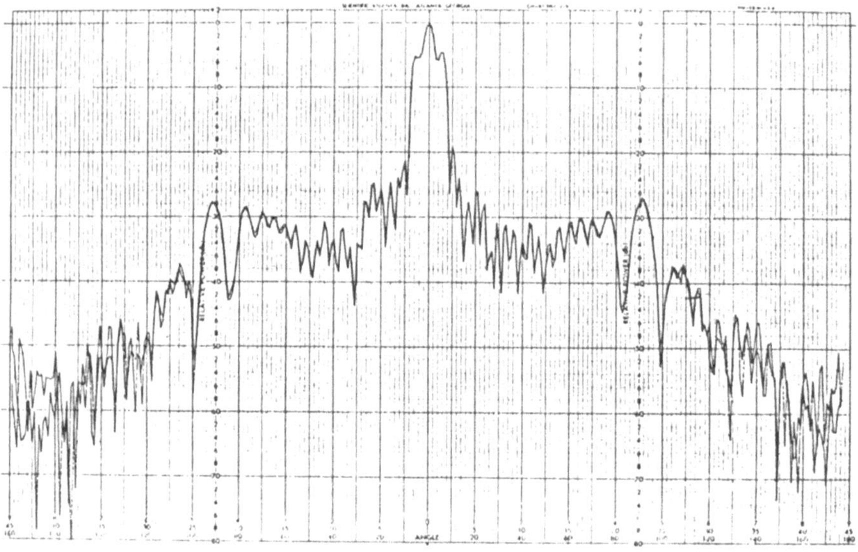

Figure 6. Comparison at R_1 = 193 cm of Measured Data versus Data Transformed from R_2 = 299 cm

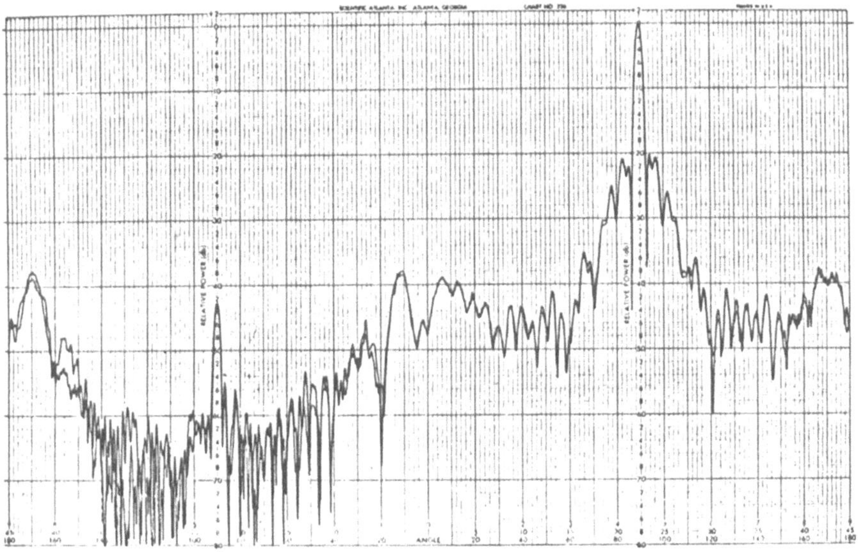

Figure 7. Comparison of R_∞ of Measured Data versus Data Transformed from $R_2' = 300$ cm

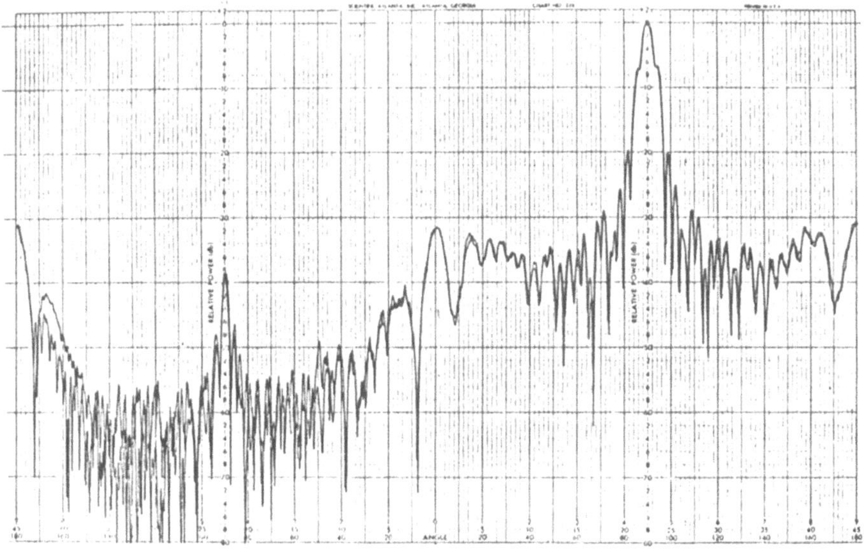

Figure 8. Comparison at $R_2' = 300$ cm of Measured Data versus Data Transformed from R_∞, the Compact Far Field

SPHERICAL NEAR-FIELD TO NEAR-FIELD TRANSFORMS RADIALLY OUTWARD AND INWARD

Measurements were made for the reflector antenna at several different near-field radii. This set of measurements permitted a comparison of the results of the measurements in the near-field region. See Figures 5 and 6.

The first comparison shows two corresponding traces from two different spherical data sets. One data set was derived by processing the measured data acquired at a close-in radius

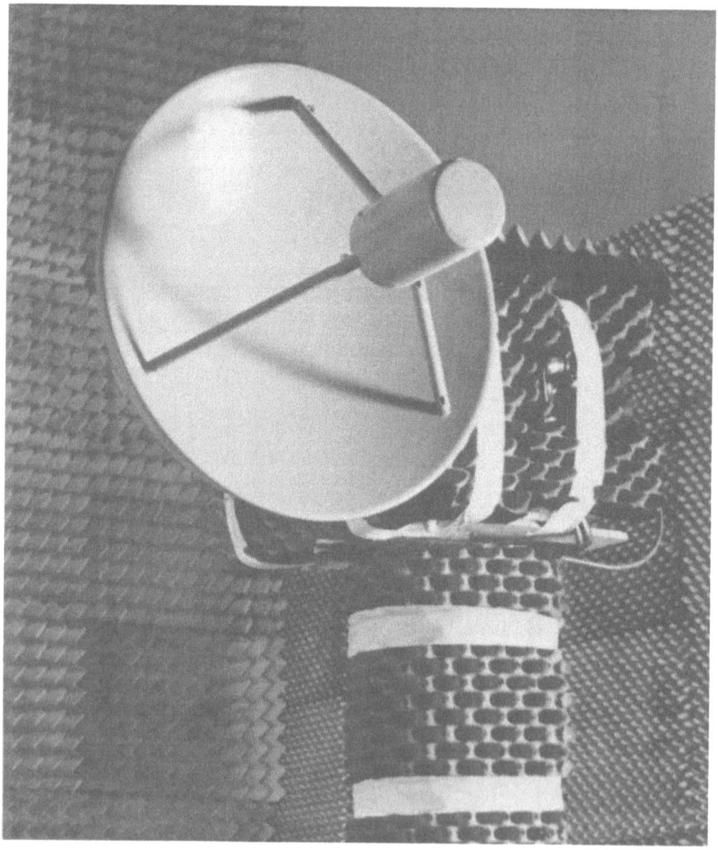

Figure 9. Photograph of the Test Antenna. A 24 (cm) Paraboloidal Reflector Mounted on the Test Positioner

R_1 = 193 cm to yield its equivalent at a larger radius R_2 = 299 cm. The other data set was not processed at all and represents raw data directly recorded at radius R_2. The agreement between R_1-to-R_2 data versus R_2 data demonstrates concept of radial near-field to near-field transforms.

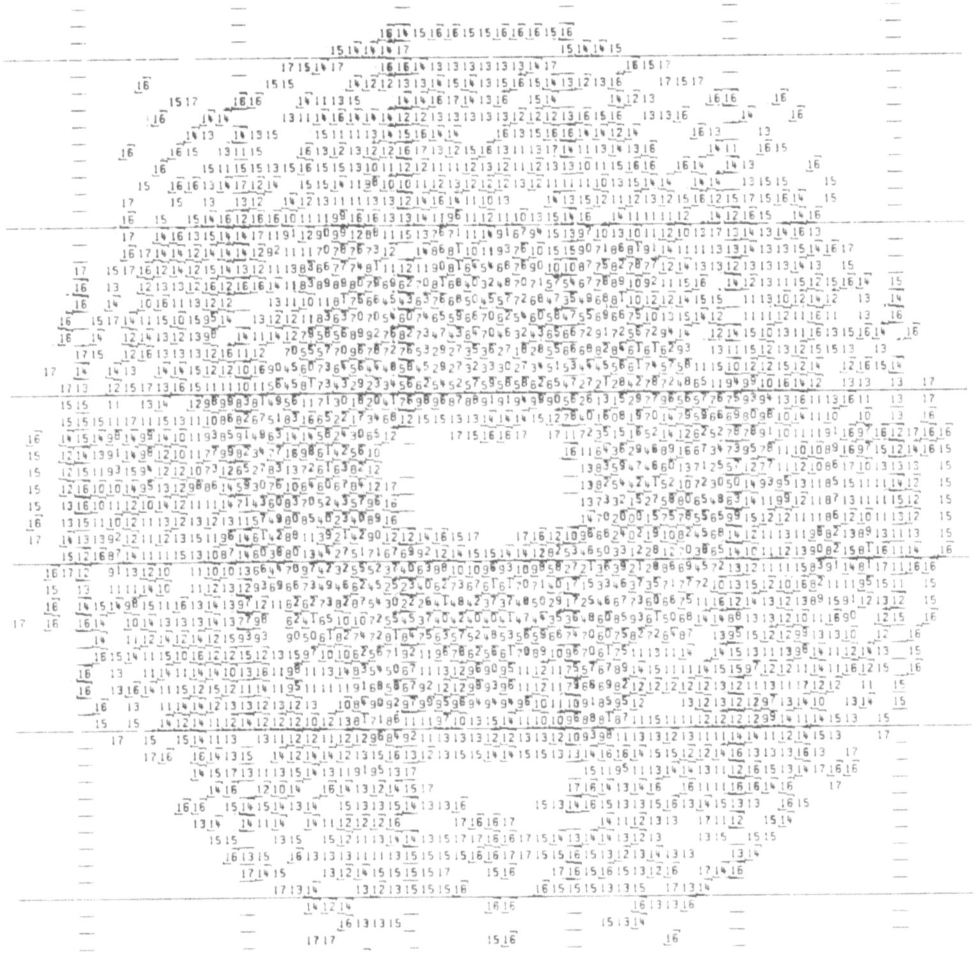

Figure 10. A Near-Field Radiation Distribution Plot at 80 cm Range Length. Note: Top and Bottom are Reversed from Photo.

The second comparison also shows two corresponding traces from different spherical data sets at the smaller radius. One data set was derived by processing the measured data acquired at a larger radius of 299 cm and processed to yield its equivalent at a smaller radius of 193 cm. The second data set is unprocessed data measured directly.

Clearly both inward and outward transforms between near-field radii are demonstrated by these results.

APERTURE IMAGING

The compact range data could also be transformed back to a sphere that virtually touched the feed structure of the antenna. Because the aperture of the antenna was set forward of the origin of the coordinate system this very close-in sphere intercepted the tube of collimated rays over approximately ±21 degrees of arc. The peak of the antenna beam lay on the equator of the coordinate system, and the pseudo rectangular character of the grid system there could be utilized to form an image with a near-field radiation distribution plot. The image displayed in Figure 10 was obtained by truncating the radiation distribution at a level of 17 dB below the peak.

(The image displayed is inverted from the photograph of Figure 9 with the top and bottom interchanged but right and left maintained.) The circular boundary of the dish and the central area blockage are clearly revealed. The blockage region in the lower portion of the image corresponds to the vertical strut at the top of the photograph. Using more aperture resolution contours would be visible.

CONCLUSION

The scattering matrix ansatz provides a transmission equation which may be considered sufficiently general for the coupled two-antenna case to permit radial transform in the free space waveguide.

REFERENCES

1. F. Jensen, Ph.D. Thesis LD-15, Technical University of Denmark (1970).

2. P. F. Wacker, National Bureau of Standards Report NBSIR75-809 (1975).

3. F. H. Larsen, Ph.D. Thesis LD-36, Technical University of Denmark (1980).

VI FINAL REPORTS of WORKING DISCUSSION GROUPS

Background:

The working discussion groups are to work on isolating unresolved problems and providing recommendations for potential future research projects which will require active interaction of engineering scientists of all NATO-member countries. The five questions (W-A to W-E) chosen define issues for which immediate answers are required. The composition of the chosen speakers, senior research scientists/engineers is well suited to tackle these problems and we do expect clear-cut resolutions and recommendations for additional near-future NATO Advanced Study Institutes and/or NATO Advanced Research Workshops.

We have also scheduled a formal Working Discussion Group (W-F) for research management representation of various research funding organizations of active NATO-member countries for the purpose of generating future close interaction and sponsorship of large-scale international research projects.

WORKING DISCUSSION GROUP TOPICS WITH BRIEF DESCRIPTIONS

VI.1: Mathematical Inversion Methods and Transient Techniques

VI.2: Numerical Instabilities in Electromagnetic Inverse Problems

VI.3: Polarization Utilization in High Resolution Target Imaging

VI.4: Definition of Image Quality and Image Resolution in Remote Sensing and Surveillance

VI.5: Holographic and Tomographic Imaging and Related Phase Problems

VI.6: Enhancement of Electromagnetic Imaging Research Interac tion of Active Laboratories within NATO-member countries

SUGGESTED DUTIES OF WORKING DISCUSSION GROUP CHAIRMEN, COORDINATORS, MODERATORS, ADVISORS, REPORTERS, AND WORKING MEMBERS

(Suggested guidelines by W-M. Boerner)

Chairmen: Will carry out the same duties as a chairman of a paper session which include:

(i) introduction of speakers and topics,

(ii) making sure that alloted time is strictly adhered to,

(iii) coordination of questions and answers.

W.-M. Boerner et al. (eds.), Inverse Methods in Electromagnetic Imaging, Part 2, 1267–1268.

Coordinators: Responsibiilties include making sure that:

(i) the discussion group pens are set-up properly.

(ii) all members assigned to a working group have been introduced to one another.

(iii) all workshop info available at the workshop hotel can be easily accessed.

Moderators: Moderators are chosen on the basis of their leadership potential in their field of expertise and their duties include assuring dynamic interaction of working group members by:

(i) stimulating discussions so that objectives of the topics to be discussed can be met.

(ii) closely interacting with the group coordinator

Advisors: Advisors are chosen on the basis of their accredited seniorleadership roles, as well as for their past NATO-ASI/ARW experience. The main duties are to:

(i) assure efficient and effective group discussion interaction,

(ii) closely work with the group reporter,

Reporters: Reporters are to collect all question statements, written notes, discussion results, i.e., acts as the recorder of discussions. The main duties are to:

(i) streamline discussion results and to direct the writing and reporting proceedings for the intermediate, final draft and print-ready group discussion reports which are limited to about five to eight pages each,

(ii) closely work with both the Advisor and the Co-Editors on the final individual group reports and the overall summary of discussion group activities.

VI.1 FINAL REPORT of WORKING DISCUSSION GROUP: W-A on MATHEMATICAL INVERSION METHODS AND TRANSIENT TECHNIQUES

Abstract: The general purposes of this working group were:

1. To review the basic mathematical properties of the various inverse methods, discussed at this ARW.
2. To evaluate the status of these methods, which include transient (time domain) as well as spectral (frequency-domain) approaches,
3. To recommend as a result of this evaluation what appear to be the most important problems and promising methods.

Coordinator: Arthur K. Jordan
(Group Leader)

Moderators: Karl Langenberg
André Roger

Advisors: Pierre C. Sabatier
Hans Blok

Reporters: Cornel Eftimiu
Arthur K. Jordan

Discussion Group Activists:
(* The names in this column weren't present at the ARW, however, their contributions were considered during the discussions.)

1. Blok, H.	13. Lüneberg, E.	* 1. Ahn, S-Y.
2. Boerner, W-M.	14. Manson, A.C.	2. Bollig, G.
3. Borden, B.H.	15. Moffatt, D.L.	3. Brück, D.
4. Chaloupka, H.	16. Moses, H.E.	4. Corones, J.P.
5. Chaudhuri, S.K.	17. Ramm, A.G.	5. Davison, M.E.
6. Crosta, G.	18. Roger, A.	6. Deschamps, G.A.
7. Dangelmayr, G.	19. Sabatier, P.C.	7. Güttinger, W.
8. Eftimiu, C.	20. Schimpf, H.	8. Hess, D.
9. Hollmann, B.F.	21. Sphicopoulos, T.	9. Jaggard, D.L.
10. Jordan, A.K.	22. Süss, H.	10. Lee, T.C.
11. Krueger, R.J.	23. Tabbara, W.	11. Nashed, M.Z.
12. Lang, H-D.	24. Wright, F.J.	12. Tijhuis, A.G.
		13. Wacker, P.F.

Synopsis

The last few years have seen significant technological progress

W.-M. Boerner et al. (eds.), Inverse Methods in Electromagnetic Imaging, Part 2, 1269–1271.

in the field of electromagnetic imaging, in spite of the fact that a unified, comprehensive theory of the electromagnetic inverse scattering is yet to be formulated and developed.

It is the general consensus of this Working Group that an important body of theoretical results pertinent to the electromagnetic inverse scattering theory currently only covers one-dimensional problems and in most cases only for mono-static instrument arrangements.

It is the foremost recomendation of this group that future research efforts be directed towards the development of an inverse electromagnetic scattering theory in two and three dimensions, addressing in particular, the polarization vector properties of the object interrogating electromagnetic wave, also for the general bistatic and/or monostatic non-reciprocal cases.

Specific Questions

The group considered a number of specific questions, which could be grouped into two classes:

1. Problems assuming the availability of complete experimental data

Such problems, of course, do not belong to the real world, yet their consideration will undoubtedly facilitate the theoretical development. Questions relating to the mathematical aspects like the existence and uniqueness of a solution of the inverse problem for bounded, non-spherical scatterers, with various types of boundaries (conductors, dielectrics) fall in this category. Also, problems relating to the analytic and asymptotic behavior of the field in the complex frequency plane, including the role of resonances or natural frequencies could provide the framework for a formulation of the inverse electromagnetic problem, by analogy with the quantum mechanical case. Finally, the special but important case of discontinuous inhomogeneities, as well as the treatment of dispersive media deserve special attention.

While there are no systematic approaches to such topics, we would like to encourage the continuing investigation of special cases and special techniques, which could provide a theoretical laboratory for the testing of more general formulations.

2. Problems taking into account the incomplete availability of experimental data.

These are the problems met within real life, but only limited progress has been made so far in this direction. Questions involving apertures, limited frequency bands, discrete sets of data and the like fall in this category. Also, the important aspects of data processing, nondestructive evaluation, numerical stability, control theory, computer simulation, etc. must be addressed to ensure the eventual practical implementation of any theory.

Multidisciplinary Approach

Electromagnetic inverse scattering theory cannot be isolated from similar efforts in fields like quantum mechanics, acoustics, electricity, seismology, oceanography, atmospherical physics, etc. Not only are the mathematical results of a common developmental basis, but the techniques employed in these fields complement and support each other. It is the consensus of this working group that Mathematical Methods in Inverse Scattering are most profitably discussed in groups in which investigators realize that this may be a pratical recommendation even for experimental work, but we believe it to be ideally suited for theoretical work, at least at the present stage of the methodology.

Final Recommendations

Given the relatively inexpensive requirements of theoretical research, and the fundamental problems it addresses, we believe that Mathematical Methods of Inverse Scattering represents a Low Risk / High Payoff endeavor par excellence. A vigorous interaction of individuals and groups, within one country as well as across the borders, represents an essential developmental condition. We strongly recommend that meetings facilitating the exchange of ideas, like the one from which this report has emerged, or Summer Schools, Advanced Institutes, etc., be organized periodically, on the basis of joint support from both governmental and private organizations.

References:

1. Sabatier, P.C. "Theoretical Considerations for Inverse Scattering", Radio Science, Vol. 18, pp. 1-18, 1983.

2. Sabatier, P.C. "Inverse Scattering Theory", Springer-Verlag, 1981.

3. Nashed, M.Z. "Operator-Theoretic Method in Inverse Scattering", IEEE Trans. A & P, Vol. 21, 1981.

VI.2 FINAL REPORT of WORKING DISCUSSION GROUP W-B on NUMERICAL INSTABILITIES IN ELECTROMAGNETIC INVERSE PROBLEMS

Abstract: Numerical instabilities are inherent in many inverse problems. The purpose of this working discussion group is to investigate:

1. The causes for numerical instability
2. The selection of optimum methods for the solution of inverse problems
3. Recommendations for further studies

Coordinator: Rudolf H. Dittel
(Group Leader)

Moderators: Tapan K. Sarkar
Mario Bertero

Advisors: Edward R. Pike
Christine DeMol

Reporters: Robert M. Bevensee
Rudolf H. Dittel
Christine DeMol

Discussion Group Activists:

(* The names in this column weren't present at the ARW, however, their contributions were considered during the discussions.)

1. Bertero, M.	6. Hawkes, H.W.	*1. Dianat, S.A.
2. Bevensee, R.M.	7. Jory, V.V.	2. Musso, G.
3. DeMol, C.	8. Pike, E.R.	3. Nashed, M-Z.
4. Dittel, R.H.	9. Sarkar, T.	4. Taflove, A.
5. Fischer, M.	10. Umashankar, K.	5. Wacker, P.F.

1. CAUSES OF NUMERICAL INSTABILITIES

The problem of numerical stability has been analysed mainly in the case of linear problems where the concept of ill-posedness is basic. This concept applies only to mathematical models formulated in functional (infinite-dimensional) spaces. The operator equation

$$Bx = y$$

is said to be well-posed (in the sense of Hadamard) if the following three conditions are satisified:

- there exists at most one solution (uniqueness)

W.-M. Boerner et al. (eds.), Inverse Methods in Electromagnetic Imaging, Part 2, 1273–1275.

- the solution exists for any y (existence)
- a small variation on the data y produces a small variation on the solution x (continuous dependence of the solution of the data)

Otherwise, the problem is said to be ill-posed (or improperly posed). The requirement of uniqueness can always be fulfilled by considering generalized solutions instead of classical solutions (Nashed, M.Z., Ed., Generalized Inverses...). The two other requirements, however, are not satisfied whenever the operator B behaves like a filter so that some "information" is lost when going from x to y. Hence, an infinite-dimensional model describing an inverse problem is ill-posed.

A practical inverse problem with sampled data can be considered as a projection of an ill-posed problem on a finite-dinensional space and therefore, when uniqueness holds, it is always well-posed in the strict mathematical sense, since it is essentially equivalent to a linear algebraic problem. It may, however, be numerically ill-conditioned so that the errors on the data, which are unavoidable in practical situations, can be considerably amplified during the solution procedure. It is the condition number of the matrix (corresponding to the original operator B) that controls error propagation from the data to the solution. If the condition number of this matrix (ratio of the maximum to the minimum eigenvalue) remains relatively stable as one increases the number of unknowns, then the problem is computationally stable. For numerically ill-conditioned problems this condition number changes as one alters the number of unknowns. Hence, a well-posed problem may also be well/ill-conditioned numerically.

Given an ill-posed problem, various schemes are available for defining as associated problem which is well-posed. This approach is referred to as regularization of the ill-posed problem. In particular, as ill-posed problem may be regularized by

a) changing the definition of what is meant be an acceptable solution,
b) changing the space to which the acceptable solution belongs,
c) revising the problem statement,
d) introducing regularizing operators, and
e) introducing probabilistic concepts so as to obtain a stochastic extension of the original deterministic problem.

The above techniques have been issustrated by examples (Sarkar et.al., IEEE, in the references). It is important to point out that the regularization method of Tikhonov of Wiener filtering are just two such approaches amongst a class of different regularizing techniques.

2. SELECTION OF OPTIMUM METHODS

Given an ill-posed problem, there is no optimum method for the solution of the problem. The choice of a particular method is dictated primarily by the amount of a-priori information available. As there is no objective way to select the best solution procedure, the choice of a particular technique depends solely on the experience of the user. The 'optimum' solution technique is generally the one which includes in its formulation all the a-priori information available based on mathematical formulation, experience of the user and educated guesses.

3. RECOMMENDATION FOR FURTHER STUDIES

As a result of the investigation of the working group and based on the experience of its members, the following subject areas for further studies are recommended:

a) investigations of relations between solutions obtained with various cost functions,

b) investigations of multidimensional problems and vector problems ('multi' refers to geometry and variables),

c) investigations concerning the optimization in the selection of sampling points,

d) calculations of singular systems for kernels of physical interest,

e) investigations of practical numerical techniques dealing with experimental kernels,

f) investigations of non-linear problems,

g) investigations on the use of high level computer languages (e.g., LISP) for the solution of inverse problems in order to prevent numerical instabilities.

References:

1. Bloom, F., Ill-posed problems for integro-differential equations mechanics and electromagnetic theory, SIAM Studies in Applied Mathematics, 3, Philadelphia, PA, 1981.

2. Nashed, M.Z., Ed., Generalized Inverses and Applications, New York, Academic Press, 1976.

3. Nashed, M.Z., IEEE Transactions on Antennas and Propagation, AP 29, 1981, pp. 220-231.

4. Sarkar, T.K., Weiner, D.D., and Jain, V.K., "Some Mathematical Considerations in Dealing with Inverse Problems", IEEE Trans. Antennas and Propagation, AP-29, pp. 373-379, 1981.

5. Tikhonov, A.N. and Arsenin, V., Solution of ill-posed problems, New York, Winston-Wiley, 1977.

VI.3 FINAL REPORT of WORKING DISCUSSION GROUP W-C on POLARIZATION UTILIZATION IN HIGH RESOLUTION IMAGING

Abstract: There is good evidence from particular problem areas that target detection and classification can be considerably improved by use of polarization information. However, the general usefulness of this technique in target detection and classification still involves many unresolved issues. Many of these issues have been identified at this ARW and initial recommendations were made for futher research.

Coordinator: (Group Leader) James W. Wright

Moderators: J. Richard Huynen
Otto Kessler

Advisors: Robert S. Raven
Dag T. Gjessing

Reporter: André J. Poelman
James W. Wright

Discussion Group Activists:

(* The names in this column weren't present at the ARW, however, their contributions were considered during the discussions.)

1. Blanchard, A.	16. Kahn, W.K.	*1. Agrawal, A.P.
2. Boerner, W-M.	17. Kessler, O.	2. Bradley, T.C.
3. Carpenter, D.	18. Lis, C.	3. Dalton, R.
4. Chaudhuri, S.K.	19. Manson, A.C.	4. Deschamps, G.A.
5. Davidovitz, M.	20. Pedersen, H.	5. Hayward,
6. Detlefsen, J.	21. Poelman, A.J.	6. Hess, D.
7. Evans, D.A.	22. Popp, R.	7. Kelly, J.D.
8. Gallagher, J.G.	23. Raven, R.S.	8. Meyer, T.H.
9. Gjessing, D.T.	24. Sillence, C.D.	9. Nespor, J.D.
10 Gniss, H.	25. Thiel, M.F.A.	10. Rackson, M.M.
11 Grüner	26. Tricoles, C.P.	11. Root, L.W.
12 Guy, J.R.F.	27. Vogel, M.	12. Rope, E.L.
13 Heath, G.J.	28. Wei, P.S.P.	13. Stock, D.
14 Hjelmstad, J.	29. Winkler, G.	14. Vannicola, V.C.
15 Huynen, J.R.	30. Wright, J.W.	15. Wacker, P.F.

W.-M. Boerner et al. (eds.), Inverse Methods in Electromagnetic Imaging, Part 2, 1277–1280.

There is good evidence from particular problem areas that target detection and classification can be considerably improved by the use of polarization information. However, the general usefulness of this technique in target detection and classification is still an unresolved issue. The following comments and recommendations are given in the context of high resolution imaging, but obviously apply in a more general context. The promising results from particular investigations clearly indicate that investigations of the benefits to particular applications are well worth pursuing.

The following unresolved issues have been identified:

- It is appparent that there is a limited (and often incom plete) amount of polarimetric measurement data available.

- There is a need to develop analytical methods (signal/data processing) which constructively exploit polarization information for specific applications.

- There is an inadequate understanding of the full implications of the scattering matrix (or matricies) and its applications.

- More careful thought is required regarding definitions and notations of parameters relating to polarization.

- Standardization of the presentation of measured polarimetric data is required in order to obtain successful exchange of data between organizations. In particular, careful characterization of the data gathering systems, methodologies, etc., should be performed and supplied with the data.

- As far as the display of polarimetric data is concerned, there seems to be no universal or ideal method, because it is so heavily dependent on the application and the analyst.

- There seems to be no universal definition for the word "image", therefore, it should be defined in every context in which it is used.

The following disciplines should be involved in the study of the usefulness of the polarization information:

- signal/data processing
- decision/estimation theory
- microwave/optical measurements
- radar/ladar engineering
- electromagnetic (vector) scattering theory
- mathematical inverse techniques
- passive sensing, including radiometry

At present it is too early to suggest a specific broadbased scientific approach for R & D on this topic. This is because of the wide variety of applications requiring different approaches (frequencies, resolution, etc.) and the lack of awareness of signal/-data processing algorithms pertaining to the exploitation of polarization information.

It is suggested that in the near future a working group should be organized to discuss analytical and measurements techniques regarding the acquisition and utilization of polarization information in relationship to specific applications. This working group should be open to all interested parties, but the primary objective is to insure that measurements for a specific application will meet the requirements of the analyst and application engineers.

The quality of data is clearly dependent upon the characteristics of the measurement instrument. Therefore, the design, construction and characterization deserves and requires more care and effort than has generally been required for conventional radar systems. This additional effort clearly implies that the division of funding on a given program will show larger proportion of the funding in the design, fabrication and hardware characterization areas. Some items of particular importance are:

- Antenna with high purity of polarization across the appropriate beam-width and bandwidth
- High power, fast polarization switches which cause minimal degradation of the antenna characteristics
- Wideband and narrow band sources with low phase noise near the carrier
- Analysis of system effects in the measurement of all polarization scattering matrix data.

A few examples of particular applications are:

- anti-armour
- anti-aircraft
- anti-ship and anti-anti-ship missiles
- remote sensing (both military and civilian applications)

The assessment of the benefits of the use of polarimetric data requires the knowledge of:

- target signatures
- clutter signatures (natural and man-made for all seasons)
- material properties
- radome characteristics (if used)
- transmission path characteristics

We believe, that the data acquisition on potential targets and clutter is one of the major requirements. These are required for many applications, a few examples of which are given above. We feel, that NATO would probably be able, to give attention to the question of measured data on clutter and on unclassified objects relevant to each of these applications and should do so now. We feel, that with regard to the targets appropriate to military applications most of the data would be classified and the exchange of data on these targets between NATO countries may be more difficult and hence such data should be aquired on a national basis.

We strongly recommend, that NATO should ensure the standardization of the definitions and terminology and of standard specification with regard to how any data which are collected should be organized. This latter standard is not intended to be restrictive, but is to insure that the user understands the organization of the data and the proper interpretation of each data item.

We believe, that a government or a quasi-governmental agency should manage any program to insure relevance, coordination, and eliminate unnecessary duplication of effort and broadest dissemination of the data to all interested parties.

Passive techniques have received only limited attention, but it does appear that there is some potential for polarimetric techniques since it is evident that the incident radiation is polarized although it appears that the polarization is a function of the time of day. Any continuing radiometric programs should incorporate the means to investigate the potential of polarimetric techniques.

Rather than recommend any particular programs in remote sensing, we believe, that clutter measurement programs be organized to provide sufficient diversity in clutter types and conditions to permit investigation of the usefulness of polarimetric techniques and provide guidance for the remote sensing community.

Full polarization radar backscatter data is the most important present requirement. A trial program should be set up by NATO to collect data from either full scale or scale model targets. The scale modelling facilities available provide, when they are applicable, a very reliable and cost effective method. The coherent scattering matrix data from targets can be combined with clutter data to investigate detection and discrimination algorithms.

VI.4 FINAL REPORT of WORKING DISCUSSION GROUP W-D on DEFINITION OF IMAGE QUALITY AND IMAGE RESOLUTION IN REMOTE SENSING AND SURVEILLANCE

Abstract: In view of an increasing committment to produce remote-sensing imagery, this sub-group devoted its attention to the problems of quality specification and calibration of images, in a sense, more general than optically dimmed images. Several specific problems were identified and discussed.

Coordinator: (Group Leader) Akira Ishimaru

Moderators: Richard W. Larson
Terje Lund

Advisors: John O. Thomas
Stephen Rotheram

Reporters: Michael E.R. Walford
Akira Ishimaru

Discussion Group Activists:

(* The names in this column weren't present at the ARW, however, their contributions were considered during the discussions.)

1. Blanchard, A.	13. Larson, R.W.	*1. Adams, M.F.
2. Boerner, W-M.	14. Lund, T.	2. Bradley, T.C.
3. Brand, H.	15. Orhaug, T.	3. DiSalvio, A.N.
4. Detlefsen, J.	16. Osterrieder, S.	4. Freeston, I.L.
5. Ferwerda, H.	17. Pedersen, H.	5. Lövhaugen, O.
6. Fung, A.K.	18. Rotheram, S.	6. Lyzenga, D.
7. Giess, S.C.	19. Schimpf, H.	7. Macklin, J.T.
8. Gniss, H.	20. von Schlachta, K.	8. Narod, B.
9. Hawkes, H.W.	21. Tabbara, W.	9. Newton, R.W.
10. Huynen, J.R.	22. Thomas, J.O.	10. Pirolo, D.G.
11. Ishimaru, A.K.	23. Walford, M.E.	11. Shuchman, R.
12. Jaques, A.F.	24. Wei, P.S.P.	12. Szu, H.

1. The Specification of Image Quality in Remote Sensing

The group wishes to emphasize the importance of specifing the quality of images produced by remote sensing systems. Quality

W.-M. Boerner et al. (eds.), Inverse Methods in Electromagnetic Imaging, Part 2, 1281–1283.

is a complex concept; a complete unique specification of the quality of an image may be possible only for a specific remote sensing system and only if the specific purpose of a particular user is considered. Even in such special cases a quantative specification may not be possible, particularly when remote sensing imagery is used for non-scientific purposes. Nevertheless it is important to establish some flexible system of measurable standards of quality which can act as a yardstick for users, producers and processors of remote sensing imagery.

In the familiar optical case an optical system is used to produce an image on film. The quality of the optical system can be quantatively described in terms of lens aberations, focal length, depth of field, numerical aperture and by some suitable statement of geometrical mapping between target and image. The quality of the recording medium can be specified in terms of speed, graininess, dynamic range, linearity, quantization levels, trichromatic spectral efficiency and dimensional stability. The quality of the image depends upon the system and the medium qualities. It can be partially specified by such measurables as spatial resolution, the range and resolution of optical density, granularity, etc., but the image quality also depends upon the users purpose in a way that is not always quantifiable and may depend upon illumination conditions, atmospheric state, target condition. Therefore, it absolutely requires that ground truth observations be made if we are to specify how much information an image carries relevant to a specific user application.

Now to generalize the concept of an image: it is any accessible, organized store of data derived from remotely sensing the physical environment. The problem now is to specify the quality of the image in any such store which is to be accessed by users, processors or producers of imagery, whether the store be tape, computer RAM, holographic plate or SAR imagery on film. This problem is far from solved. A solution must encompass the following points:

1. system calibration is more than ever necessary for "unconventional" images,

2. in establishing standards for image quality, it is necessary to consider that both human visual and machine perception may be applied to the images

3. the image quality concept must be extended to other signatures than optical and to three or more dimensional images, and is must incorporate polarimetric parameters,

4. it may be necessary to use different quality definitions for coherent and incoherent images

5. image distortion and information loss due to relative motion between object and sensor may require specification.

The group considers that these problems be addressed as a matter of urgency with due consultation with potential users, processors, and producers of imagery.

2. Calibration of SAR and Other Radar-Derived Imagery

Calibration of radar-derived images must be ensured as a matter of urgency. Ground-based reflectors, ground-based power monitors, and/or naturally occuring or man-made large flat surfaces may provide suitable means, permitting routine checks on antenna gain, power and orientation, tracking and ranging. They must be distributed over secure, well-surveyed stable locations in such a way that frequent calibrations are automatically included in the imagery itself. This is a most important, but inexpensive pratical matter on which some expertise already exists within this working group.

References:

RADAR IMAGERY, Torleiv Orhaug, Contribution S-4
INVERSE METHODS IN SLAR/SAR, Richard W. Larson, Contribution S-3

VI.5 FINAL REPORT of WORKING DISCUSSION GROUP W-E on HOLOGRAPHIC AND TOMOGRAPHIC IMAGING AND RELATED PHASE PROBLEMS

Abstract: The phase retrieval problem occurs in electromagnetic inverse scattering, optical and x-ray applications. The phase unwrapping problem is of importance in signal processing applications, ultrasound, and geophysical diffraction tomography. This working group identified problem areas and made general recommendations.

Coordinator: (Group Leader) Hedzer A. Ferwerda

Moderators: F. Alberto Grünbaum
Alan P. Anderson

Advisors: Adolf Lohmann
Harry E. Moses

Reporter: Anthony J. Devaney
Hedzer A. Ferwerda

Discussion Group Activists:

(* The names in this column weren't at the ARW, however, their contributions were considered during the discussions.)

1. Anderson, A.P.	8. Grünbaum, F.A.	*1. Gori, F.
2. Baltes, H.P.	9. Kaveh, M.	2. Hess, D.
3. Bevensee R.	10. Lohmann, A.	3. Soumekh, M.
4. Devaney, A.J.	11. Manson, A.C.	4. Szu, H.
5. Ferwerda, H.A.	12. Moses, H.E.	5. Vogelzang, E.
6. Fiddy, M.A.	13. Schwierz, G.	6. Wabnitz, P.F.
7. Gough, P.T.	14. Wei, P.S.P.	7. Yevich, D.

Introduction

Two problems of importance to the remote sensing and inverse scattering community are the "phase retrieval" and "phase unwrapping" problems. The phase retrieval problem consists of the determination of the phase of the pertinent wave function from one or more intensity distributions (which yield the absolute value of transforms of the wave function). The phase unwrapping problem occurs when the values of the reconstructed phase are restricted to an interval of width 2π, while the actual phase can vary continuously to values outside this interval. The phase retrieval problem occurs in inverse scattering, optical, and x-ray applications. The

W.-M. Boerner et al. (eds.), Inverse Methods in Electromagnetic Imaging, Part 2, 1285–1287.

phase unwrapping problem is of importance in signal processing applications, ultrasound, and geophysical diffraction tomography.

Problem Areas

This group identified a number of specific problem areas involving phase retrieval/phase unwrapping among which we include:

a) Only in very special cases it is possible to deduce the phase of a function from its modulus. As a rule the problem does not have a unique solution. It should be investigated whether prior knowledge, such as positivity can, enforce uniqueness. The mathematical mechanism behind Fienup's algorithm needs clarification, as it turns out sometimes the algorithm fails to produce the correct solution, even in those cases where the solution is known to be unique. The same remark applies to the Gerchberg-Saxton and Misell algorithm.

b) The role of higher order correlation functions in the phase retrieval problem.

c) As rapid zero finding algorithms are available, it seems possible to obtain the (complex) zeroes of the intensity distributions. All results following from zero-flipping could be checked against prior knowledge, thus reducing the ambiguity.

d) Even in cases where the phase retrieval problem is known to admit a unique solution algorithm for actually performing the retrieval are lacking.

e) Is it clear that the phase retrieval problem correctly solves the phase unwrapping problem? If this were not the case, could the minimum phase (obtained via Hilbert transformation from the modulus) be a suitable starting point for phase unwrapping?

The effect of partial coherence in the phase retrieval problem

The effect of noise of the results of phase retrieval should be systematically studied, e.g., via maximum-entropy or other approaches. This is particularly imperative for images obtained under low-dose illumination.

An interesting question directed to the group could not be openly discussed in the time available, namely "Is it possible to deduce position and extent of a VOID in an otherwise rather homogeneous material from the intensity of backscattered radiation alone?". This research is relevant to a study being conducted at the department of dentistry at the University of Groningen, where incipient caries are detected by optical methods. (H.A.F.)

Recommendations and Conclusions

As a general recommendation, the group emphasized unifying physical models with design consideration and signal processing algorithms. This unification is important, for example, in monostatic radar signal processing. By making use of the high frequency (physical optics) model for the monostatic radar return from a conducting object, it may be possible to deduce the phase of the received signal from its magnitude-a result of significant potential value to radar signal processing.

A general conclusion of the group was that progress on the phase retrieval problem would benefit from cross fertilization between the optics, electrical engineering, mathematical and crystallography communities. An example of this is the conclusion of the group that certain well accepted notions concerning the multidimensional phase retrieval problem may be incorrect. This conclusion follows from work done in the mathematical communities as was pointed out by Dr. F. Alberto Grünbaum - a mathematician.

The group recommended that its report should contain a selected compilation of references on phase retrieval and phase unwrapping problems which could serve as pivots for futher development.

Key References to Phase Retrieval/Holography

W.O. Saxton, book in the series Electron and Electro-Physics, L. Marton (Ed.), Ac Press. ca. 1980.

H.P. Baltes, (Ed.), Inverse Source Problems in Optics, Chap. 2, by H.A. Ferwerda, Springer 1978.

H.P. Baltes (Ed.), Inverse Scattering Problems in Optics, Chap. 2, by G. Ross, M.A. Fiddy, M. Nieto-Vesperinas, Springer, 1980.

Taylor, IEEE, Special Issue, Trans. A & P- 29(2).

References for 2-Dimensional Phase Problems:

Y. Bruck and L.G. Sodin, Opt. Comm. (ca 1980).

Fienup, Papers by, in Optics Letters, Optical Engineering (ca 1980).

A.M.J. Huiser and P. van Toorn, Optics Letters (ca 1980).

H.A. Ferwerda in "Optics in 4 dimensions", Proc. of AIP-meeting in Ensenada, August 1980. M. Marchado, L. Nardlucci (Eds.).

VI.6 FINAL REPORT of WORKING DISCUSSION GROUP W-F on ENHANCEMENT OF INTERACTION BETWEEN ACTIVE R & D LABS WITHIN NATO-MEMBER COUNTRIES ON THE SUBJECT OF ELECTROMAGNETIC IMAGING

Abstract: The present ARW-IMEI has successfully brought together key experts and observers from a majority of NATO-member countries allowing them to judge the current level of interaction. In this report we attempt to summarize the existing interaction paths, Substantiate why enhancement of interaction is desireable and to suggest mechanisms for achieving these goals.

Coordinator: (Group Leader) Charles J. Holland

Moderators: Martin Vogel
Leonard A. Cram

Advisors: Jerry L. Eaves
Karl Steinbach

Reporters: Roderick M. Logan
Tilo Kester

Discussion Group Activists:

(* The names in this column weren't present at the ARW, however, their contributions were considered during the discussions.)

1. Boerner, W-M.	8. Logan, R.M.	*1. Baars, E.P.
2. Brand, H.	9. Pfeiler, M.	2. Barber, B.C.
3. Cram. L.A.	10. Kester, T.	3. Holberg, K.
4. Eaves, J.L.	11. Schwierz, G.	4. Johnson, F.A.
5. Gjessing, D.T.	12. Steinbach, K.	5. Sinclair, C.
6. Holland, C.	13. Vogel, M.	6. Tischer, S.
7. Keydel, W.	14. Winkler, G.	7. Voles, R.

1. Introduction

The field of study under consideration is a very broad one, covering a rather wide spectrum of mathematical methods and signal processing techniques; not to mention the large number of diverse applications. An indication of the scope can be deduced from Table 1, presenting an overview on societies, journals, and user organizations. The topics covered by the papers presented at this NATO-ARW-IMEI-1983 give a representative sampling of this new field

W.-M. Boerner et al. (eds.), Inverse Methods in Electromagnetic Imaging, Part 2, 1289–1296.

covering all the main techniques and many, although not all, applications areas. From the breadth of the field, it follows that it is a multi-disciplinary one, in which mathematicians, physicists, engineering and computer scientists, electronic engineers and metrologists, all have a major role. Inverse processing crosses many boundaries between disciplines. To gather the participants at this workshop, the net has been cast wide both geographically and technically, allowing us to judge the current level of interaction. In this report we attempt to summarize the existing interaction paths, to indicate why enhancement is desireable in some areas and to suggest mechanisms for achieving this.

2. Existing Interaction

There are many existing paths for interaction between workers on this wide topic. Most of these are in the form of established professional societies in the NATO-member nations. These societies provide an opportunity for various sub-groups of all those involved in the total field to come together and exchange views on problems and progress. In general, these societies and meetings do not provide the ideal mechanism for interaction across national boundaries or across disciplinary boundaries. Many relevant papers are published in professional journals; but these do not provide an adequate mechanism for interaction, especially across disciplinary boundaries. These short-comings provide one of the main motivations behind setting up the present ARW IMEI. The main existing interaction paths are contained in table 1.

In addition to the items listed in table 1, the subject of inverse methods is increasingly being discussed at international conferences on specific sensor areas e.g., SPIE, RADAR 82, etc. Particularly relevant is the annual conference on Inverse Methods "R.C.P. 264, rencontre interdisciplinaire problèmes inverses", organized by Prof. Pierre Sabatier, at U.S.T.L., Place E. Bataillon, Montpellier, France, which was held on 29-30 November, 1-4 December 1983.

It would be impractical to list here all the laboratories (Government and Industry) and universities actively working in this field within NATO-member nations, but a good guide to the main areas of activity is seen by referring to the affiliations of the authors of the contributed papers in these Proceedings.

3. Need for Enhanced Interaction

Progress in inverse methods for electromagnetic imaging will lead to improvements in the performance of electromagnetic sensors for a wide range of military and civil applications. Other sensor areas, such as ultrasound, which can use similar processing techniques would also benefit. It follows that all NATO member nations would see benifits from progress in research on inverse methods and this in itself could be taken as a justification for seeking

TABLE 1. OVERVIEW OF EXISTING MECHANISMS FOR INTERACTION

Professional Societies

- EARSCL	(EU)
- Geoscience and Remote Sensing/IEEE	(US)
- Ges. f. Photogrammetrie u. Forschung	(GE)
- IEE	(UK)
- IEEE (Societies: APS, AES, MTT, RS)	(US)
- International Glaciological Society	(UK)
- Photogr. and Remote Sensing Soc.	(UK)
- Remote Sensing Society	(UK)
- Union of Radio Science....URSI	(UN)

User Organizations

- Council of Europe
- ESA
- European Community
- NATO
  -- DRG Panels
  -- AGARD Panels
- United Nations Commission

Journals

- IEE Journals
- IEEE Journals
- Journal of Optical Society of America
- International Journal of Remote Sensing
- Medical
- Geology
- Remote Sensing of the Environment
- Journal of Glaciology
- Geophysical Prospecting

---

enhanced interaction. In civil areas such as medical tomography and environmental monitoring inverse methods offer the prospect of really dramatic improvements in capability. In the military area there is as urgent need to improve the performance of electromagnetic sensors such as radar to meet the increasingly sophisticated threat. Some of the techniques considered at this meeting offer the best hope for doing this without a major escallation in costs.

For this report we have selected 3 specific examples of the need for enhanced research interaction, as given in the paragraph below. These reflect the views emerging from the other Working Group reports (Sections VI.1 to VI.5).

3.1 Data Bases and Standards

In several areas there was felt to be a need for improved experimental data bases to allow theories to be validated and system concepts to be evaluated, and for measurement standards so that measurements from one laboratory can be readily compared with those from another. The need is particularly acute in 2 areas:

a) Polarimetry; Work so far in this area based on elegant theoretical analysis and limited experimental measurements shows exciting possibilities in the area of radar target discrimination and identification. However, there is a lack of validated experimental data on targets and backgrounds and there is a need for standards for data features, gathering equipment and terms. Such a data base and standards would best be achieved by close interaction between the interested nations.

b) Synthetic Aperture Radar (SAR); It is seen from Fig. 1 that SAR techniques are being applied in many areas, and several working systems are in existence. However, there is felt to be a need for a flexible system of standards and a facility for ground truth measurements to allow image quality to be assessed and compared. The system of standards needs to be agreed between all the interested NATO nations and the ground truth facility should ideally be accessible for use by all NATO nations.

3.2 Inter-Disciplinary Interaction

By its nature the problems of inverse methods require a heavy emphasis on mathematics. There is also a need for theoretical physics to deal with some of the difficult electromagnetic scattering calculations. It became clear during the workshop that there is a gap to be bridged between the mathematical and theoretical physics work on the one hand and the direct applications work on the other. Two examples of this would be:

a) The need for mathematical working on inverse problems to interact with the applications people, especially those involved in the signal processing and instrument design.

b) The need for those involved in (radar) detection and estimation, which is normally based on simple models and a statistical approach, to interact with the physicists and mathematiicians studing the deterministic aspects of polarization vector wave propagation and scattering.

At the workshop there was evidence of useful interaction, particularly on item (a) above, but we need to consider how this interaction can be continued and enhanced in the future.

Figure 1. Summary of Techniques and Applications

Techniques * / Applications	Synthetic Aperture	Super resolution	Inverse SAR	Polarimetry	Holographic	Tomographic	Transient
Medical			-			-	
Non-destructive testing	-	-		=	o	-	o
Surveillance (radar, etc.)	+	o	+	o			o
Target acquisition	+		o	=			-
Target classification	+	o	+	=			
Target identification	+	-	+	=		-	o
Buried object detection			=	-	-	=	o
Tunnel detection			=	-	-	=	o
Counter surveillance		=	=	o			
Robotics							
Meteorology	o			o			
Geology	+			o	o		o
Agriculture	+						
Environmental	+	o					
Aviation	o			=			
Metrology				o			

+ Large activity
o Medium activity
= Strong need for enhancement
- Some need for enhancement

* Some of the mathematical inversion techniques are so general that they would fit almost every area considered, and are not listed.

3.3 New Application Areas

Interaction between research workers is especially important for new application areas, that is to say, those areas, which are not yet sufficiently well established to readily attract research funding. An area such as x-ray medical tomography is already well established, although there are some problems remaining, such as x-ray safety aspects, and probably does not need this special attention to interaction. But, it was apparent at the present ARW that the related but more general area of non-destructive testing is at an early stage and needs closer interaction. This is shown in Fig. 1, which reflects the views of those attending the present ARW.

4. Recommendations for Enhanced Interaction

4.1 Data Bases and Standards for Polarimetry

This activity will need lengthy discussions and planning followed by joint, or at least, jointly planned, trials. Since the fore-seen application areas are predominately military, we suggest that the DRG should set up an RSG (or re-direct an existing RSG) to address this topic. One of the co-directors of this present ARW would be appropriate to coordinate this RSG.

4.2 Standards for SAR

We recommend that a small Exploration Group should be set up to consider this problem further with a brief to report in about 6-12 months with a plan for setting up the necessary facility in the most efficient way. The needs of all potential application areas for airborne/spaceborne SAR should be taken into account. Dr. S.R. Brooks of Marconi Research Center would be a suitable person to organize this group since he has been involved in this topic and has contacts with the European Space Agency.

4.3 Interdisciplinary Interaction

One way of enhancing this is through gatherings such as the present ARW. Consideration should be given to setting up a future ARW or Interdisciplinary Studies Institute which is specifically designed to enhance interaction (e.g. the mathematicians would be asked, in advance, to prepare papers aimed at applications specialists and the applications specialists would be asked to present papers on areas where they need mathematical progress). We recommend that the Director of the present ARW should consider options for future interdisciplinary meetings and report to the NATO Science Committee within 12 months.

4.4 New Application Areas

Of the possible new areas the subject of non-destructive testing has been selected as being particularly appropriate for increased interaction. We recommend, therefore, that an ARW should be set up on this topic in order to examine the need for enhanced inter-

TABLE 2. SUMMARY OF MAIN RECOMMENDATIONS

FORUM	TOPIC	FORMAT	TIMING	GEOGRAPHICAL REGION	ACTORS	W-...
NATO Science Committee	Mathematical Inversion Methods	ARWs (E.M. Kennaugh) (Memmorial W.S.)	annual December annual Spring	Montpellier France Columbus U.S.A.	Prof. P.C. Sabatier U.S.T.L., Montp. E.S.L.-O.S.U. Columbus OH	A
NATO Science Committee	signal processing, including phase	ARW	end 1985	Glasgow	Prof. T.S. Durray Dept. of Electrical Univer. of Strathclyde Glasgow, U.K.	E
NATO Science Committee	Optical Applications of Inverse Scattering	ASI	1986	Italy	Dr. A.K. Jordan Naval Research Lab. Washington, D.C., USA	A E
NATO Science Committee	Numerical Inversion Methods (appl maths)*	Exploratory Group (Continuation of Workshop Group B)	present to end of 1985	Europe (probably)	Members of Working Group B, To chair group initially, Dr. R.H. Dittel, of DFVLR, Oberpfaffenhofen, FRG	B
NATO Science Committee	Non-destructive testing	ARW	As soon as possible	Ames, IO U.S.A.	Dr. Jim Corones Ames R.C., Iowa St. Prof. K. Langenberg Univ. of Kassel Kassel, FR Germany	E B A
NATO Science Committee	Basic Theory of High Resol-ution Radar Polarimetry	ARW/ ASI	Spring 1985	Chicago U.S.A.	Prof. W-M. Boerner Comm. Lab, EECS, U.I.C., Chicago	C
NATO DRG	Polarimetric measurements (radar) on targets and backgrounds	RSG	1984 - 1986	Member nations as appropriate (US and Europe)	Representatives would be nominated by member Governments from the following nations: U.S.A. (Lead nation), Holland/Germany/Norway UK, & others if appr.	C F
NATO Science Committee (+ ESA)	Calibration of SAR Imagery	Exploratory Group	1984 - 1986	Member nations as appropriate (USA and Europe)	Dr. S.R. Brooks, Marconi Research Centre, U.K., to organize	D

action and to make for future meetings as appropiate; Dr. Jim Corones, Ames Research Lab, Iowa State University, Ames, Iowa and Prof. Karl Langenberg, Universität Kassel, Kassel, FR Germany would be suitable experts to act as Directors for the workshop.

VI.7 OVERALL FINAL REPORT ON NATO-ARW-IMEI-1983

Wolfgang-M. Boerner / Hans Brand / Leonard A. Cram / Dag T. Gjessing / Wolfgang Keydel / Günter Schwierz / Martin Vogel

Concluding Remarks: (by Leonard Cram)
This was a very stimulating and successful week of work. The field of imaging by inverse processing is important and fruitful; its rapid development gained impetus from the activities of the workshop. The foresight shown by Wolfgang-M. Boerner in conceiving and by NATO in supporting this Advanced Research Workshop was shown to be fully justified.

Inverse processing crosses many boundaries between disciplines. To gather the participants at this workshop, the net has been cast wide, both geographically and technically. The recognition of the wide interdisciplinary nature of the subject is reminiscent of the growth in the 1940's and 1950's of such unified studies as information theory and cybernetics. It may be that this meeting will prove to have been as historic and epoch-making as were the inceptions of those earlier concepts. In any case, it is already clear that the application in the fields of radar imaging which occupied such a large part of the deliberations are unlikely to be long in bearing fruit.

The success of the workshop reflects credit on all of the participants and also on the workers. The hive had no drones. Everyone worked with enthusiasm, industry and co-operaton.

The high quality of the conference side of the proceedings was insured by the authors with their excellent and very pertinent papers and by the clarity and liveliness of their presentations. The close control provided by the chairmen of each session permitted time for immediate discussion of the many, good and positive questions which arose from the papers. The confernece papers and discussions also stimulated many of the participants to formulate written questions for investigation by working discussion groups.

In these working groups everyone contributed in a lively and co-operative spirit with the result that useful study reports were produced. Credit is particularly due to the officers in the six working groups, i.e., to each of the co-ordinators, moderators, advisors, and reporters. The authors of all the workshop

W.-M. Boerner et al. (eds.), Inverse Methods in Electromagnetic Imaging, Part 2, 1297–1298.

reports performed particularly important functions.

The Co-Directors, of course, played a very significant part in achieving the smooth running of the whole proceedings, not only in their activities during the proceedings, but also in their earlier planning meetings when the form of the programme was determined. Particular thanks are due to those Co-Directors from the local region who planned and organized the external expeditions which were of both technical and cultural interest.

The greatest credit for the whole stimulating and productive event is due to the Director, Wolfgang-Martin Boerner. He conceived the idea and pursued NATO to persuade them of its importance. He identified most of the participants as experts in the field and persuaded them to attend. He picked the team. He, primarily, was responsible for planning the entertainment - most of which was also educational. He impressed everyone throughout the week with his boundless mental and physical energy. Everyone appreciated the efficient way in which he organized and used their energies from dawn, with the 6:00 a.m. alarm call, to near midnight throughout the whole week; and always, he worked harder and longer than anyone else.

The event would have been a huge success just as a conference. As a workshop, it was a splendid interdisciplinary experience; it gave all participants an introduction, in a working environment, to many previously unknown colleagues engaged in the same field. Most importantly, this Advanced Research Workshop has provided a report which will accelerate the inevitable development of inverse processing in electromagnetic imaging.

FINAL

TECHNICAL PROGRAM OUTLINE

NATO-ADVANCED RESEARCH WORKSHOP
on
INVERSE METHODS IN ELECTROMAGNETIC IMAGING

Kur- und Kongresshotel Residenz
Bad Windsheim, FRG, September 18-24, 1983

in memory of

Edward Morton Kennaugh (1922 to 1983)

- an exact account of its proceedings -

SESSION OVERVIEW

Sunday, Sept. 18, 1983:		
20:00 to 21:30,	OS:	Opening Session
Monday, Sept. 19, 1983:		
08:00 to 10:00,	MI:	Microwave Imaging
10:30 to 12:30,	RS:	Radiometry/ Rough Surface Scattering
13:45 to 15:25,	NM:	Numerical Methods
Tuesday, Sept. 20, 1983:		
08:00 to 10:00,	RP:	Radar Polarimetry
10:30 to 12:30,	MM:	M-to-MM-Wave Imaging
13:45 to 15:25,	IS:	Inverse Scattering
Wednesday, Sept. 21, 1983:		
08:00 to 10:00,	TS:	Transient Analysis
10:30 to 12:30,	PT:	Projection & Propagation Tomography
13:00 to 21:30,	SCT:	Scientific/Cultural Tour
Thursday, Sept, 22, 1983:		
08:00 to 10:00,	SI:	Surveillance Imagery
10:30 to 12:30,	OI:	Optical Imaging
13:45 to 15:25,	IM:	Inverse Methods
19:30 to 21:30,	SS:	Special Session
Friday, Sept. 23, 1983:		
08:00 to 10:00,	SR:	Scatterometry & Remote Sensing
10:30 to 12:30,	SP:	Special Processing
16:00 to 17:30,	FS:	Final Session
19:00 to 21:30,	WB:	Workshop Banquet
Saturday, Sept. 24, 1983:		
08:30 to 21:30,	CST:	Cultural/Scientific Tour

NOTE: All Lectures are presented in English, and take place in the main Lecture Center, and the Banquet in the Residenz Saal. Session presentations are cross-referenced with topic and paper number found in the PROCEEDING'S TABLE OF CONTENTS, i.e., **O.2.**

SUNDAY, 18 September 1983

10:00 to 12:00 - **WORKSHOP DIRECTORS' MEETING** (Room E)
Director: WOLFGANG-M. BOERNER, Univ. of IL at Chicago, U.S.A.
Co-directors: HANS BRAND, Univ. Erlangen-Nürnberg, FR Germany
LEONARD A. CRAM, THORN EMI, Wells/Somerset, England
DAG T. GJESSING, RNCISR, ESTP, Kjeller, Norway
GÜNTER SCHWIERZ, RGG-Siemens-Med., Erlangen, FRG
MARTIN VOGEL, DFVLR, Oberpfaffenhofen, FRG

12:00 to 14:00 - **LUNCH BREAK**

14:30 - **DEPARTURE OF WORKSHOP TOUR BUS "ARW-IMEI"** from Nuremberg Airport (outside arrival gate)

15:00 - **DEPARTURE OF WORKSHOP TOUR BUS "ARW-IMEI"** from Nürnberg Hauptbahnhof (main exit towards city: Bahnhofsplatz)

14:00 to 20:00 - **REGISTRATION AND RECEPTION** (Foyer and Room E)
FRAU. MARGARETE GEIGER, Univ. Erlangen-Nürnberg
DIPL. ING. SIEGFRIED OSTERRIEDER, Univ. Erlangen-Nürnberg
PROF. WOLFGANG-M. BOERNER, Univ. of IL at Chicago
DR. MARTIN VOGEL, DFVLR, Oberpfaffenhofen, FRG

17:45 to 19:00 - **BUFFET**
MR. ROLF K. ERLENBACH, Manager, KuK Residenz Hotel
MR. HANS-DIETER ERB, Hotel Coordinator for NATO-ARW

SESSION OS (Lecture Center)

20:00 to 21:30 - **OPENING SESSION OF WORKSHOP**
Chairman: DR. DAG T. GJESSING
Environmental Surveillance Tech. Programme, Royal Norweigen Council for Industrial & Scientific Research, Kjeller, NORWAY

20:00 - OS-1: **OPENING REMARKS**
DR. MARTIN VOGEL
DFVLR, Oberpfaffenhofen, FR Germany

20:20 - OS-2: **WELCOME NOTES**
PROF. HANS BRAND: HOST INSTITUTE
Universtät Erlangen-Nürnberg, FR Germany

DR. WOLFGANG KEYDEL: HOST NATION
DFVLR, Oberpfaffenhofen, FR Germany

DR. MANFRED PFEILER: SIEMENS SPONSORSHIP
Siemens AG, Erlangen, FR Germany

20:30 - OS-3:* **DEDICATION OF WORKSHOP:**
Special Welcome for Mrs. Mary Kennaugh
THE ORGANIZING & DIRECTORS COMMITTEE

(* Substantiated in the Proceedings PREFACE, DEDICATION, and MEMORIAL PAPER - **0.1**)

20:45 - OS-4:* **TECHNICAL PROGRAM DESCRIPTION**
PROF. WOLFGANG-M. BOERNER
Univ. of Illinois at Chicago, U.S.A.

(* Expanded in the Proceedings PREFACE and its LEAD PAPER - **0.2**)

21:30 - OS-5: **GREETINGS AND WISHES FOR A SUCCESSFUL WORKSHOP**
MR. ROLF K. ERLENBACH, Manager
MR. HANS-DIETER ERB, Assistant Manager
KuK Hotel Residenz, Bad Windsheim, FR Germany

22:00 - **DAY'S END**
"Hotel at Rest" (See Information on KuK Residenz Hotel Work shop Arrangements: NOISE)

MONDAY, 19 September 1983

7:00 to 8:00 - **BREAKFAST**

SESSION MI (Lecture Center)

8:00 to 10:00 - **MICROWAVE IMAGING**
Chairman: DR. KARL VON SCHLACHTA
FFM/FGAN, Wachtberg-Werthhoven, FRG

8:00 - MI-1: Fast Millimeter Wave Imaging
PROF. HANS BRAND
IV.11 Inst. für HF-Technik, Univ. Erlangen-Nürnberg, FRG

8:30 - MI-2: Holographic and Tomographic Imaging with Microwaves and Ultrasound
V.4 PROF. ALAN P. ANDERSON
Univ. of Sheffield, U.K.
co-author: DR. M.F. ADAMS
Formerly with the Techn. Univ. München, FR Germany
Now with Eikontech, Ltd., FR Germany

9:00 - MI-3: A 4-Channel Millimeter-wave-on-line Imaging Method
PROF. JÜRGEN DETLEFSEN
IV.14 Techn. Univ. München, FR Germany

9:20 - MI-4: Phase Comparison Monopulse Side Scan Radar
PROF. PETER T. GOUGH
V.10 Univ. of Canterbury, Christchurch, New Zealand

9:40 - MI-5: Electromagnetic Imaging of Dielectric Targets
PROF. HEINZ CHALOUPKA
IV.13 Ruhr-Universität Bochum, FR Germany

10:00 to 10:30 - **COFFEE BREAK**

SESSION RS (Lecture Center)

10:30 to 12:30 - **RADIOMETRY/ROUGH SURFACE SCATTERING**
Chairman: DR. COLIN D. SILLENCE
THORN EMI, Electronics, Ltd., Wells, Somerset, U.K.

10:30 - RS-1: The Radiative Transfer Approach in Electromagnetic Imaging
IV.1 PROF. AKIRA ISHIMARU
Univ. of Washington, Seattle, WA, USA

11:00 - RS-2: Inverse Methods in Rough Surface Scattering
PROF. ADRIAN K. FUNG
IV.7 Univ. of Kansas, Lawrence, KS, USA

11:30 - RS-3: Inverse Methods for Ocean Wave Imaging by SAR
IV.6 DR. STEPHEN ROTHERAM
co-author: DR. J. T. MACKLIN
Marconi Research Center, Chelmsford, Essex, U.K.

11:50 - RS-4: Polarization-Dependence in Microwave Radiometry
DR. KONRAD GRÜNER
III.4 DFVLR, Oberpfaffenhofen, FR Germany

12:10 - RS-5: Deconvolution of Microwave Radiometry Data
DR. RUDOLF H. DITTEL
II.14 DFVLR, Oberpfaffenhofen, FR Germany

12:30 to 13:45 - **LUNCH BREAK:** Mittagstisch

SESSION NM (Lecture Center)

13:45 to 15:25 - **NUMERICAL METHODS**
Chairman: DR. RODERICK M. LOGAN
Admiralty Surface Weapons Establishment,

13:45 - NM-1: Detailed Near/Far Field Modeling of Complex, Electrically Large Three-Dimensional Strcutures
II.10 DR. ALLEN TAFLOVE
co-author: DR. KORODA UMASHANKAR
IIT Research Institute, Chicago, IL., U.S.A.

14:05 - NM-2: Stability and Resolution in Electromagnetic Inverse Scattering: Part-1
II.4 PROF. MARIO BERTERO
Univ. di Genova, Italy
co-authors: PROF. CHRISTINE DEMOL
Univ. Libre de Bruxelles, Belgium
DR. EDWARD R. PIKE
Royal Signals & Radar Establishment, Malvern, U.K.

14:25 - NM-3:* Stability and Resolution in Electromagnetic Inverse Scattering: Part-2
PROF. CHRISTINE DeMOL
II.5 Univ. Libre de Bruxelles, Belgium
co-authors: DR. EDWARD R. PIKE
Royal Signals & Radar Establishment, Malvern, U.K.
PROF. MARIO BERTERO
Univ. di Genova, Italy

(* Replacement for paper by PROF. M. ZUHAIR NASHED: Numerical Stability in Electromagnetic Scattering: **XX**)

14:45 - NM-4: Approximation Schema for Generalized Inverses: Spectral Results & Regularization of Ill-Posed Problems
II.8 DR. VIRGINIA V. JORY
Georgia Institute of Technology, Atlanta, GA., USA

15:05 - NM-5: Approximation of Impulse Response from Time Limited Output & Input: Theory & Experiment
II.7 PROF. TAPAN SARKAR
co-authors: DR. SOHEIL A. DIANAT
Rochester Institute of Tech., Rochester, NY, USA
DR. BRUCE Z. HOLLMANN
Naval Surface Weapons Center, Dahlgren, VA., U.S.A.

15:25 to 16:00 - **TEA BREAK**

MEETING W-1 (Lecture Center, Rooms A to E)

16:00 to 17:45 - **WORKING DISCUSSION GROUP MEETING**
Chairman: DR. ERNST LÜNEBURG
DFVLR, Obperfaffenhofen

16:00 - Organization of Working Discussion Groups
MR. LEONARD A. CRAM
THORN EMI, Wells, Somerset, U.K.

16:10 - Intent, Purpose, Aim of Working Discussion Group Meetings
DR. DAG T. GJESSING
RNCISR, ESTP. Kjeller, NORWAY
MR. TERJE LUND
RNCISR, ESTP, Kjeller, NORWAY

16:20 - Guidelines for Preparation of Working Discussion Group Reports
DR. TILO KESTER
NATO Publications Coordination Office, Overijse/belgium

16:30 - Separation into Six Specific Working Discussion Groups (Detailed instructions are outlined in the "Information on Working Group Discussion Programs")

W-A: Mathematical Inverse Methods and Transient Techniques
Coordinator: Dr. Arthur K. Jordan

W-B: Numerical Inversion Methods
Coordinator: Dr. Rudolf H. Dittel

W-C: Polarization Utilization in the Electromagnetic Vector Inverse Problem
Coordinator: Dr. James W. Wright

W-D: Image Quality and Image Resolution in Remote Sensing and Surveillance
Coordinator: Prof. Akira Ishimaru

W-E: Holographic and Tomographic Imaging and Related Phase Problems
Coordinator: Prof. Hedzer A. Ferwerda

W-F: Enhancement of Interactions Between Active Laboratories within Nato-member Countries on the Subject of Electromagnetic Imaging Research
Coordinator: Dr. Charles J. Holland

17:45 to 18:30 - **Supper**

18:40 to 20:00 - **Visit to Downtown Bad Windsheim**
18:40 - Departure with Guide from Kurhotel Residenz

20:10 - **Reception by First Mayor at the Historical City Hall Chamber**

22:00 - **DAY'S END:** "Hotel at Rest" (NOISE)

TUESDAY, 20 September 1983

7:00 to 8:00 - **BREAKFAST**

SESSION RP (Lecture Center)

8:00 to 10:00 - **RADAR POLARIMETRY**
Chairman: DR. JAMES W. WRIGHT
New Technology, Inc., Huntsville, AL., U.S.A.

8:00 - RP-1:* Contributions of Professors Edward M. Kennaugh and Georges A. Deschamps to Polarimetric Radar Theory **XX**
PROF. WOLFGANG-M. BOERNER
University of Illinois at Chicago, Chicago, USA

(* Replacement for paper of PROF. GEORGES A. DESCHAMPS: Polarization Descriptors-Utilization of Symmetries: **III.1**)

8:20 - RP-2: Extension of Kennaugh's Optimal Polarization Concept to the Asymmetric Matrix Case
III.6 MR. MARAT DAVIDOVITZ
co-author: PROF. WOLFGANG-M. BOERNER
Univ. of Illinois at Chicago, Illinois, U.S.A.

8:40 - RP-3: On the Null Representation of Radar Parameters
MR. ROBERT S. RAVEN
III.7 Westinghouse Defense & Space Center, Baltimore, MD, U.S.A.

9:00 - RP-4: Towards a Theory of Perception for Radar Targets With Application to the Analysis of Their Data Base
IV.2 Structures and Presentations
DR. J. RICHARD HUYNEN
Quest Research, Los Altos Hills, CA, U.S.A.

9:30 - RP-5: Polarization Information Utilization in Primary Radar: An Introduction and Update to Activities at
III.2 SHAPE Technical Centre
MR. ANDRE J. POELMAN
co-author: DR. JOHN R.F. GUY
SHAPE Technical Centre, The Hague, Netherlands

10:00 to 10:30 - **COFFEE BREAK**

SESSION MM (Lecture Center)

10:30 to 12:30 - **M-to-MM-WAVE METROLOGY**
Chairman: DR.-ING. KARL STEINBACH
U.S. Army Mobility Command, MERADCOM, Ft. Belvoir, VA., U.S.A.

10:30 - MM-1: Inverse Methods in Microwave Metrology
PROF. WALTER K. KAHN
V.3 George Washington Univ., Washington, D.C., U.S.A.

11:00 - MM-2: Polarization Measurements in Radioastronomy
PROF. MANFRED F. A. THIEL
III.3 Fachhochschule Köln, FR Germany

11:20 - MM-3:* Theory and Design of a Dual Polarization Radar for Clutter Analysis
III.8 MR. JERALD D. NESPOR
co-authors: MR. AMIT P. AGRAWAL
PROF. WOLFGANG-M. BOERNER

(* Repacement for paper of MR. LLOYD W. ROOT: Calibration of Polarimetric RCS Measurement Systems: **XX**)

11:40 - MM-4: Diagnostic Measurements and Analysis of Wave Mechanisms in Radomes
V.11 DR. CONSTANTINE TRICOLES
co-authors: DR. E.L. ROPE
DR. R.A. HAYWARD
General Dynamics, San Diego, CA., U.S.A.

12:00 - MM-5: Demands on Polarization Purity in the Measurement & Imaging of Distributed Clutter
III.13 PROF. ANDREW J. BLANCHARD
co-author: RICHARD W. NEWTON
Univ. of Texas at Arlington, Arlington, TX., U.S.A.

12:30 to 13:45 - **LUNCH BREAK:** Mittagstisch

SESSION IS (Lecture Center)

13:45 to 15:25 - **INVERSE SCATTERING**
Chairman: PROF. MICHAEL A. FIDDY
Queen Elizabeth College, London, U.K.

13:45 - IS-1: Recent Advances in the Theory of Inverse Scattering with Sparse Data
II.1 PROF. VAUGHAN H. WESTON
Purdue University, W. Lafayette, IN., U.S.A.

14:05 - IS-2: Inverse Electromagnetic Scattering for Radially Inhomogeneous Dielectric Spheres
I.9 DR. CORNEL EFTIMIU
McDonnell Douglas Res. Center, St. Louis, MO, USA

14:25 - IS-3: A Solution of the Time-Dependent Inverse Source Problem for Three-Dimensional EM Wave Propagation
I.3 PROF. HARRY E. MOSES
Univ. of Lowell, Lowell, MA., U.S.A.

14:45 - IS-4: Theoretical Study and Numerical Resolution of Inverse Problem Via the Functional Derivatives
I.5 PROF. ANDRE J. ROGER
Univ. of Marseille, France

15:05 - IS-5: Dissipative Inverse Problems in the Time Domain
I.6 J. P. CORONES
co-authors: M. E. DAVISON
Iowa State University, Ames, IA., U.S.A.
PROF. ROBERT J. KRUEGER
Univ. of Nebraska, Lincoln, NE, U.S.A.

15:25 to 16:00 - **TEA BREAK**

MEETING W-2 (Lecture Center, Rooms A to E)

16:00 to 17:30 - **WORKING DISCUSSION GROUP MEETINGS**
Various Groups Meet Separately. (Detailed Instructions provided in "Information on Working Group Discussion Programs. . .")
Moderators of Discussions:
Chairman: DR. RONALD A. EVANS
British Aerospace Corp., Warton/Lancashire, UK

Moderators: W-A, PROF. KARL LANGENBERG and PROF. ANDRE J. ROGER
W-B, PROF. TAPAN SARKAR and PROF. MARIO BERTERO
W-C, DR. J. RICHARD HUYNEN and DR. OTTO KESSLER
W-D, MR. RICHARD W. LARSON and MR. TERJE LUND
W-E, F. ALBERTO GRÜNBAUM and DR. ALAN P. ANDERSON
W-F, DR. MARTIN VOGEL and MR. LEONARD A. CRAM

17:30 to 21:15 - **VISIT TO THE FRANCONIAN OPEN-AIR MUSEUM** (REGISTRATION FOR PARTICIPATION REQUESTED)
17:30 - Departure by bus from Kurhotel Residenz

17:45 to 20:00 - Viewing with Guide of the Museum

20:00 to 21:00 - Karpfenessen im Museumsrestaurant with musical presentations by the "Bad Windsheimer Sänger".

18:00 to 19:00 - **SUPPER** (For those not attending the cultural program)

20:00 to 21:30 - **DISCUSSIONS**
Scientific Discussions on Topics of Mutual Interest

22:00 - **DAY'S END**
"Hotel at Rest" (NOISE)

WEDNESDAY, 21 September 1983

7:00 to 8:00 - **BREAKFAST**

SESSION TS (Lecture Center)

8:00 to 10:00 - **TRANSIENT ANALYSIS IN INVERSE SCATTERING**
Chairman: PROF. JOHN O. THOMAS
Imperial College of Sci., Blackett Lab, London, UK

8:00 - TS-1: Critical Analysis of the Mathematical Methods used in Electromagnetic Inverse Theories: A Quest for New Routes in the Space of Parameters
I.1
PROF. PIERRE C. SABATIER
Univ. des Sciences et Techniques du Languedoc, Montpellier, France

8:40 - TS-2: Transient Methods in Electromagnetic Imaging
I.4 PROF. KARL LANGENBERG
co-authors: MARTIN FISCHER
D. BRÜCK
G. BOLLIG
Universität Kassel, FR Germany

9:00 - TS-3: One-Dimensional Time Domain Inverse Scattering Applicable to Electromagnetic Imaging
I.8 PROF. HANS BLOK
co-author: DR. ANTON G. TIJHUIS
Delft Technical University, Netherlands

9:20 - TS-4: Time-Dependent Radar Target Signatures Synthesis and Detection of Electromagnetic Authenticity Features
II.12 PROF. DAVID L. MOFFATT
co-author: DR. TED C. LEE
The Ohio State University, Columbus, OH., U.S.A.

9:40 - TS-5: Applications of Time-Domain Inverse Scattering to Electromagnetic Waves
II.11 PROF. WALID TABBARA
Lab. des Signaux & Systems, CNRS-ESE;GIF-SUR-YVETTE, France

10:00 to 10:30 - **COFFEE BREAK**

SESSION PT (Lecture Center)

10:30 to 12:30 - **PROJECTION AND PROPAGATION TOMOGRAPHY**
Chairman: DR. MANFRED PFEILER
Siemens AG, Erlangen, FR Germany

10:30 - PT-1: Review of Tomographic Imaging Methods Applied to Medical Imaging
V.2 DR. GÜNTER SCHWIERZ
RGG-Siemens-Med., FR Germany

11:10 - PT-2: The Limited Angle Reconstruction Problem in Tomography
PROF. F. ALBERTO GRÜNBAUM
II.2 Univ. of California, Berkley, CA., U.S.A.

11:30 - PT-3: Algorithms & Error Analysis for Diffraction Tomography Using the Born and Rytov Approximations
V.6 DR. MOS KAVEH
co-author: DR. M. SOUMEKH
Univ. of Minnesota, Minneapolis, MN., U.S.A.

11:50 - PT-4: Diffraction Tomography
DR. ANTHONY J. DEVANEY
V.5 Schlumberger-Doll Research, Ridgefield, CT., U.S.A.

12:10 - PT-5: Extension of Scalar to Vector Propagation Tomography - A Computer Numerical Approach
V.7 MR. BRIAN D. JAMES
co-author: MR. CHAU-WING YANG
PROF. WOLFGANG-M. BOERNER
presented by: MR. ANTHONY C. MANSON
Univ. of Illinois at Chicago, Illinois, U.S.A.

12:30 to 13:00 - **LUNCH BREAK** (Short Sandwich Pick-Up Type Luncheon)

13:00 to 21:30 - **SCIENTIFIC/CULTURAL TOUR:** ERLANGEN-NÜRNBERG
Sponsor: DR. WERNER J. HAAS

Co-Chairmen: DR. MANFRED PFEILER
Siemens AG, Erlangen, FR Germany
PROF. HANS BRAND
Universität Erlangen-Nürnberg, FR Germany

Tour Co-ordinators:
DR. GUNTER SCHWIERZ
DIPL.-ING. GERHARD SILBERMANN
Siemens AG, Erlangen, FR Germany
DIPL.-ING. ROMAN GLOCKLER
DIPL.-ING. SIEGFRIED OSTERRIEDER
Universität Erlangen-Nürnberg, FR Germany

13:00 - **DEPARTURE FROM KuK RESIDENZ HOTEL BY TOUR BUS**

14:00 to 16:30 (i) **VISIT OF SIEMENS ZENTRAL LABORATORIUM**
Siemens AG, Erlangen, FR Germany

(ii) **VIEWING OF NUCLEAR MAGNETIC RESONANCE TOMOGRAPHIC IMAGING LAB**
Siemens Medical Research Division, Erlangen, FR Germany

16:30 - **DEPARTURE FROM ERLANGEN TO NÜRNBERG BY TOUR BUS**

17:15 to 18:45 - **VISIT OF GERMANISCHES NATIONAL MUSEUM IN NÜRNBERG**

19:00 to 20:30 - **SUPPER AT HEILIG-GEIST-SPITAL,** Nürnberg (at own expense)

20:45 - **DEPARTURE FROM NÜRNBERG TO BAD WINDSHEIM VIA TOUR BUS**

22:00 - **DAY'S END**
"Hotel at Rest" (NOISE)

THURSDAY, 22 September 1983

7:00 to 8:00 - **BREAKFAST**

SESSION SI (Lecture Center)

8:00 to 10:00 - **SURVEILLANCE IMAGERY**
Chairman: DR. OTTO KESSLER
Naval Air Development Center, Warminster, PA, USA

8:00 - SI-1: Inverse Methods Applied to Microwave Target Imaging
MR. LEONARD A. CRAM
IV.4 Thorn EMI, Electronics, Ltd., Wells/Somerset, UK

8:30 - SI-2: Optimum Detection of Surface Chemical Compounds
MR. TERJE LUND
IV.8 Environmental Surveillance Techn. Programme, Kjeller, Norway

9:00 - SI-3: Inversion Problems in SAR Imaging
IV.9 MR. RICHARD W. LARSON
co-authors: DR. DAVID LYZENGA
DR. ROBERT SHUCHMAN
Environmental Research Inst. of Mich., Ann Arbor, MI., U.S.A.

9:20 - SI-4: Radar Imagery
DR. TORLEIV ORHAUG
IV.3 National Defense Research Institute, Linköping, Sweden

9:40 - SI-5: Inverse Methods in Radar Glaciology
PROF. MICHAEL WALFORD
IV.10 Univ. of Bristol, Bristol, U.K.

10:00 to 10:30 - **COFFEE BREAK**

SESSION OI (Lecture Center)

10:30 to 12:30 - **OPTICAL IMAGING**
Chairman: DR.-ING. GUENTER WINKLER
Naval Weapons Center, China Lake, CA., U.S.A.

10:30 - OI-1: The Holographic Principle
PROF. ADOLF LOHMANN
V.1 Univ. Erlangen-Nürnberg, FR Germany

11:00 - OI-2: Synthesis & Detection of Electromagnetic Authenticity Features
V.13 PROF. HEINRICH P. BALTES
co-author: A. M. J. HUISER
Univ. of Alberta, Edmonton, Alberta, Canada

11:30 - OI-3:* Object Reconstruction from Partial Information
PROF. MICHAEL A. FIDDY
II.6 Queen Elizabeth College, London, U.K.

(* Replacement for paper by PROF. FRANCO GORI/STEFAN WABNITZ: Modifications of the Gerchberg Method Applied to Electromagnetic Imaging: **V.9**)

11:50 - OI-4: Image Processing: Analysis Beyond Matched Filtering
V.8 DR. HAROLD H. SZU
presended by: Dr. Arthur K. Jordan
Naval Research Laboratory, Washington, D.C., U.S.A.

12:10 - OI-5: Reconstruction of Refractive Index Profiles from Reflection Measurements
I.12 MR. E. VOGELZANG
co-authors: PROF. HEDZER A. FERWERDA
MR. D. YEVICK
State University of Groningen, Netherlands

12:30 to 13:45 - **LUNCH BREAK:** Mittagstisch

SESSION IM (Lecture Center)

13:45 to 15:25 - **INVERSE METHODS**
Chairman: DR. STEPHEN C. GIESS
Royal Signals & Radar Establishment, Malvern, U.K.

13:45 - IM-1:* Direct and Inverse Halfspace Scalar Diffraction: Some Models
DR. GIOVANNI CROSTA
II.3 Istituto di Cibernetica dell`Università; Milano, Italy

(* Resplacement for paper by DR. PAUL F. WACKER: Symmetry Relations in Electromagnetic Inverse Problems: **I.11**)

14:05 - IM-2: Electromagnetic Low Frequency Imaging
PROF. SUJEET K. CHAUDHURI
IV.12 Univ. of Waterloo, Waterloo, Canada

14:25 - IM-3: Applications of Almost-Periodic Functions to Inverse Scattering Theory
I.7 DR. ARTHUR K. JORDAN
co-authors: DR. SEYOUNG AHN
Naval Research Laboratory, Washington, D.C., USA
DR. D.L. JAGGARD
University of Pennsylvania, Philadelphia, PA, USA

14:45 - IM-4: Inverse Diffraction Problem
PROF. ALEXANDER RAMM
I.13 Kansas State University, Manhattan, KS., U.S.A.

15:05 - IM-5: Application of the Christoffel-Hurwitz Inversion Identity to Electromagnetic Imaging
I.14 MR. BRETT H. BORDEN
Naval Weapons Center, China Lake, CA., U.S.A.

15:25 to 16:00 - **TEA BREAK**

MEETING W-3 (Conference Center, Rooms A to E)

16:00 to 17:45 - **WORKING DISCUSSION GROUP MEETINGS:** Advice on Report Preparations. Various groups meet separately and prepare individual group reports.

chairman: DR. JOHN G. GALLAGHER

advisors: W-A, PROF. PIERRE C. SABATIER and PROF. HANS BLOK

W-B, DR. EDWARD R. PIKE and PROF. CHRISTINE DeMOL

W-C, MR ROBERT S. RAVEN and PROF. DAG T. GJESSING

W-D, PROF. JOHN O. THOMAS and DR. STEPHEN ROTHERAM

W-E, PROF. ADOLF LOHMANN and PROF. HARRY E. MOSES

W-F, MR. JERRY L. EAVES and DR. KARL STEINBACH

18:00 to 19:00 - **SUPPER**

SPECIAL SESSION SS * (Lecture Center)

19:30 to 21:30 - **EVALUATION OF EXPERIMENTAL RESULTS FROM HIGH RESOLUTION RADAR IMAGERY**

Chairman: DR. REINHARD POPP
MBB, Ottobrunn, FR Germany

19:30 - SS-1: Current Imaging Research Conducted at High Frequency Engineering Laboratories, UEN, **XX**
PROF. HANS BRAND
IHFT, University Erlangen-Nürnberg, FR Germany

19:40 - SS-2: Real-Time 70 GHz Imaging
IV.11 PROF. HANS BRAND
presented by: MR. SIEGFRIED OSTERRIEDER
IHFT, University Erlangen-Nürnberg, FR Germany

20:10 - SS-3: Advanced Concepts of Radar Target Phenomenology
DR. J. RICHARD HUYNEN
IV.2 Communications Lab. UIC, Chicago, USA

20:40 - SS-4: Interpretation of High Resolution Broadband Polarimetric Radar Target Down-range Signatures
III.12 MR. ANTHONY C. MANSON
authors group PROF. WOLFGANG-M. BOERNER
presentation: DR. J. RICHARD HUYNEN
Communications Laboratory, EECS, UIC, Chicago, USA

21:20 - SS-5: Conclusions: INTENSE GROUP DISCUSSIONS

(* Replacing Memorial Organ Concert on organ works by Johann Pachelbel (1653-1706), Johann Sebastian Bach (1685-1750), and by Richard Wagner (1813-**1883**), which was cancelled due to organ failure.)

22:00 - **DAY'S END**
"Hotel at Rest" (NOISE)

FRIDAY, 23 September 1983

7:00 to 8:00 - **BREAKFAST**

SESSION SR (Lecture Center)

8:00 to 10:00 - **SCATTEROMETRY AND REMOTE SENSING**
Chairman: MR. JERRY L. EAVES
Georgia Institute of Technology, Atlanta, GA, USA

8:00 - SR-1: Optimum Detection Techniques in Relation to Shape and Size of Objects, Motion Pattern and Material
IV.5 Composition
PROF. DAG T. GJESSING
co-authors: MR. JENS HJELMSTAD
MR. TERJE LUND
Environmental Surveillance Technology Programme, RNCISR, Kjeller, Norway

8:40 - SR-2: Topological Approach to Inverse Scattering in Remote Sensing
I.2 PROF. WERNER GÜTTINGER
co-author: DR. FRANCIS J. WRIGHT
Universitaät of Tübingen, FR Germany

9:00 - SR-3:* Singularities in Quasi-Geometrical Imaging
II.13 DR. GERHARD DANGELMAYR
co-authors: DR. FRANCIS J. WRIGHT
Universität of Tübingen, FR Germany

(* Replacement for paper by DR. DONALD J.R. STOCK: Radar Target Handling in Clutter, Consideration of the Measurement System: **III.9**)

9:20 - SR-4: Study of Two Scatterer Interference with a Polarimetric FM/CW Radar
III.10 DR. P. SAMUEL P. WEI
co-authors: MR. MICHAEL M. RACKSON/MR. THOMAS C. BRADLEY
MR. THOMAS H. MEYER/MR. JOHN D. KELLY
Boeing Aerospace Corporation, Seattle, WA., U.S.A.

9:40 - SR-5: Improvement of Image Fidelity in Microwave Imaging Through Diversity Techniques
V.13 DR.-ING. HEINRICH GNISS
co-author: DR.-ING. KLAUS MAGURA
Forschungsinstitut für HF-Physik, Wachtberg-Werthhoven, FRG

10:00 to 10:30 - **COFFEE BREAK**

SESSION SP (Lecture Center)

10:30 to 12:30 - **SPECIAL PROCESSING METHODS IN INVERSE SCATTERING/REMOTE SENSING**
Chairman: PROF. JUAN R. MOSIG
Ecole Polytechnique Fédérale de Lausanne, Switzerland

10:30 - SP-1: Maximum Entropy in Electromagnetic/Geophysical/-Ultrasonic Imaging
II.9 DR. ROBERT BEVENSEE
Lawrence Livermore Lab., Univ. of California, Livermore CA, USA

10:50 - SP-2: Polarization Vector Signal Processing for Radar Clutter Suppression
III.14 MR. VINCENT C. VANNICOLA
co-author: MR. STANLEY LIS
Rome Air Development Center-West, Rome N.Y., USA

11:10 - SP-3: Multistatic Vector Inverse Scattering
DR. GREGORY E. HEATH
III.5 MIT, Lincoln Laboratories, Bedford, MA., U.S.A.

11:50 - SP-5: Polarization Dependence in Angle Tracking Systems
DR. DANIEL CARPENTER
III.11 TRW. Inc., Redondo Beach, CA., U.S.A.

12:10 - SP-6:* Applications of the Abel Transform in Remote Sensing and Related Problems
DR.-ING. HELMUT SÜSS
I.10 DFVLR, Oberpfaffenfafen, FR Germany

(* Replacement for paper by DR. DOREN HESS: Far-Field-to-Near-Field Tranforms in Spherical Coordinates: I.12)

12:30 to 13:45 - **LUNCH BREAK:** Mittagstisch

MEETING W-R (Lecture Center)

13:45 to 15:25 - **FINAL WORKING DISCUSSION GROUP MEETING**
Submission of Final Reports by Group Reporters.
Preparation or Overall Statement on Group Reports.
Chairman: DR. RODERICK M. LOGAN
Admirality Surface Weapons Establ.,
Cosham/Portsmouth, U.K.

VI.1 reporters: W-A, DR. CORNEL EFTIMIU and DR. ARTHUR K. JORDAN
VI.2 W-B, DR. RUDOLF DITTEL and DR. ROBERT M. BEVENSEE
VI.3 W-C, IR. ANDRE J. POELMAN and DR. JAMES W. WRIGHT
VI.4 W-D, PROF. MICHAEL E.R. WALFORD and PROF. AKIRA ISHIMARU
VI.5 W-E, DR. ANTHONY J. DEVANEY and DR. HEDZER A. FERWERDA
VI.6 W-F, DR. RODERICK M LOGAN and DR. TILO KESTER

15:25 to 16:00 - **TEA BREAK** (Assembly for ARW participants' group photograph)

SESSION FS (Lecture Center)

16:00 to 17:30 - **FINAL SESSION**

16:00 - FS-1: **CONCLUDING REMARKS**
MR. LEONARD A. CRAM
THORN EMI, Somerset, England, UK

16:20 - FS-2: **PROCEDURES FOR THE PUBLICATION OF THE WORKSHOP PROCEEDINGS**
MRS. BARBARA KESTER
NATO Publications Coordination Office, Overyse, Belgium

16:40 - FS-3: **SUMMARY OF WORKING DISCUSSION GROUP REPORTS**
DR. RODERICK M. LOGAN
Admiralty Surface Weapons Establ.,
Cosham/Portsmouth, U.K.

17:00 - FS-4: **SUGGESTIONS FOR FUTURE INTERACTIONS, ADVANCED RESEARCH WORKSHOPS OR ADVANCED STUDY INSTITUTES**
DR. TILO KESTER
NATO Publications Coordination Office, Overyse, Belgium

17:15 - FS-5: **INSTRUCTIONS ON SUBMISSION OF REPORT FORMS**
PROF. WOLFGANG-M. BOERNER
Univ. of Illinois at Chicago, Illinois, U.S.A.

17:30 to 19:00 - **SETTLING OF HOTEL BILLS**

Preparation for Saturday's Departure for All Participants

WORKSHOP BANQUET (Residenz-Saal)

19:00 to 21:30 - **Host:** MR. ROLF K. ERLENBACH
Manager, KuK Residenz Hotel

19:00 to 20:00 - **DINNER**

20:00 - WB-1: **INTRODUCTION OF GUESTS OF HONOR**
DR. WOLFGANG KEYDEL
DFVLR, Oberpfaffenhofen, FR Germany

20:10 - WB-2: **SUMMARY OF WORKSHOP PROGRAMME**
DR. TILO KESTER
NATO Publications Coordination Office, Overyse, Belgium
DR. DAG T. GJESSING
Environmental Surveillance Technology Programme, RNCISR, Kjeller, Norway
MR. LEONARD A. CRAM
THORN EMI, Sommerset, U.K.

20:30 - WB-3: **THANKS TO THE HOSTS AND THE SPONSORS**
PROF. WALTER K. KAHN
The George Washington Univ., Washington, D.C., USA

20:40 - WB-4: **TILMAN RIEMENSCHNEIDER: THE SCULPTOR, ARTIST, SCHOLAR AND STATESMAN**
MR. WILLIAM McKEE WISEHART
KuK Hotel Residenz, Bad Windsheim, FR Germany

21:10 - WB-5: **DETAILED INSTRUCTIONS ON THE CULTURAL/SCIENTIFIC TOUR OF SATURDAY**
PROF. HANS BRAND
DIPL.-ING. SIEGFRIED OSTERRIEDER
Univ. Erlangen-Nürnberg

22:00 - **DAY'S END** "Hotel at Rest" (NOISE)

SATURDAY, 24 September 1983

7:00 to 8:00 - **BREAKFAST**

8:00 to 8:30 - **PREPARATIONS FOR DEPARTURE** OF ALL PARTICIPANTS

8:30 to 20:30 - **CULTURAL/SCIENTIFIC TOUR**

Tour Co-chairmen: PROF. HANS BRAND
DIPL.-ING. SIEGFRIED OSTERRIEDER
Inst. für HF-Technik, Univ.

Erlangen-Nürnberg, FR Germany

Interpreter: MR. WILLIAM McKEE WISEHART
KuK Hotel Residenz, Bad Windsheim, FRG

8:30 - **DEPARTURE** (Note; Luggage of tour participants will be taken along with the tour buses, as advised during registration) NOTE; THE HOTEL MANAGEMENT REQUESTS THE DEPARTURE OF ALL NATO-ARW PARTICIPANTS BY 9:00 (AM).

9:00 - **ARRIVAL** in **ROTHENBURG** (Visits will include sightseeing tour, shopping, St. Jakob Cathedral, City Hall)

10:30 - **DEPARTURE FROM ROTHENBURG** (Ride along picturesque Tauber Valley to Creglingen)

11:15 - **VIEWING OF MARIENALTAR** by Tilman Riemenschneider.

12:00 - **DEPARTURE FOR WÜRZBURG**

13:00 - **ARRIVAL AT WÜRZBURG HAUPTBAHNHOF**

13:30 - **LUNCH** near Käpelle with sight of Marienburg, the Bishop's Castle of Würzburg

15:00 - **DEPARTURE FROM WÜRZBURG** (Ride along the River Main, return on Autobahn to Erlangen/Nürnberg)

17:45 - **ARRIVAL AT NUREMBERG AIRPORT**

18:30 - **FARWELL AT NÜRNBERG HAUPTBAHNOF**

KEYWORD INDEX

for

PARTS 1 & 2

D

E

F

G

N

O

P

T

AUTHOR INDEX

for

Parts 1 & 2

W

Y

NATO-ARW-IMEI-1983

LIST OF REGISTERED INVITEES

11/21/83 Mon 12:25:27

(Last minute cancellation of attendance due to: * illness/ + non-attending co-authors ** other priority commitments)

Name/Address	Accompanied By	Country = Area - Tel.No.
Prof. Alan P. Anderson Electr. Engr. Dept. University of Sheffield Sheffield S1 3JD UNITED KINGDOM	Mrs. Maureen	44 = 742 - 78555
Dipl. Ing. Egon Peter Baars ** FHP/FGAN Königstrasse 2 D-5307 Wachtberg-Werthhoven FR GERMANY		49 = 228 - 852-350 49 = 228 - 8521
Prof. Henry P. Baltes H.M. Tory Professor Dept. of Electr. Engr. University of Alberta Edmonton, CANADA T6G 2E1		1 = 403 - 432-5871
Mr. Brian D. Barber ** Space & New Concepts Div. Royal Aircraft Establishment Farnborough Hants., GU14-6TD UNITED KINGDOM		44 = 252 - 24461
Prof. Mario Bertero Dipartimento di Fisica Università Di Genova Via Dodecaneso 33 I-16146 - Genova, ITALY		39 = 10 - 5993-221 39 = 10 - 330138
Dr. Robert Bevensee Elctr. Engnr. Dept. L-156 Lawrence Livermore Lab. Univ. of California P.O. Box 5504 Livermore, CA 94550 USA	Mrs. Mae	1 = 415 - 422-6787 1 = 415 - 837-0516

Prof. Andrew Blanchard 1 = 817 - 273-3497
Dept. of Elect. Engr.
University of Texas
at Arlington
Arlington, TX 76019 USA

Prof. Hans Blok Mrs. Aartje 31 = 15 - 566311
Delft Univ. of Technology
Dept. of Electr. Engr. 31 = 15 - 786291
Laboratory of Electormagnetic Research
P.O. Box 5031
2600 GA Delft
The NETHERLANDS

Prof. Wolfgang-M. Boerner, Mrs. Eileen 1 =312-996-5489
Electr. Engr. & Comp. Sci. Dept. 1 =312-498-4457
Univ. of Ill. at Chicago
851 S. Morgan St.
P.O. Box 4348, SEO-1141
Chicago, IL 60680 USA

Mr. Brett H. Borden 1 = 619 - 939-3962
Code 3814
Naval Weapons Center
China Lake, CA 93555 USA

Mr. Thomas C. Bradley ** 1 = 206 - 773-3687
RF Sensors Lab, Engineering Tech. 1 = 206 - 329-3153
Kent Space Center
Boeing Aerospace Co.
6042 33rd Ave. NE
P.O. Box 3999, MS 8H 51
Seattle, WA 98124 USA

Prof. Dr.-Ing. Hans Brand 49 = 9131 - 857215
Inst. für HF-Technik
Universität Erlangen-Nürnberg
Cauerstr. 9,
D-8520 Erlangen
FR GERMANY

Dr. Daniel Carpenter Mrs. Dixie 1 = 213 - 535-3147
Elctr. Development Oper. 1 = 213 - 376-4080
Defense & Space Systems Group of TRW, Inc.
1619 3rd Street
Manhattan Beach CA 90266 USA

Prof. Dr.-Ing. Heinz Chaloupka
Inst. für HF-Technik
Ruhr Universität Bochum
Hustadtring 28
D-4630 Bochum 1
FR GERMANY
49 = 231 - 705762

Prof. Sujeet K. Chaudhuri
Dept. of Elect. Engr.
University of Waterloo
Waterloo, Ontario
CANADA N2L 3G1
1 = 519 - 885-1211
Ext. 2843

Mr. James P. Corones +
Applied Mathematical Sciences
Ames Laboratory
Iowa State University
Ames, IA 50011 USA
1 = 515 - 294-3592

Mr. Leonard A. Cram
THORN EMI Electronics Ltd.
Wookey Hole Road
Wells, Somerset
BA5 1AA ENGLAND U.K.
44 = 749 - 72081
Ext. 209
44 = 93484 - 3222
TELEX 44254

Dr. Giovanni Crosta
Istituto di Cibernetica
University of Milano
Via Viotti, 5
I-20133 Milano ITALY
39 = 2 - 235293
39 = 2 - 3088679

Dr. Gerhard Dangelmayr
Institute for Information Sciences
University of Tübingen
Köstlinstrasse 6,
D-7400 Tübingen 1
FR GERMANY
49 = 7071 - 293 305

Mr. Marat Davidovitz
Electrical Engineering Dept.
University of Illinois at Urbana
1406 W. Green St.
Urbana IL 61801 USA
1 = 312 - 333-9365

M. E. Davison +
Applied Mathematical Sciences
Ames Laboratory
Iowa State University
Ames, IA 50011 USA
1 = 515 - 294-3592

Prof. Christine DeMol — 32 = 2 - 640 00 15 Ext. 5573; 32 = 2 - 521 34 31
Dept. de Math., Campus Plaine C.P. 217
Université Libre de Bruxelles
Blvd. du Triomphe
B-1050 Bruxelles, BELGIUM

Dr. Georges A. Deschamps * Mrs. Elsa * — 1 = 312 - 996-5489; 1 = 217 - 359-2301
Professor Emeritus
Dept. Electr. Engr.
Univ. of Ill. at Urbana
1406 W. Green St.
Urbana, IL 61820 USA

Prof. Dr.-Ing. Juergen Detlefsen — 49 = 89 - 2105 8389
Institut f. Mikrowellentechnik
Techn. Univ. Munich
Arcisstr. 21
D-8000 München 2
FR GERMANY

Dr. Anthony J. Devaney — 1 = 203 - 431-5386
Schlumberger-Doll Research
Old Quarry Road
P.O. Box 307
Ridgefield, CT. 06877 USA

Mr. Albert DiSalvio ** — 1 = 904 - 882-4631
Air Force Armament Lab.
AFATL/DLMT
Eglin AFB, FL 32542 USA

Dr. Rudolf H. Dittel — 49 = 8153 - 28-1; 49 = 8806 - 1578
Institut für HF-Technik
DFVLR-OPH/NE-HF
D-8031 Oberpfaffenhofen, Post Wessling
FR GERMANY

Mr. Jerry L. Eaves Mrs. Vila — 1 = 404 - 424-9609; 1 = 404 - 696-4913
Principal Research Engr.
EES/RAIL
Georgia Institute of Technology
Atlanta, GA 30332 USA

Dr. Cornel Eftimiu Mrs. Elena — 1 = 314 - 233-2501
Laboratory Bldg. 110
McDonnell Douglas Research
P.O. Box 516
St. Louis, MO. 63166 USA

Mr. Rolf K. Erlenbach
Mr. Hans-Dieter Erb
KuK Hotel Residenz
Erkenbrechtallee 33
8532 Bad Windsheim, FR GERMANY
49 = 09841/911
TELEX 61526 RESIDD

Dr. Ronald A. Evans
W7E
British Aerospace Corporation
Warton, Lancashire PR-1AX
UNITED KINGDOM
44 = 772 - 633333
Ext. 76

Prof. Hedzer Ferwerda Mrs. Rie
Dept. of Applied Physics
State University of Groningen
Nijenborgh 18,
NL-9747 AG Groningen
THE NETHERLANDS
31 = 50 - 115959

Prof. Michael A. Fiddy
Physics Department
Queen Elizabeth College
Campden Hill Road,
London, W8 7AH U.K.
44 = 1 - 937-5411
Ext. 239

Dr. Martin Fischer
Theor. Elektrotechnik
Universität Saarbrücken
D-6600 Saarbrücken
FR GERMANY
49 = 681 - 3023195

Prof. I.L. Freeston *
Dept. of Electr.& Electr.Engr.
University of Sheffield
Mappin Street
Sheffield, U.K. S1 3JD
44 = 742 - 78555
TELEX 54348 ULSHEFG

Prof. Adrian K. Fung
Dept. of Electrical Engineering
University of Kansas
2291 Irving Hill Drive - West Campus
Lawrence, KS 66045 USA
1 = 913 - 864-4832

Dr. John G. Gallagher
Royal Signals and Radar Est.
St. Andrews Rd., Great Malvern
Worcs WR14 3PS U.K.
44 = 6845 - 2733
Ext. 3430

Mrs. Margarete Geiger 49 = 9131 - 857241
Inst. für HF-Technik
Universität Erlangen-Nürnberg
Cauerstr. 9
D-8520 Erlangen
FR GERMANY

Dr. Stefen C. Giess 44 = 6845 - 2733 Ext.2511
Royal Signals and Radar Est.
St. Andrews Rd.
Great Malvern Worcs WR 14 3PS U. K.

Prof. Dag T. Gjessing 47 = 2 - 712660
Royal Norwegian Council for Scientific & TELEX 76528
Industrial Research 47 = 2 - 746448
Environmental Surveillance Tech. Programme
P.O. Box 25,
N-2007, Kjeller, NORWAY

Dipl. Ing. Roman Glöckler * 49 = 9131 - 857215
Inst. für HF-Technik
Universität Erlangen-Nürnberg
Cauerstr. 9
D-8520 Erlangen
FR GERMANY

Dr.-Ing. Henrich Gniss 49 = 228 - 852-239
FHP/FGAN TELEX 885589
Königstr. 2
D-5037 Wachtberg-Werthhoven
FR GERMANY

Prof. Franco Gori ** 39 = 6- 495 - 3748
Physics Department
University of Rome
P.le A. Moro, 5
I-00185 Roma ITALY

Prof. Peter T. Gough 64 = 3 - 482009 Ext. 322
Department of Electrical Engineering
University of Canterbury
Christchurch 1
NEW ZEALAND

Prof. F. Alberto Grünbaum 1 = 415 - 642-5348
Department of Mathematics
University of California
at Berkeley
Berkeley, CA. 94720 USA

Dr.-Ing. Konrad Grüner — 49 = 8153 - 28340
Inst. f. HF-Technik
DFVLR-Oberpfaffenhofen
D-8031 Post Wessling
FR GERMANY

Prof. Werner Güttinger * — 49 = 7071-296899
Inst. f. Information Sciences
Universität Tübingen
Köstlinstrasse 6
D-7400 Tübingen 1
FR GERMANY

Dr. John R.F. Guy — Mrs. Avril — 31 = 70 - 245550 Ext. 456
SHAPE Technical Centre
P.O. Box 174
2501 CD The Hague
THE NETHERLANDS

Dr. Werner J. Haas — 49 = 9131 - 84-1
Director, Tomographic Imaging
Siemens Medical Division
Siemens AG.
Henke-Strasse-127
D-8520 Erlangen, FR GERMANY

Mr. Henry William Hawkes — 44 = 242 - 21491; 44 = 242 - 602607
Government Communications Headquarters
Benhall, Cheltenham,
Glos., GL52 5AH,
UNITED KINGDOM

Dr. Gregory E. Heath — Mrs. Myrna, Gregory Jr., Michael — 1 = 617 - 863-5500 Ext. 2815
Lincoln Laboratory/KB-231
M. I. T.
244 Wood St.
Lexington, MA. 02173-0073 USA

Dr. Doren Hess * — 1 = 404 - 441-4210 Ext. 2850; TELEX 054-2898
Scientific Atlanta
Mail Stop 28I
P.O. Box 105027
Atlanta, GA. 30348 USA

Mr. Jens Hjelmstad — 47 = 2 - 712660; TELEX 76528
Royal Norwegian Council for Scientific
& Industrial Research
Environmental Surveillance
Technology Programme
P.O. Box 25,
N-2007 Kjeller
NORWAY

Mr. Karl Holberg ** Norwegian Defense Research Establishment P.O. Box 25 N-2007 Kjeller NORWAY	47 = 2 - 535603 TELEX 76528 47 = 2 - 535603
Dr. Charles J. Holland Math. Div., Code 411 Office of Naval Research Arlington, VA 22217 USA	1 = 202 - 696-4362
Dr. Bruce Z. Hollmann Code F-12 Naval Surface Weapons Center P.O. Box 427 Dahlgren, VA 22448 USA	1 = 703 - 663-8057 1 = 703 - 663-3581
Dr. J. Richard Huynen Mrs. Etty Quest Research 10531 Blandor Way Los Altos Hills, CA 94022 USA	1 = 415 - 941-2374
Prof. Akira Ishimaru Dept. of Elect. Engr. FT-10 University of Washington Seattle, WA 98195 USA	1 = 206 - 543-2169
Mr. Anthony F. Jaques Radar Division Thorn-EMI Electronics Ltd. 248 Blyth Road, UB3-1BP Hayes, Middlessex, UB3 1BP, U. K.	44 = 1 - 573-3888 Ext. 2099
Dr. Fred A. Johnson ** Chief Scientist (Royal Navy) Ministry of Defense, Main Building Whitehall, London SW1A 2HB, U.K.	44 = 1 - 218-2340 44 = 1 - 218-9000
Dr. Arthur K. Jordan Mrs. Mary Code 4110 J Space Sciences Division Naval Research Laboratory Washington, D.C. 20375 USA	1 = 202 - 767-6609
Dr. Virginia V. Jory Engineering Experiment Station Georgia Institute of Technology Atlanta, GA 30332 USA	1 = 404 - 424-9690 1 = 404 - 378 2173

Prof. Walter K. Kahn 1 = 202 - 676-7186
Electr Engr. & Comp. Sci. Dept.
School of Engr & Applied Science
George Washington University
Washington, D.C. 20052 USA

Dr. Mos Kaveh 1 = 612 - 373-2489
Dept. of Elect. Engr.
University of Minnesota
Minneapolis, MN 55455 USA

Dr. Edward Morton Kennaugh Mrs. Mary 1 = 614 - 451-4428
Professor Emeritus, OSU
4447 Lummisford Lane East
Columbus, OH 43214 USA

Dr. Otto Kessler 1 = 215 - 441-2306
Code 3022/Surveillance Radar
Naval Air Development Center
Warminster, PA 18974 USA

Dr. Tilo Kester Mrs. Barbara 32 = 2687 - 6636
NATO Publication Coordination Office
Elcerlyclaan 2
B-1900 Overijse
BELGIUM

Dr. Wolfgang Keydel 49 = 8153 - 28-305
Director, Inst. f. HF-Technik
DFVLR-Oberpfaffenhofen
D-8031 Wessling
FR GERMANY

Prof. Robert J. Krueger Mrs. Carol 1 = 515 - 294-3592
Dept. of Mathematics Statistics
University of Nebraska
Lincoln, NE 68588 USA

Mr. Dieter Lang 49 = 7071 - 29-6456
Institut für Informationsverarbeitung
Universität Tübingen
Köstlinstrasse 6
D-7400 Tübingen 1
FR GERMANY

Prof. Karl Langenberg 49 = 561 - 804-6368
Dept of Elctr.Engr./FB-16 49 = 561 - 496745
University of Kassel
Wilheimshöher Allee 71,
D-3500 Kassel
FR GERMANY

Mr. Richard W. Larson Radar & Optics Division Environmental Research Inst. of Michigan 3300 Plymouth Road P.O. Box 8618 Ann Arbor, MI 48107 USA	1 = 313 - 994-1200 Ext. 510
Mr. Stanley Lis Signal Processing/Surveillance Division Rome Air Development Center-West Griffis AFB Rome, NY 13441 USA	1 = 315 - 330-4437
Dr. Odd Lovhaugen ** Central Inst. for Indust. Research Forskningsv. 1, P.O. Box 350 Blindern, Oslo 3 NORWAY	47 = 2 - 452010
Dr. Roderick M. Logan Mrs. Dianne Admiralty Surface Weapons Est. Ministry of Defence Cosham Portsmouth, Hants., PO6-4AA, U.K.	44 = 705 - 379411 Ext. 2136 44 = 243 - 779043
Prof. Adolf Lohmann Physikalisches Institut Universität Erlangen-Nürnberg Erwin-Rommel-Str. 1 D-8520 Erlangen FR GERMANY	49 = 9131 - 857408
Dr. Ernst Lüneburg DFVLR, Inst. F. HF-Technik D-8031 Oberpfaffenhofen FR GERMANY	49 = 8153-28-343 49 = 8153-1405
Mr. Terje Lund Royal Norwegian Council for Scientific & Industrial Research Envirn. Surveillance Technology Programme P.O. Box 25 N-2007 Kjeller NORWAY	47 = 2 - 712660 47 = 2 - 777767
Dr. David R. Lyzenga + Radar Science Lab, Radar Division, Environmental Research Inst. of Michigan 3300 Plymouth Rd./P.O. Box 8618 Ann Arbor, MI 48107 USA	1 = 313 - 994-1200 Ext. 250

Dr.-Ing. Klaus Magura 49 = 228 - 852225
FHP-FGAN
Königstr. 2
D-5307 Wachtberg-Werthhoven
FR GERMANY

Mr. Anthony C. Manson Mrs. Cindy 1 = 312 - 996-5284
Communications Laboratory, 1 = 312 - 991-9762
EECS Department
Univ. of Ill. at Chicago 1139 SEO
851 S. Morgan St.
Chicago, IL 60680 USA

Prof. David L. Moffatt Mrs. Susan 1 = 614 - 422-5749
ElectroScience Laboratory 1 = 614 - 885-6710
Ohio State University
1320 Kinnear Road
Columbus, OH 43212 USA

Prof. Harry E. Moses 1 = 617 - 458-2504
College of Pure & Applied Science
University of Lowell
450 Aiken Street
Lowell, MA 01854 USA

Prof. Juan R. Mosig 41 = 21 - 47-26-78
Dept. of E E, Laboratory of TELEX 26420 EPFVD CH
Electromagnetics & Acoustics
Ecole Polytechnique
Fédérale de Lausanne
16, ch. de Bellerive
CH-1007 Lausanne, SWITZERLAND

Dr. Giorgio Musso * 39 = 10 - 60011
Elettronica San Giorgio-ELSAG S.p.A.
via Hermada, 6
I-16154 Genova-Sestri
ITALY

Prof. Barry Narod * 1 = 604 - 228-3110
Dept. of Astronomy & Geophysics
University of British Columbia
Vancouver, B.C. V6T 1W5
CANADA

Dr. M. Zuhair Nashed ** 1 = 302 - 738-2653
Dept. of Mathematical Sciences 1 = 302 - 738-6907
University of Delaware
501 Kirkbridge Office Building
Newark, DE 19711 USA

Mr. Jerald D. Nespor — 1 = 312 - 996-5140
Commun. Lab, EECS Dept.
Univ. of Ill. at Chicago,
851 S. Morgan/P.O. Box 4348
Chicago, IL 60680 USA

Dr. Richard W. Newton + — 1 = 713 - 845-5422
Remote Sensing Center,
Texas A & M University
College Station, TX 77843 USA

Dr. Torleiv Orhaug — 46 = 13 - 118157
Applied Electronics Dept. — 46 = 13 - 118000
National Defense Research Inst.
P.O. Box 1165
S-581 11 Linköping SWEDEN

Dipl. Ing. Siegfried Osterrieder — 49 = 9131 - 857219
Inst. für HF-Technik — 49 = 9131 - 302498
Universität Erlangen-Nürnberg
Cauerstr. 9
D-8520 Erlangen
FR GERMANY

Dr. Hans M. Pedersen — 47 = 2 - 452010
Central Inst. for Indust. Research
Forskningsv. 1, P.O. Box 350
Blindern, Oslo 3
NORWAY

Dr. Manfred Pfeiler — 49 = 9131 - 84 2079
Siemens AG, Bereich Medizinische Technik
Henkestrasse 127
D-8520 Erlangen, Postfach 3260
FR GERMANY

Dr. Edward R. Pike — 44 = 6845 - 2733
Royal Signals & Radar Est. — Ext. 2907
St. Andrews Road — 44 = 6845 - 4910
Great Malvern, Worcs., WR14 3PS
UNITED KINGDOM

Lt. David Pirolo ** — 1 = 904 - 882-4631
Air Force Armament Technology Lab.
DLMT
Eglin AFB, FL 32542 USA

Mr. André J. Poelman — Mrs. Carli — 31 = 70 - 245550
Sensor Branch, Command — Ext. 457
Control & Systems Div.
SHAPE-Technical Center
P.O. Box 174

Dr.-Ing. Reinhard Popp 49 = 89 - 6000 8212
MBB, Abt. AE 312
Postfach 801149
D-8000 München
FR GERMANY

Prof. Alexander G. Ramm 1 = 913 - 532-6750
Department of Mathematics 1 = 913 - 539 0959
Kansas State University
Cardwell Hall
Manhattan, KS 66506 USA

Mr. Robert S. Raven Mrs. Phylene 1 = 301 - 765-4616
Radar Development
AESD-Systems Development
Westinghouse Defense & Space Center
Friendship International Airport
P.O. Box 1693/MS-1105
Baltimore, MD 21203 USA

Dr. André Roger 33 = 91 - 98-23-44
Laboratoire d'Optique Electromàgnetique
Faculté St-Jérome
F-13397 Marseille Cedex 13
FRANCE

Mr. Lloyd W. Root ** 1 = 205 - 876-8131
DRSMI-REG
United States Army
Missile Command
Redstone Arsenal, AL 35809 USA

Dr. Stephen Rotheram Mrs. Joan 44 = 245 - 73331
Marconi Research Centre Ext. 58
West Hanningfield Road 44 = 621 - 891590
Chelmsford
Essex, CM2 8HN
UNITED KINGDOM

Prof. Pierre C. Sabatier 33 = 67 - 544850
Laboratorie de Physique Mathématique 33 = 67 - 794351
Université des Sciences
et Techniques du Languedoc
Place Eugene-Bataillon
F-34060 Montpellier Cedex FRANCE

Prof. Tapan Sarkar 1 = 716 - 475-2165
Rochester Institute of Technology
Dept. of Electr. Engnr.
One Lomb Memorial Drive
Rochester, NY 14623 USA

Dr. Hartmut Schimpf FHP/FGAN Königstr. 2 D-5037 Wachtberg-Werthhoven FR GERMANY	49 = 228 - 852255
Dr. Karl von Schlachta FFM/FGAN Königstr. 2 D-5307 Wachtberg-Werthhoven FR GERMANY	49 = 228 - 852220
Dr. Günter Schwierz Grundlagen Laboratorium RGG-Siemens-Med. Henke-Str. 172/8 D-852 Erlangen FR GERMANY	49 = 9131 - 84-2720 49 = 9134 - 1426
Dr. Robert A. Shuchman + Radar Science Laboratory Environmental Research Inst. of Michigan 3300 Plymouth Rd./P.O. Box 8618 Ann Arbor, MI 48107 USA	1 = 313 - 994 1200
Dipl. Ing. Gerhard Silbermann Siemens AG Bereich Medizinische Technik Entwicklungsleitung Vakuumtechnik Henke-str. 127 D-8520 Erlangen, FR GERMANY	49 = 9131 - 84-3956
Mr. Colin D. Sillence THORN EMI, Electronics Limited Wookey Hole Road Wells, Somerset BA5-1AA, U.K.	44 = 749 - 72081 Ext. 274 44 = 749 - 4697
Dr. Craig Sinclair ** NATO Headquarters Sientific Affairs Div. B-1110 Bruxelles BELGIUM	32 = 2 - 241-00-40 Ext. 2198
Prof. Thomas Sphicopoulos LEMA-DE Ecole Polytechnique Fédérale de Lausanne 16, ch. de Bellerive CH-1007 Lausanne, SWITZERLAND	41 = 21 - 47-26-78

Dr. Karl Steinbach 1 = 703 - 664-4970
Chief Scientist 1 = 703 - 591-5824
U.S. Army Mob. Equip. R & D Cmnd.
Attn. DRDME-HS,
Fort Belvoir, VA 22060 USA

Dr. Donald Stock ** 49 = 731 - 3920
Radio & Radar Systems
Group-Radar Division
AEG-Telefunken Al, Anlagentechnik AG
Sedanstrasse 10
D-7900 Ulm/Donau
FR GERMANY

Dr.-Ing. Helmut Süss 49 = 8153 - 28-372
Inst. f. HF-Technik
DFVLR-Oberpfaffenhofen
D-8031 Post Wessling
FR GERMANY

Dr. Harold Szu ** 1 = 202 - 767-2407
Optical Sciences Division 1 = 301 - 897-8137
Code 6530
Naval Research Laboratory
Washington, D.C. 20375 USA

Prof. Walid Tabbara 33 = 6 - 941-8040
Groupe d`Electromagnètisme
Laboratoire des Signaux et Systémes
CNRS-ESE
F-91190 GIF-SUR-YVETTE
FRANCE

Dr. Manfred Thiel 49 = 228 - 443624
Angewandte Mathematik
Maschinenbauwesen
Fachhochschule Köln
Am Reitweg 1
D-5000 Köln 1
FR GERMANY

Dr. John O. Thomas 44 = 1 - 589-5111
The Blackett Laboratory Ext. 2565
Imperial Col. of Sci. & Tech. 44 = 23 - 588-433
Prince Consort Rd.
London SW7 2BZ U. K.

Prof. Dr.-Ing. Siegfried Tischer ** 49 = 621 - 4048091
Bundesakademie für Wehrver- 49 = 6203 - 81407
waltung und Wehrtechnik
Seckenheimer Landstrasse 8-10
D-6800 Mannheim, FR GERMANY

Dr. Constantine P. (Gus) Tricoles 1 = 619 - 573-6303 OR 6308
Electronics Division
General Dynamics
P.O. Box 81127
San Diego, CA 92138 USA

Dr. Koroda Umashankar 1 = 312 - 567-4489
EM Interaction w/Complex Systems 1 = 312 - 690-1859
IIT Research Institute
10 West 35th Street
Chicago, IL 60616 USA

Dr. Martin Vogel Mrs. Elli * 49 = 8153 - 28-325
Inst. f. HF-Technik
DFVLR, Oberpfaffenhofen
D-8031 Wessling
FR GERMANY

Dr. Roger Voles ** 44 = 1 - 573-3888
Chief Scientist TELEX 22417
Thorn EMI Electronics
135 Blyth Road
Hayes, Middlessex, U.K. UB3-1BP

Mr. Stefan Wabnitz ** 39 = 6 - 5802504
Fondazione ugo Bordoni
Viale Trastevere, 108
I-00153 ROMA
ITALY

Dr. Paul F. Wacker * 1 = 213 - 957-8118
3037 Montrose Ave. Apt. 4
La Crescenta, CA 91214 USA

Dr. Michael E.R. Walford 44 = 272 - 24161
H.H. Wills Physics Laboratory Ext. 434
University of Bristol
Royal Fort, Tyndall Avenue
Bristol BS8 1TL U.K.

Dr. P. Samuel P. Wei 1 = 206 - 773-2293
Boeing Aerospace Corporation or 3083
Kent Space Center
6042 33rd Ave. NE
P.O. Box 3999, Mail Stop 8H-51
Seattle, WA 98124 USA

Prof. Vaughan H. Weston Division of Math Sciences Purdue University Mathematical Sci. Bldg. West Lafayette, IN 47907 USA	Mrs. Jacquelyn	1 = 317 - 494-1959 1 = 317 - 463-5779
Dr.-Ing. Guenter Winkler Code 381, Physics Div. Naval Weapons Center China Lake, CA 93555 USA		1 = 619 - 939-2970 or 2977
Dr. Francis J. Wright Institut für Informationsverarbeitung Universität Tübingen Köstlinstrasse 6, D-7400 Tübingen FR GERMANY		49 = 7071 - 29-6899
Dr. James W. Wright New Technology Inc. P.O. Box 5245 Huntsville, AL 35814/5247 USA	Mrs. Ruth	1 = 205 - 837-7663